TODAY'S TECHNICIAN ™

Classroom Manual for
Automotive
Electricity & Electronics

Fourth Edition

■

Barry Hollembeak

THOMSON

DELMAR LEARNING

Australia   Canada   Mexico   Singapore   Spain   United Kingdom   United States

coil

THOMSON

DELMAR LEARNING

Today's Technician/Classroom Manual for Automotive Electricity & Electronics, 4th Edition

Barry Hollembeak

Vice President, Technology and Trades ABU:
David Garza

Director of Learning Solutions:
Sandy Clark

Senior Acquisitions Editor:
David Boelio

Product Manager:
Matthew Thouin

Marketing Director:
Deborah S. Yarnell

Channel Manager:
William Lawrensen

Marketing Coordinator:
Mark Pierro

Project Editor:
Toni Hansen

Art/Design Specialist:
Cheri Plasse

Technology Project Manager:
Kevin Smith

Technology Project Specialist:
Linda Verde

Editorial Assistant:
Andrea Domkowski

Library of Congress Cataloging-in-Publication Data
Hollembeak, Barry.
    Classroom manual for automotive electricity and electronics/Barry Hollembeak. — 4th ed.
    p. cm. — (Today's technician)
Includes index.
ISBN-13: 978-1-4180-1267-0
ISBN-10: 1-4180-1267-X
    1. Automobiles—Electric equipment—Maintenance and repair. 2. Automobiles—Electronic equipment—Maintenance and repair I. Title. II. Title: Automotive electricity and electronics. III. Series.
TL272.H625 2006
629.25′40288—dc22

2006044542

NOTICE TO THE READER

Publisher does not warrant or guarantee any of the products described herein or perform any independent analysis in connection with any of the product information contained herein. Publisher does not assume, and expressly disclaims, any obligation to obtain and include information other than that provided to it by the manufacturer.

The reader is expressly warned to consider and adopt all safety precautions that might be indicated by the activities herein and to avoid all potential hazards. By following the instructions contained herein, the reader willingly assumes all risks in connection with such instructions.

The publisher makes no representation or warranties of any kind, including but not limited to, the warranties of fitness for particular purpose or merchantability, nor are any such representations implied with respect to the material set forth herein, and the publisher takes no responsibility with respect to such material. The publisher shall not be liable for any special, consequential, or exemplary damages resulting, in whole or part, from the readers' use of, or reliance upon, this material.

CONTENTS

PREFACE

Thanks to the support the Today's Technician™ series has received from those who teach automotive technology, Thomson Delmar Learning, the leader in automotive-related textbooks, is able to live up to its promise to provide new editions of the series regularly. We have listened and responded to our critics and our fans and present this new updated and revised fourth edition. By revising our series regularly, we can and will respond to changes in the industry, changes in technology, changes in the certification process, and to the ever-changing needs of those who teach automotive technology.

The Today's Technician™ series, by Thomson Delmar Learning, features textbooks that cover all mechanical and electrical systems of automobiles and light trucks (whereas the Heavy-duty Trucks portion of the series does the same for heavy-duty vehicles). Principally, the individual titles correspond to the main areas of ASE (National Institute for Automotive Service Excellence) certification. Additional titles include remedial skills and theories common to all of the certification areas and advanced or specific subject areas that reflect the latest technological trends. Each text is divided into two volumes: a Classroom Manual and a Shop Manual.

Unlike yesterday's mechanic, the technician of today and for the future must know the underlying theory of all automotive systems and be able to service and maintain those systems. Dividing the material into two volumes provides the reader with the information needed to begin a successful career as an automotive technician without interrupting the learning process by mixing cognitive and performance learning objectives into one volume.

The design of Thomson Delmar Learning's Today's Technician™ series was based on features that are known to promote improved student learning. The design was further enhanced by a careful study of survey results, in which the respondents were asked to value particular features. Some of these features can be found in other textbooks, whereas others are unique to this series.

Each Classroom Manual contains the principles of operation for each system and subsystem. The Classroom Manual also contains discussions on design variations of key components used by the different vehicle manufacturers. This volume is organized to build on basic facts and theories. The primary objective of this volume is to allow the reader to gain an understanding of how each system and subsystem operates. This understanding is necessary to diagnose the complex automobiles of today and tomorrow. Although the basics contained in the Classroom Manual provide the knowledge needed for diagnostics, diagnostic procedures appear only in the Shop Manual. An understanding of the basics is also a requirement for competence in the skill areas covered in the Shop Manual.

A spiral-bound Shop Manual covers the "how-to's." This volume includes step-by-step instructions for diagnostic and repair procedures. Photo Sequences are used to illustrate some of the common service procedures. Other common procedures are listed and are accompanied with line drawings and photos that allow the reader to visualize and conceptualize the finest details of the procedure. This volume also contains the reasons for performing the procedures as well as when that particular service is appropriate.

The two volumes are designed to be used together and are arranged in corresponding chapters. Not only are the chapters in the volumes linked together, but also the contents of the chapters are linked. This linking of content is evidenced by marginal callouts that refer the reader to the chapter and page in which that the same topic is addressed in the other volume. This feature is valuable to instructors. Users of other two-volume textbooks without this feature must search the index or table of contents to locate supporting information in the other volume. This is not only cumbersome but also creates additional work for an instructor when planning the presentation of material and when making reading assignments. It is also valuable to the students; with page references, they also know exactly where to look for supportive information.

Both volumes contain clear and thoughtfully selected illustrations, many of which are original drawings or photos specially prepared for inclusion in this series. This means that the art is a vital part of each textbook and not merely inserted to increase the numbers of illustrations.

The page layout used in the series is designed to include information that would otherwise break up the flow of information presented to the reader. The main body of the text includes all of the "need-to-know" information and illustrations. In the wide side margins of each page are many of the special features of the series. Items that are truly "nice-to-know" information include simple examples of concepts just introduced in the text, explanations or definitions of terms that will not be defined in the glossary, examples of common trade jargon used to describe a part or operation, and exceptions to the norm explained in the text. Many textbooks attempt to include this type of information and insert it in the main body of text; this tends to interrupt the thought process and cannot be pedagogically justified. By placing this information off to the side of the main text, the reader can select when to refer to it.

Classroom Manual

Features of this manual include:

Cognitive Objectives

These objectives define the contents of the chapter and define what the student should have learned upon completion of the chapter.

Each topic is divided into small units to promote easier understanding and learning.

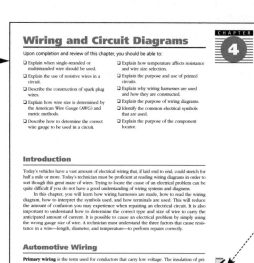

Wiring and Circuit Diagrams

CHAPTER **4**

Upon completion and review of this chapter, you should be able to:

❏ Explain when single-stranded or multistranded wire should be used.
❏ Explain the use of resistive wires in a circuit.
❏ Describe the construction of spark plug wires.
❏ Explain how wire size is determined by the American Wire Gauge (AWG) and metric methods.
❏ Describe how to determine the correct wire gauge to be used in a circuit.

❏ Explain how temperature affects resistance and wire size selection.
❏ Explain the purpose and use of printed circuits.
❏ Explain why wiring harnesses are used and how they are constructed.
❏ Explain the purpose of wiring diagrams.
❏ Identify the common electrical symbols that are used.
❏ Explain the purpose of the component locator.

Introduction

Today's vehicles have a vast amount of electrical wiring that, if laid end to end, could stretch for half a mile or more. Today's technician must be proficient at reading wiring diagrams in order to sort though this great maze of wires. Trying to locate the cause of an electrical problem can be quite difficult if you do not have a good understanding of wiring systems and diagrams.

In this chapter, you will learn how wiring harnesses are made, how to read the wiring diagram, how to interpret the symbols used, and how terminals are used. This will reduce the amount of confusion you may experience when repairing an electrical circuit. It is also important to understand how to determine the correct type and size of wire to carry the anticipated amount of current. It is possible to cause an electrical problem by simply using the wrong gauge size of wire. A technician must understand the three factors that cause resistance in a wire—length, diameter, and temperature—to perform repairs correctly.

Automotive Wiring

Primary wiring is the term used for conductors that carry low voltage. The insulation of primary wires is usually thin. **Secondary wiring** refers to wires used to carry high voltage, such as ignition spark plug wires. Secondary wires have extra thick insulation.

Most of the primary wiring conductors used in the automobile are made of several strands of copper wire wound together and covered with a polyvinyl chloride (PVC) insulation (Figure 4-1). Copper has low resistance and can be connected to easily by using crimping connectors or soldered connections. Other types of conductor materials used in automobiles include silver, gold, aluminum, and tin-plated brass.

AUTHOR'S NOTE: Copper is used mainly because of its low cost and availability.

Stranded wire means the conductor is made of several individual wires that are wrapped together. Stranded wire is used because it is very flexible and has less resistance than solid

Shop Manual
Chapter 00, page 00

91

Cross-References to the Shop Manual

Reference to the appropriate page in the Shop Manual is given whenever necessary. Although the chapters of the two manuals are synchronized, material covered in other chapters of the Shop Manual may be fundamental to the topic discussed in the Classroom Manual.

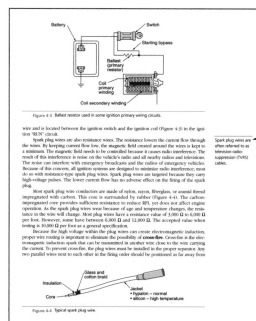

Figure 4-3 Ballast resistor used in some ignition primary wiring circuits.

wire and is located between the ignition switch and the ignition coil (Figure 4-3) in the ignition "RUN" circuit.

Spark plug wires are also resistance wires. The resistance lowers the current flow through the wires. By keeping current flow low, the magnetic field created around the wires is kept to a minimum. The magnetic field needs to be controlled because it causes radio interference. The result of this interference is noise on the vehicle's radio and all nearby radios and televisions. The noise can interfere with emergency broadcasts and the radios of emergency vehicles. Because of this concern, all ignition systems are designed to minimize radio interference; most do so with resistance-type spark plug wires. Spark plug wires are targeted because they carry high-voltage pulses. The lower current flow has no adverse effect on the firing of the spark plug.

Most spark plug wire conductors are made of nylon, rayon, fiberglass, or aramid thread impregnated with carbon. This core is surrounded by rubber (Figure 4-4). The carbon-impregnated core provides sufficient resistance to reduce RFI, yet does not affect engine operation. As the spark plug wires wear because of age and temperature changes, the resistance in the wire will change. Most plug wires have a resistance value of 3,000 Ω to 6,000 Ω per foot. However, some have between 6,000 Ω and 12,000 Ω. The accepted value when testing is 10,000 Ω per foot as a general specification.

Because the high voltage within the plug wires can create electromagnetic induction, proper wire routing is important to eliminate the possibility of **cross-fire**. Cross-fire is the electromagnetic induction spark that can be transmitted in another wire close to the wire carrying the current. To prevent cross-fire, the plug wires must be installed in the proper separator. Any two parallel wires next to each other in the firing order should be positioned as far away from

Spark plug wires are often referred to as television-radio-suppression (TVRS) cables.

Figure 4-4 Typical spark plug wire.

93

Marginal Notes

These notes add "nice-to-know" information to the discussion. They may include examples or exceptions, or may give the common trade jargon for a component.

Author's Notes

This feature includes simple explanations, stories, or examples of complex topics. These are included to help students understand difficult concepts.

Summaries

Each chapter concludes with a summary of key points from the chapter. These are designed to help the reader review the chapter contents.

Review Questions

Short-answer essay, fill-in-the-blank, and multiple-choice questions are found at the end of each chapter. These questions are designed to accurately assess the student's competence in the stated objectives at the beginning of the chapter.

A Bit of History

This feature gives the student a sense of the evolution of the automobile. This feature not only contains nice-to-know information, but also should spark some interest in the subject matter.

Terms to Know List

A list of new terms appears next to the Summary.

---

3. Longer shelf life (approximately 18 months).
4. Ability to be shipped with electrolyte installed, reducing the possibility of accidents and injury to the technician.
5. Higher cold cranking amps rating.

The major disadvantages of the maintenance-free battery include:
1. **Grid growth** when the battery is exposed to high temperatures.
2. Inability to withstand **deep cycling**.
3. Low reserve capacity.
4. Faster discharge by parasitic loads.
5. Shorter life expectancy.

Grid growth is a condition where the grid grows little metallic fingers that extend through the separators and short out the plates.

Deep cycling is to discharge the battery to a very low state of charge before recharging it.

A BIT OF HISTORY

Buick first introduced the storage battery as standard equipment in 1906.

Hybrid Batteries

AUTHOR'S NOTE: The following discussion on hybrid batteries refers to a battery type and not to the batteries that are used in hybrid electric vehicles (HEVs).

The **hybrid battery** combines the advantages of the low-maintenance and maintenance-free battery. The hybrid battery can withstand six deep cycles and still retain 100% of its original reserve capacity. The grid construction of the hybrid battery consists of approximately 2.75% antimony alloy on the positive plates and a calcium alloy on the negative plates. This allows the battery to withstand deep cycling while retaining reserve capacity for improved cranking performance. Also, the use of antimony alloys reduces grid growth and corrosion. The lead calcium has less gassing than conventional batteries.

Grid construction differs from other batteries in that the plates have a lug located near the center of the grid. In addition, the vertical and horizontal grid bars are arranged in a radial pattern (Figure 5-15). By locating the lug near the center of the grid and using the **radial grid**

Radial means branching out from a common center.

Grid with active material

100% glass separator

Grid only

Figure 5-15 Hybrid grid and separator construction.

Summary

❑ Automotive electrical horns operate on an electromagnetic principle that vibrates a diaphragm to produce a warning signal.

❑ Horn switches are either installed in the steering wheel or as a part of the multifunction switch. Most horn switches are normally open switches.

❑ Horn switches that are mounted on the steering wheel require the use of sliding contacts. The contacts provide continuity for the horn control in all steering wheel positions.

❑ The most common type of horn circuit control is to use a relay.

❑ Most two-speed windshield wiper motors use permanent magnet fields whereby the motor speed is controlled by the placement of the brushes on the commutator.

❑ Some two-speed and all three-speed wiper motors use two electromagnetic field windings: series field and shunt field. The two field coils are wound in opposite directions so that their magnetic fields will oppose each other. The strength of the total magnetic field will determine at what speed the motor will operate.

❑ Park contacts are located inside the wiper motor assembly and supply current to the motor after the switch has been turned to the PARK position. This allows the motor to continue operating until the wipers have reached the PARK position.

❑ Intermittent wiper mode provides a variable interval between wiper sweeps and is controlled by a solid-state module.

❑ Systems that have a depressed-park feature use a second set of contacts with the park switch, which are used to reverse the rotation of the motor for about 15 degrees after the wipers have reached the normal park position.

❑ Blower fan motors use a resistor block that consists of two or three helically wound wire resistors that are connected in series to control fan speed.

❑ The blower motor circuit includes the control assembly, blower switch, resistor block, and blower motor.

❑ Electric defoggers heat the rear window by means of a resistor grid.

❑ Electric defoggers may incorporate a timer circuit to prevent the high current required to operate the system from damaging the battery or charging system.

❑ The electrically controlled mirror allows the driver to position the outside mirrors by use of a switch that controls dual drive, reversible PM motors.

❑ Power windows, seats, and door locks usually use reversible PM motors, whereby motor rotational direction is determined by the direction of current flow through the switch wipers.

Terms to Know
Child safety latch
Clockspring
Depressed park
Diaphragm
Electrochromic mirrors
Grid
Park contacts
Resistor block
Sector gear
Trimotor
Window regulator

Terms to Know (continued)
Resistance
Right-hand rule
Saturation
Self-induction
Semiconductors
Series circuit
Series-parallel circuit
Shell
Static electricity
Valence ring
Voltage
Watts

❑ Voltage drop is caused by a resistance in the circuit that reduces the electrical pressure available after the resistance.

❑ Kirchhoff's voltage law states that the total voltage drop in an electrical circuit will always equal the available voltage at the source.

Review Questions

Short-Answer Essays

1. List and define the three elements of electricity.
2. Describe the use of Ohm's law.
3. List and describe the three types of circuits.
4. Explain the principle of electromagnetism.
5. Describe the principle of induction.
6. Describe the basics of electron flow.
7. Define the two types of electrical current.
8. Describe the difference between insulators, conductors, and semiconductors.
9. Explain the basic concepts of capacitance.
10. What does the measurement of "watt" represent?

Fill in the Blanks

1. _____ are negatively charged particles. The nucleus contains positively charged particles called _____ and particles that have no charge called _____.
2. A _____ allows electricity to easily flow through it. An _____ does not allow electricity to easily flow through it.
3. For the electrons to move in the same direction, there must be an _____ applied.
4. The _____ of current flow states that current flows from a positive point to a less positive point.
5. Resistance is defined as _____ to current flow and is measured in _____.
6. _____ is the ability of two conducting surfaces to store voltage.
7. Kirchhoff's voltage law states that the _____ _____ in an electrical circuit will always _____ available voltage at the source.
8. The _____ of all the resistors in series is the total resistance of that series circuit.
9. _____ is defined as an electrical pressure.
10. _____ is defined as the rate of electron flow.

Shop Manual

To stress the importance of safe work habits, the Shop Manual dedicates one full chapter to safety. Other important features of this manual include:

Performance-Based Objectives

These objectives define the contents of the chapter and define what the student should have learned upon completion of the chapter. These objectives also correspond with the list of required tasks for ASE certification. *Each ASE task is addressed.*

Although this textbook is not designed to simply prepare someone for the certification exams, it is organized around the ASE task list. These tasks are defined generically when the procedure is commonly followed and specifically when the procedure is unique for specific vehicle models. Imported- and domestic-model automobiles and light trucks are included in the procedures.

Marginal Notes

These notes add "nice-to-know" information to the discussion. They may include examples or exceptions, or may give the common trade jargon for a component.

Special Tools List

Whenever a special tool is required to complete a task, it is listed in the margin next to the procedure.

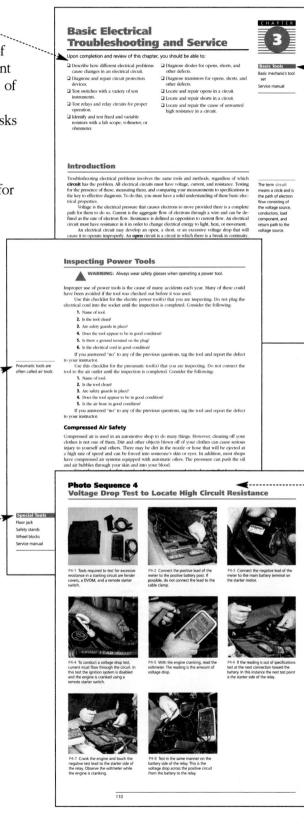

Basic Tools List

Each chapter begins with a list of the Basic Tools needed to perform the tasks included in the chapter.

Photo Sequences

Many procedures are illustrated in detailed Photo Sequences. These detailed photographs show the students what to expect when they perform particular procedures. They also can provide the student a familiarity with a system or type of equipment, which the school may not have.

Cautions and Warnings

Throughout the text, warnings are given to alert the reader to potentially hazardous materials or unsafe conditions. Cautions are given to advise the student of things that can go wrong if instructions are not followed or if a nonacceptable part or tool is used.

Service Tips

Whenever a short-cut or special procedure is appropriate, it is described in the text. These tips are generally those things commonly done by experienced technicians.

Cross-References to the Classroom Manual

Reference to the appropriate page in the Classroom Manual is given whenever necessary. Although the chapters of the two manuals are synchronized, material covered in other chapters of the Classroom Manual may be fundamental to the topic discussed in the Shop Manual.

Customer Care

This feature highlights those little things a technician can do or say to enhance customer relations.

Job Sheets

Located at the end of each chapter, the Job Sheets provide a format for students to perform procedures covered in the chapter. A reference to the ASE task addressed by the procedure is included on the Job Sheet.

Testing Circuit Protection Devices

CAUTION: Fuses and other protection devices do not wear out. They fail because something went wrong. Never replace a fuse or fusible link, or reset a circuit breaker, without finding out why it failed.

Classroom Manual
Chapter 3, page 78

Special Tools
Ohmmeter
Voltmeter
Test light

A blade type fuse is called a spade fuse.

A protection device is designed to "turn off" the system whenever excessive current or an **overload** occurs. There are three basic types of **fuses** in automotive use: cartridge, blade, and ceramic. A fuse is a replaceable element that will melt should the current passing through it exceed the fuse rating. The cartridge fuse is found on most older domestic cars and a few imports. To check this type of fuse, look for a break in the internal metal strip. Discoloration of the glass cover or glue bubbling around the metal caps is an indication of overheating. Late-model domestic vehicles and many imports use blade or spade fuses. To check the fuse, pull it from the fuse panel and look at the fuse element through the transparent plastic housing. Look for internal breaks and discoloration. The ceramic fuse is used on many older European imports. To check this type of fuse, look for a break in the contact strip on the outside of the fuse. All types of fuses can be checked with an ohmmeter or test light. If the fuse is good, there will be continuity through it.

Fuses are rated by the current at which they are designed to blow. A three-letter code is used to indicate the type and size of fuses. Blade fuses have codes ATC or ATO. All glass SFE fuses have the same diameter, but the length varies with the current rating. Ceramic fuses are available in two sizes, code GBF (small) and the more common code GBC (large). The amperage rating is also embossed on the insulator. Codes such as AGA, AGW, and AGC indicate the length and diameter of the fuse. Fuse lengths in each of these series is the same, but the current rating can vary. The code and the current rating is usually stamped on the end cap.

The current rating for blade fuses is indicated by the color of the plastic case (Table 3-1). In addition, it is usually marked on the top. The insulator on ceramic fuses is color coded to indicate different current ratings.

Fuses are located in a box or panel, usually under the dashboard, behind a panel in the foot well, or in the engine compartment. Fuses are generally numbered, and the main components abbreviated. On late-model cars there may be icons or symbols indicating which circuits they serve. This identification system is covered in more detail in the owner's and service manuals.

SERVICE TIP: To calculate the correct fuse rating, use Watt's law: watts divided by volts equals amperes. For example, if you are installing a 55-watt pair of fog lights, divide 55 by the battery voltage (12 volts) to find out how much current the circuit has to carry. Since 55 ÷ 12 = 4.58, the current is approximately 5 amperes. To allow for current surges, the correct in-line fuse should be rated slightly higher than the normal current flow. In this case, an 8- or 10-ampere fuse would do the job.

A fuse link is commonly called a fusible link.

A **fusible link** is a conductor with a special heat resistant insulation. When there is an overload in the circuit, the link melts and opens the circuit. Fusible links are used in circuits where limiting the maximum current is not extremely critical. They are often installed in the positive battery lead to the ignition switch and other circuits that are not normally fused.

A fuse link is a short a lighter gauge of wire that occur in the rest of the circuit it overheats, indicating that If the link stretches, the w burned out, check for cont

To replace a fuse lin tightly crimp or solder a n

88

To perform the regulator voltage test using a VAT-40, follow these steps:

1. Connect the large red and black cables across the battery, observing polarity.
2. Select REGULATOR.
3. Select INT 18 V.
4. Zero the ammeter.
5. Clamp the inductive pickup around the AC generator output wire.
6. Start the engine and hold between 1,500 and 2,000 rpm.
7. Allow the engine to run until the ammeter reads 10 amperes or less. This indicates the battery is fully charged.
8. Voltage should read regulated voltage (13.5–14.5 volts).
9. Load the system to between 10 and 20 amperes.
10. Voltmeter should read regulated voltage.

Special Tools
Starting/charging system tester
Tachometer

Classroom Manual
Chapter 7, page 170

Diode/Stator Test

An AC generator may have an open diode yet test close to manufacturer's specifications. If there is an open diode that is not determined in testing, a newly installed regulator may fail. In addition, an open diode can lead to the failure of other diodes. The **diode/stator test** is performed to determine the condition of the diodes. This is performed in the following manner:

1. Connect the large red and black cables across the battery, observing polarity.
2. Select the CHARGING position.
3. Select INT 18 V.
4. Zero the ammeter.
5. Clamp the inductive pickup around all of the negative battery cables.
6. Start the engine and hold between 1,500 and 2,000 rpm.
7. Adjust the load control knob to obtain an indicated charge rate of 15 amperes.
8. Set the selector to the DIODE/STATOR position while observing the red and blue DIODE/STATOR scale.
9. Return the load control knob to the OFF position.

If the meter was in the blue section of the scale, the diodes and stator are good. If the meter was in the red section of the scale, the diodes or the stator is bad. The AC generator will need to be disassembled to perform bench testing of these units.

Special Tools
Starting/charging system tester
Tachometer

This test is not valid for AC generators that failed the full field test.

Diode Pattern Testing

CUSTOMER CARE: It is good practice to check the diode pattern of the AC generator anytime an electronic component fails. Because the electronics of the vehicle cannot accept AC current, the damage to the replaced component could have been the result of a bad diode. By performing this check, it is possible to find the cause of the problem.

Set an oscilloscope on the lowest scale available. Connect the primary test leads on the AC generator output terminal and ground. Start the engine and place a moderate load on the charging system (15 to 20 amperes). Different patterns may appear. What is considered normal depends on the load placed on the system.

Classroom Manual
Chapter 7, page 172

Special Tools
Oscilloscope
Carbon pile

Job Sheet 4

4

Name _____ Date _____

Meter Symbol Interpretation

Upon completion of this job sheet, you should be able to convert commonly found symbols into numeric values.

Convert the following values into the electrical units noted:

1. 2.4 K Ω = _____ Ω
2. 954 mV = _____ V
3. 5.76 K Ω = _____ Ω
4. 2 mA = _____ A
5. 22 K Ω = _____ Ω
6. 4.5 mA = _____ A
7. 456 mA = _____ A
8. 1024 mV = _____ V
9. 0.786 K Ω = _____ Ω
10. 32 K Ω + 112 Ω = _____ Ω
11. 1400 Ω = _____ K Ω
12. 0.000235 A = _____ mA
13. 0.987 V = _____ mV
14. 5 K V = _____ mV
15. 123, 955 Ω = _____ K Ω
16. 144,000 mA = _____ A
17. 126 mV + 11.874 V = _____ V
18. 320,000 Ω = _____ Ω
19. 0.000045 A = _____ mA
20. 12,600 mV = _____ V

Instructor's Response _____
_____
_____

Case Studies

Case Studies concentrate on the ability to properly diagnose the systems. Beginning with Chapter 3, each chapter ends with a case study in which a vehicle has a problem, and the logic used by a technician to solve the problem is explained.

CASE STUDY

The vehicle owner complains that the brake lights do not light. He also says the dome light is not working. The technician verifies the problem, then checks the battery for good connections and tests the fusible links. All are in good condition.

A study of the wiring diagram indicates that the brake light and dome light circuits share the same fuse. It is also indicated that the ignition switch illumination light circuit is shared with these two circuits. A check of the ignition switch illumination light shows that it is not operating either. The technician checks the fuse that is identified in the wiring diagram. It is blown. When a replacement fuse is installed, the dome and brake lights work properly for three tests, then the fuse blows again.

Upon further testing of the shared circuits, an intermittent short to ground is located in the steering column in the ignition switch illumination circuit. The technician solders in a repair wire to replace the damaged section. After all repairs are completed, a final test indicates proper operation of all circuits.

Terms to Know

Color codes	Pull to seat	Splice clip
Crimping	Push to seat	Trouble codes
Crimping tool	Solderless connectors	Troubleshooting
Heat shrink tube	Splice	Vehicle identification number (VIN)

ASE-Style Review Questions

1. Splicing copper wire is being discussed.
   Technician A says it is acceptable to use solderless connections.
   Technician B says acid core solder should not be used on copper wires.
   Who is correct?
   A. A only   C. Both A and B
   B. B only   D. Neither A nor B

2. Use of wiring diagrams is being discussed.
   Technician A says a wiring diagram is used to help find the fault.
   Technician B says the wiring diagram will give the exact location of the components in the car.
   Who is correct?
   A. A only   C. Both A and B
   B. B only   D. Neither A nor B

3. The fuse for the parking lights is open.
   Technician A says find the cause for the blown fuse.
   Technician B says the fuse probably wore out due to age.
   Who is correct?
   A. A only   C. Both A and B
   B. B only   D. Neither A nor B

4. Repairs to a twisted/shielded wire are being discussed.
   Technician A says a twisted/shielded wire carries high current.
   Technician B says because a twisted/shielded wire carries low current, any repairs to the wire must not increase the resistance of the circuit.
   Who is correct?
   A. A only   C. Both A and B
   B. B only   D. Neither A nor B

150

Terms to Know List

Terms in this list can be found in the Glossary at the end of the manual.

ASE-Style Review Questions

Each chapter contains ASE-style review questions that reflect the performance-based objectives listed at the beginning of the chapter. These questions can be used to review the chapter as well as to prepare for the ASE certification exam.

ASE Challenge Questions

1. The fuse for an A/C blower motor circuit fails after a short period of time.
   Technician A says that the blower motor may be binding internally.
   Technician B says that the blower motor ground circuit may have excessive resistance.
   Who is correct?
   A. A only   C. Both A and B
   B. B only   D. Neither A nor B

2. The brake lights of a vehicle are inoperative. A voltmeter that is connected across the terminals of the brake light switch indicates system voltage (12 volts) regardless of whether the brake pedal is depressed.
   Technician A says that there is no power available at the brake light bulbs.
   Technician B says that the brake light switch may be open.
   Who is correct?
   A. A only   C. Both A and B
   B. B only   D. Neither A nor B

3. Relay testing is being discussed.
   Technician A says that it is acceptable during the test sequence to bypass the relay control coil terminals with a fused jumper wire.
   Technician B says that the voltage drop across the relay load contact terminals is checked with the relay control circuit de-energized.
   Who is correct?
   A. A only   C. Both A and B
   B. B only   D. Neither A nor B

4. The troubleshooting of a parallel circuit that contains three dimly lit bulbs is being discussed. A voltmeter that is placed across each of the bulbs indicates 7.2 volts.
   Technician A says that the power supply that is common to all three bulbs may be faulty.
   Technician B says that the ground terminal that is common to all three bulbs may have excessive resistance.
   Who is correct?
   A. A only   C. Both A and B
   B. B only   D. Neither A nor B

5. An A/C compressor clutch coil spike suppression diode is being tested with an analog ohmmeter. The meter indicates infinite resistance in both directions.
   Technician A says that the diode is electrically open.
   Technician B says that the use of this diode would result in the immediate failure of the circuit fuse.
   Who is correct?
   A. A only   C. Both A and B
   B. B only   D. Neither A nor B

118

ASE Challenge Questions

Each technical chapter ends with five ASE challenge questions. These are not more review questions; rather, they test the students' ability to apply general knowledge to the contents of the chapter.

Reviewers

The author and publisher would like to extend a special thanks to the following instructors for their contributions to this text:

Steve Bertram
Palomar College, San Marcos, CA

Terry Enyart
University of Northwestern Ohio, Lima, OH

C. Neel Flannagan
Aiken Technical College, Graniteville, SC

Robert Gibbens
North Central Kansas Technical College, Beloit, KS

Chris Hadfield
College of Lake County, Grayslake, IL

Don Lumsdon
Ivy Tech State College, Terre Haute, IN

Kent McCleary
University of Northwestern Ohio, Lima, OH

Dick Rogers
Lincoln Land Community College, Springfield, IL

Shane Sampson
Western Iowa Tech Community College, Sioux City, IA

Dan Wilson
Westwood College, Denver, CO

John Wood
Ranken Technical College, St. Louis, MO

Introduction to Automotive Electrical and Electronic Systems

Upon completion and review of this chapter, you should be able to:

❏ Explain the importance of learning automotive electrical systems.

❏ Describe the role of electrical systems in today's vehicles.

❏ Explain the interaction of the electrical systems.

❏ Explain the purpose of the starting system.

❏ Explain the purpose of the charging system.

❏ Explain the purpose of the ignition system.

❏ Describe the purpose of various electrical accessories.

❏ Describe the role of the computer in today's vehicles.

❏ Explain the purpose of vehicle communication networks.

❏ Explain the purpose of various electronic chassis control and accessory systems.

❏ Describe the purpose of passive restraint systems.

❏ Explain the purpose of alternate propulsion systems.

Introduction

You are probably reading this book for one of two reasons. Either you are preparing yourself to enter into the field of automotive service or you are expanding your skills to include automotive electrical systems. In either case, congratulations on selecting one of the most fast-paced segments of the automotive industry. Working with the electrical systems can be challenging, yet very rewarding; however, it can also be very frustrating at times.

For many people, learning electrical systems can be a struggle. It is my hope that I am able to present the material to you in such a manner that you will not only understand electrical systems but will excel at it. There are many ways the theory of electricity can be explained, and many metaphors can be used. Some compare electricity to a water flow, while others explain it in a purely scientific fashion. Everyone learns differently. I am presenting electrical theory in a manner that I hope will be clear and concise. If you do not fully comprehend a concept, then it is important to discuss it with your instructor. Your instructor may be able to use a slightly different method of instruction to help you to completely understand the concept. Electricity is somewhat abstract; so if you do have questions, be sure to ask your instructor.

Why Become an Electrical System Technician?

In the past it was possible for technicians to work their entire careers and be able to almost completely avoid the vehicle's electrical systems. They would specialize in engines, steering/suspension, or brakes. Today there is not a system on the vehicle that is immune to the role of electrical circuits. Engine controls, electronic suspension systems, and antilock brakes are common on today's vehicles. Even electrical systems that were once thought of as being simple have evolved to computer controls. Headlights are now pulse-width modulated using high-side drivers and will automatically brighten and dim based on the light intensity of oncoming traffic. Today's vehicles are equipped with twenty or more computers, laser-guided cruise control, sonar park assist, infrared climate control, fiber optics, and radio frequency transponders and decoders. Simple systems have become more computer reliant. For example, the horn circuit on the 2007 Chrysler 300C involves three separate control modules to function. Even the tires have computers involved, with the addition of tire pressure monitoring systems!

Today's technician must possess a full and complete electrical background to be able to succeed. The future will provide great opportunities for those technicians who have prepared themselves properly.

The Role of Electricity in the Automobile

In the past, electrical systems were basically stand-alone. For example, the ignition system was only responsible for supplying the voltage needed to fire the spark plugs. Ignition timing was controlled by vacuum and mechanical advance systems. Today there are very few electrical systems that are still independent.

Today, most manufactures **network** their electrical systems together through computers. This means that information gathered by one system can be used by another. The result may be that a faulty component may cause several symptoms. Consider the following example. The wiper system can interact with the headlight system to turn on the headlights whenever the wipers are turned on. The wipers can interact with the vehicle speed sensor to provide for speed-sensitive wiper operation. The speed sensor may provide information to the antilock brake module. The antilock brake module can then share this information with the transmission control module, and the instrument cluster can receive vehicle speed information to operate the speedometer. If the vehicle speed sensor should fail, this could result in no antilock brake operation and a warning light turned on in the dash. But it could also result in the speedometer not functioning, the transmission not shifting, and the wipers not operating properly.

A BIT OF HISTORY

Karl Benz of Mannheim, Germany, patented the world's first automobile on January 29, 1886. The vehicle was a three-wheeled automobile called the Benz Motorwagen. That same year Gottieb Daimler built a four-wheeled vehicle. It was powered by a 1.5-horsepower engine that produced 50% more power than that of the Benz Motorwagon. The first automobile to be produced for sale in the United States was the 1896 Duryea.

Introduction to the Electrical Systems

The purpose of this section is to acquaint you with the electrical systems that will be covered in this book. We will define the purpose of these systems.

AUTHOR'S NOTE: The discussion of the systems in this section of the chapter provides you with an understanding of their *main* purpose. Some systems have secondary functions. All of these will be discussed in detail in later chapters.

The Starting System

The **starting system** is a combination of mechanical and electrical parts that work together to start the engine. The starting system is designed to change the electrical energy, which is being supplied

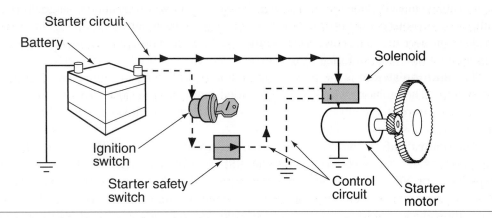

Figure 1-1 Major components of the starting system.

by the battery, into mechanical energy. For this conversion to be accomplished, a starter or cranking motor is used. The basic starting system includes the following components (Figure 1-1):

1. Battery.
2. Cable and wires.
3. Ignition switch.
4. Starter solenoid or relay.
5. Starter motor.
6. Starter drive and flywheel ring gear.
7. Starting safety switch.

The starter motor (Figure 1-2) requires large amounts of current (up to 400 amperes) to generate the torque needed to turn the engine. The conductors used to carry this amount of current

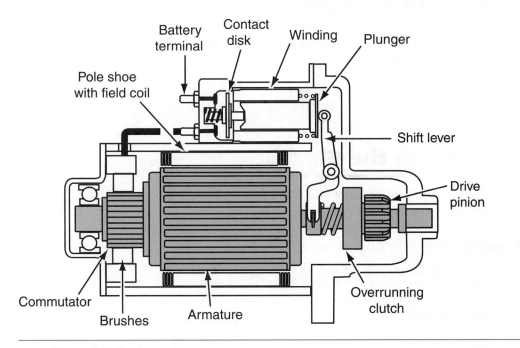

Figure 1-2 Starter motor.

(battery cables) must be large enough to handle the current with very little voltage drop. It would be impractical to place a conductor of this size into the wiring harness to the ignition switch. To provide control of the high current, all starting systems contain some type of magnetic switch. There are two basic types of magnetic switches used: the solenoid and the relay.

The **ignition switch** is the power distribution point for most of the vehicle's primary electrical systems. The ignition switch is spring loaded in the start position. This momentary contact automatically moves the contacts to the RUN position when the driver releases the key. All other ignition switch positions are detent positions.

The **neutral safety switch** is used on vehicles that are equipped with automatic transmissions. It opens the starter control circuit when the transmission shift selector is in any position except PARK or NEUTRAL. Vehicles that are equipped with automatic transmissions require a means of preventing the engine from starting while the transmission is in gear. Without this feature, the vehicle would lunge forward or backward once it was started, causing personal or property damage. The normally open neutral safety switch is connected in series into the starting system control circuit and is usually operated by the shift lever (Figure 1-3). When in the PARK or NEUTRAL position, the switch is closed, allowing current to flow to the starter circuit. If the transmission is in a gear position, the switch is opened and current cannot flow to the starter circuit.

Many vehicles that are equipped with manual transmissions use a similar type of safety switch. The start/clutch interlock switch is usually operated by movement of the clutch pedal (Figure 1-4).

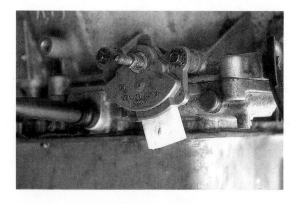

Figure 1-3 The neutral safety switch is usually attached to the transmission.

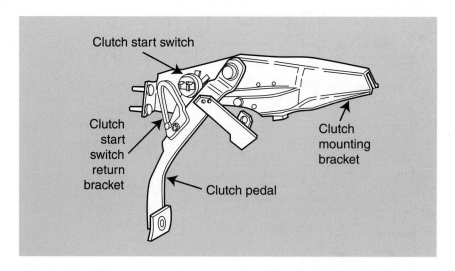

Figure 1-4 Most vehicles with a manual transmission use a clutch start switch.

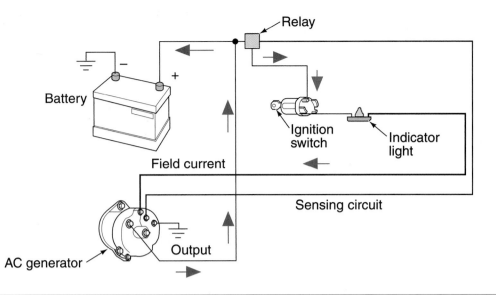

Figure 1-5 Components of the charging system.

The Charging System

The automotive storage battery is not capable of supplying the demands of the electrical systems for an extended period of time. Every vehicle must be equipped with a means of replacing the current that is being drawn from the battery. A **charging system** is used to restore to the battery the electrical power that was used during engine starting. In addition, the charging system must be able to react quickly to high load demands required of the electrical system. It is the vehicle's charging system that generates the current to operate all of the electrical accessories while the engine is running.

The purpose of the charging system is to convert the mechanical energy of the engine into electrical energy to recharge the battery and run the electrical accessories. When the engine is first started, the battery supplies all the current required by the starting and ignition systems.

As illustrated in Figure 1-5, the entire charging system consists of the following components:

1. Battery.

2. AC generator or DC generator.

3. Drive belt.

4. Voltage regulator.

5. Charge indicator (lamp or gauge).

6. Ignition switch.

7. Cables and wiring harness.

8. Starter relay (some systems).

9. Fusible link (some systems).

All charging systems use the principle of electromagnetic induction to generate the electrical power. A **voltage regulator** controls the output voltage of the AC generator, based on charging system demands, by controlling field current. The battery, and the rest of the electrical system, must be protected from excessive voltages. To prevent early battery and electrical system failure, regulation of the charging system is very important. Also, the charging system must supply enough current to run the vehicle's electrical accessories when the engine is running.

The Ignition System

One of the requirements for an efficient engine is the correct amount of heat delivered into the cylinders at the right time. This is the responsibility of the **ignition system.** The ignition system supplies

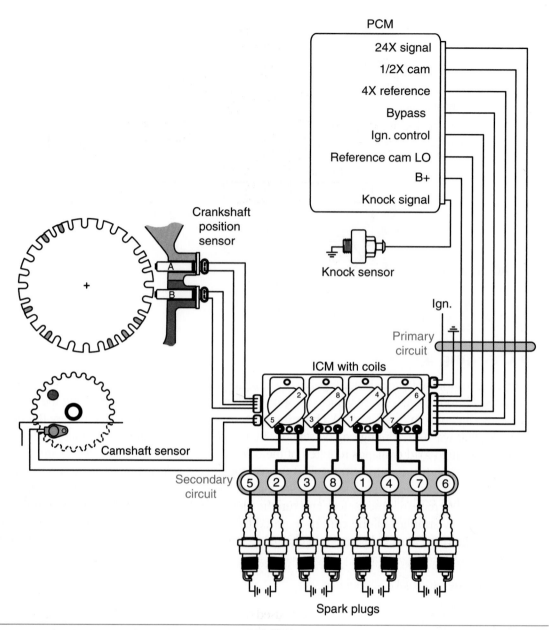

Figure 1-6 The ignition system.

properly timed high-voltage surges to the spark plugs. These voltage surges cause an arc across the electrodes of a spark plug, and this heat begins the combustion process inside the cylinder.

All ignition systems consist of two interconnected electrical circuits (Figure 1-6): a primary circuit (low voltage) and a secondary circuit (high voltage).

Depending on the type of ignition system, components in the primary circuit may include the following:

- Battery.
- Ignition switch.
- Ballast resistor or resistance wire (some systems).
- Starting bypass (some systems).
- Ignition coil primary winding.
- Triggering device.
- Switching device or control module.
- Ground.

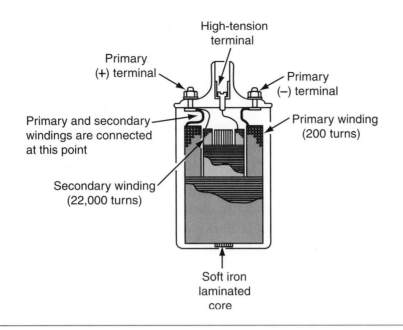

Figure 1-7 Ignition coil.

The secondary circuit may include these components:

- Ignition coil secondary winding.
- Distributor cap and rotor (some systems).
- Ignition (spark plug) cables.
- Spark plugs.

The **ignition coil** (Figure 1-7) is the heart of the ignition system. Its purpose is to build up the low battery voltage of approximately 12.6 volts to a voltage that is high enough to jump across the spark plug gap and ignite the air-fuel mixture. The coil is capable of producing approximately 30,000 to 60,000 volts.

The Lighting System

The **lighting system** consists of all of the lights used on the vehicle (Figure 1-8). This includes headlights, front and rear park lights, front and rear turn signals, side marker lights, daytime running lights, cornering lights, brake lights, back-up lights, instrument cluster backlighting, and interior lighting.

Figure 1-8 Automotive lighting system.

The lighting system of today's vehicles can consist of more than 50 light bulbs and hundreds of feet of wiring. Incorporated within these circuits are circuit protectors, relays, switches, lamps, and connectors. In addition, more sophisticated lighting systems use computers and sensors. Since the lighting circuits are largely regulated by federal laws, the systems are similar among the various manufacturers. However, there are variations that exist in these circuits.

With the addition of solid-state circuitry in the automobile, manufacturers have been able to incorporate several different lighting circuits or modify the existing ones. Some of the refinements that were made to the lighting system include automatic headlight washers, automatic headlight dimming, automatic on/off with timed-delay headlights, and illuminated entry systems. Some of these systems use sophisticated body computer–controlled circuitry and fiber optics.

Some manufacturers have included such basic circuits as turn signals into their body computer to provide for pulse-width dimming in place of a flasher unit. The body computer can also be used to control instrument panel lighting based on inputs that include if the side marker lights are on or off. By using the body computer to control many of the lighting circuits, the amount of wiring has been reduced. In addition, the use of computer control of these systems has provided a means of self-diagnosis in some applications.

In addition, high-density discharge (HID) headlamps are becoming an increasingly popular option on many vehicles. These headlights provide improved lighting over conventional headlamps.

Vehicle Instrumentation Systems

Vehicle instrumentation systems (Figure 1-9) monitor the various vehicle operating systems and provide information to the driver about their correct operation. Warning devices also provide information to the driver; however, they are usually associated with an audible signal. Some vehicles use a voice module to alert the driver to certain conditions.

Electrical Accessories

Electrical accessories provide for additional safety and comfort. There are many electrical accessories that can be installed into today's vehicles. These include safety accessories such as the horn, windshield wipers, and windshield washers. Comfort accessories include the blower motor, electric defoggers, power mirrors, power windows, power seats, and power door locks.

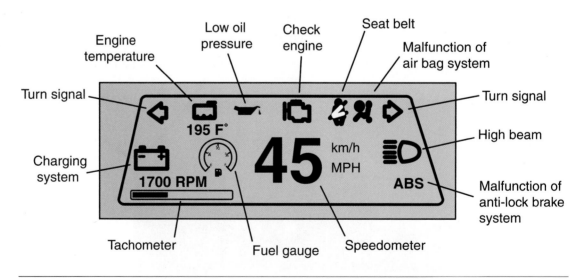

Figure 1-9 Instrument panel.

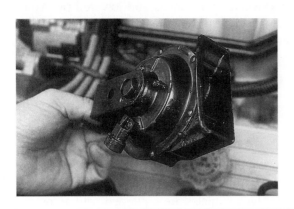

Figure 1-10 Automotive horn.

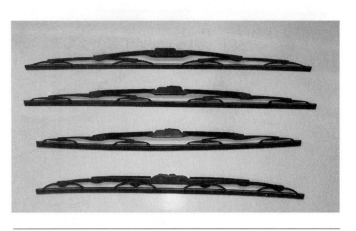

Figure 1-11 Windshield wipers.

Horns. A **horn** is a device that produces an audible warning signal (Figure 1-10). Automotive electrical horns operate on an electromagnetic principle that vibrates a diaphragm to produce a warning signal. This vibration of the diaphragm is repeated several times per second. As the diaphragm vibrates it causes a column of air that is in the horn to vibrate. The vibration of the column of air produces the sound.

Windshield Wipers. Windshield wipers are mechanical arms that sweep back and forth across the windshield to remove water, snow, or dirt (Figure 1-11). The operation of the wiper arms is through the use of a wiper motor. Most windshield wiper motors use permanent magnet fields, or electromagnetic field motors.

Electric Defoggers. Electric defoggers heat the rear window to remove ice and/or condensation. Some vehicles use the same circuit to heat the outside driver-side mirror. When electrons are forced to flow through a resistance, heat is generated. Rear window defoggers use this principle of controlled resistance to heat the glass. The resistance is through a grid that is baked on the inside of the glass (Figure 1-12). The system may incorporate a timer circuit that controls the relay.

Power Mirrors. Power mirrors are outside mirrors that are electrically positioned from the inside of the driver compartment. The electrically controlled mirror allows the driver to position the outside mirrors by use of a switch. The mirror assembly will use built-in, dual-drive, reversible permanent magnet (PM) motors.

Figure 1-12 Rear window defogger grid.

Power Windows. Power windows are windows that are raised and lowered by use of electrical motors. Many vehicle manufacturers have replaced the conventional window crank with electric motors that operate the side windows. The motor used in the power window system is a reversible PM or two-field winding motor. The power window system usually consists of the following components:

1. Master control switch.
2. Individual control switches.
3. Individual window drive motors.

Power Door Locks. Electric **power door locks** use either a solenoid or a permanent magnet reversible motor to lock and unlock the door. Many vehicles are equipped with automatic door locks that are activated when the gear shift lever is placed in the DRIVE position. The doors unlock when the selector is returned to the PARK position.

Computers

A **computer** is an electronic device that stores and processes data and is capable of operating other devices (Figure 1-13). The use of computers on automobiles has expanded to include control and operation of several functions, including climate control, lighting circuits, cruise control, antilock braking, electronic suspension systems, and electronic shift transmissions. Some of these are functions of what is known as a body computer (BCM). Some body computer–controlled systems include direction lights, rear window defogger, illuminated entry, intermittent wipers, and other systems that were once thought of as basic.

A computer processes the physical conditions that represent information (data). The operation of the computer is divided into four basic functions:

1. Input.
2. Processing.
3. Storage.
4. Output.

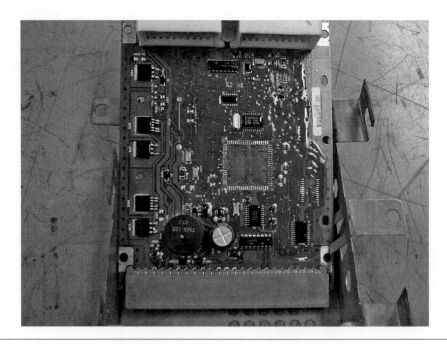

Figure 1-13 The automotive computer.

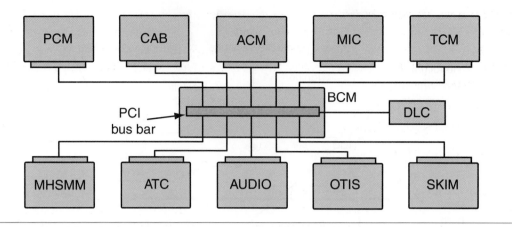

Figure 1-14 Automotive computers are networked together through multiplexing.

Vehicle Communication Networks

Most manufacturers now use a system of vehicle communications called **multiplexing** to allow control modules to share information (Figure 1-14). Multiplexing provides the ability to use a single circuit to distribute and share data between several control modules throughout the vehicle. Because the data is transmitted through a single circuit, bulky wiring harnesses are eliminated.

Vehicle manufacturers will use multiplexing (MUX) systems to enable different control modules to share information. A MUX wiring system uses **bus** data links that connect each module. The term *bus* refers to the transporting of data from one module to another. Each module can transmit and receive digital codes over the bus data links. The signal sent from a sensor can go to any one of the modules and can be shared by the other modules.

Electronic Chassis Control and Accessory Systems

With the growing use of computers, most systems can be controlled electronically. This provides for improved monitoring of the systems for proper operation and the ability to detect if a fault occurs. The systems that are covered in this book include the following:

Electronic Automatic Temperature Control. An **automatic temperature control (ATC) system** is capable of maintaining a preset level of comfort control as selected by the driver. Sensors are used to determine the present temperatures, and the system can adjust the level of heating or cooling as required.

The system uses actuators that open and close air-blend doors to achieve the desired in-vehicle temperature. Some systems control fan motor speeds to keep the temperature very close to what the driver requests.

Electronic Cruise Control Systems. Cruise control is a system that allows the vehicle to maintain a preset speed with the driver's foot off of the accelerator. Most cruise control systems are a combination of electrical and mechanical components.

Cruise control systems are also referred to as *speed control*.

Memory Seats. The **memory seat** feature allows the driver to program different seat positions that can be recalled at the push of a button. The memory seat feature is an addition to the basic power seat system. Most memory seat systems share the same basic operating principles, the difference being in programming methods and number of positions that can be programmed. Most systems provide for two seat positions to be stored in memory.

An **easy exit** feature may be an additional function of the memory seat that provides for easier entrance and exit of the vehicle by moving the seat all the way back and down. Some systems also move the steering wheel up and to full retract.

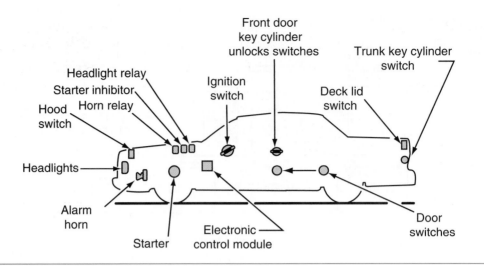

Figure 1-15 Typical components of an antitheft system.

Electronic Sunroofs. Some manufacturers have introduced electronic control of their electric sunroofs. These systems incorporate a pair of relay circuits and a timer function into the control module. Motor rotation is controlled by relays that are activated according to signals received from the slide, tilt, and limit switches.

Antitheft Systems. The **antitheft system** is a deterrent system designed to scare off would-be thieves by sounding alarms and/or disabling the ignition system. Figure 1-15 illustrates many of the common components that are used in an antitheft system. These components include:

1. An electronic control module.
2. Door switches at all doors.
3. Trunk key cylinder switch.
4. Hood switch.
5. Starter inhibitor relay.
6. Horn relay.
7. Alarm.

In addition, many systems incorporate the exterior lights into the system. The lights are flashed if the system is activated.

Some systems use ultrasonic sensors that will signal the control module if someone attempts to enter the vehicle through the door or window. The sensors can be placed to sense the parameter of the vehicle and sound the alarm if someone enters within the protected parameter distance.

Automatic Door Locks. **Automatic door locks (ADL)** use a passive system to lock all doors when the required conditions are met. Many automobile manufacturers are incorporating automatic door locks as an additional safety and convenience system. Most systems lock the doors when the gear selector is placed in DRIVE, the ignition switch in RUN, and all doors are shut. Some systems will lock the doors when the gear shift selector is passed through the REVERSE position, while others do not lock the doors unless the vehicle is moving 15 mph or faster.

The system may use the body computer or a separate controller to control the door lock relays. The controller (or body computer) takes the place of the door lock switches for automatic operation.

Keyless Entry. The **keyless entry system** allows the driver to unlock the doors or the deck lid (trunk) from outside of the vehicle without the use of a key. The main components of the

Figure 1-16 Keyless entry system keypad.

Figure 1-17 Remote keyless entry system transponder.

keyless entry system are the control module, a coded-button keypad located on the driver's door (Figure 1-16), and the door lock motors.

Some keyless entry systems can be operated remotely. Pressing a button on a hand-held transmitter will allow operation of the system from distances of 25 to 50 feet (Figure 1-17).

Electronic Heated Windshield. The **heated windshield system** is designed to melt ice and frost from the windshield three to five times faster than conventional defroster systems (Figure 1-18). The windshield undergoes a special process during manufacturing to allow for current flow through the glass without interfering with the driver's vision.

Intelligent Windshield Wipers. The **intelligent windshield wiper systems** use a monitoring system that usually consists of infrared light beams to detect if water is present on the windshield and to automatically turn on the wiper system (Figure 1-19). Once the sensing module determines the windshield requires wiper operation, it sends a request signal to the module that controls the wiper system.

Many systems will also include a feature that will automatically turn on the headlights if the wipers are activated. This is especially useful in those states and countries that have laws requiring the headlights to be turned on anytime the wipers are active.

Electronic Shift Transmissions. The use of solenoids and relays in controlling the operation of the engine has been expanded to include the drivetrain. Many of today's vehicles are equipped

Figure 1-18 The heated windshield removes ice and frost from the windshield in just a few minutes.

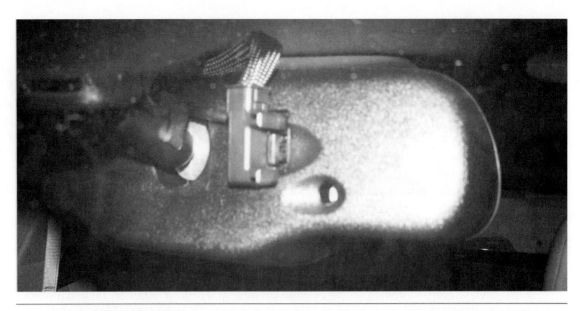

Figure 1-19 Automatic wiper system sensor and module.

with **electronic shift automatic transmissions.** The control module uses several inputs to determine torque converter clutch operation, hydraulic pressure levels, and shift points. The use of electronics within the transmission has improved shift quality, fuel economy, and towing abilities (Figure 1-20). In addition, some manual transmissions will have electronically controlled actuators to control clutch and shift operation.

Speed-Sensitive Steering. Most vehicles are available with some form of power steering, either as an option or as standard equipment. Conventional power steering systems provided for a certain degree of assist to the driver when turning the steering wheel. The disadvantage of conventional

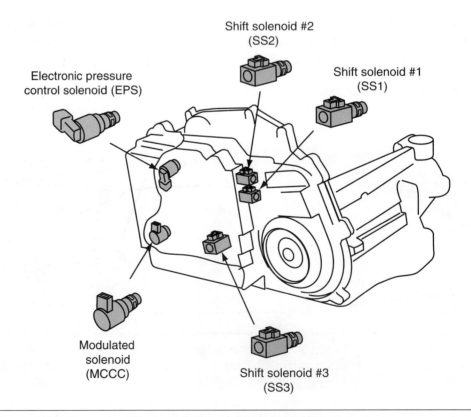

Figure 1-20 Electronic shift transmission output actuators.

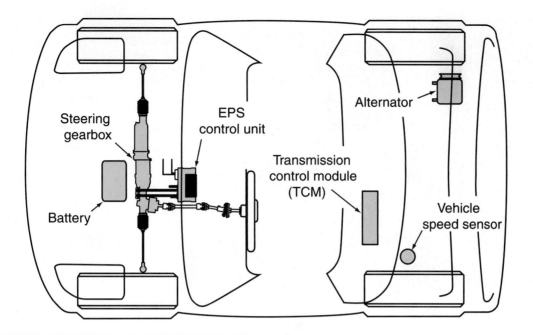

Figure 1-21 The Honda/Acura NSX electronic rack and pinion system.

power steering is the reduced road feel that the system offers during medium and high speeds. At these speeds it is desirable for a feeling of increased control and performance. Through the use of electronic controls, the advantages of high power assist and excellent road feel can be achieved.

Manufacturers have chosen different methods of accomplishing **speed-sensitive steering.** The most common is to use a means of varying the output of the power steering pump. Other methods include the use of electric motor–driven power steering pumps and electric rack and pinion steering (Figure 1-21).

Electronic Suspension Systems. Electronic suspension systems can be either adaptive or active type systems. **Adaptive suspension systems** are able to change ride characteristics by altering shock damping and ride height continuously (Figure 1-22). **Active suspension systems**

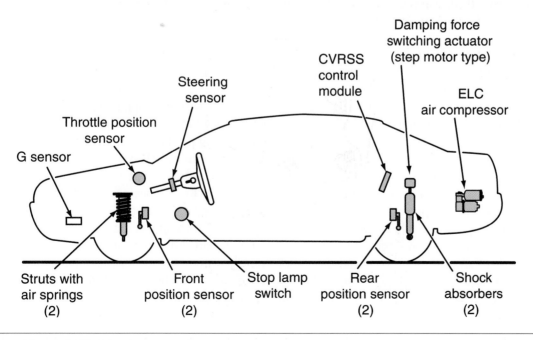

Figure 1-22 Adaptive suspension system components.

15

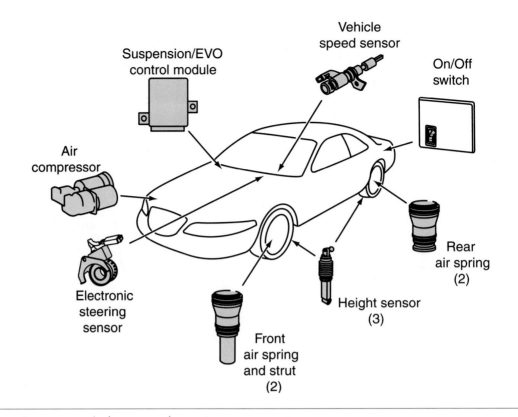

Figure 1-23 Active suspension system components.

are controlled by double-acting hydraulic cylinders or solenoids (actuators) mounted at each wheel. The actuators, instead of conventional springs or air springs, support the vehicle's weight (Figure 1-23).

Adaptive and active suspension systems provide additional benefits over conventional passive suspension systems. They are able to change ride height, shock damping, and spring rates in response to changing road and driving conditions. Electronic suspension system types can vary from basic shock-damping variations to a complex system of height and ride control that utilizes extensive computer programming.

Antilock Brake Systems. The modern brake system is more than adequate to stop the vehicle under normal conditions; however, in approximately 1% of its use it will fail to stop the vehicle safely. This failure is generally the fault of the driver, who allowed his or her vehicle to enter an uncontrollable skid. Wheel lockup during braking will increase the stopping distance. A good driver knows that "pumping" the brakes during an emergency keeps the vehicle from entering into an uncontrollable skid. A tire that is on the verge of slipping produces more friction with respect to the road than one that is locked. The **antilock brake system (ABS)** is designed to act similarly to a driver pumping the brakes, but with much more control and at a much faster rate.

The ABS system is capable of pumping each brake up to 15 times per second. Usually, the control module can pulse the two front brakes separately and the rear brakes as a pair. ABS automatically stops the vehicle in the shortest possible distance, without locking a wheel. In addition, ABS maintains directional control on almost any type of road surface or condition.

There are several versions and generations of ABS. The basic categories are integrated (Figure 1-24) and nonintegrated (Figure 1-25), and 2-wheel and 4-wheel ABS. System operation of electronically controlled ABS is similar; the differences are basically in component design.

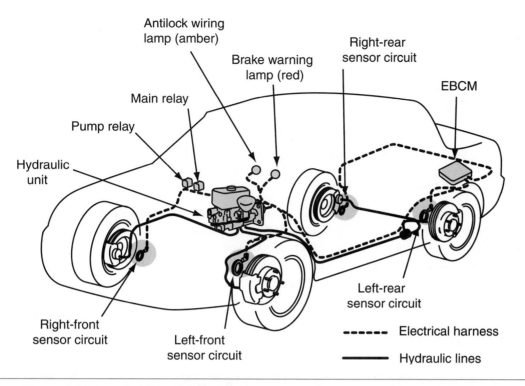

Figure 1-24 Integrated ABS braking system.

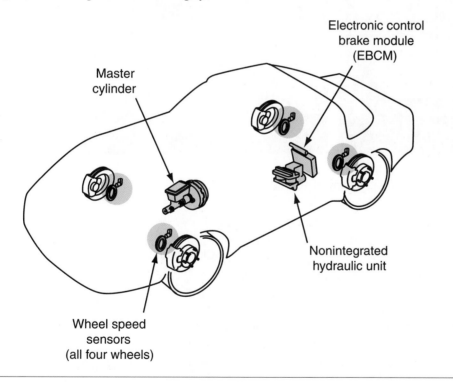

Figure 1-25 Nonintegrated ABS system.

Automatic Traction Control. **Automatic traction control** limits the amount of tire spin on slippery road conditions by applying the brakes automatically. The same technology that is used for ABS systems is also applied to automatic traction control. An electronic control monitor monitors the wheel speed sensors. If it determines that one of the drive wheels is spinning faster than the other, it will automatically apply the brakes to the spinning wheel. With the brake applied, it requires a greater amount of torque to spin the wheel. Since a differential delivers equal torque to each drive wheel, the greater torque requirement is also transferred to the stationary or slower-moving wheel. This allows the wheel that has the greatest amount of traction ability to move the vehicle.

Electronic Stability Control. The next evolution of the antilock brake and traction control system is the incorporation of **electronic stability control.** This system uses additional sensors and inputs to determine if the vehicle is actually moving in the direction intended by the driver, as indicated by steering wheel position sensors, yaw sensors, and lateral sensors. If the actual path is not the intended path, the module will apply the appropriate brake to bring the vehicle back onto the correct path.

Passive Restraint Systems

Federal regulations have mandated the use of automatic **passive restraint systems** in all vehicles sold in the United States after 1990. Passive restraints are ones that operate automatically, with no action required on the part of the driver or occupant.

Air bag systems are on all of today's vehicles. The need to supplement the existing restraint system during frontal collisions has led to the development of the supplemental inflatable restraint (SIR) or air bag systems (Figure 1-26).

A typical air bag system consists of sensors, a diagnostic module, a clock spring, and an air bag module. Figure 1-27 illustrates the typical location of the common components of the SIR system.

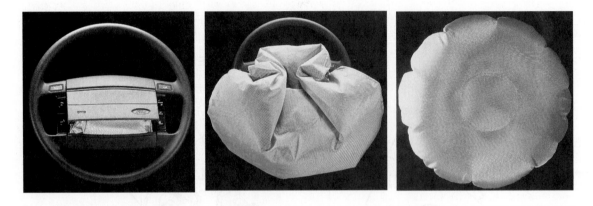

Figure 1-26 Air bag deployment sequence.

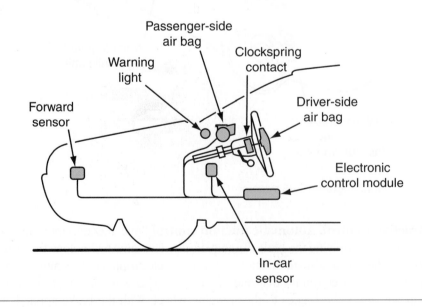

Figure 1-27 Typical location of components of the air bag system.

Additional Electrical Features

Additional systems that will be discussed in this book include vehicle audio entertainment, navigational, hands-free cell phone, and park assist systems.

Alternate Propulsion Systems

Due to the increase in regulations concerning emissions and the public's desire to become less dependent on foreign oil, most major automotive manufacturers have developed alternative fuel or alternate power vehicles. Since the 1990s, most major automobile manufacturers have developed an **electric vehicle (EV).** The primary advantage of an EV is a drastic reduction in noise and emission levels. General Motors introduced the EV1 electric car to the market in 1996. The original battery pack in this car contained twenty-six 12-volt batteries that delivered electrical energy to a three-phase 102-kilowatt (kW) AC electric motor. The electric motor is used to drive the front wheels. The driving range is about 70 miles (113 km) of city driving or 90 miles (145 km) of highway driving.

EV battery limitation was a major stumbling block to most consumers. One method of improving the electric vehicle resulted in the addition of an on-board power generator that is assisted by an internal combustion engine, resulting in the **hybrid electric vehicle (HEV).**

Basically, the hybrid electric vehicle relies on power from the electric motor, the engine, or both (Figure 1-28). When the vehicle moves from a stop and has a light load, the electric motor moves the vehicle. Power for the electric motor comes from stored electricity in the battery pack. During normal driving conditions, the engine is the main power source. Engine power is also used to rotate a generator that recharges the storage batteries. The output from the generator may also be used to power the electric motor, which is run to provide additional power to the powertrain. A computer controls the operation of the electric motor, depending on the power needs of the vehicle. During full throttle or heavy load operation, additional electricity from the battery is sent to the motor to increase the output of the powertrain.

Fuel cell–powered vehicles have a very good chance of becoming the drives of the future. They combine the reach of conventional internal combustion engines with high efficiency, low fuel consumption, and minimal or no pollutant emission. At the same time, they are extremely quiet. Because they work with regenerative fuel such as hydrogen, they reduce the dependence on crude oil and other fossil fuels.

A fuel cell–powered vehicle (Figure 1-29) is basically an electric vehicle. Like the electric vehicle, it uses an electric motor to supply torque to the drive wheels. The difference is that the fuel cell produces and supplies electric power to the electric motor instead of batteries. Most of

Figure 1-28 HEV power system. (Reprinted with permission)

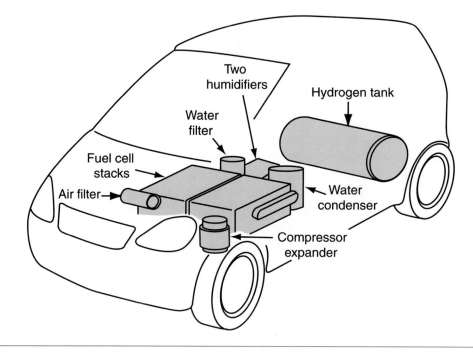

Figure 1-29 Fuel cell vehicle components.

the vehicle manufacturers and several independent laboratories are involved in fuel cell research and development programs. A number of prototype fuel cell vehicles have been produced, with many being placed in fleets in North America and Europe.

Summary

❏ The starting system is a combination of mechanical and electrical parts that work together to start the engine.

❏ The charging system replaces the electrical power used by the battery and to provide current to operate all of the electrical accessories while the engine is running.

❏ The ignition system must supply properly timed high-voltage surges to the spark plugs.

❏ The lighting system consists of all of the lights used on the vehicle.

❏ Vehicle instrumentation systems monitor the various vehicle operating systems and provide information to the driver.

❏ Electrical accessories provide additional safety and comfort.

❏ Many of the basic electrical accessory systems have electronic controls added to them to provide additional features and enhancement.

❏ Computers are electronic devices that gather, store, and process data.

❏ Most vehicles use a multiplexing system to share information between computer systems.

❏ The memory seat feature allows the driver to program different seat positions that can be recalled at the push of a button.

❏ Some manufacturers have introduced electronic control of their electric sunroofs. These systems incorporate a pair of relay circuits and a timer function into the control module.

❏ Antitheft systems are deterrent systems designed to scare off would-be thieves by sounding alarms and/or disabling the ignition system.

❏ Automatic door locks is a passive system used to lock all doors when the required conditions are met. Many automobile manufacturers are incorporating the system as an additional safety and convenience feature.

❏ The heated windshield system is designed to melt ice and frost from the windshield three to five times faster than conventional defroster systems.

❏ The use of electronics within the transmission has improved shift quality and fuel economy.

❏ In speed-sensitive steering systems, steering effort is controlled based on vehicle speed and rate of steering wheel rotation.

❏ Adaptive and active suspension systems provide additional benefits over conventional passive suspension systems by varying ride height, shock damping, and spring rates in response to changing road and driving conditions.

❏ The antilock braking system (ABS) is designed to act like a driver pumping the brakes, but with much more control and at a much faster rate.

❏ Automatic traction control limits the amount of tire spin on slippery road conditions by applying the brakes automatically.

❏ Passive restraints operate automatically, with no action required on the part of the driver or occupant.

❏ Electric vehicles powered by an electric motor run off a battery pack.

❏ The hybrid electric vehicle relies on power from the electric motor, the engine, or both.

❏ A fuel cell–powered vehicle is basically an electric vehicle, except that the fuel cell produces and supplies electric power to the electric motor instead of batteries.

Review Questions

Short-Answer Essays

1. Describe your level of comfort concerning automotive electrical systems.
2. Explain why you feel it is important to understand the operation of the automotive electrical system.
3. Explain how the use of computers has changed the automotive electrical system.
4. What advantages does the electronic shift transmission offer?
5. What is the purpose of antilock braking systems?
6. Describe the purpose of automatic traction control systems.
7. What safety benefits can be achieved from the automatic door lock system?
8. What is the purpose of the starting system?
9. What is the purpose of the charging system?
10. What is the function of the air bag system?

Fill in the Blanks

1. The _____ _____ is the heart of the ignition system.
2. Vehicle instrumentation systems _____ the various operating systems and provide information to the driver about their correct operation.

Terms to Know
(continued)
Electrical accessories
Electronic shift automatic transmissions
Electronic stability control
Electronic suspension systems
Fuel cell
Heated windshield system
Horn
Hybrid electric vehicle (HEV)
Ignition coil
Ignition switch
Ignition system
Intelligent windshield wiper systems
Keyless entry system
Lighting system
Memory seat
Multiplexing
Network
Neutral safety switch
Passive restraint systems
Power door locks
Power mirrors
Power windows
Speed-sensitive steering
Starting system
Vehicle instrumentation systems
Voltage regulator
Windshield wipers

3. A _____ is an electronic device that stores and processes data.

4. Automatic traction control limits the amount of tire spin on slippery road conditions by _____ the brakes automatically.

5. The _____ _____ feature is an additional function of the memory seat that provides for easier entrance and exit of the vehicle.

6. The antilock brake system (ABS) is designed to act in a similar manner as a driver _____ the brakes, but with much more control and at a much faster rate.

7. The _____ system is a deterrent system.

8. The intelligent windshield wiper system uses a monitoring system that usually consists of _____ _____ beams to detect if water is present.

9. _____ restraints operate automatically with no action required on the part of the driver.

10. The _____ _____ _____ uses an on-board power generator that is assisted by an internal combustion engine.

Multiple Choice

1. Electric vehicles power the motor by:
   A. A generator.
   B. A battery pack.
   C. An engine.
   D. None of the above.

2. Automatic traction control:
   A. Automatically speeds up the wheel that is too slow.
   B. Automatically applies the brake for the wheel that is spinning.
   C. Is only active after 50 mph.
   D. Increases the torque output of the engine to move the vehicle.

3. The memory seat system:
   A. Operates separately of the power seat system.
   B. Requires the vehicle to be moving before the seat position can be recalled.
   C. Allows for the driver to program different seat positions that can be recalled at the push of a button.
   D. Can only be equipped on vehicles with manual position seats.

4. The following are true about the easy exit feature EXCEPT:
   A. It is an additional function of the memory seat.
   B. The driver's door is opened automatically.
   C. The seat is moved all the way back and down.
   D. The system may move the steering wheel up.

5. The following are components of the starting system EXCEPT:
   A. The flywheel ring gear.
   B. Neutral safety switch.

C. Harmonic balancer.
D. Battery.

6. The purpose of the ignition coil is to:
   A. Build up low voltage to a voltage high enough to jump the spark plug gap.
   B. Determine engine speed.
   C. Locate top dead center of the compression stroke.
   D. Provide low-voltage protection to the computer.

7. Automotive horns operate on the principle of:
   A. Induced voltage.
   B. Depletion zone bonding.
   C. Frequency modulation.
   D. Electromagnetism.

8. The purpose of multiplexing is to:
   A. Increase circuit loads to a sensor.
   B. Prevent electromagnetic interference.
   C. Allow computers to share information.
   D. Prevent multiple system failures from occurring.

9. The following are true about the air bag system EXCEPT:
   A. It is an active system.
   B. It is a supplemental system.
   C. It is mandated by the federal government.
   D. Deployment is automatic.

10. Alternate propulsion systems include:
    A. Electric vehicles.
    B. Hybrid vehicles.
    C. Fuel cell vehicles.
    D. All of the above.

Basic Theories

Upon completion and review of this chapter, you should be able to:

❑ Explain the theories and laws of electricity.

❑ Describe the difference between insulators, conductors, and semiconductors.

❑ Define voltage, current, and resistance.

❑ Define and use Ohm's law correctly.

❑ Explain the basic concepts of capacitance.

❑ Explain the difference between AC and DC currents.

❑ Define and illustrate series, parallel, and series-parallel circuits and the electrical laws that govern them.

❑ Explain the theory of electromagnetism.

❑ Explain the principles of induction.

Introduction

The electrical systems used in today's vehicles can be very complicated (Figure 2-1). However, through an understanding of the principles and laws that govern electrical circuits, technicians can simplify their job of diagnosing electrical problems. In this chapter, you will learn the laws that dictate electrical behavior, how circuits operate, the difference between types of circuits, and how to apply Ohm's law to each type of circuit. You will also learn the basic theories of semiconductor construction. Because magnetism and electricity are closely related, a study of electromagnetism and induction is included in this chapter.

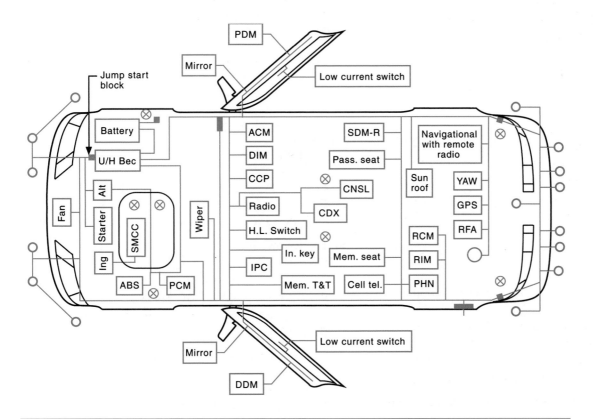

Figure 2-1 The electrical system of today's vehicle can be complicated.

Basics of Electron Flow

Because electricity is an energy form that cannot be seen, some technicians regard the vehicle's electrical system as being more complicated than it is. These technicians approach the vehicle's electrical system with some reluctance. It is important for today's technician to understand that electrical behavior is confined to definite laws that produce predictable results and effects. To facilitate the understanding of the laws of electricity, a short study of atoms is presented.

Atomic Structure

An **atom** is the smallest part of a chemical element that still has all the characteristics of that element. An atom is constructed of a fixed arrangement of **electrons** in orbit around a **nucleus**—much like planets orbiting the sun (Figure 2-2). **Electrons** are negatively charged particles. The nucleus contains positively charged particles called **protons** and particles that have no charge, which are called **neutrons.** The protons and neutrons that make up the nucleus are tightly bound together. The electrons are free to move within their orbits at fixed distances around the nucleus. The attraction between the negative electrons and the positive protons causes the electrons to orbit the nucleus. All of the electrons surrounding the nucleus are negatively charged, so they repel each other when they get too close. The electrons attempt to stay as far away from each other as possible without leaving their orbits.

Atoms attempt to have the same number of electrons as there are protons in the nucleus. This makes the atom **balanced** (Figure 2-3). To remain balanced, an atom will shed an electron

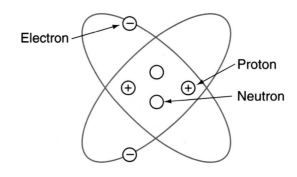

Figure 2-2 The basic construction of an atom.

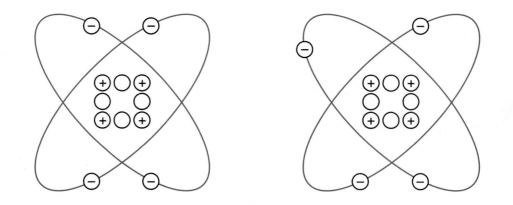

Figure 2-3 If the number of electrons and protons in an atom are the same, the atom is balanced.

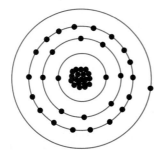

Figure 2-4 Basic structure of a copper atom.

or attract an electron from another atom. A specific number of electrons are in each of the electron orbit paths. The orbit closest to the nucleus has room for 2 electrons; the second orbit holds up to 8 electrons; the third holds up to 18; and the fourth and fifth hold up to 32 each. The number of orbits depends on the number of electrons the atom has. For example, a copper atom contains 29 electrons; 2 in the first orbit, 8 in the second, 18 in the third orbit, and 1 in the fourth (Figure 2-4). The outer orbit, or **shell** as it is sometimes called, is referred to as the **valence ring.** This is the orbit we care about in our study of electricity.

In studying the laws of electricity, the only concern is with the electrons that are in the valence ring. Since an atom seeks to be balanced, an atom that is missing electrons in its valence ring will attempt to gain other electrons from neighboring atoms. Also, if the atom has an excess amount of electrons in its valence ring, it will try to pass them on to neighboring atoms.

Like charges repel each other; unlike charges attract each other.

Conductors and Insulators

To help explain why you need to know about these electrons and their orbits, let's continue to look at the atomic structure of copper. Copper is a metal and is the most commonly used **conductor** of electricity. A conductor is something that supports the flow of electricity through it. As stated earlier, the copper atom has 29 electrons and 29 protons, but there is only 1 electron in the valence ring. For the valence ring to be completely filled, it would require 32 electrons. Since there is only 1 electron, it is loosely tied to the atom and can be easily removed, making it a good conductor.

Copper, silver, gold, and other good conductors of electricity have only one or two electrons in their valence ring. These atoms can be made to give up the electrons in their valence ring with little effort.

Since electricity is the movement of electrons from one atom to another, atoms that have one to three electrons in their valence ring support electricity. They allow the electron to easily move from the valence ring of one atom to the valence ring of another atom. Therefore, if we have a wire made of millions of copper atoms, we have a good conductor of electricity. To have electricity, we simply need to add one electron to one of the copper atoms. That atom will shed the electron it had to another atom, which will shed its original electron to another, and so on. As the electrons move from atom to atom, a force is released. This force is what we use to light lamps, run motors, and so on. As long as we keep the electrons moving in the conductor, we have electricity.

Insulators are materials that don't allow electrons to flow through them easily. Insulators are atoms that have five to eight electrons in their valence ring. The electrons are held tightly around the atom's nucleus and they can't be moved easily. Insulators are used to prevent electron flow or to contain it within a conductor. Insulating material covers the outside of most conductors to keep the moving electrons within the conductor.

In summary, the number of electrons in the valence ring determines whether an atom is a good conductor or insulator. Some atoms are not good insulators or conductors; these are called **semiconductors.** In short:

1. Three or fewer electrons—conductor.

2. Five or more electrons—insulator.

3. Four electrons—semiconductor.

A BIT OF HISTORY

Electricity was discovered by the Greeks over 2,500 years ago. They noticed that when amber was rubbed with other materials it was charged with an unknown force that had the power to attract objects, such as dried leaves and feathers. The Greeks called amber "elektron." The word *electric* is derived from this word and means "to be like amber."

Electricity Defined

Random movement of electrons is not electric current; the electrons must move in the same direction.

The speed of light is 186,000 miles per second (299,000 kilometers per second).

An *E* can be used for the symbol to designate voltage (electromotive force). A *V* is also used as a symbol for voltage.

Electricity is the movement of electrons from atom to atom through a conductor (Figure 2-5). Electrons are attracted to protons. Since we have excess electrons on the other end of the conductor, we have many electrons being attracted to the protons. This attraction sort of pushes the electrons toward the protons. This push is normally called electrical pressure. The amount of electrical pressure is determined by the number of electrons that are attracted to protons. The electrical pressure or **electromotive force (EMF)** attempts to push an electron out of its orbit and toward the excess protons. If an electron is freed from its orbit, the atom acquires a positive charge because it now has one more proton than it has electrons. The unbalanced atom or **ion** attempts to return to its balanced state so it will attract electrons from the orbit of other **balanced atoms.** This starts a chain reaction as one atom captures an electron and another releases an electron. As this action continues to occur, electrons will flow through the conductor. A stream of free electrons forms and an electrical current is started. This does not mean a single electron travels the length of the insulator; it means the overall effect is electrons moving in one direction. All this happens at the speed of light. The strength of the electron flow is dependent on the potential difference or voltage.

The three elements of electricity are voltage, current, and resistance. How these three elements interrelate governs the behavior of electricity. Once the technician comprehends the laws that govern electricity, understanding the function and operation of the various

Conductor

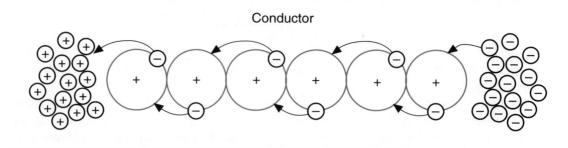

Figure 2-5 As electrons flow in one direction from one atom to another, an electrical current is developed.

automotive electrical systems is an easier task. This knowledge will assist the technician in diagnosis and repair of automotive electrical systems.

Voltage

Voltage can be defined as an electrical pressure (Figure 2-6) and is the electromotive force (EMF) that causes the movement of the electrons in a conductor. In Figure 2-5, voltage is the force of attraction between the positive and negative charges. An electrical pressure difference is created when there is a mass of electrons at one point in the circuit, and a lack of electrons at another point in the circuit. In the automobile, the battery or generator is used to apply the electrical pressure.

The amount of pressure applied to a circuit is stated in the number of volts. If a voltmeter is connected across the terminals of an automobile battery, it may indicate 12.6 volts. This is actually indicating that there is a difference in potential of 12.6 volts. There is 12.6 volts of electrical pressure between the two battery terminals.

In a circuit that has current flowing, voltage will exist between any two points in that circuit (Figure 2-7). The only time voltage does not exist is when the potential drops to zero. In Figure 2-7 the voltage potential between points A and C and between points B and C is 12.6 volts. However, between points A and B the pressure difference is zero and the voltmeter will indicate 0 volts.

Shop Manual
Chapter 2, page 39

One volt (V) is the amount of pressure required to move one ampere of current through one ohm of resistance.

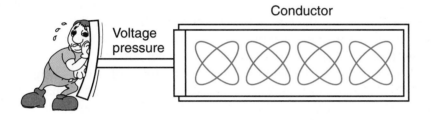

Figure 2-6 Voltage is the pressure that causes the electrons to move.

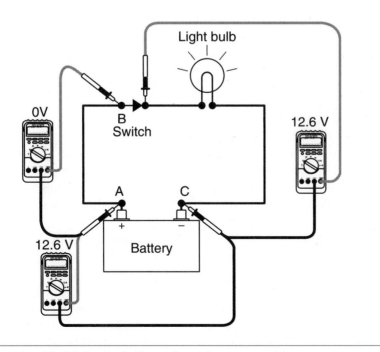

Figure 2-7 A simplified light circuit illustrating voltage potential.

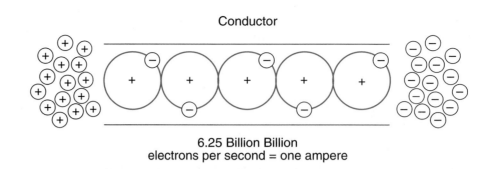

Conductor

6.25 Billion Billion
electrons per second = one ampere

Figure 2-8 The rate of electron flow is called current and is measured in amperes.

Shop Manual
Chapter 2, page 45

One **ampere** (A) represents the movement of 6.25 billion billion electrons (or one coulomb) past one point in a conductor in one second.

The symbol for current is I.

Current

Current can be defined as the *rate* of electron flow (Figure 2-8) and is measured in **amperes.** Current is a measurement of the electrons passing any given point in the circuit in one second. Because the flow of electrons is at the speed of light, it would be impossible to physically see electron flow. However, the rate of electron flow can be measured. Current will increase as pressure or voltage is increased—provided circuit resistance remains constant.

An electrical current will continue to flow through a conductor as long as the electromotive force is acting on the conductor's atoms and electrons. If a potential exists in the conductor, with a build up of excess electrons at the end of the conductor farthest from the EMF and there is a lack of electrons at the EMF side, current will flow. The effect is called electron drift and accounts for the method in which electrons flow through a conductor.

An electrical current can be formed by the following forces: friction, chemical reaction, heat, pressure, and magnetic induction. Whenever electrons flow or drift in mass, an electrical current is formed. There are six laws that regulate this electrical behavior:

1. Like charges repel each other.
2. Unlike charges attract each other.
3. A voltage difference is created in the conductor when an EMF is acting on the conductor.
4. Electrons flow only when a voltage difference exists between the two points in a conductor.
5. Current tends to flow to ground in an electrical circuit as a return to source.
6. **Ground** is defined as the common negative connection of the electrical system and is the point of lowest voltage. The ground circuit used in most automotive systems is through the vehicle chassis and/or engine block.

So far we have described current as the movement of electrons through a conductor. Electrons move because of a potential difference. This describes one of the common theories about current flow. The **electron theory** states that since electrons are negatively charged, current flows from the most negative to the most positive point within an electrical circuit. In other words, current flows from negative to positive. This theory is widely accepted by the electronic industry.

Another current flow theory is called the **conventional theory.** This states that current flows from positive to negative. The basic idea behind this theory is simply that although electrons move toward the protons, the energy or force that is released as the electrons move begins at the point where the first electron moved to the most positive charge. As electrons continue to move in one direction, the released energy moves in the opposite direction. This theory is the oldest theory and serves as the basis for most electrical diagrams.

Trying to make sense of it all may seem difficult. It is also difficult for scientists and engineers. In fact, another theory has been developed to explain the mysteries of current flow. This theory is called the hole-flow theory and is actually based on both electron theory and the conventional theory.

As a technician, you will find references to all of these theories. Fortunately, it really doesn't matter as long as you know what current flow is and what affects it. From this understanding, you will be able to figure out how the circuit basically works, how to test it, and how to repair it. In this text, we will present current flow as moving from positive to negative and electron flow as moving from negative to positive. Remember that current flow is the result of the movement of electrons, regardless of the theory.

The ampere is named after André Ampère, who in the late 1700s worked with magnetism and current flow to develop some foundations for understanding the behavior of electricity.

Shop Manual
Chapter 2, page 43

Resistance

The third component in electricity is **resistance.** Resistance is the opposition to current flow and is measured in **ohms.** In a circuit, resistance controls the amount of current. The size, type, length, and temperature of the material used as a conductor will determine its resistance. Devices that use electricity to operate (motors and lights) have a greater amount of resistance than the conductor.

A complete electrical **circuit** consists of the following: (1) a power source, (2) a load or resistance unit, and (3) conductors. Resistance (load) is required to change electrical energy to light, heat, or movement. There is resistance in any working device of a circuit, such as a lamp, motor, relay, coil, or other load component.

The ohm (Ω) is the unit of measurement for resistance of a conductor such that a constant current of 1 ampere in it produces a voltage of 1 volt between its ends.

The symbol for resistance is *R.*

There are five basic characteristics that determine the amount of resistance in any part of a circuit:

1. The atomic structure of the material: The higher the number of electrons in the outer valence ring, the higher the resistance of the material.

2. The length of the conductor: The longer the conductor, the higher the resistance.

3. The diameter of the conductor: The smaller the cross-sectional area of the conductor, the higher the resistance.

4. Temperature: Normally an increase of temperature of the conductor causes an increase in the resistance.

5. Physical condition of the conductor: If the conductor is damaged by nicks or cuts, the resistance will increase because the conductor's diameter is decreased by these.

There may be unwanted resistance in a circuit. This could be in the form of a corroded connection or a broken conductor. In these instances, the resistance may cause the load component to operate at reduced efficiency or to not operate at all.

It does not matter if the resistance is from the load component or from unwanted resistance. There are certain principles that dictate its impact in the circuit:

1. Voltage always drops as current flows through the resistance.

2. An increase in resistance causes a decrease in current.

3. All resistances change the electrical energy into heat energy to some extent.

Voltage Drop Defined

Voltage drop occurs when current flows through a load component or resistance. Voltage drop is the amount of electrical pressure lost or consumed as it pushes current flow through a resistance. Electricity is an energy. Energy cannot be created or destroyed but it can be changed. As electrical energy flows through a resistance, it is converted to some other form of energy, usually heat energy. The amount of voltage drop over a resistance or load device is an indication of how much electrical energy was converted to another energy form. After a resistance, the voltage is lower than it was before the resistance.

Voltage drop can be measured by using a voltmeter (Figure 2-9). With current flowing through a circuit, the voltmeter may be connected in parallel over the resistor, wire, or component to measure voltage drop. The voltmeter indicates the amount of voltage potential between two points in the circuit. The voltmeter reading indicates the difference between the amount of voltage available to the resistor and the amount of voltage after the resistor.

There must be a voltage present for current to flow through a resistor. Kirchhoff's law basically states that the sum of the voltage drops in an electrical circuit will always equal source voltage. In other words, all of the source's voltage is used by the circuit.

A BIT OF HISTORY

Gustav Kirchhoff was a German scientist who, in the 1800s, discovered two facts about the characteristics of electricity. One is called his voltage law, which states "The sum of the voltage drops across all resistances in a circuit must equal the voltage of the source." His law on current states "The sum of the currents flowing into any point in a circuit equals the sum of the currents flowing out of the same point." These laws describe what happens when electricity is applied to a load. Voltage drops while current remains constant; current does not drop.

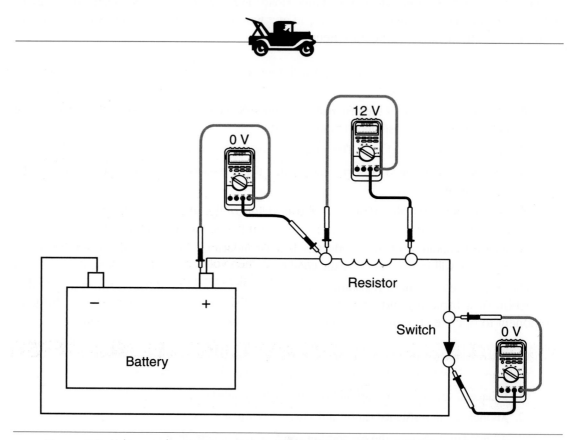

Figure 2-9 Using a voltmeter to measure voltage drop in different locations of a circuit.

Electrical Laws

Electricity is governed by well-defined laws. The most fundamental of these are Ohm's law and Watt's law. Today's technician must understand these laws in order to completely grasp electrical theory.

Ohm's Law

Understanding **Ohm's law** is the key to understanding how electrical circuits work. Ohm's law defines the relationship between current, voltage, and resistance. The law states that it takes one volt of electrical pressure to push one ampere of electrical current through one ohm of electrical resistance. This law can be expressed mathematically as:

1 Volt = 1 Ampere × 1 Ohm

This formula is most often expressed as: $V = A \times R$. V stands for voltage (electrical pressure), A stands for ampere (current), and R represents resistance. This formula is often used to find the amount of one electrical characteristic when the other two are known. As an example: if we have 2 amps of current and 6 ohms of resistance in a circuit, we must have 12 volts of electrical pressure.

V = 2 Amps × 6 Ohms   V = 2 × 6   V = 12 Volts

If we know the voltage and resistance but not the current of a circuit, we can quickly calculate it by using Ohm's law. Since $V = A \times R$, A would equal V divided by R. Let's supply some numbers to this. If we have a 12-volt circuit with 6 ohms of resistance, we can determine the amount of current in this way:

$$A = \frac{V}{R} \quad \text{or} \quad \frac{12 \text{ Volts}}{6 \text{ Ohms}} \quad \text{or } A = 2 \text{ Amps}$$

The same logic is used to calculate resistance when voltage and current are known. $R = V/A$. One easy way to remember the formulas of Ohm's law is to draw a circle and divide it into three parts as shown in Figure 2-10. Simply cover the value you want to calculate. The formula you need to use is all that shows.

To show how easily this works, consider the 12-volt circuit in Figure 2-11. This circuit contains a 3-ohm light bulb. To determine the current in the circuit, cover the A in the circle to expose the formula $A = V/R$. Then plug in the numbers, $A = 12/3$. Therefore, the circuit current is 4 amperes.

To further explore how Ohm's law works, refer to Figure 2-12 of a simple circuit. If the battery voltage is 12 volts and the amperage is 24 amperes, the resistance of the lamp can be determined using $R = V/A$. In this instance, the resistance is .5 Ω. For another example, if the resistance of the bulb is 2 Ω and the amperage is 12, then voltage can be found using $V = A \times R$. In this instance, voltage would equal 24 volts. Now, if the circuit had 12 volts and the resistance of the bulb was 2 Ω, then amperage could be determined by $A = V/R$. In this case, amperage would be 6 amperes.

> The resistance of an actual lamp in an automotive application will change when current passes through it because its temperature changes.

A BIT OF HISTORY

Georg S. Ohm was a German scientist in the 1800s who discovered that all electrical quantities are proportional to each other and therefore have a mathematical relationship.

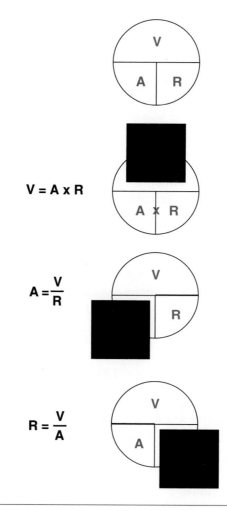

$$V = A \times R$$

$$A = \frac{V}{R}$$

$$R = \frac{V}{A}$$

Figure 2-10 The mathematical formula for Ohm's law using a circle to help understand the different formulas that can be derived from it. To expose the formula to use, cover the unknown value.

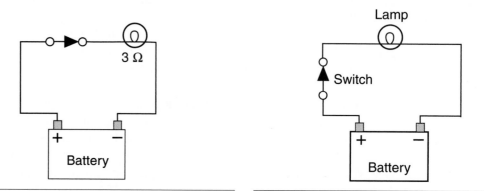

Figure 2-11 Simplified light circuit with 3 ohms of resistance in the lamp.

Figure 2-12 Simple lamp circuit to help use Ohm's law.

Ohm's law is the basic law of electricity. It states that the amount of current in an electric circuit is inversely proportional to the resistance of the circuit and it is directly proportional to the voltage in the circuit. For example, if the resistance decreases and the voltage remains constant, the amperage will increase. If the resistance stays the same and the voltage increases, the amperage will also increase.

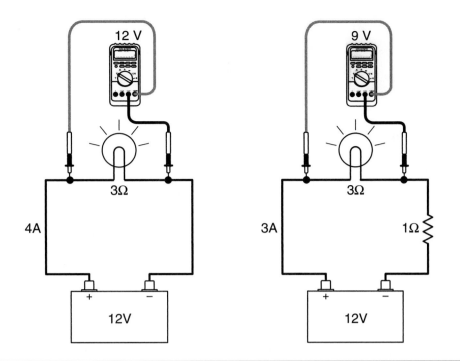

Figure 2-13 The light circuit in Figure 2-11 shown with normal circuit values and with added resistance in series.

For example, refer to Figure 2-13; on the left side is a 12-volt circuit with a 3-ohm light bulb. This circuit will have 4 amps of current flowing through it. If a 1-ohm resistor is added to the same circuit (as shown to the right in Figure 2-13), total resistance is now 4 ohms. Because of the increased resistance, current dropped to 3 amps. The light bulb will be powered by less current and will be less bright than it was before the additional resistance was added.

Another point to consider is voltage drop. Before adding the 1-ohm resistor, the source voltage (12 volts) was dropped by the light bulb. With the additional resistance, the voltage drop of the light bulb decreased to 9 volts. The remaining 3 volts were dropped by the 1-ohm resistor. This can be proven by using Ohm's law. When the circuit current was 4 amps, the light bulb had 3 ohms of resistance. To find the voltage drop, we multiply the current by the resistance.

$$V = A \times R \quad \text{or} \quad V = 4 \times 3 \quad \text{or} \quad V = 12$$

When the extra resistor was added to the circuit, the light bulb still had 3 ohms of resistance, but the current in the circuit decreased to 3 amps. Again voltage drop can be determined by multiplying the current by the resistance.

$$V = A \times R \quad \text{or} \quad V = 3 \times 3 \quad \text{or} \quad V = 9$$

The voltage drop of the additional resistor is calculated in the same way: $V = A \times R$ or $V = 3$ volts. The total voltage drop of the circuit is the same for both circuits. However, the voltage drop at the light bulb changed. This also would cause the light bulb to be dimmer. Ohm's law and its application will be discussed in greater detail later.

Watt's Law

Power (P) is the rate of doing electrical work. Power is expressed in **watts.** A watt is equal to 1 volt multiplied by 1 ampere. There is another mathematical formula that expresses the

It is possible to convert horsepower ratings to electrical power rating using the conversion factor: 1 horsepower equals 746 watts.

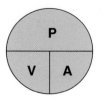

Figure 2-14 The mathematical formula for Watt's law.

relationship between voltage, current, and power. It is simply: P = V × A (Figure 2-14). Power measurements are measurements of the rate at which electricity is doing work.

The best examples of power are light bulbs. Household light bulbs are sold by wattage. A 100-watt bulb is brighter and uses more electricity than a 60-watt bulb.

Referring back to Figure 2-13, the light bulb in the circuit on the left had a 12-volt drop at 4 amps of current. We can calculate the power the bulb uses by multiplying the voltage and the current.

$$P = V \times A \quad \text{or} \quad P = 12 \times 4 \quad \text{or} \quad P = 48$$

The power output of the bulb is 48 watts. When the resistor was added to the circuit, the bulb dropped 9 volts at 3 amps of current. The power of the bulb is calculated in the same way as before.

$$P = V \times A \quad \text{or} \quad P = 9 \times 3 \quad \text{or} \quad P = 27$$

This bulb produced 27 watts of power, a little more than half of the original. It would be almost half as bright. The key to understanding what happened is to remember the light bulb didn't change; the circuit changed.

Another example of using Watt's law is to determine the amperage if an additional accessory is added to the vehicle's electrical system. If the accessory is rated at 75 watts, the amperage draw would be:

$$A = P / V = 75 / 12 = 6.25 \text{ amps}$$

This tells the technician that this circuit will probably require a 10-amp rated fuse.

Types of Current

There are two classifications of electrical current flow: direct current (DC) and alternating current (AC). The type of current flow is determined by the direction it flows and by the type of voltage that drives it.

Direct Current

DC voltage is created and stored in the automotive battery.

Direct current (DC) can only be produced by a chemical reaction (such as in a battery) and has a current that is the same throughout the circuit and flows in the same direction (Figure 2-15). Voltage and current are constant if the switch is turned on or off. Most of the electrically controlled units in the automobile require direct current.

Alternating Current

Alternating current (AC) is produced anytime a conductor moves through a magnetic field. In an alternating current circuit, voltage and current do not remain constant. Alternating

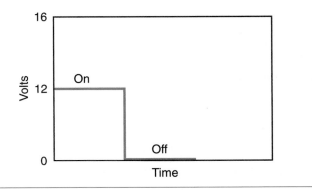

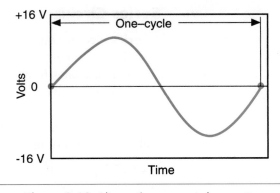

Figure 2-15 Direct current flow is in the same direction and remains constant on or off.

Figure 2-16 Alternating current does not remain constant.

current changes directions from positive to negative. The voltage in an AC circuit starts at zero and rises to a positive value. Then it falls back to zero and goes to a negative value. Finally it returns to zero (Figure 2-16). The AC voltage, shown in Figure 2-16, is called a sine wave. Figure 2-16 shows one **cycle.**

Electrical Circuits

The electrical term *continuity* refers to the circuit being continuous. For current to flow, the electrons must have a continuous path from the source voltage to the load component and back to the source. A simple automotive circuit is made up of three parts:

1. Battery (power source).
2. Wires (conductors).
3. Load (light, motor, etc.).

The basic circuit shown (Figure 2-17) includes a switch to turn the circuit on and off, a protection device (fuse), and a load. When the switch is turned to the ON position, the circuit is referred to as a **closed circuit.** When the switch is in the OFF position, the circuit is referred to as an **open circuit.** In this instance, with the switch closed, current flows from the positive terminal of the battery through the light and returns to the negative terminal of the battery. To have a complete circuit, the switch must be closed or turned on. The effect of opening and closing the switch to control electrical flow would be the same if the switch was installed on the ground side of the light.

There are three different types of electrical circuits: (1) the series circuit, (2) the parallel circuit, and (3) the series-parallel circuit.

The unrectified current produced within a generator is the most common example of alternating current found in the automobile. Generator circuits convert AC current to DC current.

The portion of the circuit from the positive side of the source to the load component is called the insulated side or "hot" side of the circuit. The portion of the circuit that is from the load component to the negative side of the source is called the ground side of the circuit.

The electrical term **closed circuit** means that there are no breaks in the path and current will flow. **Open circuit** is used to mean that current flow is stopped. By opening the circuit, the path for electron flow is broken.

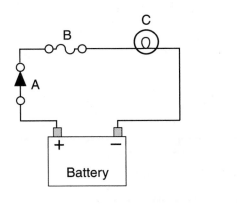

Figure 2-17 A basic electrical circuit including (A) a switch, (B) a fuse, and (C) a lamp.

Series Circuit

A **series circuit** consists of one or more resistors (or loads) with only one path for current to flow. If any of the components in the circuit fails, the entire circuit will not function. All of the current that comes from the positive side of the battery must pass through each resistor, then back to the negative side of the battery.

The total resistance of a series circuit is calculated by simply adding the resistances together. As an example, refer to Figure 2-18. Here is a series circuit with three light bulbs; one bulb has 2 ohms of resistance and the other two have 1 ohm each. The total resistance of this circuit is 2 + 1 + 1 or 4 ohms.

The characteristics of a series circuit are:

1. The total resistance is the sum of all resistances.

2. The current is the same at all points of the circuit (Figure 2-19).

3. The voltage drop across each resistance will be different if the resistance values are different (Figure 2-20).

4. The sum of all voltage drops equals the source voltage.

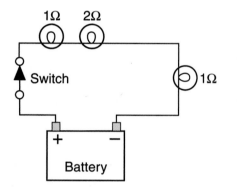

Figure 2-18 The total resistance in a series circuit is the sum of all resistances in the circuit.

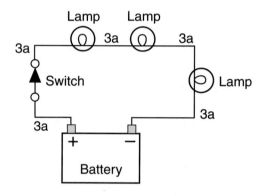

Figure 2-19 Regardless of where it is measured, amperage is the same at all points in a series circuit.

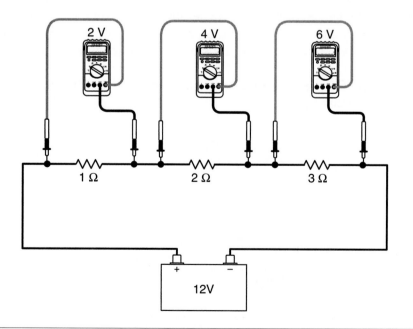

Figure 2-20 The voltage drop across each resistor in series will be different if the resistance values of each are different.

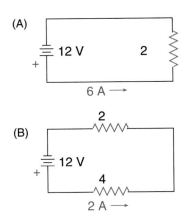

(A)

12 V   2

6 A →

(B)

2

12 V

4

2 A →

Figure 2-21 Circuit resistance controls, or determines, the amount of current flow.

To illustrate the laws of the series circuit, refer to Figure 2-21. The illustration labeled (A) is a simple 12-volt series circuit with a 2 Ω resistor. Using Ohm's law, it can be determined that the current is 6 amperes (A = V/R). Since the 2 Ω resistor is the only one in the circuit, all 12 volts are dropped across this resistor (V = A × R).

In Figure 2-21(B), an additional 4 Ω resistor is added in series to the existing 2 Ω resistor. Battery voltage is still 12 volts. Since this is a series circuit, total resistance is the sum of all of the resistance. In this case, total resistance is 6 Ω (4 Ω + 2 Ω). Using Ohm's law, total current through this circuit is 2 amperes (A = V/R = 12/6 = 2). In a series circuit, current is the same at all points of the circuit. No matter where current was measured in this example, the meter would read 2 amperes. This means that 2 amperes of current is flowing through each of the resistors. By comparing the amperage flow of the two circuits in Figure 2-21, you can see that circuit resistance controls (or determines) the amount of current flow. Understanding this concept is critical to performing diagnostics. Since this is a series circuit, adding resistance will decrease amperage draw.

Using Ohm's law, voltage drop over each resistor in the circuit can be determined. In this instance, V is the unknown value. Using V = A × R will determine the voltage drop over a resistance. Remember that amperage is the same throughout the circuit (2 amperes). The resistance value is the resistance of the resistor we are determining voltage drop for. For the 2 Ω resistor, the voltage drop would be A × R = 2 × 2 = 4 volts. Since the battery provides 12 volts and 4 volts are dropped over the 2 Ω resistor, 8 volts are left to be dropped by the 4 Ω resistor. To confirm this, V = A × R = 2 × 4 = 8. The sum of the voltage drops must equal the source voltage. Source voltage is 12 volts and the sum of the voltage drops is 4 volts + 8 volts = 12 volts.

These calculations work for all series circuits, regardless of the number of resistances in the circuit. Refer to Figure 2-22 for an example of a series circuit with four resistors. Total resistance

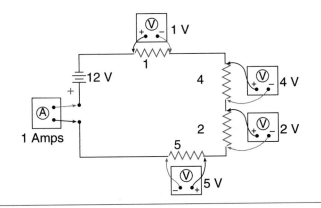

Figure 2-22 A series circuit used to demonstrate Ohm's law and voltage drop.

is 12 Ω (1 Ω + 4 Ω + 2 Ω + 5 Ω). Total amperage is 1 amp (A = V/R = 12 volts/12 Ω = 1 amp). Voltage drop over each resistor would be calculated as follows:

1. Voltage drop over the 1 Ω resistor = A × R = 1 × 1 = 1 volt
2. Voltage drop over the 4 Ω resistor = A × R = 1 × 4 = 4 volts
3. Voltage drop over the 2 Ω resistor = A × R = 1 × 2 = 2 volts
4. Voltage drop over the 5 Ω resistor = A × R = 1 × 5 = <u>5 volts</u>

Total voltage drop = 12 volts

Parallel Circuit

In a **parallel circuit,** each path of current flow has separate resistances that operate either independently or in conjunction with each other (depending on circuit design). In a parallel circuit, current can flow through more than one parallel leg at a time (Figure 2-23). In this type of circuit, failure of a component in one parallel leg does not affect the components in other legs of the circuit.

The characteristics of a parallel circuit are:

1. The voltage applied to each parallel leg is the same.

2. The voltage dropped across each parallel leg will be the same; however, if the leg contains more than one resistor, the voltage drop across each of them will depend on the resistance of each resistor in that leg.

3. The total resistance of a parallel circuit will always be less than the resistance of any of its legs.

4. The current flow through the legs will be different if the resistance is different.

5. The sum of the current in each leg equals the total current of the parallel circuit.

Figuring total resistance is a bit more complicated for a parallel circuit than a series circuit. Total resistance in a parallel circuit is always less than the lowest individual resistance because current has more than one path to follow. The method used to calculate total resistance depends on how many parallel branches are in the circuit, the resistance value of each branch, and personal preferences. Several methods of calculating total resistance are discussed. Choose the ones that work best for you.

If all resistances in the parallel circuit are equal, use the following formula to determine total resistance:

$$R_T = \frac{\text{Value of one resistor}}{\text{Total number of branches}}$$

For example, if the parallel circuit, shown in Figure 2-23, had a 120 Ω resistor for R_1 and R_2, total circuit resistance would be:

$$R_T = 120 \ \Omega / 2 \text{ branches} = 60 \ \Omega$$

The legs of a **parallel circuit** are also called parallel branches or shunt circuits.

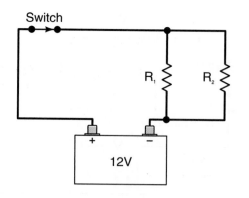

Figure 2-23 In a parallel circuit, there are multiple paths for current flow.

Note that total resistance is less than any of the resistances of the branches. If a third branch was added in parallel that also had 120 Ω resistance, total circuit resistance would be:

$$R_T = 120\ \Omega / 3\ \text{branches} = 40\ \Omega$$

Note what has happened when the third parallel branch was added. If more parallel resistors are added, more circuits are added, and the total resistance will decrease. With a decrease of total resistance, total amperage draw increases. The total resistance of a parallel circuit with two legs or two paths for current flow can be calculated by using this formula:

$$R_T = \frac{R_1 \times R_2}{R_1 + R_2}$$

If the value of R_1 in Figure 2-23 was 3 ohms and R_2 had a value of 6 ohms, the total resistance can be found.

$$R_T = \frac{R_1 \times R_2}{R_1 + R_2} \quad \text{or} \quad R_T = \frac{3 \times 6}{3 + 6} \quad \text{or} \quad R_T = \frac{18}{9} \quad \text{or} \quad R_T = 2$$

Based on this calculation, we can determine that the total circuit current is 6 amps (12 volts divided by 2 ohms). Using basic Ohm's law and a basic understanding of electricity, we can quickly determine other things about this circuit.

Each leg of the circuit has 12 volts applied to it; therefore, each leg must drop 12 volts. So the voltage drop across R_1 is 12 volts, and the voltage drop across R_2 is also 12 volts. Using the voltage drops, we can quickly find the current that flows through each leg. Since R_1 has 3 ohms and drops 12 volts, the current through it must be 4 amps. R_2 has 6 ohms and drops 12 volts and its current is 2 amps (A = V/R). The total current flow through the circuit is 4 + 2 or 6 amps. To calculate current in a parallel circuit, each shunt branch is treated as an individual circuit. To determine the branch current, simply divide the source voltage by the shunt branch resistance:

$$A = V \div R$$

Referring to Figure 2-24, the total resistance of a circuit with more than two legs can be calculated with the following formula:

$$R_T = \frac{1}{\dfrac{1}{R_1} + \dfrac{1}{R_2} + \dfrac{1}{R_3} \cdots \dfrac{1}{R_n}}$$

Using Figure 2-24 and its resistance values, total resistance would be calculated by:

$$R_T = \frac{1}{1/4 + 1/6 + 1/8}$$

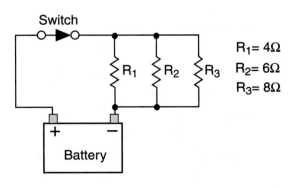

Figure 2-24 A parallel circuit with different resistances in each branch.

This means total resistance is equal to the reciprocal of the sum of 1/4 + 1/6 + 1/8. The next step is to add 1/4 + 1/6 + 1/8. To do this, the least common denominator must be found. In this case, the least common denominator is 24 so the formula now looks like this:

$$\frac{1}{6/24 + 4/24 + 3/24}$$

Now the fractions can be added together (remember to add only the numerator):

$$\frac{1}{13/24}$$

Since we are working with reciprocals, the formula now looks like this:

$$1 \times \frac{24}{13} = 1.85 \ \Omega$$

Often it is much easier to calculate total resistance of a parallel circuit by using total current. Begin by finding the current through each leg of the parallel circuit; then add them together to find total current. Use basic Ohm's law to calculate the total resistance.

First, using the circuit illustrated in Figure 2-24, calculate the current through each branch:
1. Current through R_1 = V/R = 12/4 = 3 amperes
2. Current through R_2 = V/R = 12/6 = 2 amperes
3. Current through R_3 = V/R = 12/8 = 1.5 amperes

Add all of the current flow through the branches together to get the total current flow:

Total amperage = 3 + 2 + 1.5 = 6.5 amperes

Since this is a 12-volt system and total current is 6.5 amperes, total resistance is:

R = 12 volts/6.5 amps = 1.85 Ω.

Series-Parallel Circuits

The **series-parallel circuit** has some loads that are in series with each other and some that are in parallel (Figure 2-25). To calculate the total resistance in this type of circuit, calculate the **equivalent series loads** of the parallel branches first. Next, calculate the series resistance and add it to the equivalent series load. For example, if the parallel portion of the circuit has two branches with 4 Ω resistance each and the series portion has a single load of 10 Ω, use the following method to calculate the equivalent resistance of the parallel circuit:

$$R_T = \frac{R_1 \times R_2}{R_1 + R_2} \quad \text{or} \quad \frac{4 \times 4}{4 + 4} \quad \text{or} \quad \frac{16}{8} \quad \text{or 2 ohms}$$

The **equivalent series load,** or equivalent resistance, is the equivalent resistance of a parallel circuit plus the resistance in series and is equal to the resistance of a single load in series with the voltage source.

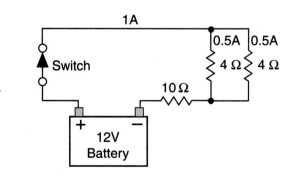

Figure 2-25 A series-parallel circuit with known resistance values.

40

Then add this equivalent resistance to the actual series resistance to find the total resistance of the circuit.

2 ohms + 10 ohms = 12 ohms

With the total resistance now known, total circuit current can be calculated. Because the source voltage is 12 volts, 12 is divided by 12 ohms.

A = V/R   or   A = 12/12   or   A = 1 amp

The current flow through each parallel leg is calculated by using the resistance of each leg and voltage drop across that leg. To do this, you must first find the voltage drops. Since all 12 volts are dropped by the circuit, we know that some are dropped by the parallel circuit and the rest by the resistor in series. We also know that the circuit current is 1 amp, the equivalent resistance value of the parallel circuit is 2 ohms, and the resistance of the series resistor is 10. Using Ohm's law we can calculate the voltage drop of the parallel circuit:

V = A × R   or   V = 1 × 2   or   V = 2

Two volts are dropped by the parallel circuit. This means 2 volts are dropped by each of the 4-ohm resistors. Using our voltage drop, we can calculate our current flow through each parallel leg.

A = V/R   or   A = 2/4   or   A = 0.5 amps

Since the resistance on each leg is the same, each leg has 0.5 amps through it. If we did this right, the sum of the amperages will equal the current of the circuit. It does: 0.5 + 0.5 = 1.

A slightly different series-parallel circuit is illustrated in Figure 2-26. In this circuit, a 2 Ω resistor is in series to a parallel circuit containing a 6 Ω and a 3 Ω resistor. To calculate total resistance, first find the resistance of the parallel portion of the circuit using (6 × 3)/(6 + 3) = 18/9 = 2 ohms of resistance. Add this amount to the series resistance; 2 + 2 = 4 ohms of total circuit resistance. Now total current can be calculated by A = V/R = 12/4 = 3 amperes. This means that 3 amperes is flowing through the series portion of the circuit. To figure how much amperage is in each of the parallel branches, the amount of applied voltage to each branch must be calculated. Since the series circuit has 3 amperes going through a 2 Ω resistor, the voltage drop over this resistor can be figured using V = A × R = 3 × 2 = 6 volts. Since the source voltage is 12 volts, this means that 6 volts are applied to each of the resistors in parallel (12 − 6 = 6 volts) and that 6 volts are dropped over each of these resistors. Current through each branch can now be calculated:

Current through the 6 Ω branch is A = V/R = 6/6 = 1 ampere

Current through the 3 Ω branch is A = V/R = 6/3 = 2 amperes

The sum of the current flow through the parallel branches should equal total current flow (1 + 2 = 3 amperes).

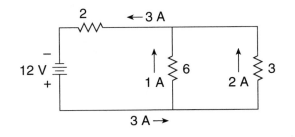

Figure 2-26 In a series-parallel circuit, the sum of the currents through the legs will equal the current through the series portion of the circuit.

Based on what was just covered, the characteristics of a series-parallel circuit can be summarized as follows:

1. Total resistance is the sum of the resistance value of the parallel portion and the series resistance.

2. Voltage drop over the parallel branch resistance is determined by the resistance value of the series resistor.

3. Total amperage is the sum of the current flow through each parallel branch.

4. The amperage through each parallel branch is determined by the resistance in the branch.

It is important to realize that the actual or measured values of current, voltage, and resistance may be somewhat different than the calculated values. The change is caused by the effects of heat on the resistances. As the voltage pushes current through a resistor, the resistor heats up. The resistor changes the electrical energy into heat energy. This heat may cause the resistance to increase or decrease depending on the material it is made of. The best example of a resistance changing electrical energy into heat energy is a light bulb. A light bulb gives off light because the conductor inside the bulb heats up and glows when current flows through it.

Using Ohm's Law

The primary importance of being able to use Ohm's law is to predict what will happen if something else happens. Technicians use electrical meters to measure current, voltage, and resistance. When a measured value is not within specifications, you should be able to determine why. Ohm's law is used to do that.

Most automotive electrical systems are wired in parallel. Actually, the system is made up of a number of series circuits wired in parallel. This allows each electrical component to work independently of the others. When one component is turned on or off, the operation of the other components should not be affected.

Figure 2-27 illustrates a 12-volt circuit with one 3-ohm light bulb. The switch controls the operation of the light bulb. When the switch is closed, current flows and the bulb is lit. Four amps will flow through the circuit and the bulb.

$$A = V/R \quad \text{or} \quad A = 12/3 \quad \text{or} \quad A = 4 \text{ amps}$$

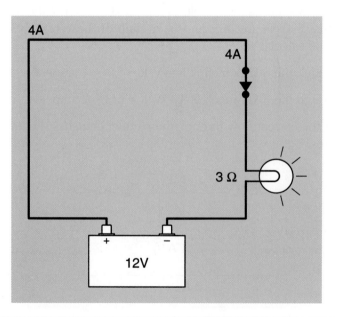

Figure 2-27 A simple light circuit.

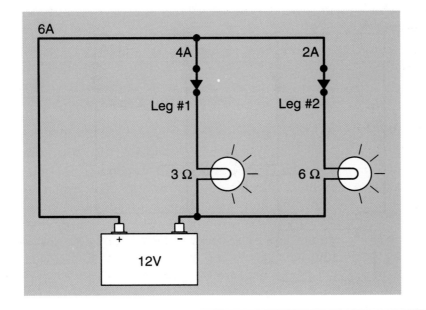

Figure 2-28 Two light bulbs wired in parallel.

Figure 2-28 illustrates the same circuit with a 6-ohm light bulb added in parallel to the 3-ohm light bulb. With the switch for the new bulb closed, 2 amps will flow through that bulb. The 3-ohm bulb is still receiving 12 volts and has 4 amps flowing through it. It will operate in the same way and with the same brightness as it did before we added the 6-ohm light bulb. The only thing that changed was circuit current, which is now 6 (4 + 2) amps.

Leg #1   A = V/R   or   A = 12/3   or   A = 4 amps
Leg #2   A = V/R   or   A = 12/6   or   A = 2 amps

If the switch to the 3-ohm bulb is opened (Figure 2-29), the 6-ohm bulb works in the same way and with the same brightness as it did before we opened the switch. In this case, two things happened: the 3-ohm bulb no longer is lit and the circuit current dropped to 2 amps.

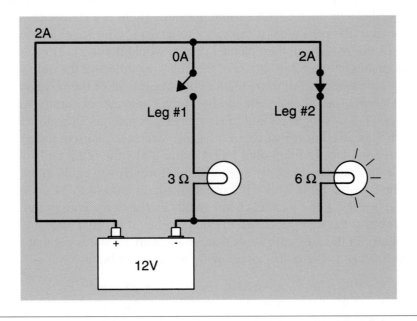

Figure 2-29 Two light bulbs wired in parallel; one switched on, the other switched off.

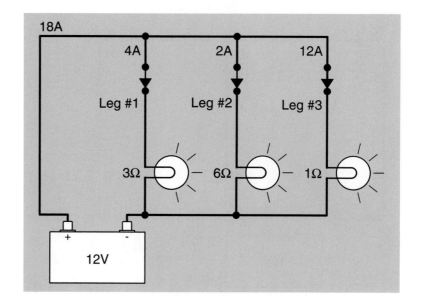

Figure 2-30 Three light bulbs wired in parallel.

Figure 2-30 is the same circuit as Figure 2-29 except a 1-ohm light bulb and switch was added in parallel to the circuit. With the switch for the new bulb closed, 12 amps will flow through that circuit. The other bulbs are working in the same way and with the same brightness as before. Again, total circuit resistance decreases so the total circuit current increases. Total current is now 18 amps:

Leg #1	A = V/R	or	A = 12/3	or	A = 4 amps
Leg #2	A = V/R	or	A = 12/6	or	A = 2 amps
Leg #3	A = V/R	or	A = 12/1	or	A = 12 amps

Total current = 4 + 2 + 12 or 18 amps

When the switch for any of these bulbs is opened or closed, the only things that happen are the bulbs either turn off or on and the total current through the circuit changes. Notice as we add more parallel legs, total circuit current goes up. There is a commonly used statement, "Current always takes the path of least resistance to ground." This statement is not totally correct. If this were a true statement, then parallel circuits would not work. However, as illustrated in the previous circuits, current flows to all of the bulbs regardless of the bulb's resistance. The resistances with lower values will draw higher currents, but all of the resistances will receive the current they allow. The statement should be, "Larger amounts of current will flow through lower resistances." This is very important to remember when diagnosing electrical problems.

From Ohm's law, we know that when resistance decreases, current increases. If we put a 0.6-ohm light bulb in place of the 3-ohm bulb (Figure 2-31), the other bulbs will work in the same way and with the same intensity as they did before. However, 20 amps of current will flow through the 0.6-ohm bulb. This will raise total circuit current to 34 amps. Lowering the resistance on the one leg of the parallel circuit greatly increases the current through the circuit. This high current may damage the circuit or components. It is possible that high current can cause wires to burn. In this case, the wires that would burn are the wires that would carry the 34 amps or the 20 amps to the bulb, not the wires to the other bulbs.

Leg #1	A = V/R	or	A = 12/0.6	or	A = 20 amps
Leg #2	A = V/R	or	A = 12/6	or	A = 2 amps
Leg #3	A = V/R	or	A = 12/1	or	A = 12 amps

Total current = 20 + 2 + 12 or 34 amps

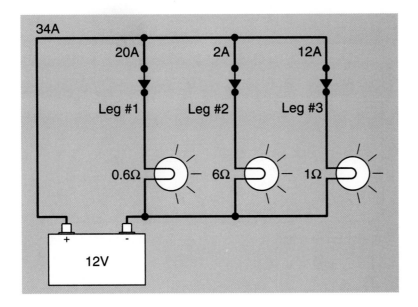

Figure 2-31 Parallel light circuit.

Let's see what happens when we add resistance to one of the parallel legs. An increase in resistance should cause a decrease in current. In Figure 2-32, a 1-ohm resistor was added after the 1-ohm light bulb. This resistor is in series with the light bulb and the total resistance of that leg is now 2 ohms. The current through that leg is now 6 amps. Again, the other bulbs were not affected by the change. The only change to the whole circuit was in total circuit current, which now drops to 12 amps. The added resistance lowered total circuit current and changed the way the 1-ohm bulb works. This bulb will now drop only 6 volts. The remaining 6 volts will be dropped by the added resistor. The 1-ohm bulb will be much dimmer than before; its power rating dropped from 144 watts to 36 watts. Additional resistance causes the bulb to be dimmer. The bulb itself wasn't changed, only the resistance of that leg changed. The dimness is caused by the circuit, not the bulb.

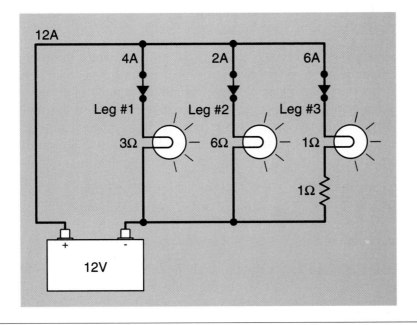

Figure 2-32 A parallel light circuit with one leg having a series resistance added.

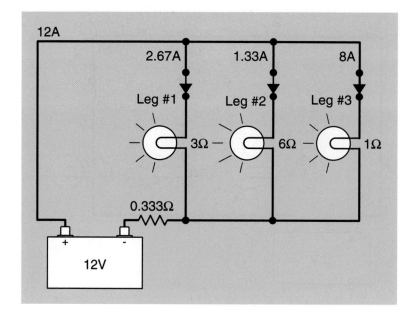

Figure 2-33 A parallel light circuit with a resistance in series to the whole circuit.

Leg #1	A = V/R	or	A = 12/3	or	A = 4 amps		
Leg #2	A = V/R	or	A = 12/6	or	A = 2 amps		
Leg #3	A = V/R	or	A = 12/1+1	or	A = 12/2	or	A = 6 amps

Total current = 4 + 2 + 6 or 12 amps

Now let's see what happens when we add a resistance that is common to all of the parallel legs. In Figure 2-33, we added a 0.333-ohm resistor (0.333 was chosen to keep the math simple!) to the negative connection at the battery. This will cause the circuit's current to decrease; it will also change the operation of the bulbs in the circuit. The total resistance of the bulbs in parallel is 0.667 ohms.

$$R_T = \cfrac{1}{\frac{1}{3} + \frac{1}{6} + \frac{1}{1}} \quad \text{or} \quad R_T = \cfrac{1}{0.333 + 0.167 + 1} \quad \text{or} \quad R_T = \frac{1}{1.5} \quad \text{or} \quad R_T = 0.667$$

The total resistance of the circuit is 1 ohm (0.667 + 0.333), which means the circuit current is now 12 amps. Because there will be a voltage drop across the 0.333-ohm resistor, each of the parallel legs will drop less than source voltage. To find the amount of voltage dropped by the parallel circuit, we multiply the amperage by the resistance. Twelve amps multiplied by 0.667 equals 8. So 8 volts will be dropped by the parallel circuit; the remaining 4 volts will be dropped by the 0.333 resistor. The amount of current through each leg can be calculated by taking the voltage drop and dividing it by the resistance of the leg.

Leg #1	A = V/R	or	A = 8/3	or	A = 2.667 amps
Leg #2	A = V/R	or	A = 8/6	or	A = 1.333 amps
Leg #3	A = V/R	or	A = 8/1	or	A = 8 amps

Total circuit current = 2.667 + 1.333 + 8 or 12 amps

The added resistance affected the operation of all the bulbs, because it was added to a point that was common to all of the bulbs. All of the bulbs would be dimmer and circuit current would be lower.

Figure 2-34 Capacitors that can be used in automotive electrical circuits.

Capacitance

Some automotive electrical systems will use a capacitor or condenser to store electrical charges (Figure 2-34). A capacitor uses the theory of **capacitance** to temporarily store electrical energy. Capacitance (C) is the ability of two conducting surfaces to store voltage. The two surfaces must be separated by an insulator.

A capacitor does not consume any power. All of the voltage stored in the capacitor is returned to the circuit when the capacitor discharges. Because the capacitor stores voltage, it will also absorb voltage changes in the circuit. By providing for this storage of voltage, damaging voltage spikes can be controlled. They are also used to reduce radio noise.

A capacitor is made by wrapping two conductor strips around an insulating strip. The insulating strip, or **dielectric,** prevents the plates from coming in contact while keeping them very close to each other. The dielectric can be made of insulator material such as ceramic, glass, paper, plastic, or even the air between the two plates. A capacitor blocks direct current. A small amount of current enters the capacitor and charges it.

Most capacitors are connected in parallel across the circuit (Figure 2-35). Capacitors operate on the principle that opposite charges attract each other and that there is a potential voltage between any two oppositely charged points. When the switch is closed, the protons at the positive battery terminal will attract some of the electrons on one plate of the capacitor away from the area near the dielectric material. As a result, the atoms of the **positive plate** are unbalanced because there are more protons than electrons in the atom. This plate now has a positive charge because of the shortage of electrons (Figure 2-36). The positive charge of this plate will

The insulator in a capacitor is called a **dielectric.** The dielectric can be made of some insulator material such as ceramic, glass, paper, plastic, or even the air between the two plates.

The plate connected to the positive battery terminal is the **positive plate.**

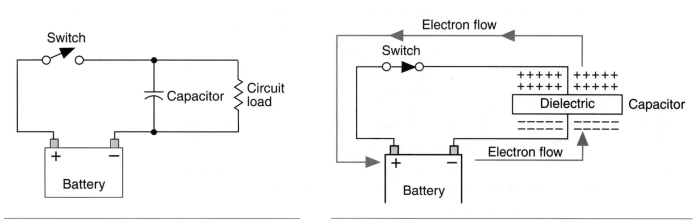

Figure 2-35 A capacitor connected to a circuit.

Figure 2-36 The positive plate sheds its electrons.

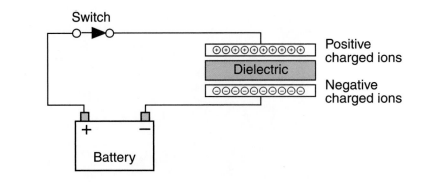

Figure 2-37 The electrons will be stored on the negative plate.

attract electrons on the other plate. The dielectric keeps the electrons on the negative plate from crossing over to the positive plate, resulting in a storage of electrons on the negative plate (Figure 2-37). The movement of electrons to the negative plate and away from the positive plate is an electrical current.

Current will flow "through" the capacitor until the voltage charges across the capacitor and across the battery are equalized. Current flow through a capacitor is only the effect of the electron movement onto the negative plate and away from the positive plate. Electrons do not actually pass through the capacitor from one plate to another. The charges on the plates do not move through the **electrostatic field.** They are stored on the plates as **static electricity.**

When the charges across the capacitor and battery are equalized, there is no potential difference and no more current will flow "through" the capacitor (Figure 2-38). Current will now flow through the load components in the circuit (Figure 2-39).

When the switch is opened, current flow from the battery through the resistor is stopped. However, the capacitor has a storage of electrons on its negative plate. Because the negative plate of the capacitor is connected to the positive plate through the resistor, the capacitor acts as the source. The capacitor will discharge the electrons through the resistor until the atoms of the positive plate and negative plate return to a balanced state (Figure 2-40).

In the event that a high-voltage spike occurs in the circuit, the capacitor will absorb the additional voltage before it is able to damage the circuit components. A capacitor can also be used to stop current flow quickly when a circuit is opened (such as in the ignition system). It can also store a high-voltage charge and then discharge it when a circuit needs the voltage (such as in some air bag systems).

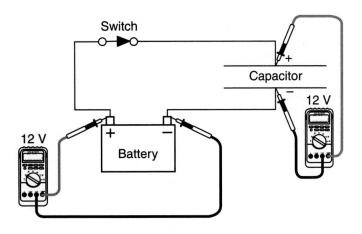

Figure 2-38 A capacitor when it is fully charged.

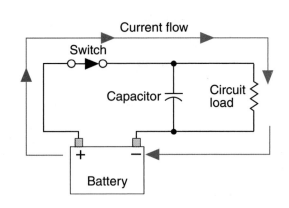

Figure 2-39 Current flow with a fully charged capacitor.

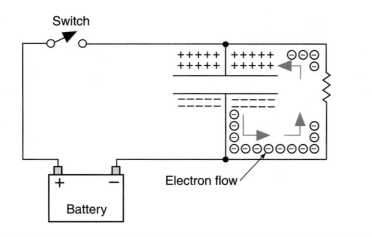

Figure 2-40 Current flow with the switch open and the capacitor discharging.

Capacitors are rated in units called farads. A one-farad capacitor connected to a one-volt source will store 6.28×10^{18} electrons. A farad is a large unit and most commonly used capacitors are rated in picofarad (a trillionth of a farad) or microfarads (a millionth of a farad). In addition, the capacitor has a voltage rating that is determined by how much voltage can be applied to it without the dielectric breaking down. The maximum voltage rating and capacitance determine the amount of energy a capacitor holds. The voltage rating is related to the strength and thickness of the dielectric. The voltage rating increases with increasing dielectric strength and the thickness of the dielectric. The capacitance increases with the area of the plates and decreases with the thickness of the dielectric.

Magnetism Principles

Magnetism is a force that is used to produce most of the electrical power in the world. It is also the force used to create the electricity to recharge a vehicle's battery, make a starter work, and produce signals for various operating systems. A magnet is a material that attracts iron, steel, and a few other materials. Because magnetism is closely related to electricity, many of the laws that govern electricity also govern magnetism.

There are two types of magnets used on automobiles, permanent magnets and electromagnets. Permanent magnets are magnets that do not require any force or power to keep their magnetic field. Electromagnets depend on electrical current flow to produce and, in most cases, keep their magnetic field.

| A   BIT   OF   HISTORY |

The force of a magnet was first discovered over 2,000 years ago by the Greeks. They noticed that a type of stone, now called magnetite, was attracted to iron. During the Dark Ages, people believed evil spirits caused the strange powers of magnetite.

Magnets

All magnets have polarity. A magnet that is allowed to hang free will align itself north and south. The end facing north is called the north-seeking pole and the end facing south is called the south-seeking pole. Like poles will repel each other and unlike poles will attract each other. These principles are shown in Figure 2-41. The magnetic attraction is the strongest at the poles.

Magnetic flux density is a concentration of the lines of force (Figure 2-42). A strong magnet produces many lines of force and a weak magnet produces fewer lines of force. Invisible lines of force leave the magnet at the north pole and enter again at the south pole. While inside the magnet, the lines of force travel from the south pole to the north pole (Figure 2-43).

The field of force (or magnetic field) is all the space, outside the magnet, that contains lines of magnetic force. Magnetic lines of force penetrate all substances; there is no known insulation against magnetic lines of force. The lines of force may be deflected only by other magnetic materials or by another magnetic field.

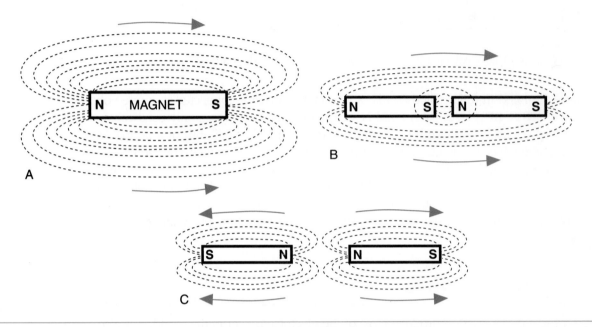

Figure 2-41 Magnetic principles: (A) All magnets have poles, (B) unlike poles attract each other, and (C) like poles repel.

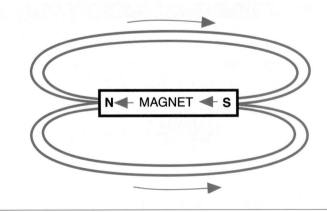

Figure 2-42 Iron filings indicate the lines of magnetic flux.

Figure 2-43 Lines of force through the magnet.

Electromagnetism

Electromagnetism uses the theory that whenever an electrical current flows through a conductor, a magnetic field is formed around the conductor (Figure 2-44). The number of lines of force and the strength of the magnetic field produced will be in direct proportion to the amount of current flow.

The direction of the lines of force is determined by the **right-hand rule.** Using the conventional theory of current flow being from positive to negative, the right hand is used to grasp the wire, with the thumb pointing in the direction of current flow. The fingers will point in the direction of the magnetic lines of force (Figure 2-45).

André Marie Ampère noted that current flowing in the same direction through two nearby wires will cause the wires to attract one another. Also, he observed that if current flow in one of the wires is reversed, the wires will repel one another. In addition, he found that if a wire is coiled with current flowing through the wire, the same magnetic field that surrounds a straight wire combines to form one larger magnetic field. This magnetic field has true north and south poles (Figure 2-46). Looping the wire doubles the flux density where the wire is running parallel to itself. The illustration (Figure 2-47) shows how these lines of force will join and add to each other.

The north pole can be determined in the coil by use of the right-hand rule. Grasp the coil with the fingers pointing in the direction of current flow (+ to −) and the thumb will point toward the north pole (Figure 2-48).

Figure 2-44 A magnetic field surrounds a conductor that has current flowing through it.

Figure 2-45 Right-hand rule to determine direction of magnetic lines of force.

Figure 2-46 Looping the conductor increases the magnetic field.

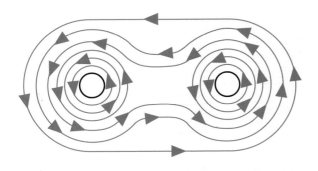

Figure 2-47 Lines of force join together and attract each other.

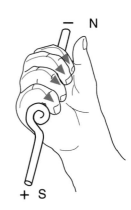

Figure 2-48 Right-hand rule to determine magnetic poles.

As more loops are added, the fields from each loop will join and increase the flux density (Figure 2-49). To make the magnetic field even stronger, an iron core can be placed in the center of the coil (Figure 2-50). The soft iron core has high **permeability** and low **reluctance,** which provides an excellent conductor for the magnetic field to travel through the center of the wire coil.

The strength of an electromagnetic coil is affected by the following factors:

1. The amount of current flowing through the wire.
2. The number of windings or turns.
3. The size, length, and type of core material.
4. The direction and angle at which the lines of force are cut.

The strength of the magnetic field is measured in ampere-turns:

$$\text{ampere-turns} = \text{amperes} \times \text{number of turns}$$

The magnetic field strength is measured by multiplying the current flow in amperes through a coil by the number of complete turns of wire in the coil. For example, in the illustration (Figure 2-51), a 1,000-turn coil with 1 ampere of current would have a field strength of 1,000

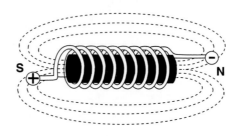

Figure 2-49 Adding more loops of wire increases the magnetic flux density.

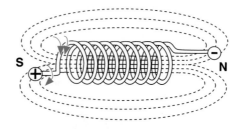

Figure 2-50 The addition of an iron core concentrates the flux density.

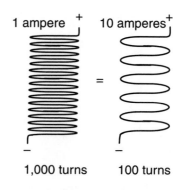

Figure 2-51 Magnetic field strength is determined by the amount of amperage and the number of coils.

ampere-turns. This coil would have the same field strength as a coil with 100 turns and 10 amperes of current.

Theory of Induction

Electricity can be produced by magnetic **induction.** Magnetic induction occurs when a conductor is moved through the magnetic lines of force (Figure 2-52) or when a magnetic field is moved across a conductor. A difference of potential is set up between the ends of the conductor and a voltage is induced. This voltage exists only when the magnetic field or the conductor is in motion.

The induced voltage can be increased by either increasing the speed in which the magnetic lines of force cut the conductor or by increasing the number of conductors that are cut. It is this principle that is behind the operation of all ignition systems, starter motors, and charging systems.

A common induction device is the ignition coil. As the current increases, the coil will reach a point of **saturation.** This is the point at which the magnetic strength eventually levels off and where current will no longer increase as it passes through the coil. The magnetic lines of force, which represent stored energy, will collapse when the applied voltage is removed. When the lines of force collapse, the magnetic energy is returned to the wire as electrical energy.

A desirable induction is called a **mutual induction.**

Mutual induction is used in ignition coils where a rapidly changing magnetic field in the primary windings creates a voltage in the secondary winding (Figure 2-53).

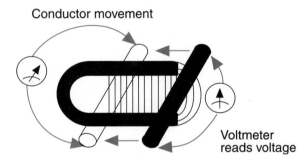

Figure 2-52 Moving a conductor through a magnetic field induces an electrical potential difference.

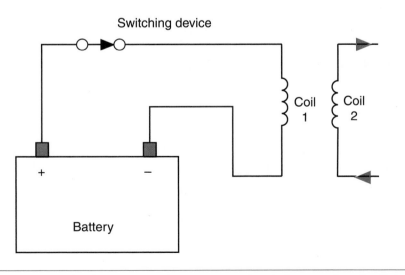

Figure 2-53 A mutual induction is used to create an electrical current in coil 2 if the current flow in coil 1 is turned off.

If voltage is induced in the wires of a coil when current is first connected or disconnected, it is called **self-induction.** The resulting current is in the opposite direction of the applied current and tends to reduce the magnetic force. Self-induction is governed by Lenz's law, which states:

An induced current flows in a direction opposite the magnetic field that produced it.

Self-induction is generally not wanted in automotive circuits. For example, when a switch is opened, self-induction tends to continue to supply current in the same direction as the original current because as the magnetic field collapses, it induces voltage in the wire. According to Lenz's law, voltage induced in a conductor tends to oppose a change in current flow. Self-induction can cause an electrical arc to occur across an opened switch. The arcing may momentarily bypass the switch and allow the circuit that was turned off to operate for a short period of time. The arcing will also burn the contacts of the switch.

Self-induction is commonly found in electrical components that contain a coil or an electric motor. To help reduce the arc across contacts, a capacitor or clamping diode may be connected to the circuit. The capacitor will absorb the high-voltage arcs and prevent arcing across the contacts. Diodes are semiconductors that allow current flow in only one direction. A clamping diode can be connected in parallel to the coil and will prevent current flow from the self-induction coil to the switch.

Magnetic induction is also the basis for a generator and many of the sensors on today's vehicles. In a generator, a magnetic field rotates inside a set of conductors. As the magnetic field crosses the wires, a voltage is induced. The amount of voltage induced by this action depends on the speed of the rotating field, the strength of the field, and the number of conductors the field cuts through. This principle will be discussed in greater detail in Chapter 7.

Magnetic sensors are used to measure speeds, such as engine, vehicle, and shaft speeds. These sensors typically use a permanent magnet. Rotational speed is determined by the passing of blades or teeth in and out of the magnetic field. As a tooth moves in and out of the magnetic field, the strength of the magnetic field is changed and a voltage signal is induced. This signal is sent to a control device, where it is interpreted. This principle is discussed in greater detail in Chapter 12.

EMI Suppression

Electromagnetic interference (EMI) is an undesirable creation of electromagnetism whenever current is switched on and off. As manufacturers began to increase the number of electronic components and systems in their vehicles, the problem of EMI had to be controlled. The low-power integrated circuits used on modern vehicles are sensitive to the signals produced as a result of EMI. EMI is produced as current in a conductor is turned on and off. EMI is also caused by static electricity that is created by friction. The friction is a result of tires contacting the road, or of fan belts contacting the pulleys.

EMI can disrupt the vehicle's computer systems by inducing false messages to the computer. The computer requires messages to be sent over circuits in order to communicate with other computers, sensors, and actuators. If any of these signals are disrupted, the engine and/or accessories may turn off.

EMI can be suppressed by any one of the following methods:

1. Adding a resistance to the conductors. This is usually done to high-voltage systems, such as the secondary circuit of the ignition system.

2. Connecting a capacitor in parallel and a choke coil in series with the circuit.

3. Shielding the conductor or load components with a metal or metal-impregnated plastic.

4. Increasing the number of paths to ground by using designated ground circuits. This provides a clear path to ground that is very low in resistance.

5. Adding a clamping diode in parallel to the component.

6. Adding an isolation diode in series to the component.

Summary

❏ An atom is constructed of a complex arrangement of electrons in orbit around a nucleus. If the number of electrons and protons are equal, the atom is balanced or neutral.

❏ A conductor allows electricity to easily flow through it.

❏ An insulator does not allow electricity to easily flow through it.

❏ Electricity is the movement of electrons from atom to atom. In order for the electrons to move in the same direction, an electromotive force (EMF) must be applied to the circuit.

❏ The electron theory defines electron flow as motion from negative to positive.

❏ The conventional theory of current flow states that current flows from a positive point to a less positive point.

❏ Voltage is defined as an electrical pressure and is the difference between the positive and negative charges.

❏ Current is defined as the rate of electron flow and is measured in amperes. Amperage is the amount of electrons passing any given point in the circuit in one second.

❏ Resistance is defined as opposition to current flow and is measured in ohms (Ω).

❏ Ohm's law defines the relationship between current, voltage, and resistance. It is the basic law of electricity and states that the amount of current in an electric circuit is inversely proportional to the resistance of the circuit and is directly proportional to the voltage in the circuit.

❏ Wattage represents the measure of power (P) used in a circuit. Wattage is measured by using the power formula, which defines the relationship between amperage, voltage, and wattage.

❏ Capacitance is the ability of two conducting surfaces to store voltage.

❏ Direct current results from a constant voltage and a current that flows in one direction.

❏ In an alternating-current circuit, voltage and current do not remain constant. AC current changes direction from positive to negative and negative to positive.

❏ For current to flow, the electrons must have a complete path from the source voltage to the load component and back to the source.

❏ The series circuit provides a single path for current flow from the electrical source through all the circuit's components and back to the source.

❏ A parallel circuit provides two or more paths for current to flow.

❏ A series-parallel circuit is a combination of the series and parallel circuits.

❏ The equivalent series load is the total resistance of a parallel circuit plus the resistance of the load in series with the voltage source.

Terms to Know

Alternating current

Ampere

Atom

Balanced

Balanced atom

Capacitance

Circuit

Closed circuit

Conductor

Conventional theory

Current

Cycle

Dielectric

Direct current (DC)

Electromagnetic interference (EMI)

Electromagnetism

Electromotive force (EMF)

Electron theory

Electrons

Electrostatic field

Equivalent series load

Ground

Induction

Insulator

Ion

Magnetic flux density

Mutual induction

Neutrons

Nucleus

Ohms

Ohm's law

Open circuit

Parallel circuit

Permeability

Positive plate

Power

Protons

Reluctance

❑ Voltage drop is caused by a resistance in the circuit that reduces the electrical pressure available after the resistance.

❑ Kirchhoff's voltage law states that the total voltage drop in an electrical circuit will always equal the available voltage at the source.

Review Questions

Short-Answer Essays

1. List and define the three elements of electricity.
2. Explain the basic principles of Ohm's law.
3. List and describe the three types of circuits.
4. Explain the principle of electromagnetism.
5. Describe the principle of induction.
6. Describe the basics of electron flow.
7. Define the two types of electrical current.
8. Describe the difference between insulators, conductors, and semiconductors.
9. Explain the basic concepts of capacitance.
10. What does the measurement of "watt" represent?

Fill in the Blanks

1. _____ are negatively charged particles. The nucleus contains positively charged particles called _____ and particles that have no charge called _____.

2. A _____ allows electricity to easily flow through it. An _____ does not allow electricity to easily flow through it.

3. For the electrons to move in the same direction, there must be an _____ applied.

4. The _____ _____ of current flow states that current flows from a positive point to a less positive point.

5. Resistance is defined as _____ to current flow and is measured in _____.

6. _____ is the ability of two conducting surfaces to store voltage.

7. Kirchhoff's voltage law states that the _____ _____ _____ in an electrical circuit will always _____ available voltage at the source.

8. The _____ of all the resistors in series is the total resistance of that series circuit.

9. _____ is defined as an electrical pressure.

10. _____ is defined as the rate of electron flow.

Multiple Choice

1. Which of the following methods can be used to form an electrical current?
   A. Magnetic induction.
   B. Chemical reaction.
   C. Heat.
   D. All of the above.
   E. None of the above.

2. In a series circuit:
   A. Total resistance is the sum of all of the resistances in the circuit.
   B. Total resistance is less than the lowest resistor.
   C. Amperage will increase as more resistance is added.
   D. All of the above.

3. All of the following concerning voltage drop are true EXCEPT:
   A. All of the voltage from the source must be dropped before it returns to the source.
   B. Corrosion is not a contributor to voltage drop.
   C. Voltage drop is the conversion of electrical energy into another energy form.
   D. Voltage drop can be measured with a voltmeter.

4. All of the following concerning voltage are true EXCEPT:
   A. Voltage is the electrical pressure that causes electrons to move.
   B. Voltage will exist between any two points in a circuit unless the potential drops to zero.
   C. Voltage is A × R.
   D. In a series circuit, voltage is the same at all points in the circuit.

5. Wattage is:
   A. A measure of the total electrical work being performed per unit of time.
   B. Expressed as P = R × A.
   C. Both A and B.
   D. Neither A nor B.

6. A capacitor:
   A. Consumes electrical power.
   B. Induces voltage.
   C. Both A and B.
   D. Neither A nor B.

7. Which statement about electrical currents is correct?
   A. Alternating current can be stored in a battery.
   B. Alternating current is produced from a voltage and current that remain constant and flow in the same direction.
   C. Direct current is used for most electrical systems on the automobile.
   D. Direct current changes directions from positive to negative.

8. Induction:
   A. Is the magnetic process of producing a current flow in a wire without any actual contact to the wire.
   B. Exists when the magnetic field or the conductor is in motion.
   C. All of the above.
   D. None of the above.

9. All of the following statements are true EXCEPT:
   A. If the resistance increases and the voltage remains constant, the amperage will increase.
   B. Ohm's law can be stated as A = V ÷ R.
   C. If voltage is increased, amperage will increase.
   D. An open circuit does not allow current flow.

10. Which of the following statements is correct?
    A. An insulator is capable of supporting the flow of electricity through it.
    B. A conductor is not capable of supporting the flow of electricity.
    C. All of the above.
    D. None of the above.

Electrical Components

Upon completion and review of this chapter, you should be able to:

❑ Describe the common types of electrical system components used and how they affect the electrical system.

❑ Explain the operation of the electrical controls, including switches, relays, and variable resistors.

❑ Describe the basic operating principles of electronic components.

❑ Explain the use of electronic components in the circuit.

❑ Explain the purpose of a circuit protection device. Describe the most common types in use.

❑ Define circuit defects, including opens, shorts, grounds, and excessive resistance.

❑ Explain the effects that each type of circuit defect has on the operation of the electrical system.

Introduction

In this chapter you will be introduced to electrical and electronic components. These components include circuit protection devices, switches, relays, variable resistors, diodes, and different forms of transistors. Today's technician must comprehend the operation of these components and the ways they affect electrical system operation. With this knowledge, the technician will be able to accurately and quickly diagnose many electrical failures.

To be able to properly diagnose the components and circuits, the technician must be able to use the test equipment that is designed for electrical system diagnosis. In this chapter you will learn about the various types of test equipment used for diagnosing electrical systems. You will learn the appropriate equipment to use to locate the fault based on the symptoms. In addition, the various types of defects that cause the system to operate improperly are discussed.

Electrical Components

Electrical circuits require different components depending on the type of work they do and how they are to perform it. A light may be wired directly to the battery, but it will remain on until the battery drains. A switch will provide for control of the light circuit. However, if variable dimming of the light is required, a rheostat is also needed.

There are several electrical components that may be incorporated into a circuit to achieve the desired results from the system. These components include switches, relays, buzzers, and various types of resistors.

Switches

Shop Manual
Chapter 3, page 92

A switch is the most common means of providing control of electrical current flow to an accessory (Figure 3-1). A switch can control the on/off operation of a circuit or direct the flow of current through various circuits. The contacts inside the switch assembly carry the current when they are closed. When they are open, current flow is stopped.

Figure 3-1 Common types of switches used in the automotive electrical system.

Figure 3-2 A simplified illustration of an SPST switch.

A **normally open (NO)** switch will not allow current flow when it is in its rest position. The contacts are open until they are acted on by an outside force that closes them to complete the circuit. A **normally closed (NC)** switch will allow current flow when it is in its rest position. The contacts are closed until they are acted on by an outside force that opens them to stop current flow.

The simplest type of switch is the single-**pole**, single-**throw** (SPST) switch (Figure 3-2). This switch controls the on/off operation of a single circuit. The most common type of SPST switch design is the hinged pawl. The pawl acts as the contact and changes position as directed to open or close the circuit.

Some SPST switches are designed to be a momentary contact switch. This switch usually has a spring that holds the contacts open until an outside force is applied and closes them. The horn button on most vehicles is of this design.

Some electrical systems may require the use of a single-pole, double-throw switch (SPDT). The dimmer switch used in the headlight system is usually an SPDT switch. This switch has one input circuit with two output circuits. Depending on the position of the contacts, voltage is applied to the high-beam circuit or to the low-beam circuit (Figure 3-3).

The term **pole** refers to the number of input circuits. The term **throw** refers to the number of output circuits.

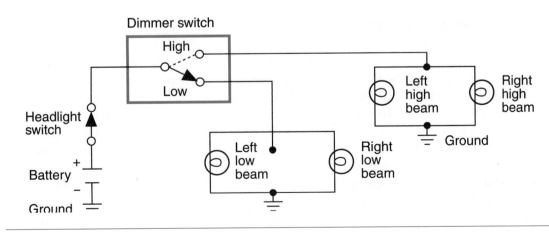

Figure 3-3 A simplified schematic of a headlight system using an SPDT dimmer switch.

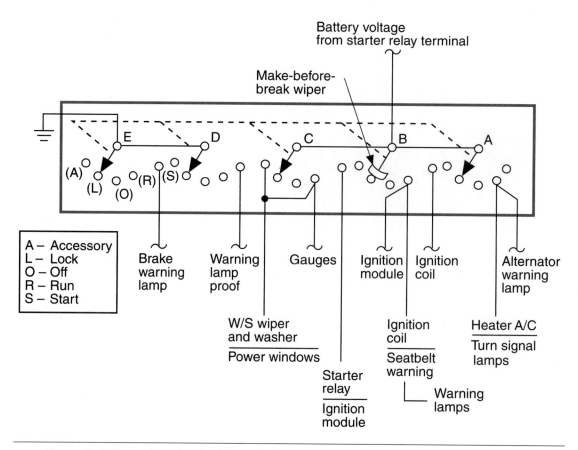

Figure 3-4 Illustration of an ignition switch.

One of the most complex switches is the **ganged switch.** This type of switch is commonly used as an ignition switch. In Figure 3-4, the five wipers are all ganged together and will move together. Battery voltage is applied to the switch from the starter relay terminal. When the ignition key is turned to the START position, all wipers move to the "S" position. Wipers D and E will complete the circuit to ground to test the instrument panel warning lamps. Wiper B provides battery voltage to the ignition coil. Wiper C supplies battery voltage to the starter relay and the ignition module. Wiper A has no output.

 AUTHOR'S NOTE: The dotted lines used in the switch symbol indicate that the wipers of the switch move together.

Once the engine starts, the wipers are moved to the RUN position. Wipers D and E are moved out of contact with any output terminals. Wiper A supplies battery voltage to the comfort controls and turn signals, wiper B supplies battery voltage to the ignition coil and other accessories, and wiper C supplies battery voltage to other accessories. The jumper wire between terminals A and R of wiper C indicate that those accessories listed can be operated with the ignition switch in the RUN or ACC position.

Mercury switches are used by many vehicle manufacturers to detect motion. This switch uses a capsule that is partially filled with mercury and has two electrical contacts located at one end. If the switch is constructed as a normally open switch, the contacts are located above the mercury level (Figure 3-5). Mercury is an excellent conductor of electricity. If the capsule is moved so the mercury touches both of the electrical contacts, the circuit is completed (Figure 3-6). This type of switch is used to illuminate the engine compartment when the hood is opened. While the hood is shut, the capsule is tilted in a position such that the

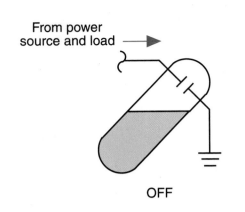

From power
source and load →

OFF

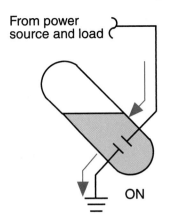

From power
source and load

ON

Figure 3-5 A mercury switch in the open position. The mercury is not covering the points.

Figure 3-6 When the mercury switch is tilted, the mercury covers the points and closes the circuit.

mercury is not able to complete the circuit. Once the hood is opened, the capsule tilts with the hood, the mercury completes the circuit, and the light turns on.

Relays

Some circuits utilize electromagnetic switches called **relays** (Figure 3-7). The coil in the relay has a very high resistance, thus it will draw very low current. This low current is used to produce a magnetic field that will close the contacts. Normally open relays have their points closed by the electromagnetic field, and normally closed relays have their points opened by the magnetic field. The contacts are designed to carry the high current required to operate the load component. When current is applied to the coil, the contacts close and heavy battery current flows to the load component that is being controlled.

The illustration (Figure 3-8) shows a relay application in a horn circuit. Battery voltage is applied to the coil. Because the horn button is a normally open–type switch, the current flow to ground is open. Pushing the horn button will complete the circuit, allowing current flow through the coil. The coil develops a magnetic field, which closes the contacts. With the contacts closed, battery voltage is applied to the horn (which is grounded). Used in this manner, the horn relay becomes a control of the high current necessary to blow the horn. The control circuit may be

Shop Manual
Chapter 3, page 94

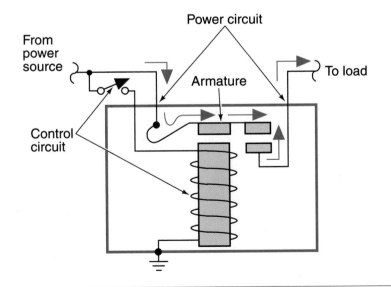

Power circuit

From power source

Control circuit

Armature

To load

Figure 3-7 A relay uses electrical current to create a magnetic field to draw the contact point closed.

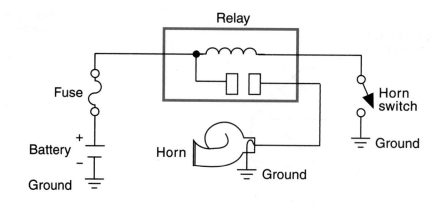

Figure 3-8 A relay can be used in the horn circuit to reduce the required size of the conductors installed in the steering column.

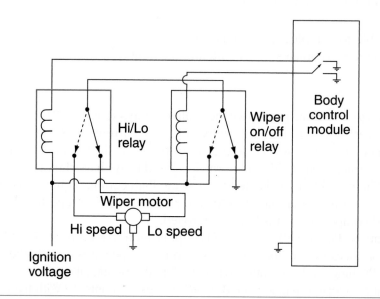

Figure 3-9 Using a relay as a diverter to control HI/LO wiper operation. The HI/LO relay diverts current to the different brushes of the wiper motor.

wired with very thin wire because it will have low current flowing through it. The control unit may have only 0.25 ampere flowing through it, and the horn may require 24 or more amperes.

Relays can also be used as a circuit diverter (Figure 3-9). In this example, the HI/LO wiper relay will direct current flow to either the high-speed brush or low-speed brush of the wiper motor to control wiper speeds.

ISO Relays.   **ISO relays** confirm to the specifications of the International Standards Organization (ISO) for common size and terminal patterns (Figure 3-10). The terminals are identified as 30, 87A, 87, 86, and 85. Terminal 30 is usually connected to battery voltage. This source voltage can be either switched (on or off by some type of switch) or connected directly to the battery. Terminal 87A is connected to terminal 30 when the relay is de-energized. Terminal 87 is connected to terminal 30 when the relay is energized. Terminal 86 is connected to battery voltage (switched or unswitched) to supply current to the electromagnet. Finally, terminal 85 provides ground for the electromagnet. Once again, the ground can be switched or unswitched.

Solenoids

A **solenoid** is an electromagnetic device and operates in the same way as a relay; however, a solenoid uses a movable iron core. Solenoids can do mechanical work, such as switching

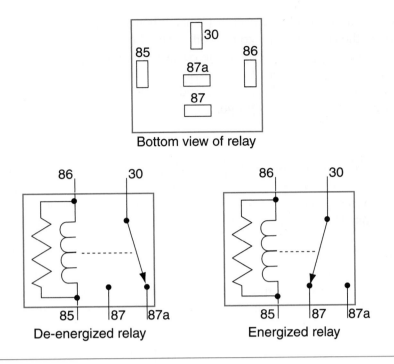

Bottom view of relay

86   30   86   30

85   87   87a   85   87   87a

De-energized relay   Energized relay

Figure 3-10 ISO relay terminal identification.

electrical, vacuum, and liquid circuits. The iron core inside the coil of the solenoid is spring loaded. When current flows through the coil, the magnetic field created around the coil attracts the core and moves it into the coil. To do work, the core is attached to a mechanical linkage, which causes something to move. When current flow through the coil stops, the spring pushes the core back to its original position. Some power door locks use solenoids to work the locking devices. Solenoids may also switch a circuit on or off, in addition to causing a mechanical action. Such is the case with some starter solenoids. These devices move the starter gear in and out of mesh with the flywheel. At the same time, they complete the circuit from the battery to the ignition circuit. Both of these actions are necessary to start an engine.

Buzzers

A **buzzer,** or **sound generator,** is sometimes used to warn the driver of possible safety hazards by emitting an audio signal (such as when the seat belt is not buckled). A buzzer is similar in construction to a relay except for the internal wiring (Figure 3-11). The coil is supplied

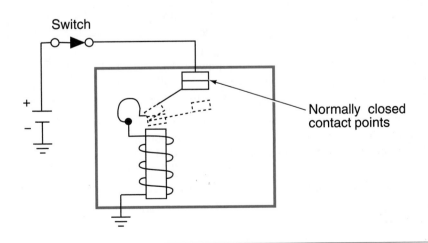

Figure 3-11 A buzzer reacts to the current flow to open and close rapidly, creating a noise.

current through the normally closed contact points. When voltage is applied to the buzzer, current flows through the contact points to the coil. When the coil is energized, the contact arm is attracted to the magnetic field. As soon as the contact arm is pulled down, the current flow to the coil is opened, and the magnetic field is dissipated. The contact arm then closes again, and the circuit to the coil is closed. This opening and closing action occurs very rapidly. It is this movement that generates the vibrating signal.

Resistors

All circuits require resistance in order to operate. If the resistance performs a useful function, it is referred to as the **load device.** However, resistance can also be used to control current flow and as sensing devices for computer systems. There are several types of resistors that may be used within a circuit. These include fixed resistors, stepped resistors, and variable resistors.

Fixed Resistors. **Fixed resistors** are usually made of carbon or oxidized metal (Figure 3-12). These resistors have a set resistance value and are used to limit the amount of current flow in a circuit. The resistance value can be determined by the color bands on the protective shell (Figure 3-13). Usually there are four or five color bands. When there are four bands, the first two

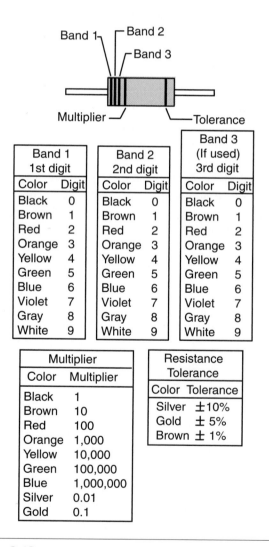

Band 1 1st digit		Band 2 2nd digit		Band 3 (If used) 3rd digit	
Color	Digit	Color	Digit	Color	Digit
Black	0	Black	0	Black	0
Brown	1	Brown	1	Brown	1
Red	2	Red	2	Red	2
Orange	3	Orange	3	Orange	3
Yellow	4	Yellow	4	Yellow	4
Green	5	Green	5	Green	5
Blue	6	Blue	6	Blue	6
Violet	7	Violet	7	Violet	7
Gray	8	Gray	8	Gray	8
White	9	White	9	White	9

Multiplier	
Color	Multiplier
Black	1
Brown	10
Red	100
Orange	1,000
Yellow	10,000
Green	100,000
Blue	1,000,000
Silver	0.01
Gold	0.1

Resistance Tolerance	
Color	Tolerance
Silver	±10%
Gold	± 5%
Brown	± 1%

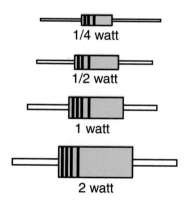

1/4 watt

1/2 watt

1 watt

2 watt

Figure 3-12 Fixed resistors.

Figure 3-13 Resistor color code chart.

are the digit bands, the third is the "multiplier," and the fourth is the tolerance. On a resistor with five bands, the first three are digit bands.

For example, if the resistor has four color bands of yellow, black, brown, and gold, the resistance value is determined as follows:

The first color band (yellow) gives the first digit value of 4.

The second color band (black) gives the second digit value of 0.

The digit value is now 40. Multiply this by the value of the third band. In this case, brown has a value of 10 so the resistor should have 400 ohms of resistance (40 × 10 = 400).

The last band gives the tolerance. Gold equals a tolerance range of ± 5%.

Stepped Resistors. A **stepped resistor** has two or more fixed resistor values. The stepped resistor can have an integral switch or have a switch wired in series. A stepped resistor is commonly used to control electrical motor speeds (Figure 3-14). By changing the position of the switch, resistance is increased or decreased within the circuit. If the current flows through a low resistance, then higher current flows to the motor and its speed is increased. If the switch is placed in the low-speed position, additional resistance is added to the circuit. Less current flows to the motor, which causes it to operate at a reduced speed.

Shop Manual
Chapter 3, page 96

A stepped resistor is also used to convert digital to analog signals in a computer circuit. This is accomplished by converting the on/off digital signals into a continuously variable analog signal.

Variable Resistors. **Variable resistors** provide for an infinite number of resistance values within a range. The most common types of variable resistors are rheostats and potentiometers. A **rheostat** is a two-terminal variable resistor used to regulate the strength of an electrical current.

Shop Manual
Chapter 3, page 97

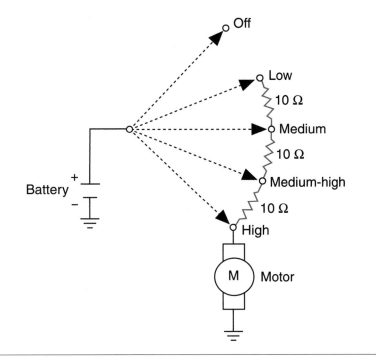

Figure 3-14 A stepped resistor is commonly used to control motor speeds. The total resistance of the switch is 30 Ω in the low position, 20 Ω in the medium position, 10 Ω in the medium-high position, and 0 Ω in the high position.

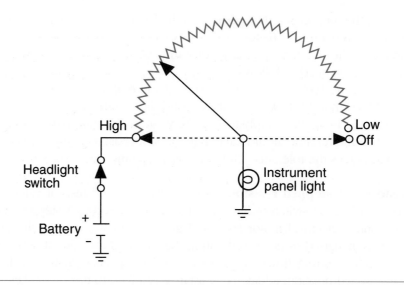

Figure 3-15 A rheostat can be used to control the brightness of a lamp.

A rheostat has one terminal connected to the fixed end of a resistor and a second terminal connected to a moveable contact called a **wiper** (Figure 3-15). By changing the position of the wiper on the resistor, the amount of resistance can be increased or decreased. The most common use of the rheostat is in the instrument panel lighting switch. As the switch knob is turned, the instrument lights dim or brighten depending on the resistance value.

A **potentiometer** is a three-wire variable resistor that acts as a voltage divider to produce a continuously variable output signal proportional to a mechanical position. When a potentiometer is installed into a circuit, one terminal is connected to a power source at one end of the resistor. The second wire is connected to the opposite end of the resistor and is the ground return path. The third wire is connected to the wiper contact (Figure 3-16). The wiper senses a variable voltage drop as it is moved over the resistor. Because the current always flows through the same amount of resistance, the total voltage drop measured by the potentiometer is very stable. For this reason, the potentiometer is a common type of input sensor for the vehicle's onboard computers.

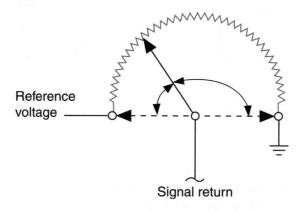

Figure 3-16 A potentiometer is used to send a signal voltage from the wiper.

Electronic Components

Because a semiconductor material can operate as both a conductor and an insulator, it is very useful as a switching device. How a semiconductor material works depends on the way current flows, or tries to flow, through it.

As discussed in Chapter 2, electrical materials are classified as conductors, insulators, or semiconductors. Semiconductors include diodes, transistors, and silicon-controlled rectifiers. These semiconductors are often called solid-state devices because they are constructed of a solid material. The most common materials used in the construction of semiconductors are silicon or germanium. Both of these materials are classified as a **crystal,** since they have a definite atom structure.

Silicon and germanium have four electrons in their outer orbits. Because of their crystal-type structure, each atom shares an electron with four other atoms (Figure 3-17). As a result of this **covalent bonding,** each atom will have eight electrons in its outer orbit. All the orbits are filled and there are no free electrons, thus the material (as a category of matter) falls somewhere between conductor and insulator.

Perfect crystals are not used for manufacturing semiconductors. They are doped with impurity atoms. This doping adds a small percentage of another element to the crystal. The doping element can be arsenic, antimony, phosphorous, boron, aluminum, or gallium.

If the crystal is doped by using arsenic, antimony, or phosphorous, the result is a material with free electrons (Figure 3-18). Materials such as arsenic have five electrons, which leaves one electron left over. This doped material becomes negatively charged and is referred to as an **N-type material.** Under the influence of an EMF, it will support current flow.

If boron, aluminum, or gallium are added to the crystal, a P-type material is produced. Materials like boron have three electrons in their outermost orbit. Because there is one fewer

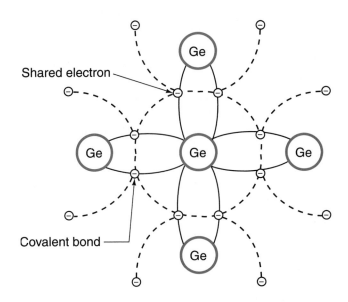

Figure 3-17 Crystal structure of germanium.

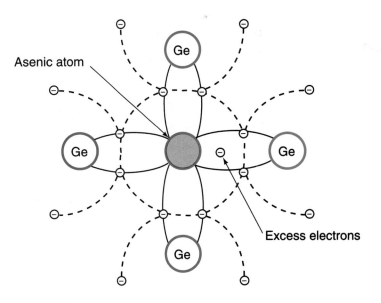

Figure 3-18 Germanium crystal doped with an arsenic atom to produce an N-type material.

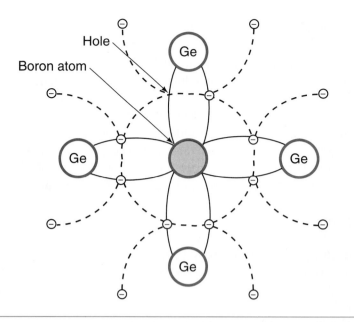

Figure 3-19 Germanium crystal doped with a boron atom to produce a P-type material.

electron, there is an absence of an electron that produces a **hole** (Figure 3-19) and becomes positively charged.

By putting N-type and P-type materials together in a certain order, solid-state components are built that can be used for switching devices, voltage regulators, electrical control, and so on.

Diodes

A **diode** is an electrical one-way check valve that will allow current to flow in one direction only. A diode is the simplest semiconductor device. It is formed by joining P-type semiconductor material with N-type material. The N (negative) side of a diode is called the **cathode** and the P (positive) side, the **anode** (Figure 3-20). The point where the cathode and anode join together is called the PN junction. The outer shell of the diode will have a stripe painted around it. This stripe designates which end of the diode is the cathode.

When a diode is made, the positive holes from the P region and the negative charges from the N region are drawn toward the junction. Some charges cross over and combine with opposite charges from the other side. When the charges cross over, the two halves are no longer balanced and the diode builds up a network of internal charges opposite to the charges at the PN junction. The internal EMF between the opposite charges limits the further diffusion of charges across the junction.

Shop Manual
Chapter 3, page 99

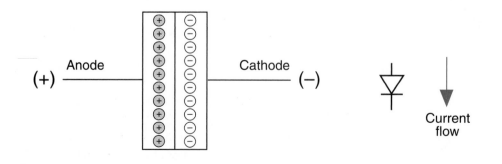

Figure 3-20 A diode and its symbol.

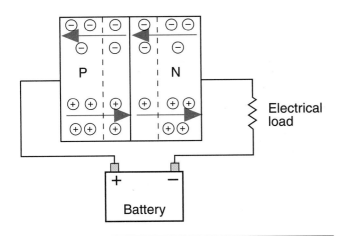

Figure 3-21 Forward-biased voltage causes current flow.

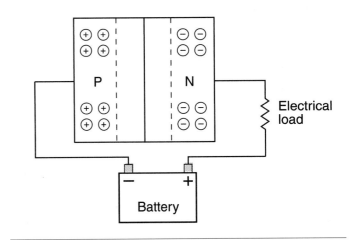

Figure 3-22 Reverse-biased voltage prevents current flow.

When the diode is incorporated within a circuit and a voltage is applied, the internal characteristics change. If the diode is **forward biased,** there will be current flow (Figure 3-21). In this state, the negative region will push electrons across the barrier as the positive region pushes holes across. When forward-biased, the diode acts as a conductor.

If the diode is **reverse biased,** there will be no current flow (Figure 3-22). The negative region will attract the positive holes away from the junction and the positive region will attract electrons away. This makes the diode act as an insulator.

When the diode is forward biased, it will have a small voltage drop across it. On standard silicon diodes, this voltage is usually about 0.6 volts. This is referred to as the **turn-on voltage.**

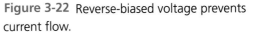

A BIT OF HISTORY

The first diodes were vacuum tube devices (also known as thermionic valves). Arrangements of electrodes were surrounded by a vacuum within a glass envelope that appeared similar to light bulbs. This arrangement of a filament and plate to create a diode was invented by John Ambrose Fleming in 1904. Current flow through the filament results in generation of heat. When heated, electrons are emitted into the vacuum. These electrons are electrostatically drawn to a positively charged outer metal plate (anode). Since the plate is not heated, the electrons will not return to the filament, even if the charge on the plate is made negative.

Zener Diodes

As stated, if a diode is reverse biased it will not conduct current. However, if the reverse voltage is increased, a voltage level will be reached at which the diode will conduct in the reverse direction. This voltage level is referred to as **zener voltage.** Reverse current can destroy a simple PN-type diode. But the diode can be doped with materials that will withstand reverse current.

A **zener diode** is designed to operate in reverse bias at the breakdown region. At the point that breakdown voltage is reached, a large current flows in reverse bias. This prevents the voltage from climbing any higher. This makes the zener diode an excellent component for regulating voltage. If the zener diode is rated at 15 volts, it will not conduct in reverse bias when the voltage is below 15 volts. At 15 volts it will conduct and the voltage will not increase over 15 volts.

Forward-biased means that a positive voltage is applied to the P-type material and negative voltage to the N-type material.

Reverse-biased means that positive voltage is applied to the N-type material and negative voltage is applied to the P-type material.

Shop Manual Chapter 3, page 99

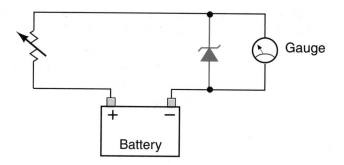

Figure 3-23 Simplified instrument gauge circuit that uses a zener diode to maintain a constant voltage to the gauge. Note the symbol used for a zener diode.

The illustration (Figure 3-23) shows a simplified circuit that has a zener diode in it to provide a constant voltage level to the instrument gauge. In this example, the zener diode is connected in series with the resistor and in parallel to the gauge. If the voltage to the gauge must be limited to 7 volts, the zener diode used would be rated at 7 volts. The zener diode maintains a constant voltage drop, and the total voltage drop in a series circuit must equal the amount of source voltage, thus voltage that is greater than the zener voltage must be dropped over the resistor. Even though source voltage may vary (as a normal result of the charging system), causing different currents to flow through the resistor and zener diode, the voltage that the zener diode drops remains the same.

The zener breaks down when system voltage reaches 7 volts. At this point, the zener diode conducts reverse current, causing an additional voltage drop across the resistor. The amount of voltage to the instrument gauge will remain at 7 volts because the zener diode "makes" the resistor drop the additional voltage to maintain this limit.

Here we see the difference between the standard diode and the zener diode. When the zener diode is reverse biased, the zener holds the available voltage to a specific value.

Avalanche Diodes

Avalanche diodes are diodes that conduct in the reverse direction when the reverse-bias voltage exceeds the breakdown voltage, similar to zener diodes in operation. However, breakdown is done by the avalanche effect. This occurs when the reverse electric field moves across the PN junction and causes a wave of ionization (like an avalanche), leading to a large current. Avalanche diodes are designed to break down at a well-defined reverse voltage without being destroyed. The reverse breakdown voltage is about 6.2 volts or higher. Avalanche diodes are commonly used in automobile AC generators (alternators).

Light-Emitting Diodes

Shop Manual
Chapter 3, page 100

A **light-emitting diode (LED)** is similar in operation to the diode, except the LED emits light when it is forward biased. An LED has a small lens built into it so that light can be seen when current flows through it (Figure 3-24). When the LED is forward biased, the holes and electrons combine and current is allowed to flow through it. The energy generated is released in the form of

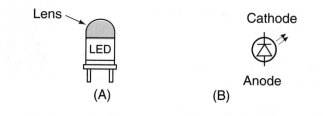

Figure 3-24 (A) A light-emitting diode uses a lens to emit the generated light. (B) Symbol for LED.

light. The light from an LED is not heat energy as is the case with other lights. It is electrical energy. Because of this, LEDs last longer than light bulbs. It is the material used to make the LED that will determine the color of the light emitted, and the turn-on voltage.

Similar to standard silicon diodes, the LED has a constant turn-on voltage. However, this turn-on voltage is usually higher than standard diodes. The turn-on voltage defines the color of the light; 1.2 volts corresponds to red, 2.4 volts to yellow.

Photo Diodes

A **photo diode** also allows current to flow in one direction only. However, the direction of current flow is opposite a standard diode. Reverse current flow only occurs when the diode receives a specific amount of light. These types of diodes can be used in automatic headlight systems.

Clamping Diodes

Whenever the current flow through a coil (such as used in a relay or solenoid) is discontinued, a voltage surge or spike is produced. This surge results from the collapsing of the magnetic field around the coil. The movement of the field across the windings induces a very high voltage spike, which can damage electronic components as it flows through the system. In some circuits, a capacitor can be used as a shock absorber to prevent component damage from this surge. In today's complex electronic systems, a **clamping diode** is commonly used to prevent the voltage spike. By installing a clamping diode in parallel with the coil, a bypass is provided for the electrons during the time that the circuit is open (Figure 3-25).

An example of the use of clamping diodes is on some air conditioning compressor clutches. Because the clutch operates by electromagnetism, opening the clutch coil circuit produces a voltage spike. If this voltage spike was left unchecked, it could damage the vehicle's onboard computers. The installation of the clamping diode prevents the voltage spike from reaching the computers. The clamping diode must be connected to the circuit in reverse bias.

Relays may also be equipped with a clamping diode. However, some use a resistor to dissipate the voltage spike. The two types of relays are not interchangeable.

A **clamping diode** is nothing more than a standard diode; the term *clamping* refers to its function.

Transistors

A **transistor** is a three-layer semiconductor. It is used as a very fast switching device. The word *transistor* is a combination of two words, *transfer* and *resist*. The transistor is used to control current flow in the circuit (Figure 3-26). It can be used to allow a predetermined amount of current flow or to resist this flow.

Shop Manual
Chapter 3, page 101

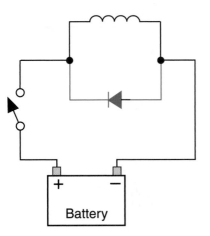

Figure 3-25 A clamping diode in parallel to a coil prevents voltage spikes when the switch is opened.

Figure 3-26 Transistors that are used in automotive applications.

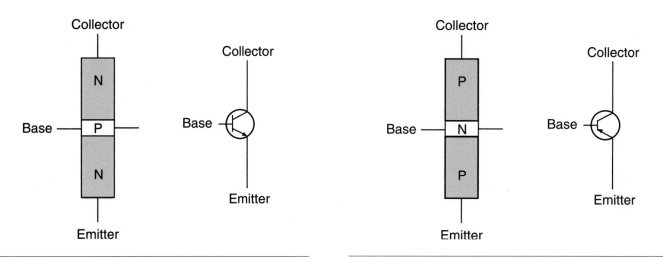

Figure 3-27 An NPN transistor and its symbol.

Figure 3-28 A PNP transistor and its symbol.

Transistors are made by combining P-type and N-type materials in groups of three. The two possible combinations are NPN (Figure 3-27) and PNP (Figure 3-28).

The three layers of the transistor are designated as **emitter, collector,** and **base.** The emitter is the outside layer of the forward-biased diode that has the same polarity as the circuit side to which it is applied. The arrow on the transistor symbol refers to the emitter lead and points in the direction of positive current flow and to the N material. The collector is the outside layer of the reverse-biased diode. The base is the shared middle layer. Each of these different layers has its own lead for connecting to different parts of the circuit. In effect, a transistor is two diodes that share a common center layer. When a transistor is connected to the circuit, the emitter-base junction will be forward biased and the collector-base junction will be reverse biased.

In the NPN transistor, the emitter conducts current flow to the collector when the base is forward biased. The transistor cannot conduct unless the voltage applied to the base leg exceeds the emitter voltage by approximately 0.7 volt. This means both the base and collector must be positive with respect to the emitter. With less than 0.7 volt applied to the base leg (compared to the voltage at the emitter), the transistor acts as an opened switch. When the voltage difference is greater than 0.7 volt at the base, compared to the emitter voltage, the transistor acts as a closed switch (Figure 3-29).

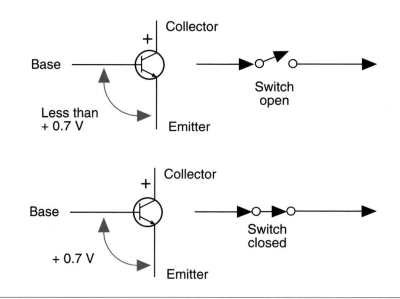

Figure 3-29 NPN transistor action.

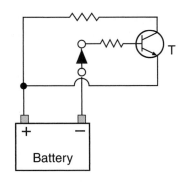

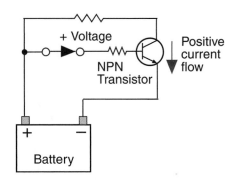

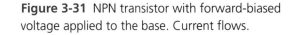

Battery

Figure 3-30 NPN transistor with reverse-biased voltage applied to the base. No current flow.

Figure 3-31 NPN transistor with forward-biased voltage applied to the base. Current flows.

When an NPN transistor is used in a circuit, it normally has a reverse bias applied to the base-collector junction. If the emitter-base junction is also reverse biased, no current will flow through the transistor (Figure 3-30). If the emitter-base junction is forward biased (Figure 3-31), current flows from the emitter to the base. Because the base is a thin layer and a positive voltage is applied to the collector, electrons flow from the emitter to the collector.

In the PNP transistor, current will flow from the emitter to the collector when the base leg is forward biased with a voltage that is more negative than that at the emitter (Figure 3-32). For current to flow through the emitter to the collector, both the base and the collector must be negative in respect to the emitter.

AUTHOR'S NOTE: Current flow through transistors is always based on hole and/or electron flow.

Current can be controlled through a transistor. Thus transistors can be used as a very fast electrical switch. It is also possible to control the amount of current flow through the collector. This is because the output current is proportional to the amount of current through the base leg.

A transistor has three operating conditions:

1. **Cutoff:** When reverse-biased voltage is applied to the base leg of the transistor. In this condition the transistor is not conducting and no current will flow.

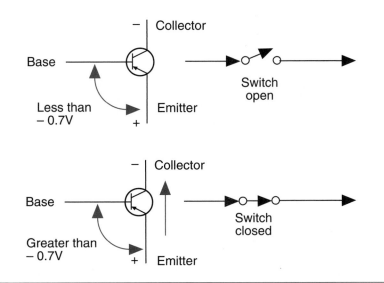

Figure 3-32 PNP transistor action.

2. **Conduction:** Bias voltage difference between the base and the emitter has increased to the point that the transistor is switched on. In this condition the transistor is conducting. Output current is proportional to that of the current through the base.

3. **Saturation:** This occurs when the collector to emitter voltage is reduced to near zero by a voltage drop across the collector's resistor.

These types of transistors are called **bipolar** because they have three layers of silicon; two of these layers are the same. Another type of transistor is the **field-effect transistor (FET).** The FET's leads are listed as source, drain, and gate. The source supplies the electrons and is similar to the emitter in the bipolar transistor. The drain collects the current and is similar to the collector. The gate creates the electrostatic field that allows electron flow from the source to the drain. It is similar to the base.

The FET transistor does not require a constant bias voltage. A voltage needs to be applied to the gate terminal to get electron flow from the source to the drain. The source and drain are constructed of the same type of doped material. They can be either N-type or P-type materials. The source and drain are separated by a thin layer of either N-type or P-type material opposite the gate and drain.

Using the illustration (Figure 3-33), if the source voltage is held at 0 volts and 6 volts are applied to the drain, no current will flow between the two. However, if a lower positive voltage is applied to the gate, the gate forms a capacitive field between the channel and itself. The voltage of the capacitive field attracts electrons from the source, and current will flow through the channel to the higher positive voltage of the drain.

This type of FET is called an **enhancement-type FET** because the field effect improves current flow from the source to the drain. This operation is similar to that of a normally open switch. A **depletion-type FET** is like a normally closed switch, whereas the field effect cuts off current flow from the source to the drain.

While electrons are flowing from the source to the drain (electron theory), positive charges are flowing from the drain to the source (conventional theory).

A BIT OF HISTORY

The transistor was developed by a team of three American physicists: Walter Houser Brattain, John Bardeen, and William Bradford Shockley. They announced their achievement in 1948. These physicists won the Nobel Prize in physics for this development in 1956.

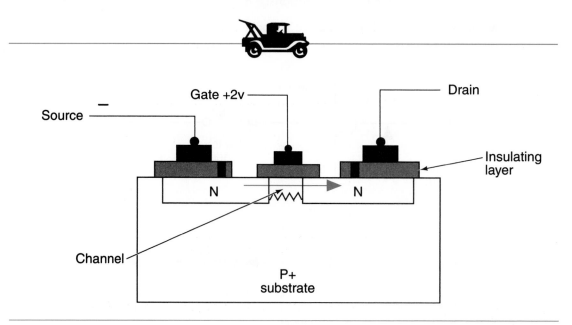

Figure 3-33 An FET uses a positive voltage to the gate terminal to create a capacitive field to allow electron flow.

Transistor Amplifiers

A transistor can be used in an amplifier circuit to amplify the voltage. This is useful when using a very small voltage for sensing computer inputs but needing to boost that voltage to operate an accessory (Figure 3-34). The waveform showing the small signal voltage that is applied to the base leg of a transistor may look like that shown (Figure 3-35A). The waveform showing the corresponding signal through the collector will be inverted (Figure 3-35B). Three things happen in an amplified circuit:

Shop Manual
Chapter 3, page 103

1. The amplified voltage at the collector is greater than that of the base voltage.

2. The input current increases.

3. The pattern has been inverted.

Some amplifier circuits use a **Darlington pair,** which is two transistors that are connected together. The first transistor in a Darlington pair is used as a preamplifier to produce a large

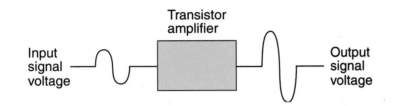

Figure 3-34 A simplified amplifier circuit.

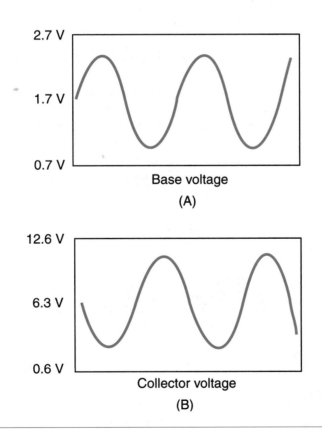

Figure 3-35 The voltage applied to the base (A) is amplified and inverted through the collector (B).

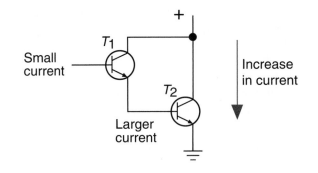

Figure 3-36 A Darlington pair used to amplify current. T1 acts as a preamplifier that creates a larger base current for T2, which is the final amplifier that creates a larger current.

current to operate the second transistor (Figure 3-36). The second transistor is isolated from the control circuit and is the final amplifier. The second transistor boosts the current to the amount required to operate the load component. The Darlington pair is utilized by most control modules used in electronic ignition systems.

Phototransistors

A **phototransistor** is a transistor that is sensitive to light. In a phototransistor, a small lens is used to focus incoming light onto the sensitive portion of the transistor (Figure 3-37). When light strikes the transistor, holes and free electrons are formed. These increase current flow through the transistor according to the amount of light. The stronger the light intensity, the more current that will flow. This type of phototransistor is often used in automatic headlight dimming circuits.

Thyristors

A **thyristor** is a semiconductor switching device composed of alternating N and P layers. It can be used to rectify current from AC to DC, and to control power to light dimmers, motor speed controls, solid-state relays, and other applications where power control is needed.

The most common type of thyristor used in automotive applications is the silicon-controlled rectifier (SCR). Like the transistor, the SCR has three legs. However, it consists of four regions arranged PNPN (Figure 3-38). The three legs of the SCR are called the anode (or P-terminal), the cathode (or N-terminal), and the gate (one of the center regions).

The SCR requires only a trigger pulse (not a continuous current) applied to the gate to become conductive. Current will continue to flow through the anode and cathode as long as the voltage remains high enough, or until gate voltage is reversed.

The SCR can be connected into a circuit in either the forward or reverse direction. Using Figure 3-38 of a forward-direction connection, the P-type anode is connected to the positive

Figure 3-37 Phototransistor.

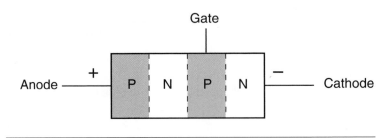

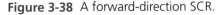

Figure 3-38 A forward-direction SCR.

side of the circuit and the N-type cathode is connected to the negative side. The center PN junction blocks current flow through the anode and cathode.

Once a positive voltage pulse is applied to the gate, the SCR turns on. Even if the positive voltage pulse is removed, the SCR will continue to conduct. If a negative voltage pulse is applied to the gate, the SCR will no longer conduct.

The SCR will also block any reverse current from flowing from the cathode to the anode. Because current can flow in only one direction through the SCR, it can rectify AC current to DC current.

Integrated Circuits

An **integrated circuit (IC)** is a complex circuit of thousands of transistors, diodes, resistors, capacitors, and other electronic devices that are formed onto a tiny silicon chip (Figure 3-39). As many as 30,000 transistors can be placed on a chip that is 1/4 inch (6.35 mm) square.

Integrated circuits are constructed by photographically reproducing circuit patterns onto a silicon wafer. The process begins with a large-scale drawing of the circuit. This drawing can be room size. Photographs of the circuit drawing are reduced until they are the actual size of the circuit. The reduced photographs are used as a mask. Conductive P-type and N-type materials, along with insulating materials, are deposited onto the silicon wafer. The mask is placed over the wafer and selectively exposes the portion of material to be etched away or the portions requiring selective deposition. The entire process of creating an integrated circuit chip takes over 100 separate steps. Out of a single wafer 4 inches (101.6 mm) in diameter, thousands of integrated circuits can be produced.

The small size of the integrated chip has made it possible for the vehicle manufacturers to add several computer-controlled systems to the vehicle without taking up much space. Also, a single computer is capable of performing several functions.

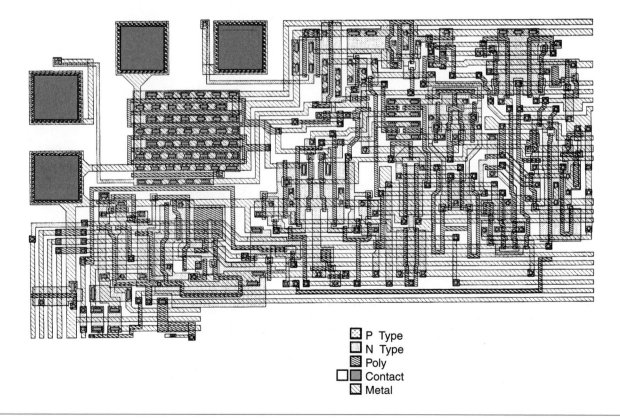

- ▦ P Type
- ▢ N Type
- ▨ Poly
- ▢▨ Contact
- ▧ Metal

Figure 3-39 An enlarged illustration of an integrated circuit with thousands of transistors, diodes, resistors, and capacitors. Actual size can be less than 1/4-inch (6.35 mm) square.

Circuit Protection Devices

Most automotive electrical circuits are protected from high current flow that would exceed the capacity of the circuit's conductors and/or loads. Excessive current results from a decrease in the circuit's resistance. Circuit resistance will decrease when too many components are connected in parallel or when a component or wire becomes shorted. A short is an undesirable, low-resistance path for current flow. When the circuit's current reaches a predetermined level, most circuit **protection devices** open and stop current flow in the circuit. This action prevents damage to the wires and the circuit's components.

Fuses

The most commonly used circuit protection device is the **fuse** (Figure 3-40). A fuse is a replaceable element that contains a metal strip that will melt when the current flowing through it exceeds its rating. The thickness of the metal strip determines the rating of the fuse. When the metal strip melts, excessive current is indicated. The cause of the **overload** must be found and repaired; then a new fuse of the same rating should be installed. The most commonly used automotive fuses are rated from 3 to 30 amps.

There are three basic types of fuses: glass or ceramic fuses, blade-type fuses, and bullet or cartridge fuses. Glass and ceramic fuses are found mostly on older vehicles. Sometimes, however, you can find them in a special holder connected in series with a circuit. Glass fuses are small glass cylinders with metal caps. The metal strip connects the two caps. The rating of the fuse is normally marked on one of the caps.

Blade-type fuses are flat plastic units and are available in three different physical sizes: mini, standard, and maxi (Figure 3-41). The plastic housing is formed around two male

> Excess current flow in a circuit is called an **overload**.

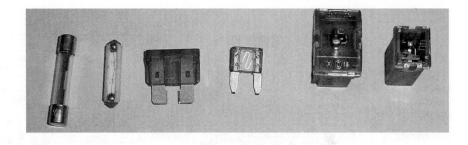

Figure 3-40 Common fuses: (A) glass cartage, (B) ceramic, (C) blade (auto), mini, maxi, and F type.

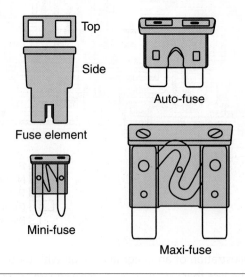

Figure 3-41 Types of blade fuses.

blade-type connectors. The metal strip connects these connectors inside the plastic housing. The rating of these fuses is on top of the plastic housing and the plastic is color coded (Figure 3-42).

Cartridge-type fuses are used in many European vehicles. These fuses are made of plastic or ceramic material. They have pointed ends and the metal strip rounds from end to end. This type of fuse is much like a glass fuse except the metal strip is not enclosed.

Fuses are typically located in a central **fuse block** or power distribution box. However, fuses may also be found in relay boxes and electrical junction boxes. Power distribution boxes are normally located in the engine compartment and house fuses and relays. A common location for a fuse box is under the instrumental panel (Figure 3-43). The fuse box may also be located behind kick panels, in the glove box, in the engine compartment, or in a variety of other places on the vehicle. Fuse ratings and the circuits they protect are normally marked on the cover of

Auto-fuse

Current Rating	Color
3	Violet
5	Tan
7.5	Brown
10	Red
15	Blue
20	Yellow
25	Natural
30	Green

Maxi-fuse

Current Rating	Color
20	Yellow
30	Green
40	Amber
50	Red
60	Blue
70	Brown
80	Natural

Mini-fuse

Current Rating	Color
5	Tan
7.5	Brown
10	Red
15	Blue
20	Yellow
25	White
30	Green

Figure 3-42 Color coding for blade-type fuses. An auto-fuse is a standard blade-type fuse.

Figure 3-43 Fuse boxes are normally located under the dash or in the engine compartment.

the fuse or power distribution box. Of course, this information can also be found in the vehicle's owner's manual and the service manual.

A fuse is connected in series with the circuit. Normally the fuse is located before all of the loads of the circuit (Figure 3-45). However, it may be placed before an individual load (Figure 3-46).

When adding accessories to the vehicle, the correct fuse rating must be selected. Use the power formula to determine the correct fuse rating (watts ÷ volts = amperes). The fuse selected should be rated slightly higher than the actual current draw to allow for current surges (5% to 10%).

Fusible Links

Fusible links are made of meltable conductor material with a special heat-resistant insulation. When there is an overload in the circuit, the conductor link melts and opens the circuit. To properly test a fusible link, use an ohmmeter or continuity tester. A vehicle may have one or

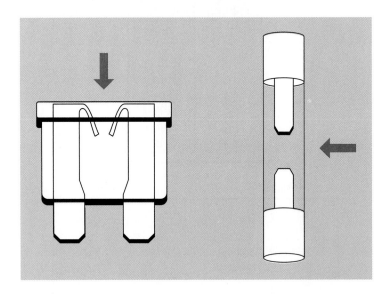

Figure 3-44 Blown fuses.

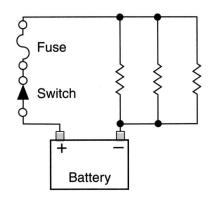

Figure 3-45 One fuse to protect the entire parallel circuit.

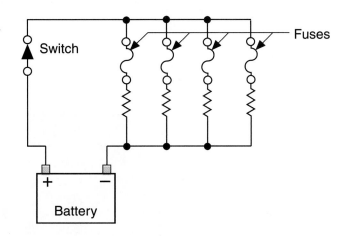

Figure 3-46 Fuses used to protect each branch of a parallel circuit.

Figure 3-47 Fusible links located near the battery.

several fusible links to provide protection for the main power wires before they are divided into smaller circuits at the fuse box. The fusible links are usually located at a main connection near the battery or starter solenoid (Figure 3-47). The current capacity of a fusible link is determined by its size. A fusible link is usually four wire sizes smaller (four numbers larger) than the circuit it protects. The smaller the wire, the larger its number. A circuit that uses 14-gauge wire would require an 18-gauge fusible link for protection.

 AUTHOR'S NOTE: Some GM vehicles have the fusible link located at the main connection near the starter motor.

 AUTHOR'S NOTE: A "blown" fusible link is usually identified by bubbling of the insulator material around the link.

Maxi-Fuses

In place of fusible links, many manufacturers use a **maxi-fuse.** A maxi-fuse looks similar to a blade-type fuse except it is larger and has a higher current capacity. It is also referred to as a cartridge fuse. By using maxi-fuses, manufacturers are able to break down the electrical system into smaller circuits. If a fusible link burns out, many of the vehicle's electrical systems may be affected. By breaking down the electrical system into smaller circuits and installing maxi-fuses, the consequence of a circuit defect will not be as severe as it would have been with a fusible link. In place of a single fusible link, there may be many maxi-fuses, depending on how the circuits are divided. This makes the technician's job of diagnosing a faulty circuit much easier.

Maxi-fuses are used because they are less likely to cause an underhood fire when there is an overload in the circuit. If the fusible link is burned in two, it is possible that the "hot" side of the fuse can come into contact with the vehicle frame and the wire can catch on fire.

Today many manufacturers are replacing maxi-fuses with "F"-type fuses. These are smaller versions of the maxi-fuses.

Circuit Breakers

A circuit that is susceptible to an overload on a routine basis is usually protected by a **circuit breaker.** A circuit breaker uses a **bimetallic strip** that reacts to excessive current (Figure 3-48). When an overload or circuit defect occurs that causes an excessive amount of current draw, the current flowing through the bimetallic strip causes it to heat. As the strip heats, it bends and opens the contacts. Once the contacts are opened, current can no longer flow. With no current flowing, the strip cools and closes again. If the excessive current cause is still in the circuit, the breaker will open again. The circuit breaker will continue to open and close as long as the overload is in the circuit. This type of circuit breaker is self-resetting or "cycled." Some circuit breakers require manual resetting by pressing a button, while others must be removed from the power to reset (Figure 3-49).

An example of the use of a circuit breaker is in the power window circuit. Because the window is susceptible to jams due to ice buildup on the window, a current overload is possible. If this should occur, the circuit breaker will heat up and open the circuit before the window motor is damaged. If the operator continues to attempt to operate the power window, the circuit breaker will open and close until the cause of the jam is removed.

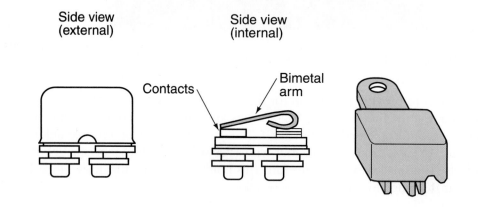

Figure 3-48 The circuit breaker uses a bimetallic strip that opens if current draw is excessive.

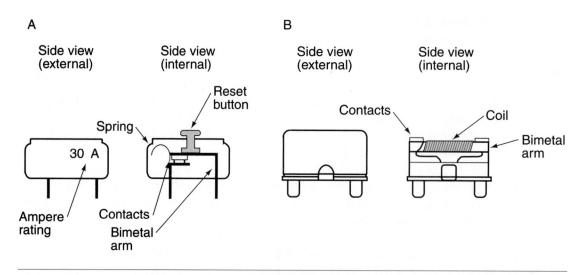

Figure 3-49 Noncycling circuit breakers. (A) can be reset by pressing the button, while (B) requires being removed from the power to be reset.

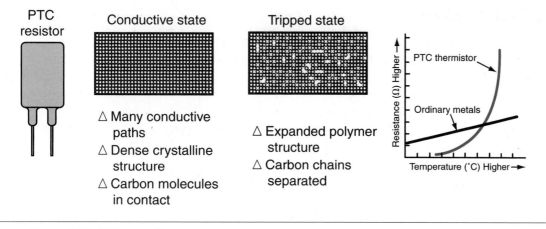

- △ Many conductive paths
- △ Dense crystalline structure
- △ Carbon molecules in contact

- △ Expanded polymer structure
- △ Carbon chains separated

Figure 3-50 PTC operation.

PTCs as Circuit Protection Devices

Automotive engineers are faced with conflicting needs to provide reliable circuit protection against shorts to ground or other overload conditions yet at the same time to reduce vehicle weight and cost. Traditionally, fuses are used to protect multiple circuits. However, this results in large, heavy, and complex wiring assemblies. The use of polymer, **positive temperature coefficient (PTC)** resistors provides a means of meeting these needs. A PTC resistor increases in resistance as temperature increases. Because of its design, a PTC resistor has the ability to trip (increase resistance to the point it becomes the load device in the circuit) during an overcurrent condition and reset after the fault is no longer present (Figure 3-50).

Conductive polymers consist of specially formulated plastics and various conductive materials. At normal temperatures, the plastic materials form a crystalline structure. The structure provides a low-resistance conductive chain. The resistance is so low that it does not affect the operation of the circuit. However, if the current flow increases above the trip threshold, the additional heat causes the crystalline structure to change to an amorphous state. In this condition, the conductive paths separate, causing a rapid increase in the resistance of the PTC. The increased resistance reduces the current flow to a safe level.

Circuit Defects

All electrical problems can be classified as being one of three types of problems: an open, short, or high resistance. Each one of these will cause a component to operate incorrectly or not at all. Understanding what each of these problems will do to a circuit is the key to proper diagnosis of any electrical problem.

Open

An **open** is simply a break in the circuit (Figure 3-51). An open is caused by turning a switch off, a break in a wire, a burned-out light bulb, a disconnected wire or connector, or anything that opens the circuit. When a circuit is open, current does not flow and the component doesn't work. Because there is no current flow, there are no voltage drops in the circuit. Source voltage is available everywhere in the circuit up to the point at which it is open. Source voltage is even available after a load, if the open is after that point.

Opens caused by a blown fuse will still cause the circuit not to operate, but the cause of the problem is the excessive current that blew the fuse. Nearly all other opens are caused by a break in the continuity of the circuit. These breaks can occur anywhere in the circuit.

Shop Manual
Chapter 3, page 103

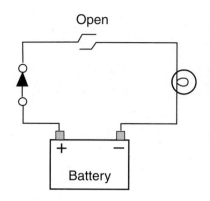

Figure 3-51 An open circuit stops all current flow.

Shorts

Shop Manual
Chapter 3, page 106

A **short** results from an unwanted path for current. **Shorted circuits** cause an increase in current flow by bypassing part of the normal circuit path. This increased current flow can burn wires or components.

An example of a shorted circuit could be found in a faulty coil. The windings within a coil are insulated from each other; however, if this insulation breaks down, a copper-to-copper contact is made between the turns. Since part of the windings will be bypassed, this reduces the number of windings in the coil through which current will flow. This results in the effectiveness of the coil being reduced. Also, since the current bypasses a portion of the normal circuit resistance, current flow is increased and excess heat can be generated.

Another example of a shorted circuit is if the insulation of two adjacent wires breaks down and allows a copper-to-copper contact (Figure 3-52). If the short is between points A and B, light 1 would be on all the time. If the short is between points B and C, both lights would illuminate when either switch is closed.

Shop Manual
Chapter 3, page 107

Another example is shown in Figure 3-53. With the two wires shorted together, the horn will sound every time the brake pedal is depressed. Also, if the horn button is pressed, the brake lights will come on.

Another type of electrical defect is a short to ground. A short to ground allows current to flow an unintentional path to ground (Figure 3-54). To see what happens in a circuit that has a

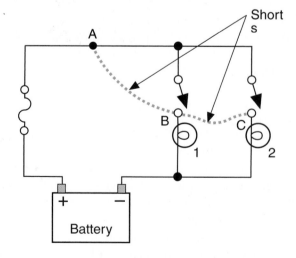

Figure 3-52 A short circuit can be a copper-to-copper contact between two adjacent wires.

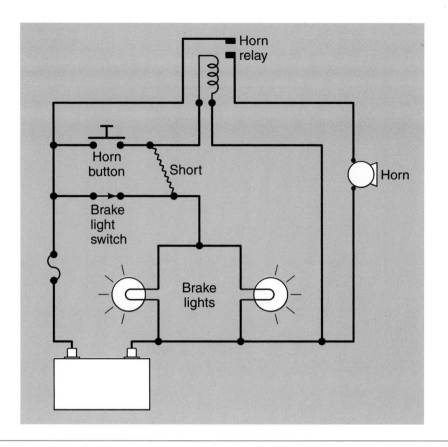

Figure 3-53 A wire-to-wire short.

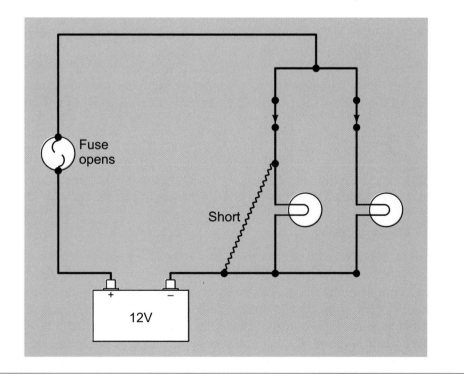

Figure 3-54 A grounded circuit.

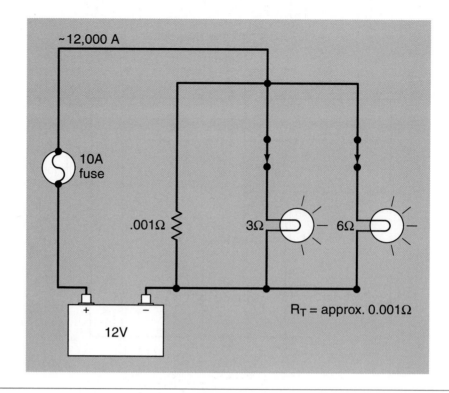

Figure 3-55 Ohm's law applied to Figure 3-54.

short to ground, refer to Figure 3-55. If normal resistance of the two bulbs is 3 ohms and 6 ohms, since they are in parallel, the total circuit resistance is 2 ohms. The short makes a path from the power side of one bulb to the return path, and to the battery. The short creates a low-resistance path. If the low-resistance path has a resistance value of 0.001 ohms, it is possible to calculate what would happen to the current in this circuit.

The short becomes another leg in the parallel circuit. Since the total resistance of a parallel circuit is always lower than the lowest resistance, we know the total resistance of the circuit is now less than 0.001 ohms. Using Ohm's law we can calculate the current flow through the circuit.

A = V/R   or   A = 12/.001   or   A = 12,000 Amps

Needless to say, it would take a large wire to carry that kind of amperage. Our 10-amp fuse would melt quickly when the short occurred. This would protect the wires and light bulbs.

High Resistance

Shop Manual
Chapter 3, page 109

High-resistance problems occur when there is unwanted resistance in the circuit. The high resistance can come from a loose connection, corroded connection, corrosion in the wire, wrong size wire, and so on. Since the resistance becomes an additional load in the circuit, the effect is that the load component, with reduced voltage and current applied, operates with reduced efficiency. An example would be a taillight circuit with a load component (light bulb) that is rated at 50 watts. To be fully effective, this bulb must draw 4.2 amperes at 12 volts (A = P ÷ V). This means a full 12 volts should be applied to the bulb. If resistance is present at other points in the circuit, some of the 12 volts will be dropped. With less voltage (and current) being available to the light bulb, the bulb will illuminate with less intensity.

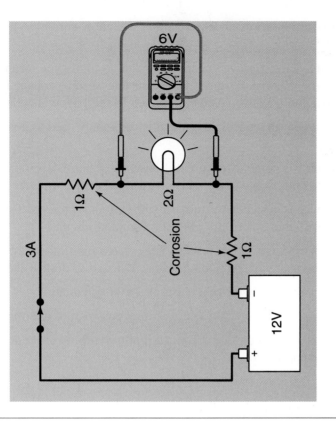

Figure 3-56 A simple light circuit with unwanted resistance.

Figure 3-56 illustrates a light circuit with unwanted resistance at the power feed for the bulb and at the negative battery terminal. When the circuit is operating properly, the 2-ohm light bulb will have 6 amps of current flowing through it and drop 12 volts. With the added resistance, the current is reduced to 3 amps and the bulb drops only 6 volts. As a result, the bulb's illumination is very dim.

Summary

- ❏ A switch can control the on/off operation of a circuit or direct the flow of current through various circuits.
- ❏ A normally open switch will not allow current flow when it is in its rest position. A normally closed switch will allow current flow when it is in its rest position.
- ❏ A relay is a device that uses low current to control a high-current circuit.
- ❏ A buzzer is sometimes used to warn the driver of possible safety hazards by emitting an audio signal (such as when the seat belt is not buckled).
- ❏ A stepped resistor has two or more fixed resistor values. It is commonly used to control electrical motor speeds.
- ❏ A variable resistor provides for an infinite number of resistance values within a range. A rheostat is a two-terminal variable resistor used to regulate the strength of an electrical current. A potentiometer is a three-wire variable resistor that acts as a voltage divider to produce a continuously variable output signal proportional to a mechanical position.

Terms to Know

Anode

Avalanche diodes

Base

Bimetallic strip

Bipolar

Buzzer

Cathode

Circuit breaker

Clamping diode

Collector

Covalent bonding

Crystal

Darlington pair

Depletion-type FET

Diode

❏ A diode is an electrical one-way check valve that will allow current to flow in one direction only.

❏ Forward bias means that a positive voltage is applied to the P-type material and negative voltage to the N-type material. Reverse bias means that positive voltage is applied to the N-type material and negative voltage is applied to the P-type material.

❏ A transistor is a three-layer semiconductor that is commonly used as a very fast switching device.

❏ An integrated circuit is a complex circuit of thousands of transistors, diodes, resistors, capacitors, and other electronic devices that are formed onto a tiny silicon chip.

❏ The protection device is designed to "turn off" the system it protects. This is done by creating an open (like turning off a switch) to prevent a complete circuit.

❏ Fuses are rated by amperage. Never install a larger rated fuse into a circuit than the one that was designed by the manufacturer. Doing so may damage or destroy the circuit.

❏ An open circuit is a circuit in which there is a break in continuity.

❏ A shorted circuit is a circuit that allows current to bypass part of the normal path.

❏ A short to ground is a condition that allows current to return to ground before it has reached the intended load component.

Review Questions

Short-Answer Essays

1. Describe the use of three types of semiconductors.

2. What types of mechanical variable resistors are used on automobiles?

3. Define what is meant by opens, shorts, grounds, and excessive resistance.

4. Explain the effects that each type of circuit defect will have on the operation of the electrical system.

5. Explain the purpose of a circuit protection device.

6. Describe the most common types of circuit protection devices.

7. Describe the common types of electrical system (non-electric) components used and how they affect the electrical system.

8. Describe the difference between a rheostat and a potentiometer.

9. Explain the difference between normally open (NO) and normally closed (NC) switches.

10. Explain the differences between forward biasing and reverse biasing a diode.

Fill in the Blanks

1. Never install a larger rated _____ into a circuit than the one that was designed by the manufacturer.

2. A _____ can control the on/off operation of a circuit or direct the flow of current through various circuits.

3. A normally _____ switch will not allow current flow when it is in its rest position. A normally _____ switch will allow current flow when it is in its rest position.

4. An _____ _____ is a complex circuit of many transistors, diodes, resistors, capacitors, and other electronic devices that are formed onto a tiny silicon chip.

5. When a _____ voltage is applied to the P-material of a diode and _____ voltage is applied to the N-material, the diode is reverse biased. When a _____ voltage is applied to the N-material of a diode and _____ voltage is applied to the P-material, the diode is forward biased.

6. A _____ is used in electronic circuits as a very fast switching device.

7. A _____ is an electrical one-way check valve that will allow current to flow in one direction only.

8. A _____ is an electromechanical device that uses low current to control a high-current circuit.

9. A _____ is a three-wire variable resistor that acts as a voltage divider. A _____ is a two-terminal variable resistor used to regulate the strength of an electrical current.

10. The _____ requires only a trigger pulse applied to the gate to become conductive.

Terms to Know
(continued)
Throw
Thyristor
Transistor
Turn-on voltage
Variable resistors
Wiper
Zener diode
Zener voltage

Multiple Choice

1. All of the following are true concerning electrical shorts, EXCEPT:
   A. A short can add a parallel leg to the circuit, which lowers the entire circuit's resistance.
   B. A short can result in a blown fuse.
   C. A short decreases amperage in the circuit.
   D. A short bypasses the circuit's intended path.

2. Which statement is true concerning circuit protection devices?
   A. A fuse automatically resets after the cause of the overload is repaired.
   B. Circuit protection devices create an open when an overload occurs.
   C. An open circuit can cause a blown fuse.
   D. Fuses are rated according to the voltage limits.

3. All of the statements concerning circuit components are true, EXCEPT:
   A. A switch can control the on/off operation of a circuit.
   B. A switch can direct the flow of current through various circuits.
   C. A relay can be an SPDT-type switch.
   D. A potentiometer changes voltage drop due to the function of temperature.

4. Which of the following statements is/are correct?
   A. A zener diode is an excellent component for regulating voltage.
   B. A reverse-biased diode lasts longer than a forward-biased diode.
   C. The switches of a transistor last longer than those of a relay.
   D. Both a and c.

5. The light-emitting diode (LED):
   A. Emits light when it is reverse biased.
   B. Has a variable turn-on voltage.
   C. Has a light color that is defined by the materials used to construct the diode.
   D. Has a turn-on voltage that is usually less than standard diodes.

6. Which of the following is the correct statement?
   A. An open means that there is continuity in the circuit.
   B. A short bypasses a portion of the circuit.
   C. High amperage draw indicates an open circuit.
   D. High resistance in a circuit increases current flow.

7. Transistors:
   A. Can be used to control the switching on/off of a circuit.
   B. Can be used to amplify voltage.
   C. Control high current with low current.
   D. All of the above.

8. All of the following are true, EXCEPT:
   A. Voltage drop can cause a lamp in a parallel circuit to burn brighter than normal.
   B. Excessive voltage drop may appear on either the insulated or grounded return side of a circuit.
   C. Increased resistance in a circuit decreases current.
   D. A diode is used as an electrical one-way check valve.

9. Which statement is correct concerning diodes?
   A. Diodes are aligned to allow current flow in one direction only.
   B. Diodes can be used to rectify DC voltages into AC voltages.
   C. The stripe is on the anode side of the diode.
   D. Normal turn-on voltage of a standard diode is 1.5 volts.

10. A "blown" fusible link is identified by:
   A. A burned-through metal wire in the capsule.
   B. A bubbling of the insulator material around the link.
   C. All of the above.
   D. None of the above.

Wiring and Circuit Diagrams

Upon completion and review of this chapter, you should be able to:

- ❏ Explain when single-stranded or multistranded wire should be used.
- ❏ Explain the use of resistive wires in a circuit.
- ❏ Describe the construction of spark plug wires.
- ❏ Explain how wire size is determined by the American Wire Gauge (AWG) and metric methods.
- ❏ Describe how to determine the correct wire gauge to be used in a circuit.

- ❏ Explain how temperature affects resistance and wire size selection.
- ❏ Explain the purpose and use of printed circuits.
- ❏ Explain why wiring harnesses are used and how they are constructed.
- ❏ Explain the purpose of wiring diagrams.
- ❏ Identify the common electrical symbols that are used.
- ❏ Explain the purpose of the component locator.

Introduction

Today's vehicles have a vast amount of electrical wiring that, if laid end to end, could stretch for half a mile or more. Today's technician must be proficient at reading wiring diagrams in order to sort through this great maze of wires. Trying to locate the cause of an electrical problem can be quite difficult if you do not have a good understanding of wiring systems and diagrams.

In this chapter, you will learn how wiring harnesses are made, how to read the wiring diagram, how to interpret the symbols used, and how terminals are used. This will reduce the amount of confusion you may experience when repairing an electrical circuit. It is also important to understand how to determine the correct type and size of wire to carry the anticipated amount of current. It is possible to cause an electrical problem by simply using the wrong gauge size of wire. A technician must understand the three factors that cause resistance in a wire—length, diameter, and temperature—to perform repairs correctly.

Automotive Wiring

Primary wiring is the term used for conductors that carry low voltage. The insulation of primary wires is usually thin. **Secondary wiring** refers to wires used to carry high voltage, such as ignition spark plug wires. Secondary wires have extra-thick insulation.

Most of the primary wiring conductors used in the automobile are made of several strands of copper wire wound together and covered with a polyvinyl chloride (PVC) insulation (Figure 4-1). Copper has low resistance and can be connected to easily by using crimping connectors or soldered connections. Other types of conductor materials used in automobiles include silver, gold, aluminum, and tin-plated brass.

Shop Manual
Chapter 4, page 133

> **AUTHOR'S NOTE:** Copper is used mainly because of its low cost and availability.

Stranded wire means the conductor is made of several individual wires that are wrapped together. Stranded wire is used because it is very flexible and has less resistance than solid

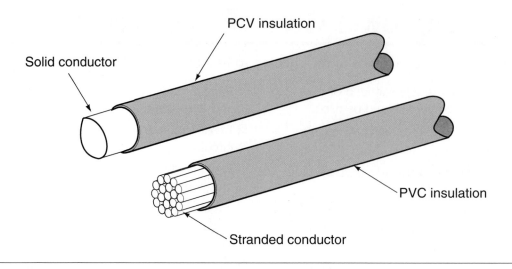

Figure 4-1 Comparison between solid and stranded primary wire.

wire. This is because electrons tend to flow on the outside surface of conductors. Since there is more surface area exposed in a stranded wire (each strand has its own surface), there is less resistance in the stranded wire than in the solid wire (Figure 4-2). The PVC insulation is used because it can withstand temperature extremes and corrosion. PVC insulation is also capable of withstanding battery acid, antifreeze, and gasoline. The insulation protects the wire from shorting to ground and from corrosion.

AUTHOR'S NOTE: General Motors has used single-stranded aluminum wire in limited applications where no flexing of the wire is expected. For example, it is used in the taillight circuits.

A **ballast resistor** was used by some manufacturers to protect the ignition primary circuit from excessive voltage. It reduces the current flow through the coil's primary windings and provides a stable voltage to the coil. This increases the life of the coil. The resistance value of the ballast resistor is usually between 0.8 and 1.2 ohms. Some automobiles use a **resistance wire** in the ignition system instead of a ballast resistor. Resistance wire is designed with a certain amount of resistance per foot. This wire is called the ballast resistor

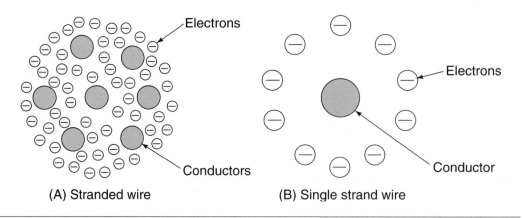

Figure 4-2 Stranded wire provides flexibility and more surface area for electron flow than a single-strand solid wire.

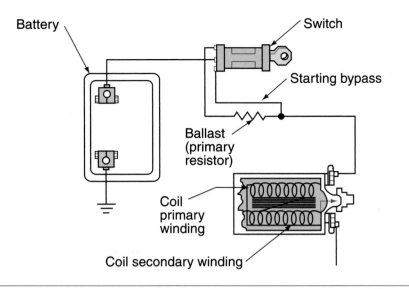

Figure 4-3 Ballast resistor used in some ignition primary wiring circuits.

wire and is located between the ignition switch and the ignition coil (Figure 4-3) in the ignition RUN circuit.

Spark plug wires are also resistance wires. The resistance lowers the current flow through the wires. By keeping current flow low, the magnetic field created around the wires is kept to a minimum. The magnetic field needs to be controlled because it causes radio interference. The result of this interference is noise on the vehicle's radio and all nearby radios and televisions. The noise can interfere with emergency broadcasts and the radios of emergency vehicles. Because of this concern, all ignition systems are designed to minimize radio interference; most do so with resistance-type spark plug wires. Spark plug wires are targeted because they carry high-voltage pulses. The lower current flow has no adverse effect on the firing of the spark plug.

Most spark plug wire conductors are made of nylon, rayon, fiberglass, or aramid thread impregnated with carbon. This core is surrounded by rubber (Figure 4-4). The carbon-impregnated core provides sufficient resistance to reduce RFI, yet does not affect engine operation. As the spark plug wires wear because of age and temperature changes, the resistance in the wire will change. Most plug wires have a resistance value of 3,000 Ω to 6,000 Ω per foot. However, some have between 6,000 Ω and 12,000 Ω. The accepted value when testing is 10,000 Ω per foot as a general specification.

Because the high voltage within the plug wires can create electromagnetic induction, proper wire routing is important to eliminate the possibility of **cross-fire.** Cross-fire is the electromagnetic induction spark that can be transmitted in another wire close to the wire carrying the current. To prevent cross-fire, the plug wires must be installed in the proper separator. Any two parallel wires next to each other in the firing order should be positioned as far away from

Spark plug wires are often referred to as television-radio-suppression (TVRS) cables.

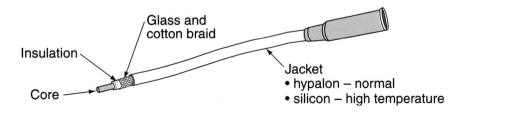

Figure 4-4 Typical spark plug wire.

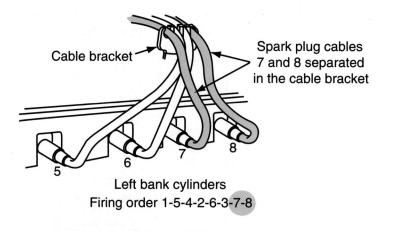

Cable bracket

Spark plug cables 7 and 8 separated in the cable bracket

5 6 7 8

Left bank cylinders
Firing order 1-5-4-2-6-3-7-8

Figure 4-5 Proper spark plug wire routing to prevent cross-fire.

each other as possible (Figure 4-5). When induction cross-fire occurs, no spark is jumped from one wire to the other. The spark is the result of induction from another field. Cross-fire induction is most common in two parallel wires that fire one after the other in the firing order.

Wire Sizes

An additional amount of consideration must be given for some margin of safety when selecting wire size. There are three major factors that determine the proper size of wire to be used:

1. The wire must have a large enough diameter, for the length required, to carry the necessary current for the load components in the circuit to operate properly.

2. The wire must be able to withstand the anticipated vibration.

3. The wire must be able to withstand the anticipated amount of heat exposure.

Wire size is based on the diameter of the conductor. The larger the diameter, the less the resistance. There are two common size standards used to designate wire size: American Wire Gauge (AWG) and metric.

The AWG standard assigns a **gauge** number to the wire based on its diameter. The higher the number, the smaller the wire diameter. For example, 20-gauge wire is smaller in diameter than 10-gauge wire (Figure 4-6). Most electrical systems in the automobile use 14-, 16-, or 18-gauge wire. Some high-current circuits will also use 10- or 12-gauge wire. Most battery cables are 2-, 4-, or 6-gauge cable.

Both wire diameter and wire length affect resistance. Sixteen-gauge wire is capable of conducting 20 amperes for 10 feet with minimal voltage drop. However, if the current is to be carried for 15 feet, 14-gauge wire would be required. If 20 amperes were required to be carried for 20 feet, then 12-gauge wire would be required. The additional wire size is needed to prevent voltage drops in the wire. The illustration (Figure 4-7) lists the wire size required to carry a given amount of current for different lengths.

Another factor to wire resistance is temperature. An increase in temperature creates a similar increase in resistance. A wire may have a known resistance of 0.03 ohms per 10 feet at 70°F. When exposed to temperatures of 170°F, the resistance may increase to 0.04 ohms per 10 feet. Wires that are to be installed in areas that experience high temperatures, as in the engine compartment, must be of a size such that the increased resistance will not affect the operation of the load component. Also, the insulation of the wire must be capable of withstanding the high temperatures.

American Wire Gauge Sizes

Gauge size	Conductor diameter (inches)
20	0.032
18	0.040
16	0.051
14	0.064
12	0.081
10	0.102
8	0.128
6	0.162
4	0.204
2	0.258
1	0.289
0	0.325
2/0	0.365
4/0	0.460

Figure 4-6 Gauge and wire size chart.

Total Approximate Circuit Amperes	Wire Gauge (for Length in Feet)								
12 V	3	5	7	10	15	20	25	30	40
1.0	18	18	18	18	18	18	18	18	18
1.5	18	18	18	18	18	18	18	18	18
2	18	18	18	18	18	18	18	18	18
3	18	18	18	18	18	18	18	18	18
4	18	18	18	18	18	18	18	16	16
5	18	18	18	18	18	18	18	16	16
6	18	18	18	18	18	18	16	16	16
7	18	18	18	18	18	18	16	16	14
8	18	18	18	18	18	16	16	16	14
10	18	18	18	18	16	16	16	14	12
11	18	18	18	18	16	16	14	14	12
12	18	18	18	18	16	16	14	14	12
15	18	18	18	18	14	14	12	12	12
18	18	18	16	16	14	14	12	12	10
20	18	18	16	16	14	12	10	10	10
22	18	18	16	16	12	12	10	10	10
24	18	18	16	16	12	12	10	10	10
30	18	16	16	14	10	10	10	10	10
40	18	16	14	12	10	10	8	8	6
50	16	14	12	12	10	10	8	8	6
100	12	12	10	10	6	6	4	4	4
150	10	10	8	8	4	4	2	2	2
200	10	8	8	6	4	4	2	2	1

Note: 18 AWG as indicated above this line could be 20 AWG electrically.
18 AWG is recommended for mechanical strength.

Figure 4-7 The distance the current must be carried is a factor in determining the correct wire gauge to use.

Metric Size (mm²)	AWG (Gauge) Size	Ampere Capacity
0.5	20	4
0.8	18	6
1.0	16	8
2.0	14	15
3.0	12	20
5.0	10	30
8.0	8	40
13.0	6	50
19.0	4	60

Figure 4-8 Approximate AWG to metric equivalents.

In the metric system, wire size is determined by the cross-sectional area of the wire. Metric wire size is expressed in square millimeters (mm²). In this system the smaller the number, the smaller the wire conductor. The approximate equivalent wire size of metric to AWG is shown in Figure 4-8.

Terminals and Connectors

Shop Manual
Chapter 4, page 135

To perform the function of connecting the wires from the voltage source to the load component reliably, terminal connections are used. Today's vehicles can have as many as 500 separate circuit connections. The terminals used to make these connections must be able to perform with very low voltage drop. Terminals are constructed of either brass or steel. Steel terminals usually have a tin or lead coating. A loose or corroded connection can cause an unwanted voltage drop that results in poor operation of the load component. For example, a connector used in a light circuit that has as little as 10% voltage drop (1.2 V) may result in a 30% loss of lighting efficiency.

Terminals can be either crimped or soldered to the conductor. The terminal makes the electrical connection, and it must be capable of withstanding the stress of normal vibration. The illustration (Figure 4-9) shows several different types of terminals used in the automotive electrical system. In addition, the following connectors are used on the automobile:

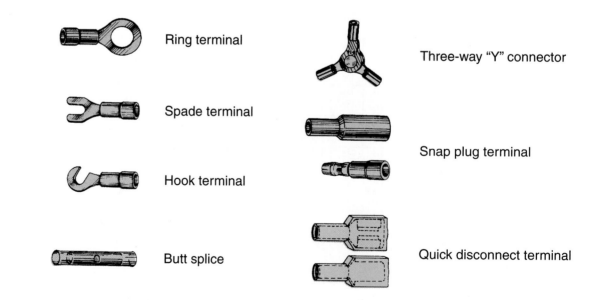

Ring terminal

Three-way "Y" connector

Spade terminal

Snap plug terminal

Hook terminal

Butt splice

Quick disconnect terminal

Figure 4-9 Primary wire terminals used in automotive applications.

Figure 4-10 Molded connectors cannot be disassembled to replace damaged terminals or to test.

1. **Molded connector:** These connectors usually have one to four wires that are molded into a one-piece component (Figure 4-10). Although the connector halves separate, the connector itself cannot be taken apart.

2. **Multiple-wire, hard-shell connector:** These connectors usually have a hard, plastic shell that holds the connecting terminals of separate wires (Figure 4-11). The wire terminals can be removed from the shell to be repaired.

3. **Bulkhead connectors:** These connectors are used when several wires must pass through the bulkhead (Figure 4-12).

4. **Weather-Pack Connectors:** These connectors have rubber seals on the terminal ends and on the covers of the connector half (Figure 4-13). They are used on computer circuits to protect the circuit from corrosion, which may result in a voltage drop.

5. **Metri-Pack Connectors:** These are like the weather-pack connectors but do not have the seal on the cover half (Figure 4-14).

6. **Heat Shrink Covered Butt Connectors:** Recommended for air bag applications by some manufacturers. Other manufacturers allow NO repairs to the circuitry, while still others require silver-soldered connections.

Shop Manual
Chapter 4, page 142

Shop Manual
Chapter 4, page 142

Shop Manual
Chapter 4, page 144

Shop Manual
Chapter 4, page 146

To reduce the number of connectors in the electrical system, a **common connection** can be used (Figure 4-15). Common connections are used to share a source of power or a common ground and are often called a splice. If there are several electrical components that are physically close to each other, a single common connection (splice) eliminates using a separate connector for each wire.

Printed Circuits

Printed circuit boards are used to simplify the wiring of the circuits they operate. Other uses of printed circuit boards include the inside of radios, computers, and some voltage regulators. Most instrument panels use printed circuit boards as circuit conductors. A printed circuit is made of a thin phenolic or fiberglass board that copper (or some other conductive material) has been deposited on. Portions of the conductive metal are then etched or eaten away by acid. The remaining strips of conductors provide the circuit path for the instrument panel illumination

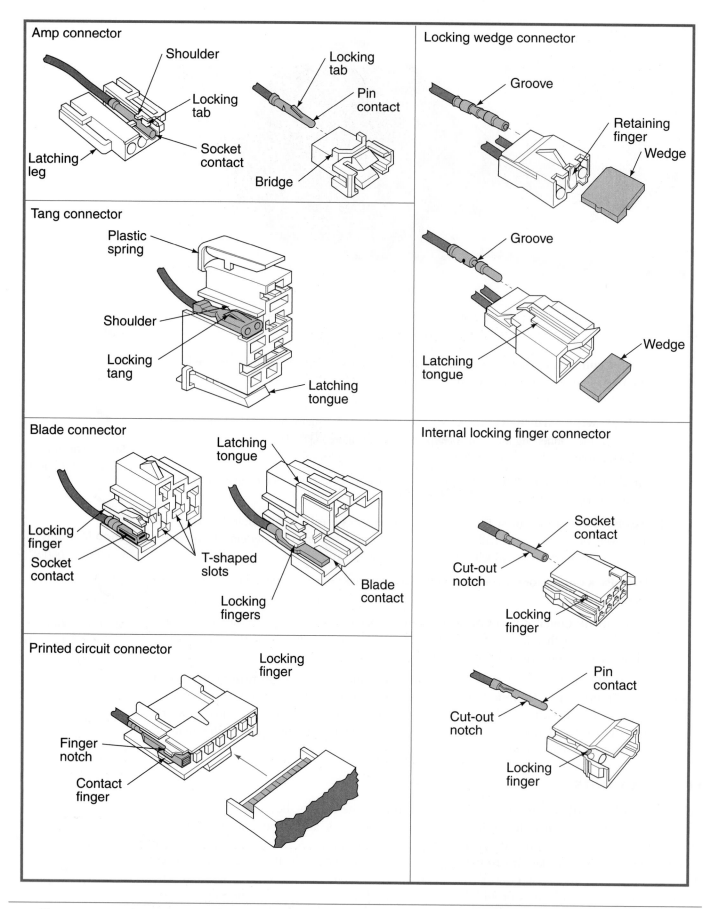

Figure 4-11 Multiple-wire, hard-shell connectors.

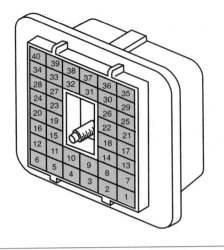

Figure 4-12 Bulkhead connector.

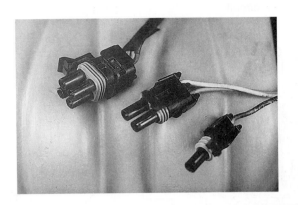

Figure 4-13 Weather-pack connector is used to prevent connector corrosion.

Figure 4-14 Metri-pack connector.

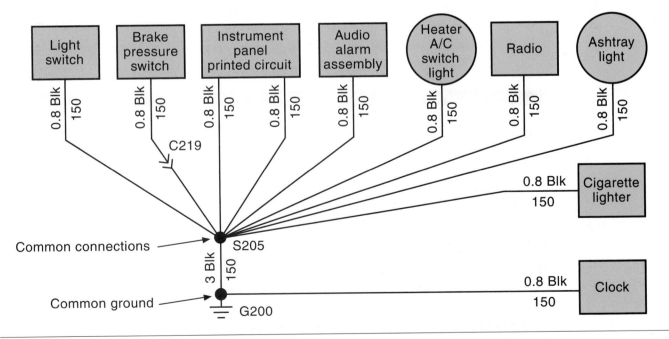

Figure 4-15 Common connections (splices) are used to reduce the amount of wire and connectors.

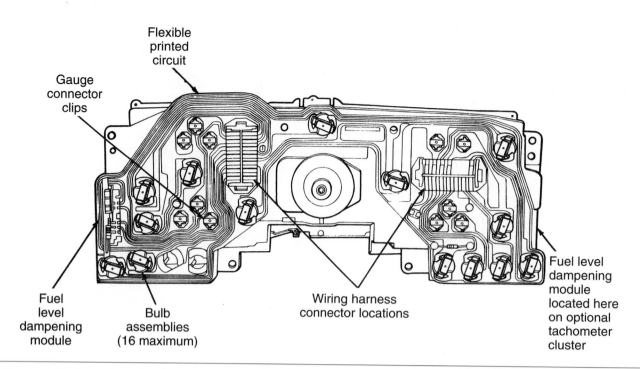

Gauge connector clips

Flexible printed circuit

Fuel level dampening module

Bulb assemblies (16 maximum)

Wiring harness connector locations

Fuel level dampening module located here on optional tachometer cluster

Figure 4-16 Printed circuits eliminate bulky wires behind the instrument panel.

lights, warning lights, indicator lights, and gauges of the instrument panel (Figure 4-16). The printed circuit board is attached to the back of the instrument panel housing. An edge connector joins the printed circuit board to the vehicle wiring harness.

Whenever it is necessary to perform repairs on or around the printed circuit board, it is important to follow these precautions:

1. When replacing light bulbs, be careful not to cut or tear the surface of the printed circuit board.

2. Do not touch the surface of the printed circuit with your fingers. The acid present in normal body oils can damage the surface.

3. If the printed circuit board needs to be cleaned, use a commercial cleaning solution designed for electrical use. If this solution is not available, it is possible to clean the board by *lightly* rubbing the surface with an eraser.

A BIT OF HISTORY

The printed circuit board was developed in 1947 by the British scientist J. A. Sargrove to simplify the production of radios.

Wiring Harness

Most manufacturers use **wiring harnesses** to reduce the number of loose wires hanging under the hood or dash of an automobile. The wiring harness provides for a safe path for the wires of the vehicle's lighting, engine, and accessory components. The wiring harness is made by grouping insulated wires and wrapping them together. The wires are bundled into separate harness assemblies that are joined together by connector plugs. The multiple-pin connector plug may have more than 60 individual wire terminals.

There are several complex wiring harnesses in a vehicle, in addition to the simple harnesses. The engine compartment harness and the under-dash harness are examples of complex harnesses (Figure 4-17). Lighting circuits usually use a more simple harness (Figure 4-18).

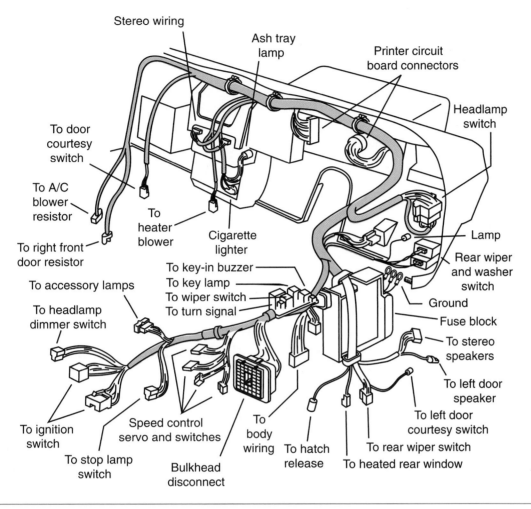

Figure 4-17 Complex wiring harness.

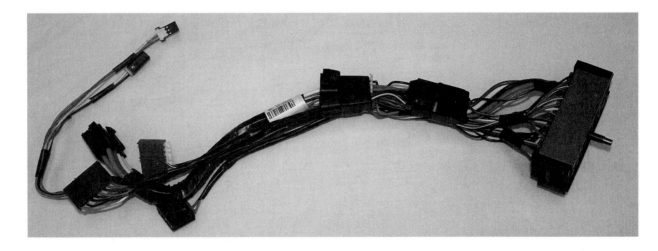

Figure 4-18 Simple wiring harness.

Figure 4-19 Flexible conduit used to make wiring harnesses.

A complex harness serves many circuits. The simple harness services only a few circuits. Some individual circuit wires may branch out of a complex harness to other areas of the vehicle.

Most wiring harnesses now use a flexible conduit to provide for quick wire installation (Figure 4-19). The conduit has a seam that can be opened to accommodate the installation or removal of wires from the harness. The seam will close once the wires are installed, and will remain closed even if the conduit is bent.

The conduit is commonly referred to as the *wire loom* or *corrugated loom*.

Wiring Protective Devices

Often overlooked, but very important to the electrical system, are proper wire protection devices (Figure 4-20). These devices prevent damage to the wiring by maintaining proper wire routing and retention. Special clips, retainers, straps, and supplementary insulators provide additional protection to the conductor over what the insulation itself is capable of providing. Whenever the technician must remove one of these devices to perform a repair, it is important that the device be reinstalled to prevent additional electrical problems.

Whenever it is necessary to install additional electrical accessories, try to support the primary wire in at least 1-foot intervals. If the wire must be routed through the frame or body, use rubber grommets to protect the wire.

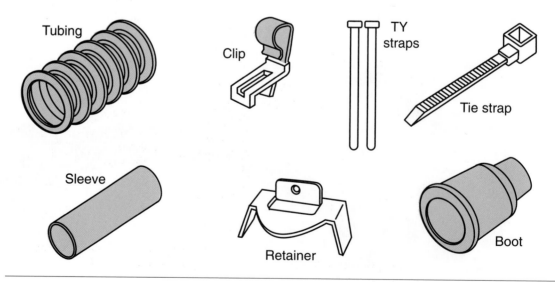

Figure 4-20 Typical wire protection devices.

Wiring Diagrams

One of the most important tools for diagnosing and repairing electrical problems is a **wiring diagram.** A wiring diagram is an electrical schematic that shows a representation of actual electrical or electronic components (by use of symbols) and the wiring of the vehicle's electrical systems. These diagrams identify the wires and connectors from each circuit on a vehicle. They also show where different circuits are interconnected, where they receive their power, where the ground is located, and the colors of the different wires. All of this information is critical to proper diagnosis of electrical problems. Some wiring diagrams also give additional information that helps you understand how a circuit operates and how to identify certain components (Figure 4-21). Wiring diagrams do not explain how the circuit works; this is where your knowledge of electricity comes in handy.

Shop Manual
Chapter 4, page 147

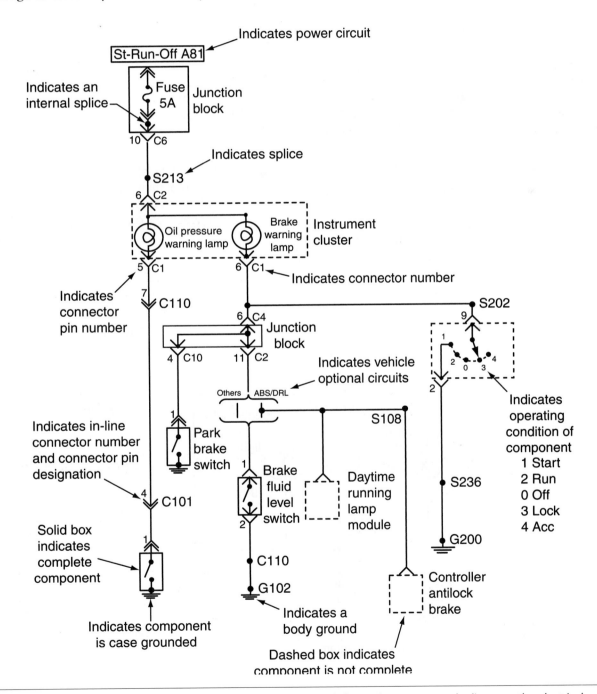

Figure 4-21 Wiring diagrams provide the technician with necessary information to accurately diagnose the electrical systems.

A wiring diagram can show the wiring of the entire vehicle or a single circuit (Figure 4-22). These single-circuit diagrams are also called block diagrams. Wiring diagrams of the entire vehicle tend to look more complex and threatening than block diagrams. However, once you simplify the diagram to only those wires, connectors, and components that belong to an individual circuit, they become less complex and more valuable.

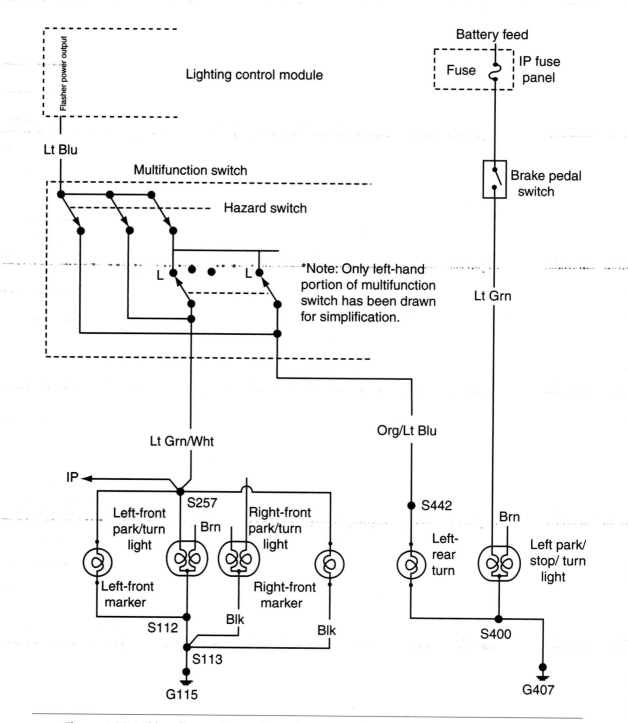

Figure 4-22 Wiring diagram illustrating only one specific circuit for easier reference. This is also known as a block diagram.

Wiring diagrams show the wires, connections to switches and other components, and the type of connector used throughout the circuit. Total vehicle wiring diagrams are normally spread out over many pages of a service manual. Some are displayed on a single large sheet of paper that folds out of the manual. A system wiring diagram is actually a portion of the total vehicle diagram. The system and all related circuitry are shown on a single page. System diagrams are often easier to use than vehicle diagrams simply because there is less information to sort through.

Remember that electrical circuits need a complete path in order to work. A wiring diagram shows the insulated side of the circuit and the point of ground. Also, when lines (or wires) cross on a wiring diagram, this does not mean they connect. If wires are connected, there will be a connector or a dot at the point where they cross. Most wiring diagrams do not show the location of the wires, connectors, or components in the vehicle. Some have location reference numbers displayed by the wires. After studying the wiring diagram, you will know what you are looking for. Then you move to the car to find it.

In addition to entire vehicle and system-specific wiring diagrams, there are other diagrams that may be used to diagnose electricity problems. An electrical **schematic** shows how the circuit is connected. It does not show the colors of the wires or their routing. Schematics are what have been used so far in this book. They display a working model of the circuit. These are especially handy when trying to understand how a circuit works. Schematics are typically used to show the internal circuitry of a component or to simplify a wiring diagram. One of the troubleshooting techniques used by good electrical technicians is to simplify a wiring diagram into a schematic.

Installation diagrams show where and how electrical components and wiring harnesses are installed in the vehicle. These are helpful when trying to locate where a particular wire or component may be in the car. These diagrams also may show how the component or wiring harness is attached to the vehicle (Figure 4-23).

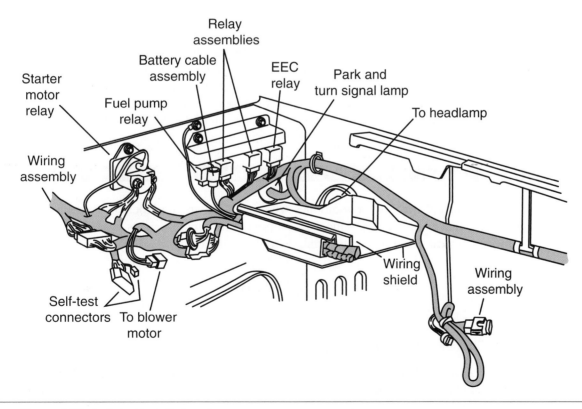

Figure 4-23 A typical installation diagram.

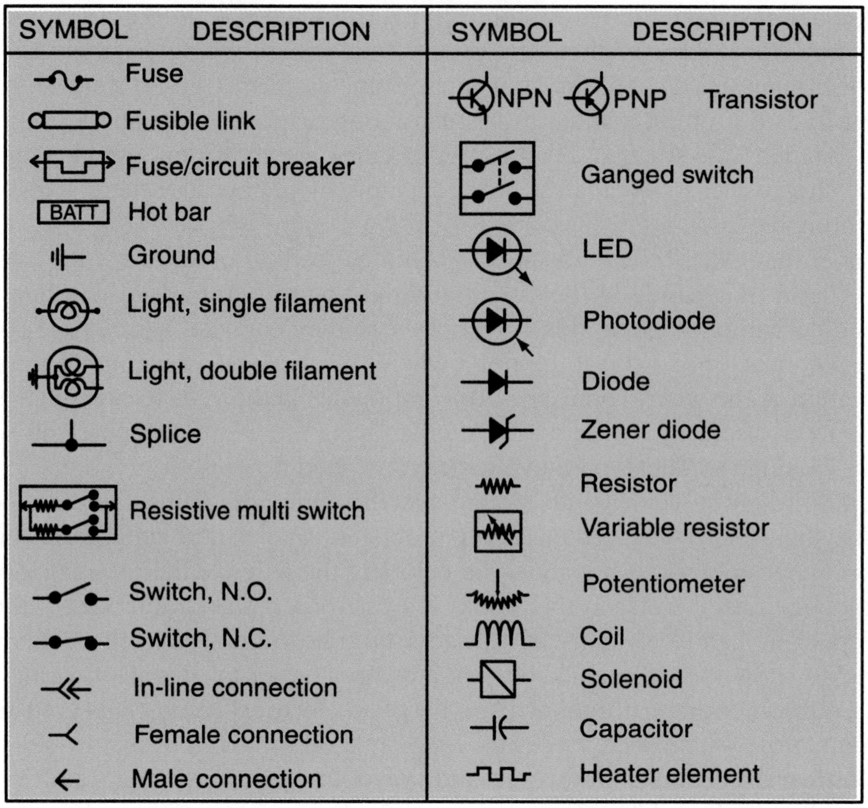

SYMBOL	DESCRIPTION	SYMBOL	DESCRIPTION
	Fuse		NPN / PNP Transistor
	Fusible link		Ganged switch
	Fuse/circuit breaker		LED
BATT	Hot bar		Photodiode
	Ground		Diode
	Light, single filament		Zener diode
	Light, double filament		Resistor
	Splice		Variable resistor
	Resistive multi switch		Potentiometer
	Switch, N.O.		Coil
	Switch, N.C.		Solenoid
	In-line connection		Capacitor
	Female connection		Heater element
	Male connection		

Figure 4-24 Common electrical symbols used in wiring diagrams.

Electrical Symbols

Most wiring diagrams do not show an actual drawing of the components. Rather, they use **electrical symbols** to represent the components. Often the symbol displays the basic operation of the component. Many different symbols have been used in wiring diagrams through the years. Figure 4-24 shows some of the commonly used symbols. You need to be familiar with all of the symbols; however, you don't need to memorize all of the variations. Wiring diagram manuals include a "legend" that helps you interpret the symbols.

A BIT OF HISTORY

The service manuals for early automobiles were hand drawn and labeled. They also had drawings of the actual components. As more and more electrical components were added to cars, this became impractical. Soon schematic symbols replaced the component drawings.

Color Codes and Circuit Numbering

Nearly all of the wires in an automobile are covered with colored insulation. These colors are used to identify wires and electrical circuits. The color of the wires is indicated on a wiring diagram. Some wiring diagrams also include circuit numbers. These numbers, or letters and numbers, help identify a specific circuit. Both types of coding make it easier to diagnose electrical

Color	Abbreviations		
Aluminum	AL		
Black	BLK	BK	B
Blue (Dark)	BLU DK	DB	DK BLU
Blue	BLU	B	L
Blue (Light)	BLU LT	LB	LT BLU
Brown	BRN	BR	BN
Glazed	GLZ	GL	
Gray	GRA	GR	G
Green (Dark)	GRN DK	DG	DK GRN
Green (Light)	GRN LT	LG	LT GRN
Maroon	MAR	M	
Natural	NAT	N	
Orange	ORN	O	ORG
Pink	PNK	PK	P
Purple	PPL	PR	
Red	RED	R	RD
Tan	TAN	T	TN
Violet	VLT	V	
White	WHT	W	WH
Yellow	YEL	Y	YL

Figure 4-25 Common color codes used in automotive applications.

problems. Unfortunately, not all manufacturers use the same method of wire identification. Figure 4-25 shows common color codes and their abbreviations. Most wiring diagrams list the appropriate color coding used by the manufacturer. Make sure you understand what color the code is referring to before looking for a wire.

In most color codes, the first group of letters designates the base color of the insulation. If a second group of letters is used, it indicates the color of the **tracer.** For example, a wire designated as WH/BLK would have a white base color with a black tracer. A tracer is a thin or dashed line of a different color than the base color of the insulation.

Ford uses four methods of color coding its wires (Figure 4-26):

1. Solid color.

2. Base color with a stripe (tracer).

3. Base color with hash marks.

4. Base color with dots.

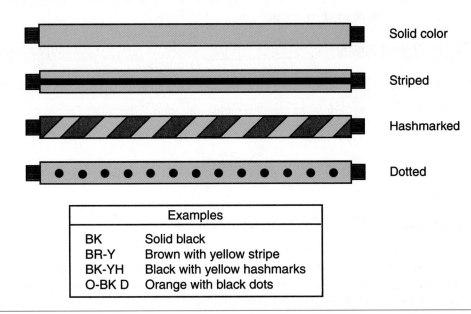

Solid color

Striped

Hashmarked

Dotted

Examples	
BK	Solid black
BR-Y	Brown with yellow stripe
BK-YH	Black with yellow hashmarks
O-BK D	Orange with black dots

Figure 4-26 Four methods that Ford uses to color code their wires.

DaimlerChrysler uses a numbering method to designate the circuits on the wiring diagram (Figure 4-27). The circuit identification, wire gauge, and color of the wire are included in the wire number. DaimlerChrysler identifies the main circuits by using a main circuit identification code that corresponds to the first letter in the wire number (Figure 4-28).

General Motors uses numbers that include the wire gauge in metric millimeters, the wire color, the circuit number, splice number, and ground identification (Figure 4-29). In this example, the circuit is designated as 100, the wire size is 0.8 mm^2, the insulation color is black, the splice is numbered S114, and the ground is designated as G117.

Most manufacturers also number connectors and terminals for identification.

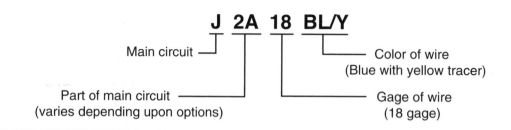

Figure 4-27 DaimlerChrysler Motors' wiring code identification.

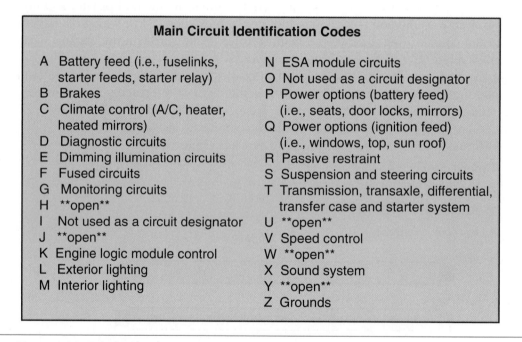

Figure 4-28 DaimlerChrysler Motors' circuit identification codes.

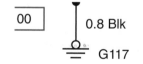

Figure 4-29 GM's method of circuit and wire identification.

Standardized Wiring Designations. The Society of Automotive Engineers (SAE) is attempting to standardize the circuit diagrams used by the various manufacturers. The system that is developed may be similar to the **DIN** used by import manufacturers. DIN is the abbreviation for Deutsche Institut füer Normung (German Institute for Standardization) and is the recommended standard for European manufacturers to follow. DIN assigns certain color codes to a particular circuit as follows:

- ❑ Red wires are used for direct battery-powered circuits and also ignition-powered circuits.
- ❑ Black wires are also powered circuits controlled by switches or relays.
- ❑ Brown wires are usually the grounds.
- ❑ Green wires are used for ignition primary circuits.

A combination of wire colors is used to identify subcircuits. The base color still identifies the circuit's basic purpose. In addition to standardized color coding, DIN attempts to standardize terminal identification and circuit numbering.

Component Locators

The wiring diagrams in most service manuals may not indicate the exact physical location of the components of the circuit. In another section of the service manual, or in a separate manual, a **component locator** is provided to help find where a component is installed in the vehicle. The component locator may use both drawings and text to lead the technician to the desired component (Figure 4-30).

Many electrical components may be hidden behind kick panels, dashboards, fender wells, and under seats. The use of a component locator will save the technician time in finding the suspected defective unit.

Summary

- ❑ Most of the primary wiring conductors used in the automobile are made of several strands of copper wire wound together and covered with a polyvinyl chloride (PVC) insulation.

- ❑ Stranded wire is used because of its flexibility and current flows on the surface of the conductors. Because there is more surface area exposed in a stranded wire, there is less resistance in the stranded wire than in the solid wire.

- ❑ There are three major factors that determine the proper size of wire to be used: (1) The wire must be large enough diameter—for the length required—to carry the necessary current for the load components in the circuit to operate properly, (2) the wire must be able to withstand the anticipated vibration, and (3) the wire must be able to withstand the anticipated amount of heat exposure.

- ❑ Wire size is based on the diameter of the conductor.

- ❑ Factors that affect the resistance of the wire include the conductor material, wire diameter, wire length, and temperature.

- ❑ Terminals can be either crimped or soldered to the conductor. The terminal makes the electrical connection and it must be capable of withstanding the stress of normal vibration.

- ❑ Printed circuit boards are used to simplify the wiring of the circuits they operate. A printed circuit is made of a thin phenolic or fiberglass board that copper (or some other conductive material) has been deposited on.

Terms to Know
Ballast resistor
Common connection
Component locator
Cross-fire
DIN
Electrical symbols
Gauge
Installation diagrams
Primary wiring
Printed circuit boards

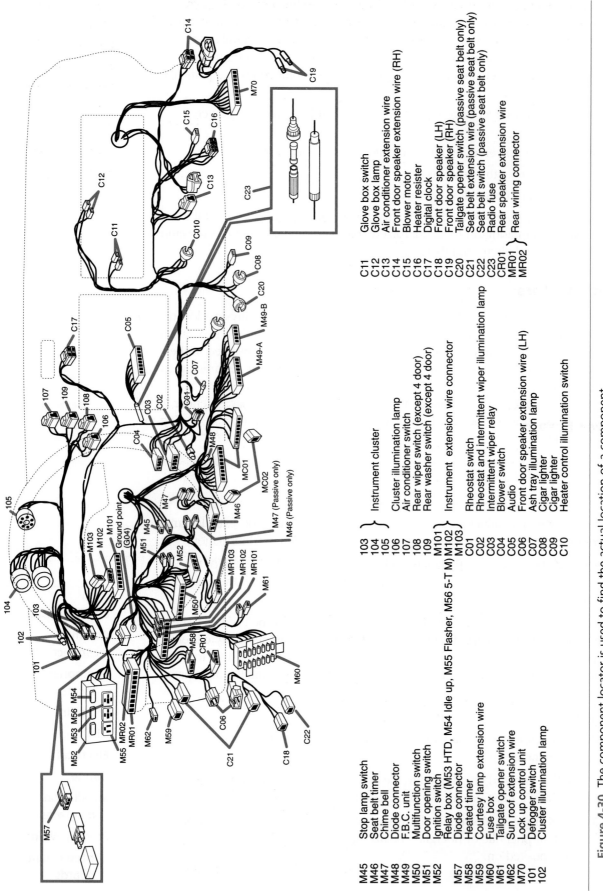

M45 Stop lamp switch
M46 Seat belt timer
M47 Chime bell
M48 Diode connector
M49 F.B.C. unit
M50 Multifunction switch
M51 Door opening switch
M52 Ignition switch
M53 } Relay box (M53 HTD, M54 Idle up, M55 Flasher, M56 5-T M)
M57 Diode connector
M58 Heated timer
M59 Courtesy lamp extension wire
M60 Fuse box
M61 Tailgate opener switch
M62 Sun roof extension wire
M70 Lock up control unit
101 Defogger switch
102 Cluster illumination lamp

103 } Instrument cluster
104 }
105 }
106 Cluster illumination lamp
107 Air conditioner switch
108 Rear wiper switch (except 4 door)
109 Rear washer switch (except 4 door)
M101 } Instrument extension wire connector
M103 }
C01 Rheostat switch
C02 Rheostat and intermittent wiper illumination lamp
C03 Intermittent wiper relay
C04 Blower switch
C05 Audio
C06 Front door speaker extension wire (LH)
C07 Ash tray illumination lamp
C08 Cigar lighter
C09 Cigar lighter
C10 Heater control illumination switch

C11 Glove box switch
C12 Glove box lamp
C13 Air conditioner extension wire
C14 Front door speaker extension wire (RH)
C15 Blower motor
C16 Heater resister
C17 Digital clock
C18 Front door speaker (LH)
C19 Front door speaker (RH)
C20 Tailgate opener switch (passive seat belt only)
C21 Seat belt extension wire (passive seat belt only)
C22 Seat belt switch (passive seat belt only)
C23 Radio fuse
CR01 Rear speaker extension wire
MR01 } Rear wiring connector
MR02 }

Figure 4-30 The component locator is used to find the actual location of a component.

110

❏ A wire harness is an assembled group of wires that branch out to the various electrical components. It is used to reduce the number of loose wires hanging under the hood or dash. It provides for a safe path for the wires of the vehicle's lighting, engine, and accessory components.

❏ The wiring harness is made by grouping insulated wires and wrapping them together. The wires are bundled into separate harness assemblies that are joined together by connector plugs.

❏ A wiring diagram shows a representation of actual electrical or electronic components and the wiring of the vehicle's electrical systems.

❏ The technician's greatest helpmate in locating electrical problems is the wiring diagram. Correct use of the wiring diagram will reduce the amount of time a technician needs to spend tracing the wires in the vehicle.

❏ In place of actual pictures, a variety of electrical symbols are used to represent the components in the wiring diagram.

❏ Color codes and circuit numbers are used to make tracing wires easier.

❏ In most color codes, the first group of letters designates the base color of the insulation. If a second group of letters is used, it indicates the color of the tracer.

❏ A component locator is used to determine the exact location of several of the electrical components.

Terms to Know
(continued)
Resistance wire
Schematic
Secondary wiring
Stranded wire
Tracer
Wiring harness
Wiring diagram

Review Questions

Short-Answer Essays

1. Explain the purpose of wiring diagrams.

2. Explain how wire size is determined by the American Wire Gauge (AWG) and metric methods.

3. Explain the purpose and use of printed circuits.

4. Explain the purpose of the component locator.

5. Explain when single-stranded or multistranded wire should be used.

6. Explain how temperature affects resistance and wire size selection.

7. List the three major factors that determine the proper size of wire to be used.

8. List and describe the different types of terminal connectors used in the automotive electrical system.

9. What is the difference between a complex and a simple wiring harness?

10. Describe the methods the three domestic automobile manufacturers use for wiring code identification.

Fill in the Blanks

1. There is _____ resistance in the stranded wire than in the solid wire.

2. _____ _____ is the electromagnetic induction spark that can be transmitted in another wire that is close to the wire carrying the current.

3. Wire size is based on the _____ of the conductor.

4. In the AWG standard, the _____ the number, the smaller the wire _____.

5. An increase in temperature creates a similar _____ in resistance.

6. _____ connectors are used when several wires must pass through the bulkhead.

7. _____ _____ _____ are used to prevent damage to the wiring by maintaining proper wire routing and retention.

8. A wiring diagram is an electrical schematic that shows a _____ of actual electrical or electronic components (by use of symbols) and the _____ of the vehicle's electrical systems.

9. In most color codes, the first group of letters designates the _____ _____ of the insulation. The second group of letters indicates the color of the _____.

10. A _____ _____ is used to determine the exact location of several of the electrical components.

Multiple Choice

1. Automotive wiring is being discussed.
   Technician A says most primary wiring is made of several strands of copper wire wound together and covered with an insulation.
   Technician B says the types of conductor materials used in automobiles include copper, silver, gold, aluminum, and tin-plated brass.
   Who is correct?
   A. A only   **C.** Both A and B
   B. B only   **D.** Neither A nor B

2. Stranded wire use is being discussed.
   Technician A says there is less exposed surface area for electron flow in a stranded wire.
   Technician B says there is more resistance in the stranded wire than in the same gauge solid wire.
   Who is correct?
   A. A only   **C.** Both A and B
   B. B only   **D.** Neither A nor B

3. Spark plug wires are being discussed.
   Technician A says RFI is controlled by using resistances in the conductor of the spark plug wire.
   Technician B says all late-model ignition systems use resistance wires to control RFI.
   Who is correct?
   A. A only   **C.** Both A and B
   B. B only   **D.** Neither A nor B

4. Spark plug wire installation is being discussed.
   Technician A says there is little that can be done to prevent cross-fire.
   Technician B says the spark plug wires must be installed in the proper separator and any two parallel wires next to each other in the firing order should be positioned as far away from each other as possible.
   Who is correct?
   A. A only   **C.** Both A and B
   B. B only   **D.** Neither A nor B

5. The selection of the proper size of wire to be used is being discussed.
   Technician A says the wire must be large enough, for the length required, to carry the amount of current necessary for the load components in the circuit to operate properly.
   Technician B says temperature has little effect on resistance and it is not a factor in wire size selection.
   Who is correct?
   A. A only   **C.** Both A and B
   B. B only   **D.** Neither A nor B

6. Terminal connectors are being discussed.
   Technician A says good terminal connections will resist corrosion.
   Technician B says the terminals can be either crimped or soldered to the conductor.
   Who is correct?
   A. A only   **C.** Both A and B
   B. B only   **D.** Neither A nor B

7. Wire routing is being discussed.
   Technician A says to install additional electrical accessories it is necessary to support the primary wire in at least 10-foot intervals.
   Technician B says if the wire must be routed through the frame or body, use metal clips to protect the wire.
   Who is correct?
   A. A only   **C.** Both A and B
   B. B only   **D.** Neither A nor B

8. Printed circuit boards are being discussed.
   Technician A says printed circuit boards are used to simplify the wiring of the circuits they operate.
   Technician B says care must be taken not to touch the board with bare hands.
   Who is correct?
   A. A only   **C.** Both A and B
   B. B only   **D.** Neither A nor B

9. Wiring harnesses are being discussed.
   Technician A says a wire harness is an assembled group of wires that branches out to the various electrical components.
   Technician B says most under-hood harnesses are simple harnesses.
   Who is correct?
   A. A only   **C.** Both A and B
   B. B only   **D.** Neither A nor B

10. Wiring diagrams are being discussed.
    Technician A says wiring diagrams give the exact location of the electrical components.
    Technician B says a wiring diagram will indicate what circuits are interconnected, where circuits receive their voltage source, and what color of wires are used in the circuit.
    Who is correct?
    A. A only   **C.** Both A and B
    B. B only   **D.** Neither A nor B

Automotive Batteries

Upon completion and review of this chapter, you should be able to:

❑ Explain the purposes of the battery.

❑ Describe the construction of conventional, maintenance-free, hybrid, and recombination batteries.

❑ Define the main elements of the battery.

❑ Explain the chemical action that occurs to produce current in a battery.

❑ Explain the chemical reaction that occurs in the battery during cycling.

❑ Describe the differences, advantages, and disadvantages between different types of batteries.

❑ Describe the different types of battery terminals used.

❑ Describe the methods used to rate batteries.

❑ Determine the correct battery to be installed into a vehicle.

❑ Explain the effects of temperature on battery performance.

❑ Describe the different loads or demands placed upon a battery during different operating conditions.

❑ Explain the major reasons for battery failure.

❑ Define battery-related terms such as deep cycle, electrolyte solution, and gassing.

Shop Manual
Chapter 5, page 171

Introduction

An automotive battery (Figure 5-1) is an **electrochemical** device capable of storing and producing electrical energy. Electrochemical refers to the chemical reaction of two dissimilar materials in a chemical solution that results in electrical current. When the battery is connected to an external load, such as a starter motor, an energy conversion occurs that results in an electrical current flowing through the circuit. Electrical energy is produced in the battery by the chemical reaction that occurs between two dissimilar plates that are immersed in an electrolyte solution. The automotive battery produces direct current (DC) electricity that flows in only one direction.

When discharging the battery (current flowing from the battery), the battery changes chemical energy into electrical energy. It is through this change that the battery releases stored energy. During charging (current flowing through the battery from the charging system), electrical energy is converted into chemical energy. As a result, the battery can store energy until it is needed.

Figure 5-1 Typical automotive 12-volt battery.

The automotive battery has several important functions, including:

1. It operates the starting motor, ignition system, electronic fuel injection, and other electrical devices for the engine during cranking and starting.

2. It supplies all the electrical power for the vehicle accessories whenever the engine is not running or when the vehicle's charging system is not working.

3. It furnishes current for a limited time whenever electrical demands exceed charging system output.

4. It acts as a stabilizer of voltage for the entire automotive electrical system.

5. It stores energy for extended periods of time.

 AUTHOR'S NOTE: The battery does not store energy in electrical form. The battery stores energy in chemical form.

The largest demand placed on the battery occurs when it must supply current to operate the starter motor. The amperage requirements of a starter motor may be over several hundred amperes. This requirement is also affected by temperatures, engine size, and engine condition.

After the engine is started, the vehicle's charging system works to recharge the battery and to provide the current to run the electrical systems. Most AC generators have a maximum output of 60 to 120 amperes. This is usually enough to operate all of the vehicle's electrical systems and meet the demands of these systems. However, under some conditions (such as the engine running at idle) generator output is below its maximum rating. If there are enough electrical accessories turned on during this time (heater, wipers, headlights, and radio) the demand may exceed the AC generator output. The total demand may be 20 to 60 amperes. During this time, the battery must supply the additional current.

Even with the ignition switch turned off, there are electrical demands placed on the battery. Clocks, memory seats, engine computer memory, body computer memory, and electronic sound system memory are all examples of key-off loads. The total current draw of key-off loads is usually less than 30 milliamperes.

Electrical loads that are still present when the ignition switch is in the OFF position are called key-off or parasitic loads.

In the event that the vehicle's charging system fails, the battery must supply all of the current necessary to run the vehicle. The amount of time a battery can be discharged at a certain current rate until the voltage drops below a specified value is referred to as **reserve capacity.** Most batteries will supply a reserve capacity of 25 amperes for approximately 120 minutes before discharging too low to keep the engine running.

The amount of electrical energy that a battery is capable of producing depends on the size, weight, active area of the plates, and the amount of sulfuric acid in the electrolyte solution.

In this chapter, you will study the design and operation of different types of batteries currently used in automobiles. These include conventional batteries, maintenance-free batteries, hybrid batteries, and recombination batteries.

Conventional Batteries

The conventional battery is constructed of seven basic components:

1. Positive plates.
2. Negative plates.
3. Separators.
4. Case.
5. Plate straps.
6. Electrolyte.
7. Terminals.

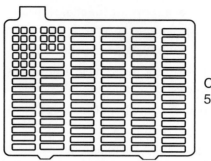

Conventional
5% antimony

Figure 5-2 Conventional battery grid.

The condition of the battery should be the first test performed on a vehicle with an electrical problem. Without proper battery performance, the entire electrical system is affected.

The lead peroxide is composed of small grains of particles. This gives the plate a high degree of porosity, allowing the electrolyte to penetrate the plate.

Lead peroxide is also called lead dioxide.

Usually, negative plate groups contain one more plate than positive plate groups to help equalize the chemical activity.

The difference between "3-year" and "5-year" batteries is the quantity of **material expanders** used in the construction of the plates and the number of plates used to build a cell. Material expanders are fillers that can be used in place of the active materials. They are used to keep the manufacturing costs low.

A plate, either positive or negative, starts with a **grid.** Grids are generally made of lead alloys, usually antimony. About 5% to 6% antimony is added to increase the strength of the grid. The grid is the frame structure with connector tabs at the top. The grid has horizontal and vertical grid bars that intersect at right angles (Figure 5-2). An active material made from ground lead oxide, acid, and material expanders is pressed into the grid in paste form. The positive plate is given a "forming charge" that converts the lead oxide paste into lead peroxide. The negative plate is given a "forming charge" that converts the paste into sponge lead.

The negative and positive plates are arranged alternately in each **cell element** (Figure 5-3). Each cell element can consist of 9 to 13 plates. The positive and negative plates are insulated from each other by separators made of microporous materials. The construction of the element is completed when all of the positive plates are connected to each other and all of the negative plates are connected to each other. The connection of the plates is by plate straps (Figure 5-4).

A typical 12-volt automotive battery is made up of six cells connected in series (Figure 5-5). This means the positive side of a cell element is connected to the negative side of the next

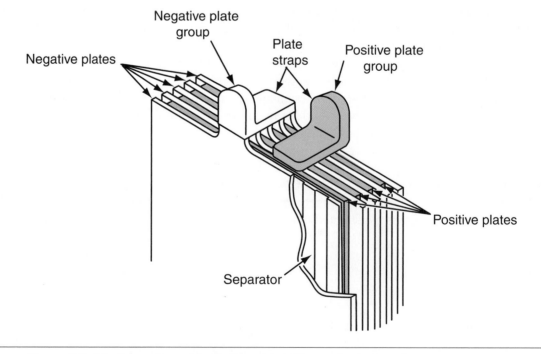

Figure 5-3 A battery cell consists of alternate positive and negative plates.

116

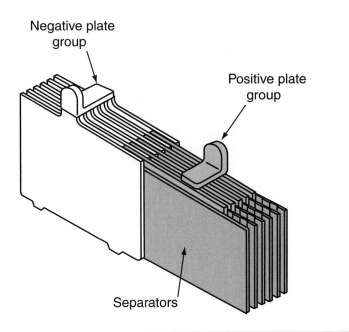

Negative plate group

Positive plate group

Separators

FiFigure 5-4 Construction of a battery element.

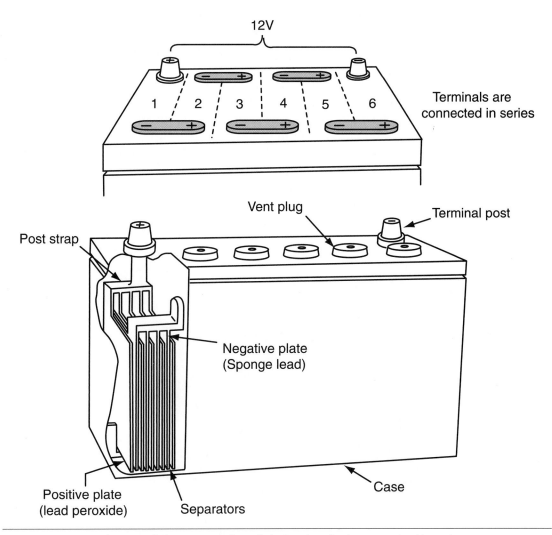

12V

1 2 3 4 5 6

Terminals are connected in series

Vent plug

Terminal post

Post strap

Negative plate (Sponge lead)

Positive plate (lead peroxide)

Separators

Case

Figure 5-5 The 12-volt battery consists of six 2-volt cells that are wired in series.

Intercell connections

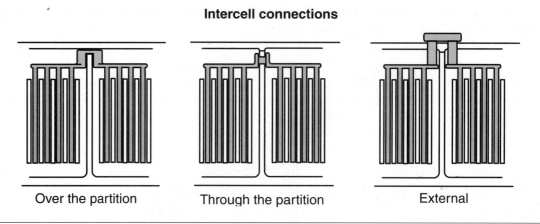

Over the partition | Through the partition | External

Figure 5-6 The cell elements can be connected using one of three intercell connection methods.

Many batteries have envelope-type separators that retain active materials near the plates.

The most common connection used to connect cell elements is through the partition. It provides the shortest path and the least resistance.

Lead acid batteries are also called flooded batteries.

Shop Manual
Chapter 5, pages 175, 180

cell element. This is repeated throughout all six cells. By connecting the cells in series, the current capacity of the cell and cell voltage remain the same. The six cells produce 2.1 volts each. Wiring the cells in series produces the 12.6 volts required by the automotive electrical system. The plate straps provide a positive cell connection and a negative cell connection. The cell connection may be one of three types: through the partition, over the partition, or external (Figure 5-6). The cell elements are submerged in a cell case filled with **electrolyte** solution. Electrolyte consists of sulfuric acid diluted with water. The electrolyte solution used in automotive batteries consists of 64% water and 36% sulfuric acid, by weight. Electrolyte is both conductive and reactive.

The battery case is made of polypropylene, hard rubber, and plastic base materials. The battery case must be capable of withstanding temperature extremes, vibration, and acid absorption. The cell elements sit on raised supports in the bottom of the case. By raising the cells, chambers are formed at the bottom of the case that trap the sediment that flakes off the plates. If the sediment was not contained in these chambers, it could cause a conductive connection across the plates and short the cell. The case is fitted with a one-piece cover.

Because the conventional battery releases hydrogen gas when it is being charged, the case cover will have vents. The vents are located in the cell caps of a conventional battery (Figure 5-7).

Chemical Action

Activation of the battery is through the addition of electrolyte. This solution causes the chemical actions to take place between the lead peroxide of the positive plates and the sponge lead

Figure 5-7 The vents of a conventional battery allow the release of gases.

of the negative plates. The electrolyte is also the carrier that moves electric current between the positive and negative plates through the separators.

The automotive battery has a fully charged **specific gravity** of 1.265 corrected to 80°F (27°C). Therefore, a specific gravity of 1.265 for electrolyte means it is 1.265 times heavier than an equal volume of water. As the battery discharges, the specific gravity of the electrolyte decreases because the electrolyte becomes more like water. The specific gravity of a battery can give you an indication of how charged a battery is.

Fully charged: 1.265 specific gravity
75% charged: 1.225 specific gravity
50% charged: 1.190 specific gravity
25% charged: 1.155 specific gravity
Discharged: 1.120 or lower specific gravity

These specific gravity values may vary slightly according to the design of the battery. However, regardless of the design, the specific gravity of the electrolyte in all batteries will decrease as the battery discharges. Temperature of the electrolyte will also affect its specific gravity. All specific gravity specifications are based on a standard temperature of 80°F (27°C). When the temperature is above that standard, the specific gravity is lower. When the temperature is below that standard, the specific gravity increases. Therefore, all specific gravity measurements must be corrected for temperature. A general rule to follow is to add 0.004 for every 10°F (5.5°C) above 80°F (27°C) and subtract 0.004 for every 10°F (5.5°C) below 80°F (27°C).

In operation, the battery is being partially discharged and then recharged. This represents an actual reversing of the chemical action that takes place within the battery. The constant cycling of the charge and discharge modes slowly wears away the active materials on the cell plates. This action eventually causes the battery plates to sulfate. The battery must be replaced once the sulfation of the plates has reached the point that there is insufficient active plate area.

In the charged state, the positive plate material is essentially pure lead peroxide, PbO_2. The active material of the negative plates is spongy lead, Pb. The electrolyte is a solution of sulfuric acid, H_2SO_4, and water. The voltage of the cell depends on the chemical difference between the active materials.

The illustration (Figure 5-8) shows what happens to the plates and electrolyte during discharge. The lead (Pb) from the positive plate combines with sulfate (SO_4) from the acid, forming lead sulfate ($PbSO_4$). While this is occurring, oxygen (O_2) in the active material of the

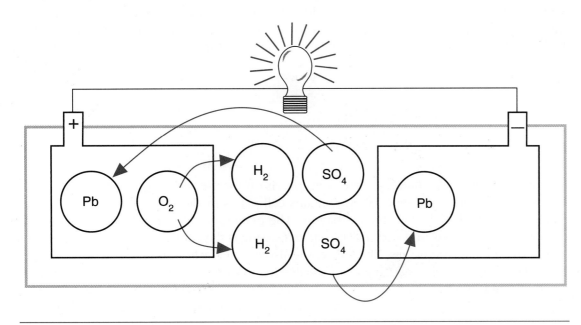

Figure 5-8 Chemical action that occurs inside of the battery during the discharge cycle.

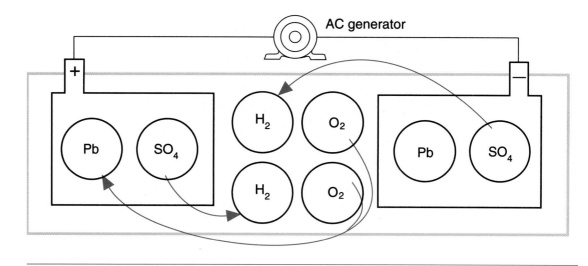

Figure 5-9 Chemical action inside of the battery during the charge cycle.

positive plate joins with the hydrogen (H_2) from the electrolyte forming water (H_2O). This water dilutes the acid concentration.

A similar reaction is occurring in the negative plate. Lead (Pb) is combining with sulfate (SO_4), forming lead sulfate ($PbSO_4$). The result of discharging is changing the positive plate from lead dioxide into lead sulfate and changing the negative plate into lead sulfate. Discharging a cell makes the positive and negative plates the same. Once they are the same, the cell is discharged.

The charge cycle is exactly the opposite (Figure 5-9). The lead sulfate ($PbSO_4$) in both plates is split into its original forms of lead (Pb) and sulfate (SO_4). The water in the electrolyte splits into hydrogen and oxygen. The hydrogen (H_2) combines with the sulfate to become sulfuric acid again (H_2SO_4). The oxygen combines with the positive plate to form the lead peroxide. This now puts the plates and the electrolyte back in their original form and the cell is charged.

A BIT OF HISTORY

Lead acid batteries date back to 1859. Alexander Graham Bell used a primitive battery to make his first local call in 1876. Once it was learned how to recharge the lead acid batteries, they were installed into the automobile. These old-style batteries could not hold a charge very well and it was believed that placing the battery on a concrete floor made them discharge faster. Although this fable has no truth to it, the idea has hung on for years.

Maintenance-Free Batteries

Maintenance-free batteries are also called lead calcium batteries.

In a **maintenance-free battery** there is no provision for the addition of water to the cells. The battery is sealed. It (Figure 5-10) contains cell plates made of a slightly different compound than what is in a conventional battery. The plate grids contain calcium, cadmium, or strontium to reduce **gassing** and self-discharge. Gassing is the conversion of the battery water into hydrogen and oxygen gas. This process is also called electrolysis. The antimony used in conventional batteries is not used in maintenance-free batteries because it increases the breakdown of water into hydrogen and oxygen and because of its low resistance to overcharging. The use of calcium, cadmium, or strontium reduces the amount of vaporization that takes

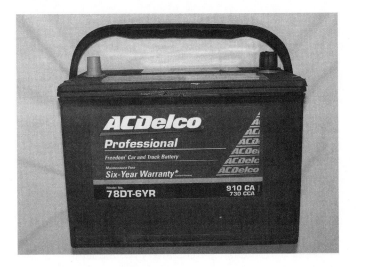

Figure 5-10 Maintenance-free batteries.

place during normal operation. The grid may be constructed with additional supports to increase its strength and to provide a shorter path, with less resistance, for the current to flow to the top tab (Figure 5-11).

Each plate is wrapped and sealed on three sides by an envelope design separator. The envelope is made from microporous plastic. By enclosing the plate in an envelope, the plate is insulated and reduces the shedding of the active material from the plate.

The battery is sealed except for a small vent so the electrolyte and vapors cannot escape (Figure 5-12). An expansion or condensation chamber allows the water to condense and drain back into the cells. Because the water cannot escape from the battery, it is not necessary to add water to the battery on a periodic basis. Containing the vapors also reduces the possibility of corrosion and discharge through the surface because of electrolyte on the surface of the battery. Vapors leave the case only when the pressure inside the battery is greater than atmospheric pressure.

AUTHOR'S NOTE: If electrolyte and dirt are allowed to accumulate on the top of the battery case, it may create a conductive connection between the positive and negative terminals, resulting in a constant discharge on the battery.

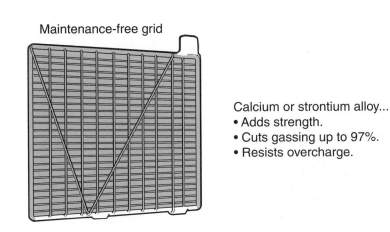

Figure 5-11 Maintenance-free battery grids with support bars give increased strength and faster electrical delivery.

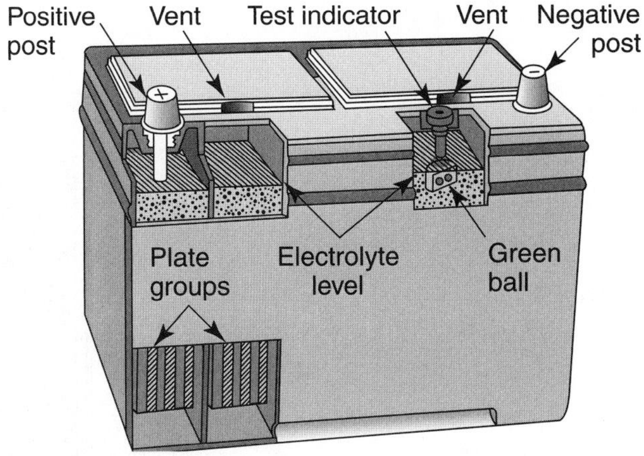

Figure 5-12 Construction of a maintenance-free battery.

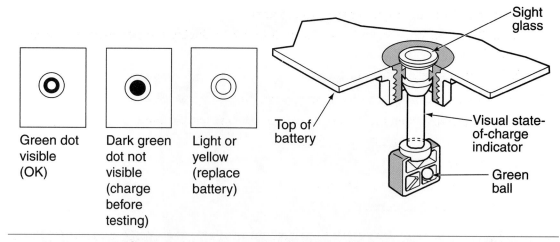

Green dot visible (OK)	Dark green dot not visible (charge before testing)	Light or yellow (replace battery)

Figure 5-13 One cell of a maintenance-free battery has a built-in hydrometer, which indicates overall battery condition.

Some maintenance-free batteries have a built-in **hydrometer** that shows the state of charge (Figure 5-13). A hydrometer is a test instrument that is used to check the specific gravity of the electrolyte to determine the battery's state of charge. If the dot that is at the bottom of the hydrometer is green, then the battery is fully charged (more than 65% charged). If the dot is black, the battery state of charge is low. If the battery does not have a built-in hydrometer, it cannot be tested with a hydrometer because the battery is sealed.

 AUTHOR'S NOTE: It is important to remember that the built-in hydrometer is only an indication of the state of charge for one of the six cells of the battery.

Many manufacturers have revised the maintenance-free battery to a "low maintenance-battery," in that the caps are removable for testing and electrolyte level checks. Also, the grid construction contains about 3.4% antimony. To decrease the distance and resistance of the path, that current flows in the grid, and to increase its strength, the horizontal and vertical grid bars do not intersect at right angles (Figure 5-14).

The advantages of maintenance-free batteries over conventional batteries include:

1. A larger reserve of electrolyte above the plates.
2. Increased resistance to overcharging.

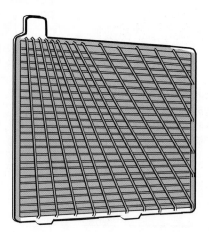

Low maintenance
3.4% or less antimony

Figure 5-14 Low-maintenance battery grid with vertical grid bars intersecting at an angle.

3. Longer shelf life (approximately 18 months).

4. Ability to be shipped with electrolyte installed, reducing the possibility of accidents and injury to the technician.

5. Higher cold cranking amps rating.

The major disadvantages of the maintenance-free battery include:

1. **Grid growth** when the battery is exposed to high temperatures.

2. Inability to withstand **deep cycling.**

3. Low reserve capacity.

4. Faster discharge by parasitic loads.

5. Shorter life expectancy.

A BIT OF HISTORY

Buick first introduced the storage battery as standard equipment in 1906.

Hybrid Batteries

AUTHOR'S NOTE: The following discussion on hybrid batteries refers to a battery type and not to the batteries that are used in hybrid electric vehicles (HEVs).

The **hybrid battery** combines the advantages of the low-maintenance and maintenance-free battery. The hybrid battery can withstand six deep cycles and still retain 100% of its original reserve capacity. The grid construction of the hybrid battery consists of approximately 2.75% antimony alloy on the positive plates and a calcium alloy on the negative plates. This allows the battery to withstand deep cycling while retaining reserve capacity for improved cranking performance. Also, the use of antimony alloys reduces grid growth and corrosion. The lead calcium has less gassing than conventional batteries.

Grid construction differs from other batteries in that the plates have a lug located near the center of the grid. In addition, the vertical and horizontal grid bars are arranged in a **radial** pattern (Figure 5-15). By locating the lug near the center of the grid and using the **radial grid**

Grid growth is a condition where the grid grows little metallic fingers that extend through the separators and short out the plates.

Deep cycling is to discharge the battery to a very low state of charge before recharging it.

Radial means branching out from a common center.

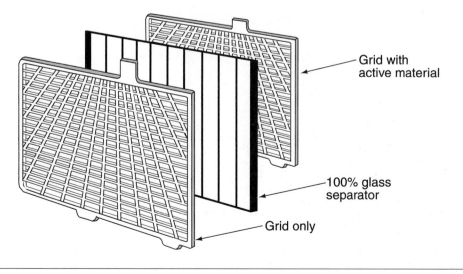

Grid with active material

100% glass separator

Grid only

Figure 5-15 Hybrid grid and separator construction.

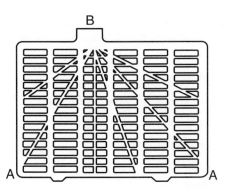

Figure 5-16 The hybrid battery grid construction allows for faster current delivery. Electrical energy at point A has a shorter distance to travel to get to the tab at point B.

Some manufacturers are using fiberglass separators.

design, the current has less resistance and a shorter path to follow to the lug (Figure 5-16). This means the battery is capable of providing more current at a faster rate.

The separators used are constructed of glass with a resin coating. The glass separators offer low electrical resistance with high resistance to chemical contamination. This type of construction provides for increased cranking performance and battery life.

Recombination Batteries

One of the most recent variations of the automobile battery is the **recombination battery** (Figure 5-17). The recombination battery is sometimes called a gel-cell battery. It does not use a liquid electrolyte. Instead, it uses separators that hold a gel-type material. The separators are placed between the grids and have very low electrical resistance. The spiral design provides a larger plate

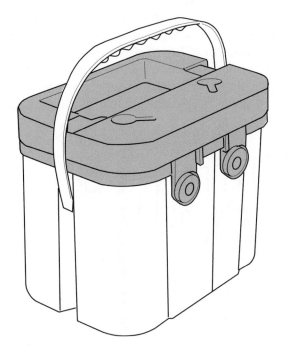

Figure 5-17 The recombination battery is one of the most recent advances in the automotive battery.

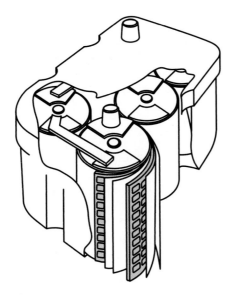

Figure 5-18 Construction of the recombination battery cells.

surface area than that in conventional batteries (Figure 5-18). In addition, the close plate spacing results in decreased resistance. Because of this design, output voltage and current are higher than in conventional batteries. The extra amount of available voltage (approximately 0.6 V) assists in cold-weather starting. Also, gassing is virtually eliminated and the battery can recharge faster.

The following are some other safety features and advantages of the recombination battery:

1. Contains no liquid electrolyte. If the case is cracked, no electrolyte will spill.
2. Can be installed in any position, including upside down.
3. Is corrosion free.
4. Has very low maintenance because there is no electrolyte loss.
5. Can last as much as four times longer than conventional batteries.
6. Can withstand deep cycling without damage.
7. Can be rated over 800 cold cranking amperes.

Recombination batteries recombine the oxygen gas that is normally produced on the positive plates with the hydrogen given off by the negative plates. This recombination of oxygen and hydrogen produces water (H_2O) and replaces the moisture in the battery. The electrolyte solution of the recombination battery is absorbed into the separators.

The oxygen produced by the positive plates is trapped in the cell by special pressurized sealing vents. The oxygen gases then travel to the negative plates through small fissures in the gelled electrolyte. There are between one and six one-way safety valves in the top of the battery. The safety valves are necessary for maintaining a positive pressure inside of the battery case. This positive pressure prevents oxygen from the atmosphere from entering the battery and causing corrosion. Also, the safety valves must release excessive pressure that may be produced if the battery is overcharged.

Due to the use of valves in recombination batteries, these are also called valve-regulated lead acid (VRLA) batteries.

Battery Terminals

Terminals provide a means of connecting the battery plates to the vehicle's electrical system. All automotive batteries have two terminals. One terminal is a positive connection; the other is

Shop Manual
Chapter 5, page 178

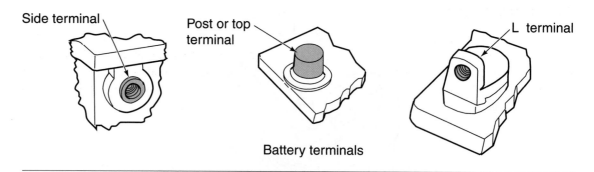

Side terminal

Post or top terminal

L terminal

Battery terminals

Figure 5-19 The most common types of automotive battery terminals.

a negative connection. The battery terminals extend through the cover or the side of the battery case. The following are the most common types of battery terminals (Figure 5-19):

1. **Post or top terminals:** Used on most automotive batteries. The positive post will be larger than the negative post to prevent connecting the battery in reverse polarity.

2. **Side terminals:** Positioned in the side of the container near the top. These terminals are threaded and require a special bolt to connect the cables. Polarity identification is by positive and negative symbols.

3. **L terminals:** Used on specialty batteries and some imports.

Battery Ratings

Battery capacity ratings are established by the Battery Council International (BCI) in conjunction with the Society of Automotive Engineers (SAE). Battery cell voltage depends on the types of materials used in the construction of the battery. Current capacity depends on several factors:

1. The size of the cell plates. The larger the surface area of the plates, the more chemical action that can occur. This means a greater current is produced.

2. The weight of the positive and negative plate active materials.

3. The weight of the sulfuric acid in the electrolyte solution.

The battery's current capacity rating is an indication of its ability to deliver cranking power to the starter motor and of its ability to provide reserve power to the electrical system. The commonly used current capacity ratings are explained in the following sections.

Ampere-Hour Rating

The **ampere-hour rating** is the amount of steady current that a fully charged battery can supply for 20 hours at 80°F (26.7°C) without the terminal voltage falling below 10.5 volts. For example, if a battery can be discharged for 20 hours at a rate of 4.0 amperes before its terminal voltage reads 10.5 volts, it would be rated at 80 ampere-hours. This method of battery rating is no longer widely used.

Cold Cranking Rating

Cold cranking
rating is also called
cold cranking amps
(CCA).

Cold cranking rating is the most common method of rating automotive batteries. It is determined by the load, in amperes, that a battery is able to deliver for 30 seconds at 0°F (−17.7°C) without terminal voltage falling below 7.2 volts for a 12-volt battery. The cold cranking rating is given in total amperage and is identified as 300 CCA, 400 CCA, 500 CCA, and so on. Some batteries are rated as high as 1,100 CCA.

Cranking Amps

Cranking Amps (CA) is an indication of the battery's ability to provide a cranking amperage at 32°F (0°C). This rating uses the same test procedure as the cold cranking rating or CCA discussed earlier, except it uses a higher temperature. To convert CA to CCA, divide the CA by 1.25. For example, a 650-CCA-rated battery is the same as 812 CA. It is important that the technician does not misread the rating and think the battery is rated as CCA instead of CA.

Reserve-Capacity Rating

The **reserve-capacity rating** is determined by the length of time, in minutes, that a fully charged battery can be discharged at 25 amperes before battery voltage drops below 10.5 volts. This rating gives an indication of how long the vehicle can be driven, with the headlights on, if the charging system should fail.

Battery Size Selection

Some of the aspects that determine the battery rating required for a vehicle include engine size, engine type, climatic conditions, vehicle options, and so on. The requirement for electrical energy to crank the engine increases as the temperature decreases. Battery power drops drastically as temperatures drop below freezing (Figure 5-20). The engine also becomes harder to crank due to the tendency of oils to thicken when cold, which results in increased friction. As a general rule, it takes 1 ampere of cold cranking power per cubic inch of engine displacement. Therefore, a 200-cubic-inch displacement (CID) engine should be fitted with a battery of at least 200 CCA. To convert this into metric, it takes 1 amp of cold cranking power for every 16 cm^3 of engine displacement. A 1.6-liter engine should require at least a battery rated at 100 CCA. This rule may not apply to vehicles that have several electrical accessories. The best method of determining the correct battery is to refer to the manufacturer's specifications.

The battery that is selected should fit the battery holding fixture and the holddown must be able to be installed. It is also important that the height of the battery not allow the terminals to short across the vehicle hood when it is shut. BCI group numbers are used to indicate the physical size and other features of the battery (Figure 5-21). This group number does not indicate the current capacity of the battery.

36-Volt Batteries

Currently automotive manufacturers are researching the use of 36-volt electrical systems. The reason for this development is due to the increased number of electrical circuits and wiring used on today's vehicles. With the use of higher voltages, current flow can be reduced and allow for smaller conductors to be used. In turn, this will reduce the size and weight of the wiring harnesses and save costs. To support this new electrical system, the charging system will also need to be converted to a 42-volt system. As with the 36-volt battery, the higher voltage output of the AC generator (alternator) means the current output can be reduced. This allows for the construction of smaller AC generators than those being used today.

Temperature	% of Cranking Power
80°F (26.7°C)	100
32°F (0°C)	65
0°F (−17.8°C)	40

Figure 5-20 The effect temperature has on the cranking power of the battery.

BCI Group Number	Maximum Overall Dimensions Including Terminals					
	Millimeters			Inches		
	L	W	H	L	W	H
Passenger Car and Light Commercial Batteries 12-V (6 Cells)						
21	208	173	222	8-3/16	6-13/16	8-3/4
22F	241	175	211	9-1/2	6-7/8	8-5/16
22HF	241	175	229	9-1/2	6-7/8	9
22NF	240	140	227	9-7/16	5-1/2	8-15/16
22R	229	175	211	9	6-7/8	8-5/16
24	260	173	225	10-1/4	6-13/16	8-7/8
24F	273	173	229	10-3/4	6-13/16	9
24H	260	173	238	10-1/4	6-13/16	9-3/8
24R	260	173	229	10-1/4	6-13/16	9
24T	260	173	248	10-1/4	6-13/16	9-3/4
25	230	175	225	9-1/16	6-7/8	8-7/8
26	208	173	197	8-3/16	6-13/16	7-3/4
26R	208	173	197	8-3/16	6-13/16	7-3/4
27	306	173	225	12-1/16	6-13/16	8-7/8
27F	318	173	227	12-1/2	6-13/16	8-15/16
27H	298	173	235	11-3/4	6-13/16	9-1/4
29NF	330	140	227	13	5-1/2	8-15/16
33	338	173	238	13-5/16	6-13/16	9-3/8
34	260	173	200	10-1/4	6-13/16	7-7/8
34R	260	173	200	10-1/4	6-13/16	7-7/8
35	230	175	225	9-1/16	6-7/8	8-7/8
36R	263	183	206	10-3/8	7-1/4	8-1/8
40R	277	175	175	10-15/16	6-7/8	6-7/8
41	293	175	175	11-3/16	6-7/8	6-7/8
42	243	173	173	9-5/16	6-13/16	6-13/16
43	334	175	205	13-1/8	6-7/8	8-1/16
45	240	140	227	9-7/16	5-1/2	8-15/16
46	273	173	229	10-3/4	6-13/16	9
47	246	175	190	9-11/16	6-7/8	7-1/2
48	306	175	192	12-1/16	6-7/8	7-9/16
49	381	175	192	15	6-7/8	7- 3/16
50	343	127	254	13-1/2	5	10
51	238	129	223	9-3/8	5-1/16	8-13/16
51R	238	129	223	9-3/8	5-1/16	8-13/16

Figure 5-21 BCI battery group numbers indicate the size and features of a battery.

52	186	147	210	7-5/16	5-13/16	8-1/4
53	330	119	210	13	4-11/16	8-1/4
54	186	154	212	7-5/16	6-1/16	8-3/8
55	218	154	212	8-5/8	6-1/16	8-3/8
56	254	154	212	10	6-1/16	8-3/8
57	205	183	177	8-1/16	7-3/16	6-15/16
58	255	183	177	10-1/16	7-3/16	6-15/16
58R	255	183	177	10-1/16	7-3/16	6-15/16
59	255	193	196	10-1/16	7-5/8	7-3/4
60	332	160	225	13-1/16	6-5/16	8-7/8
61	192	162	225	7-9/16	6-3/8	8-7/8
62	225	162	225	8-7/8	6-3/8	8-7/8
63	258	162	225	10-3/16	6-3/8	8-7/8
64	296	162	225	11-11/16	6-3/8	8-7/8
65	306	190	192	12-1/16	7-1/2	7-9/16
70	208	179	196	8-3/16	7-1/16	7-11/16
71	208	179	216	8-3/16	7-1/16	8-1/2
72	230	179	210	9-1/16	7-1/16	8-1/4
73	230	179	216	9-1/16	7-1/16	8-1/2
74	260	184	222	10-1/4	7-1/4	8-3/4
75	230	179	196	9-1/16	7-1/16	7-11/16
76	334	179	216	13-1/8	7-1/16	8-1/2
78	260	179	196	10-1/4	7-1/16	7-11/16
85	230	173	203	9-1/16	6-13/16	8
86	230	173	203	9-1/16	6-13/16	8
90	246	175	175	9-11/16	6-7/8	6-7/8
91	280	175	175	11	6-7/8	6-7/8
92	317	175	175	12-1/2	6-7/8	6-7/8
93	354	175	175	15	6-7/8	6-7/8
95R	394	175	190	15-9/16	6-7/8	7-1/2
96R	242	173	175	9-9/16	6-13/16	6-7/8
97R	252	175	190	9-15/16	6-7/8	7-1/2
98R	283	175	190	11-3/16	6-7/8	7-1/2

Figure 5-21 *(Continued)*

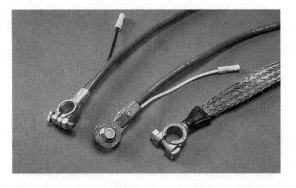

Figure 5-22 The battery cable is designed to carry the high current required to start the engine and supply the vehicle's electrical systems.

Battery Cables

Battery cables are high-current conductors that connect the battery to the vehicle's electrical system. Battery cables must be of a sufficient capacity to carry the current required to meet all electrical demands (Figure 5-22). Normal 12-volt cable size is usually 4 or 6 gauge. Various forms of clamps and terminals are used to assure a good electrical connection at each end of the cable. Connections must be clean and tight to prevent arcing, corrosion, and high-voltage resistance.

The positive cable is usually red (but not always), and the negative cable is usually black. The positive cable will fasten to the starter solenoid or relay. The negative cable fastens to ground on the engine block. Some manufacturers use a negative cable with no insulation. Sometimes the negative battery cable may have a body grounding wire to help assure that the vehicle body is properly grounded.

AUTHOR'S NOTE: It is important to properly identify the positive and negative cables when servicing, charging, or jumping the battery. Do not rely on the color of the cable for this identification; use the markings on the battery case.

Battery Holddowns

All batteries must be secured in the vehicle to prevent damage and the possibility of shorting across the terminals if the battery tips. Normal vibrations cause the plates to shed their active materials. **Holddowns** reduce the amount of vibration and help increase the life of the battery (Figure 5-23).

In addition to holddowns, many vehicles may have a heat shield surrounding the battery (Figure 5-24). This heat shield is usually made of plastic and prevents under-hood temperatures from damaging the battery.

A BIT OF HISTORY

The storage battery on early automobiles was mounted under the car. It wasn't until 1937 that the battery was located under the hood for better accessibility. Today, with the increased use of maintenance-free batteries, some manufacturers have "buried" the battery again. For example, to access the battery on some vehicles, you must remove the left front wheel and work through the wheel well. Also, some batteries are now located in the trunk area.

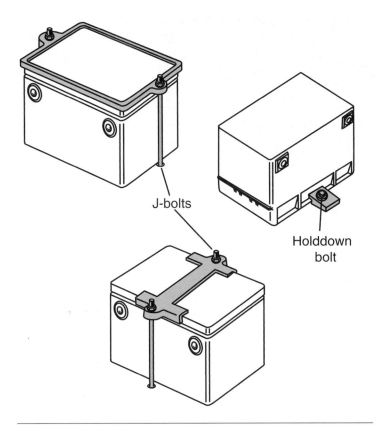

J-bolts

Holddown
bolt

Figure 5-23 Different types of battery holddowns.

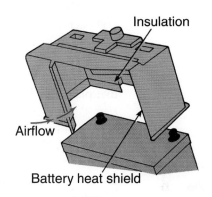

Insulation

Airflow

Battery heat shield

Figure 5-24 Some vehicles are equipped with a heat shield to protect the battery from excessive heat.

Battery Failure

Whenever battery failure occurs, first perform some simple visual inspections. Check the case for cracks, check the electrolyte level in each cell (if possible), and check the terminals for corrosion. The sulfuric acid that vents out with the battery gases attacks the battery terminals and battery cables. As the sulfuric acid reacts with the lead and copper, deposits of lead sulfate and copper sulfate are created (Figure 5-25). These deposits are resistive to electron flow and limit the amount of current that can be supplied to the electrical and starting systems. If the deposits are bad enough, the resistance can increase to a level that prevents the starter from cranking the engine.

Shop Manual
Chapter 5, page 179

Figure 5-25 Corroded battery terminals reduce the efficiency of the battery.

The sulfate that is not converted back to H_2SO_4 hardens on the plates. This results in battery sulfation, which permanently damages the battery.

One common cause of early battery failure is overcharging. If the charging system is supplying a voltage level over 15.5 volts, the plates may become warped. Warping of the plates results from the excess heat that is generated as a result of overcharging. Overcharging also causes the active material to disintegrate and shed off of the plates.

If the charging system does not produce enough current to keep the battery charged, the lead sulfate can become crystallized on the plates. If this happens, the sulfate is difficult to remove and the battery will resist recharging. The recharging process converts the sulfate on the plates. If there is an undercharging condition, the sulfate is not converted and it will harden on the plates.

Vibration is another common reason for battery failure. If the battery is not secure, the plates will shed the active material as a result of excessive vibration. If enough material is shed, the sediment at the bottom of the battery can create an electrical connection between the plates. The shorted cell will not produce voltage, resulting in a battery that will have only 10.5 volts across the terminals. With this reduced amount of voltage, the starter usually will not be capable of starting the engine. To prevent this problem, make sure that proper holddown fixtures are used.

During normal battery operation, the active materials on the plates will shed. The negative plate also becomes soft. Both of these events will reduce the effectiveness of the battery.

Terms to Know

Ampere-hour rating

Battery cables

Cell element

Cold cranking rating

Cranking amps (CA)

Deep cycling

Electrochemical

Electrolyte

Gassing

Grid

Grid growth

Holddowns

Hybrid batteries

Hydrometer

Maintenance-free battery

Material expanders

Radial

Radial grid

Recombination batteries

Reserve capacity

Reserve-capacity rating

Specific gravity

Terminals

Summary

❑ An automotive battery is an electrochemical device that provides for and stores electrical energy.

❑ Electrical energy is produced in the battery by the chemical reaction that occurs between two dissimilar plates that are immersed in an electrolyte solution.

❑ An automotive battery has the following important functions:

1. It operates the starting motor, ignition system, electronic fuel injection, and other electrical devices for the engine during cranking and starting.

2. It supplies all the electrical power for the vehicle accessories whenever the engine is not running or at low idle.

3. It furnishes current for a limited time whenever electrical demands exceed charging system output.

4. It acts as a stabilizer of voltage for the entire automotive electrical system.

5. It stores energy for extended periods of time.

❑ Electrical loads that are still placed on the battery when the ignition switch is in the OFF position are called key-off or parasitic loads.

❑ The amount of electrical energy that a battery is capable of producing depends on the size, weight, and active area of the plates and the specific gravity of the electrolyte solution.

❑ The conventional battery is constructed of seven basic components:

1. Positive plates.

2. Negative plates.

3. Separators.

4. Case.

5. Plate straps.

6. Electrolyte.

7. Terminals.

❑ Electrolyte solution used in automotive batteries consists of 64% water and 36% sulfuric acid by weight.

❑ The electrolyte solution causes the chemical actions to take place between the lead dioxide of the positive plates and the sponge lead of the negative plates. The electrolyte is also the carrier that moves electric current between the positive and negative plates through the separators.

❑ The automotive battery has a fully charged specific gravity of 1.265 corrected to 80°F.

❑ Grid growth is a condition where the grid grows little metallic fingers that extend through the separators and short out the plates.

❑ Deep cycling is discharging the battery almost completely before recharging it.

❑ In a conventional battery, the positive plate is covered with lead peroxide and the negative plate is covered with sponge lead.

❑ In maintenance-free batteries, the cell plates contain calcium, cadmium, or strontium to reduce gassing and self-discharge.

❑ The grid construction of the hybrid battery consists of approximately 2.75% antimony alloy on the positive plates and a calcium alloy on the negative plates.

❑ The recombination battery uses separators that hold a gel-type material in place of liquid electrolyte.

❑ The three most common types of battery terminals are:

1. Post or top terminals: Used on most automotive batteries. The positive post will be larger than the negative post to prevent connecting the battery in reverse polarity.

2. Side terminals: Positioned in the side of the container near the top. These terminals are threaded and require a special bolt to connect the cables. Polarity identification is by positive and negative symbols.

3. L terminals: Used on specialty batteries and some imports.

❑ The most common methods of battery rating are cold cranking, cranking amps, reserve capacity, and ampere-hour.

Review Questions

Short-Answer Essays

1. Explain the purposes of the battery.

2. Describe how a technician can determine the correct battery to be installed into a vehicle.

3. Describe the methods used to rate batteries.

4. Describe the different types of battery terminals used.

5. Explain the effects that temperature has on battery performance.

6. Describe the different loads or demands that are placed on a battery during different operating conditions.

7. List and describe the seven main elements of the conventional battery.

8. What are the major reasons that a battery fails?

9. List at least three safety concerns associated with working on or near the battery.

10. Describe the difference in construction of the hybrid battery as compared to the conventional battery.

Fill in the Blanks

1. An automotive battery is an _____ device that provides for and stores _____ energy.

2. When discharging the battery, it changes _____ energy into _____ energy.

3. The assembly of the positive plates, negative plates, and separators is called the _____ _____.

4. The electrolyte solution used in automotive batteries consists of _____% water and _____% sulfuric acid.

5. A fully charged automotive battery has a specific gravity of _____ corrected to 80°F (26.7°C).

6. _____ _____ is a condition where the grid grows little metallic fingers that extend through the separators and short out the plates.

7. The _____ _____ rating indicates the battery's ability to deliver a specified amount of current to start an engine at low ambient temperatures.

8. The electrolyte solution causes the chemical actions to take place between the lead peroxide of the _____ plates and the _____ _____ of the _____ plates.

9. Some of the aspects that determine the battery rating required for a vehicle include engine _____, engine _____, _____ conditions, and vehicle _____.

10. Electrical loads that are still present when the ignition switch is in the OFF position are called _____ loads.

Multiple Choice

1. *Technician A* says the battery provides electricity by releasing free electrons.
   Technician B says the battery stores energy in chemical form.
   Who is correct?
   A. A only
   B. B only
   C. Both A and B
   D. Neither A nor B

2. *Technician A* says the largest demand on the battery is when it must supply current to operate the starter motor.
   Technician B says the current requirements of a starter motor may be over 100 amperes.
   Who is correct?
   A. A only
   B. B only
   C. Both A and B
   D. Neither A nor B

3. *Technician A* says even with the ignition switch turned off, there are electrical demands placed on the battery.
Technician B says after the engine is started, the vehicle's charging system works to recharge the battery and to provide the current to run the electrical systems.
Who is correct?
   A. A only
   B. B only
   C. Both A and B
   D. Neither A nor B

4. The current capacity rating of the battery is being discussed.
Technician A says the amount of electrical energy that a battery is capable of producing depends on the size, weight, and active area of the plates.
Technician B says the current capacity rating of the battery depends on the types of materials used in the construction of the battery.
Who is correct?
   A. A only
   B. B only
   C. Both A and B
   D. Neither A nor B

5. The construction of the battery is being discussed.
Technician A says the 12-volt battery consists of positive and negative plates connected in parallel.
Technician B says the 12-volt battery consists of six cells wired in series.
Who is correct?
   A. A only
   B. B only
   C. Both A and B
   D. Neither A nor B

6. Maintenance of a conventional battery is being discussed.
Technician A says, in a battery that is completely charged, the positive and negative plates are the same material.
Technician B says, in a battery that is completely discharged, the positive and negative plates are different materials.
Who is correct?
   A. A only
   B. B only
   C. Both A and B
   D. Neither A nor B

7. Battery terminology is being discussed.
Technician A says grid growth is a condition where the grid grows little metallic fingers that extend through the separators and short out the plates.
Technician B says deep cycling is discharging the battery almost completely before recharging it.
Who is correct?
   A. A only
   B. B only
   C. Both A and B
   D. Neither A nor B

8. Battery rating methods are being discussed.
Technician A says the ampere-hour is determined by the load in amperes a battery is able to deliver for 30 seconds at 0°F (−17.7°C) without terminal voltage falling below 7.2 volts for a 12-volt battery.
Technician B says the cold cranking rating is the amount of steady current that a fully charged battery can supply for 20 hours at 80°F (26.7°C) without battery voltage falling below 10.5 volts.
Who is correct?
   A. A only
   B. B only
   C. Both A and B
   D. Neither A nor B

9. The hybrid battery is being discussed.
Technician A says the hybrid battery can withstand six deep cycles and still retain 100% of its original reserve capacity.
Technician B says the grid construction of the hybrid battery consists of approximately 2.75% antimony alloy on the positive plates and a calcium alloy on the negative plates.
Who is correct?
   A. A only
   B. B only
   C. Both A and B
   D. Neither A nor B

10. *Technician A* says battery polarity must be observed when connecting the battery cables.
Technician B says the battery must be secured in the vehicle to prevent internal damage and the possibility of shorting across the terminals if it tips.
Who is correct?
   A. A only
   B. B only
   C. Both A and B
   D. Neither A nor B

Starting Systems

Upon completion and review of this chapter, you should be able to:

❑ Explain the purpose of the starting system.

❑ List and identify the components of the starting system.

❑ Explain the principle of operation of the DC motor.

❑ Describe the purpose and operation of the armature.

❑ Describe the purpose and operation of the field coil.

❑ Explain the differences between the types of magnetic switches used.

❑ Identify and explain the differences between starter drive mechanisms.

❑ Describe the differences between the positive engagement and solenoid shift starter.

❑ Describe the operation and features of the permanent magnet starter.

Introduction

Shop Manual
Chapter 6, page 209

The internal combustion engine must be rotated before it will run under its own power. The starting system is a combination of mechanical and electrical parts that work together to start the engine. The starting system is designed to change the electrical energy, which is being stored in the battery, into mechanical energy. To accomplish this conversion, a starter or cranking motor is used. The starting system includes the following components:

1. Battery.

2. Cable and wires.

3. Ignition switch.

4. Starter solenoid or relay.

5. Starter motor.

6. Starter drive and flywheel ring gear.

7. Starter safety switch.

Components in a simplified cranking system circuit are shown (Figure 6-1). This chapter examines both this circuit and the fundamentals of electric motor operation.

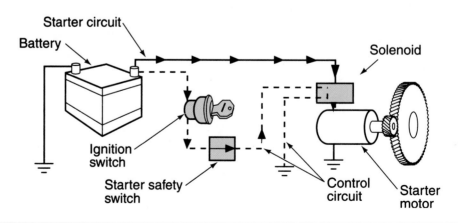

Figure 6-1 Major components of the starting system. The solid line represents the starting (cranking) circuit and the dashed line indicates the starter control circuit.

In the early days of the automobile, the vehicle did not have a starter motor. The operator had to use a starting crank to turn the engine by hand. Charles F. Kettering invented the first electric "self-starter," which was developed and built by the Delco Electrical Plant. The self-starter first appeared on the 1912 Cadillac and was actually a combination starter and generator.

Motor Principles

DC motors use the interaction of magnetic fields to convert the electrical energy into mechanical energy. Magnetic lines of force flow from the north pole to the south pole of a magnet (Figure 6-2). If a current-carrying conductor is placed within the magnetic field, two fields will be present (Figure 6-3). On the left side of the conductor, the lines of force are in the same direction. This will concentrate the flux density of the lines of force on the left side. This will produce a strong

Shop Manual
Chapter 6, page 216

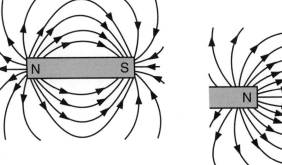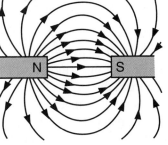

Figure 6-2 Magnetic lines of force flow from the north pole to the south pole.

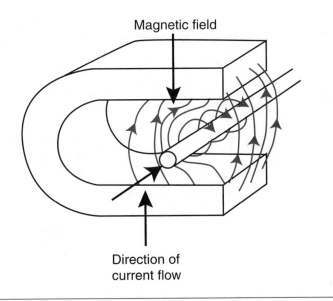

Figure 6-3 Interaction of two magnetic fields.

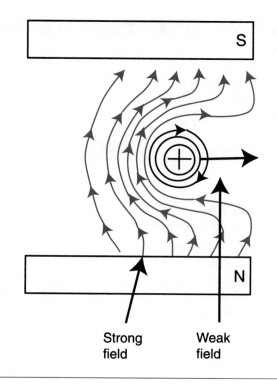

Figure 6-4 Conductor movement in a magnetic field.

magnetic field because the two fields will reinforce each other. The lines of force oppose each other on the right side of the conductor. This results in a weaker magnetic field. The conductor will tend to move from the strong field to the weak field (Figure 6-4). This principle is used to convert electrical energy into mechanical energy in a starter motor by electromagnetism.

A simple electromagnet-style starter motor is shown (Figure 6-5). The inside windings are called the **armature.** The armature is the moveable component of the motor that consists of a conductor wound around a laminated iron core. It is used to create a magnetic field. The armature rotates within the stationary outside windings, called the **field coils,** which has windings

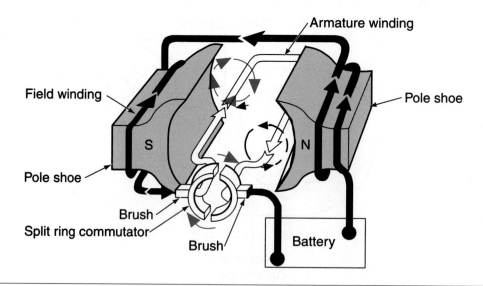

Figure 6-5 Simple electromagnetic motor.

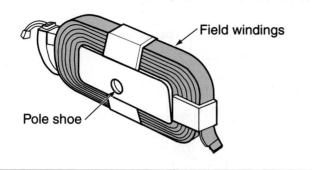

Field windings

Pole shoe

Figure 6-6 Field coil wound around a pole shoe.

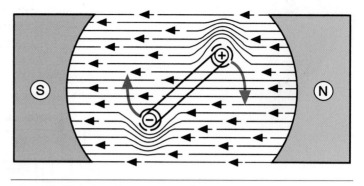

Figure 6-7 Rotation of the conductor is in the direction of the weaker field.

coiled around **pole shoes** (Figure 6-6). Field coils are heavy copper wire wrapped around an iron core to form an electromagnet. Pole shoes are made of high–magnetic permeability material to help concentrate and direct the lines of force in the field assembly.

When current is applied to the field coils and the armature, both produce magnetic flux lines (Figure 6-7). The direction of the windings will place the left pole at a south polarity and the right side at a north polarity. The lines of force move from north to south in the field. In the armature, the flux lines circle in one direction on one side of the loop and in the opposite direction on the other side. Current will now set up a magnetic field around the loop of wire, which will interact with the north and south fields and put a turning force on the loop. This force will cause the loop to turn in the direction of the weaker field. However, the armature is limited in how far it is able to turn. When the armature is halfway between the shoe poles, the fields balance one another. The point at which the fields are balanced is referred to as the **static neutral point.**

For the armature to continue rotating, the current flow in the loop must be reversed. To accomplish this, a split-ring **commutator** is in contact with the ends of the armature loops. The commutator is a series of conducting segments located around one end of the armature. Current enters and exits the armature through a set of **brushes** that slide over the commutator's sections. Brushes are electrically conductive sliding contacts, usually made of copper and carbon. As the brushes pass over one section of the commutator to another, the current flow in the armature is reversed. The position of the magnetic fields are the same. However, the direction of current flow through the loop has been reversed. This will continue until the current flow is turned off.

A single-loop motor would not produce enough torque to rotate an engine. Power can be increased by the addition of more loops or pole shoes. An armature with its many windings, with each loop attached to corresponding commutator sections, is shown (Figure 6-8). In a

Shop Manual
Chapter 6, page 234

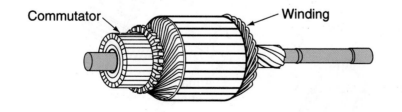

Commutator

Winding

Figure 6-8 Starter armature.

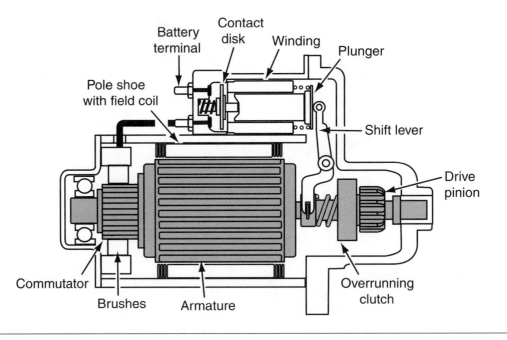

Figure 6-9 Starter and solenoid components.

typical starter motor (Figure 6-9) there are four brushes that make the electrical connections to the commutator. Two of the brushes are grounded to the starter motor frame and two are insulated from the frame. Also, the armature is supported by bushings at both ends.

Armature

Shop Manual
Chapter 6, page 232

The armature is constructed with a laminated core made of several thin iron stampings that are placed next to each other (Figure 6-10). **Laminated construction** is used because, in a solid iron core, the magnetic fields would generate **eddy currents.** These are counter voltages induced in a core. They cause heat to build up in the core and waste energy. By using laminated construction, eddy currents in the core are minimized.

The slots on the outside diameter of the laminations hold the armature windings. The windings loop around the core and are connected to the commutator. Each commutator

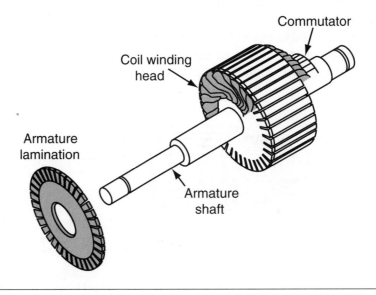

Figure 6-10 Lamination construction of a typical motor armature.

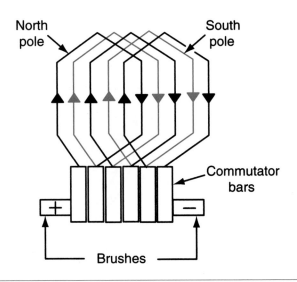

Figure 6-11 Lap-winding diagram.

segment is insulated from the adjacent segments. A typical armature can have more than 30 commutator segments.

A steel shaft is fitted into the center hole of the core laminations. The commutator is insulated from the shaft.

Two basic winding patterns are used in the armature: lap winding and wave winding. In the lap winding, the two ends of the winding are connected to adjacent commutator segments (Figure 6-11). In this pattern, the wires passing under a pole field have their current flowing in the same direction.

In the wave-winding pattern, each end of the winding connects to commutator segments that are 90 or 180 degrees apart (Figure 6-12). In this pattern design, some windings will have no current flow at certain positions of armature rotation. This occurs because the segment ends of the winding loop are in contact with brushes that have the same polarity. The wave-winding pattern is the most commonly used due to its lower resistance.

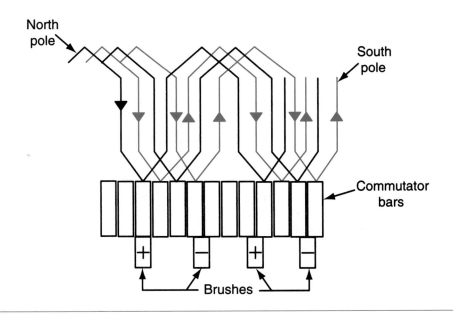

Figure 6-12 Wave-wound armature.

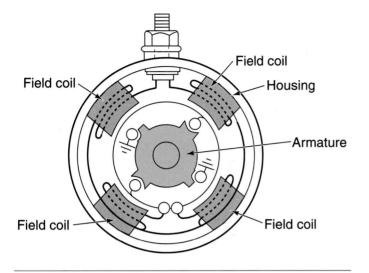

Figure 6-13 Field coils mounted to the inside of starter housing.

Figure 6-14 Magnetic fields in a 4-pole starter motor.

Shop Manual
Chapter 6, page 232

Field Coils

The field coils are electromagnets constructed of wire ribbons or coils wound around a pole shoe. The pole shoes are constructed of heavy iron. The field coils are attached to the inside of the starter housing (Figure 6-13). Most starter motors use four field coils. The iron pole shoes and the iron starter housing work together to increase and concentrate the field strength of the field coils (Figure 6-14).

When current flows through the field coils, strong stationary electromagnetic fields are created. The fields have a north and south magnetic polarity based on the direction the windings are wound around the pole shoes. The polarity of the field coils alternate to produce opposing magnetic fields.

In any DC motor, there are three methods of connecting the field coils to the armature: in series, in parallel (shunt), and a compound connection that uses both series and shunt coils.

Series-Wound Motors

Most starter motors are series-wound with current flowing first to the field windings, then to the brushes, through the commutator and the armature winding contacting the brushes at that time, then through the grounded brushes back to the battery source (Figure 6-15). This design permits all of the current that passes through the field coils to also pass through the armature.

A series-wound motor will develop its maximum torque output at the time of initial start. As the motor speed increases, the torque output of the motor will decrease. This decrease of torque output is the result of **counter electromotive force (CEMF)** caused by self-induction. Since a starter motor has a wire loop rotating within a magnetic field, it will generate an electrical voltage as it spins. This induced voltage will be opposite the battery voltage that is pushing the current through the starter motor. The faster the armature spins, the greater the amount of induced voltage that is generated. This results in less current flow through the starter from the battery as the armature spins faster. Figure 6-16 shows the relationship between starter motor speed and CEMF. Notice that, at 0 (zero) rpm, CEMF is also at 0 (zero). At this time, maximum current flow from the battery through the starter motor will be possible. As the motor spins faster, CEMF increases and current decreases. Since current decreases, the amount of rotating force (torque) also decreases.

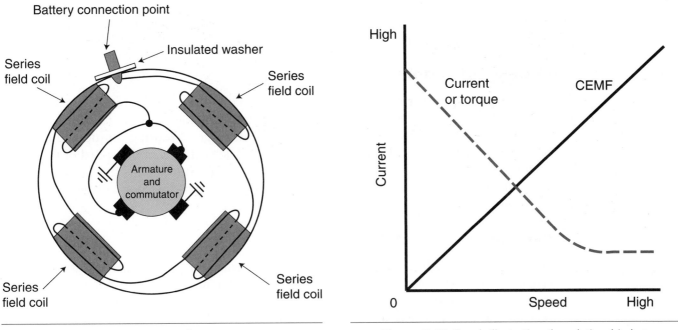

Figure 6-15 A series-wound starter motor.

Figure 6-16 Graph illustrating the relationship between CEMF, starter motor speed, and current draw. As speed increases so does CEMF, reducing current draw and torque.

Shunt-Wound Motors

Electric motors, or **shunt** motors, have the field windings wired in parallel across the armature (Figure 6-17). *Shunt* means there is more than one path for current to flow. A shunt-wound field is used to limit the speed that the motor can turn. A shunt motor does not decrease in its torque output as speeds increase. This is because the CEMF produced in the armature does not

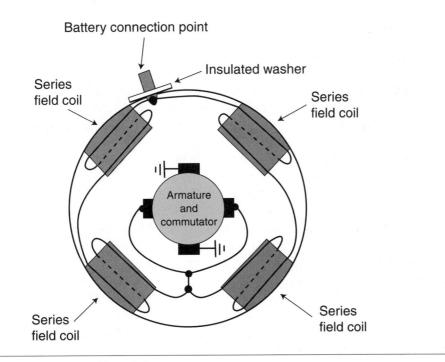

Figure 6-17 A shunt-wound (parallel) starter motor.

decrease the field coil strength. Due to a shunt motor's inability to produce high torque, it is not typically used as a starter motor. However, shunt motors may be found as wiper motors, power window motors, power seat motors, and so on.

Compound Motors

In a **compound motor** most of the field coils are connected to the armature in series and one field coil is connected in parallel with the battery and the armature (Figure 6-18). This configuration allows the compound motor to develop good starting torque and constant operating speeds. The field coil that is shunt wound is used to limit the speed of the starter motor. Also, on Ford's positive engagement starters, the shunt coil is used to engage the starter drive. This is possible because the shunt coil is energized as soon as battery voltage is sent to the starter.

Permanent Magnet Motors

Most newer vehicles have starter motors that use permanent magnets in place of the field coils (Figure 6-19). These motors are also used in many different applications. When a permanent magnet is used instead of coils, there is no field circuit in the motor. By eliminating this circuit, potential electrical problems are also eliminated, such as field-to-housing shorts. Another advantage to using permanent magnets is weight savings; the weight of a typical starter motor is reduced by 50%. Most permanent magnet starters are gear-reduction-type starters.

Multiple permanent magnets are positioned in the housing around the armature. These permanent magnets are an alloy of boron, neodymium, and iron. The field strength of these magnets is much greater than typical permanent magnets. The operation of these motors is the same as other electric motors, except there is no field circuit or windings.

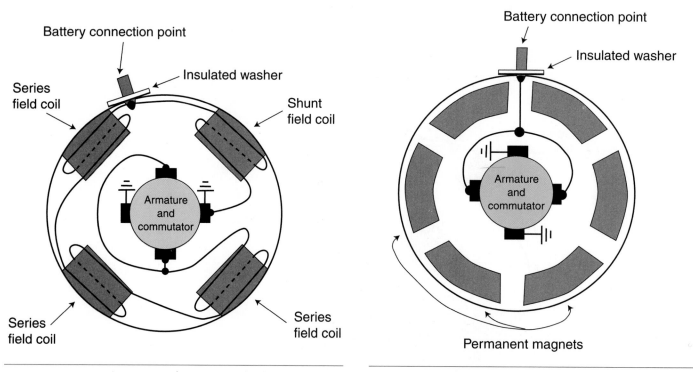

Figure 6-18 A compound motor uses both series and shunt coils.

Figure 6-19 A permanent magnet motor has only an armature circuit, as the field is created by strong permanent magnets.

Starter Drives

The **starter drive** is the part of the starter motor that engages the armature to the engine flywheel ring gear. A starter drive includes a pinion gear set that meshes with the flywheel ring gear on the engine's crankshaft (Figure 6-20). To prevent damage to the pinion gear or the ring gear, the pinion gear must mesh with the ring gear before the starter motor rotates. To help assure smooth engagement, the ends of the pinion gear teeth are tapered (Figure 6-21). Also, the action of the armature must always be from the motor to the engine. The engine must not be allowed to spin the armature. The **ratio** of the number of teeth on the ring gear and the starter drive pinion gear is usually between 15:1 and 20:1. This means the starter motor is rotating 15 to 20 times faster than the engine. The ratio of the starter drive is determined by dividing the number of teeth on the drive gear (pinion gear) into the number of teeth on the driven gear (flywheel). Normal cranking speed for the engine is about 200 rpm. If the starter drive had a ratio of 18:1, the starter would be rotating at a speed of 3,600 rpm. If the engine started and was accelerated to 2,000 rpm, the starter speed would increase to 36,000 rpm. This would destroy the starter motor if it was not disengaged from the engine.

The most common type of starter drive is the **overrunning clutch.** The overrunning clutch is a roller-type clutch that transmits torque in one direction only and freewheels in the other direction. This allows the starter motor to transmit torque to the ring gear but prevents the ring gear from transferring torque to the starter motor.

Shop Manual
Chapter 6, page 235

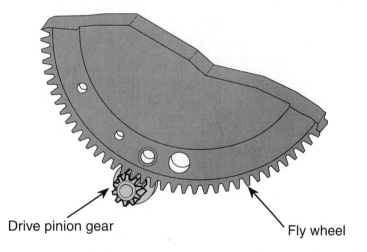

Drive pinion gear

Fly wheel

Figure 6-20 Starter drive pinion gear is used to turn the engine's flywheel.

Figure 6-21 The pinion gear teeth are tapered to allow for smooth engagement.

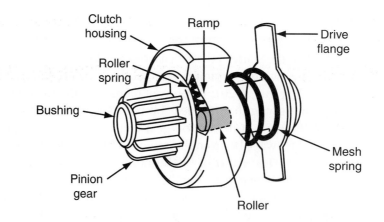

Figure 6-22 Overrunning clutch starter drive.

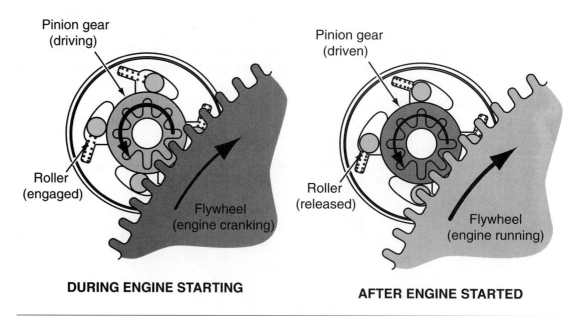

DURING ENGINE STARTING

AFTER ENGINE STARTED

Figure 6-23 When the armature turns, it locks the rollers into the tapered notch.

In a typical overrunning-type clutch (Figure 6-22), the clutch housing is internally splined to the starter armature shaft. The drive pinion turns freely on the armature shaft within the clutch housing. When torque is transmitted through the armature to the clutch housing, the spring-loaded rollers are forced into the small ends of their tapered slots (Figure 6-23). They are then wedged tightly against the pinion barrel. The pinion barrel and clutch housing are now locked together; torque is transferred through the starter motor to the ring gear and engine.

When the engine starts and is running under its own power, the ring gear attempts to drive the pinion gear faster than the starter motor. This unloads the clutch rollers and releases the pinion gear to rotate freely around the armature shaft.

Cranking Motor Circuits

The starting system of the vehicle consists of two circuits: the starter control circuit and the motor feed circuit. These circuits are separate but related. The control circuit consists of the starting portion of the ignition switch, the starting safety switch (if applicable), and the wire conductor to connect these components to the relay or solenoid. The motor feed circuit consists of heavy battery cables from the battery to the relay and the starter or directly to the solenoid if the starter is so equipped.

Starter Control Circuit Components

Shop Manual
Chapter 6, page 220

Magnetic Switches

The starter motor requires large amounts of current (up to 300 amperes) to generate the torque needed to turn the engine. The conductors used to carry this amount of current (battery cables) must be large enough to handle the current with very little voltage drop. It would be impractical to place a conductor of this size into the wiring harness to the ignition switch. To provide control of the high current, all starting systems contain some type of magnetic switch. There are two basic types of magnetic switches used: the solenoid and the relay.

Starter-Mounted Solenoids. As discussed in Chapter 3, a solenoid is an electromagnetic device that uses the movement of a plunger to exert a pulling or holding force. In the solenoid-actuated starter system, the solenoid is mounted directly on top of the starter motor (Figure 6-24). The solenoid switch on a starter motor performs two functions: It closes the circuit between the battery and the starter motor. Then it shifts the starter motor pinion gear into mesh with the ring gear. This is accomplished by a linkage between the solenoid plunger and the shift lever on the starter motor. In the past, the most common method of energizing the

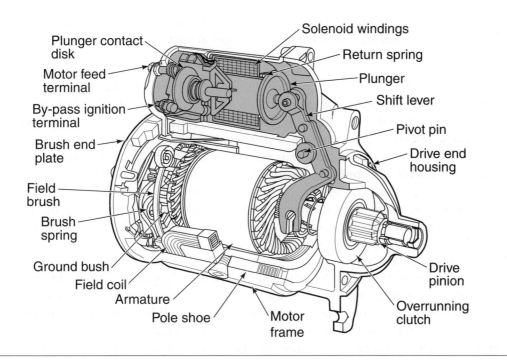

Figure 6-24 Solenoid operated starter has the solenoid mounted directly on top of the motor.

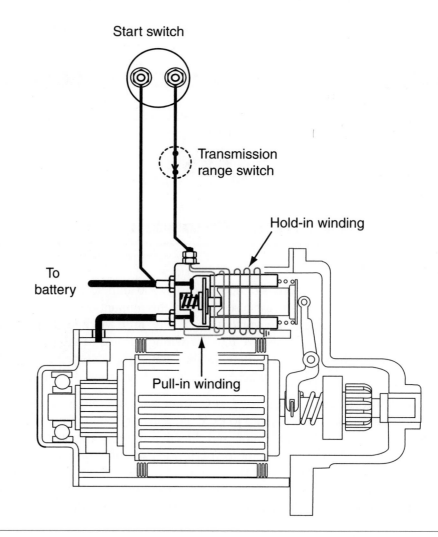

Start switch

Transmission
range switch

Hold-in winding

To
battery

Pull-in winding

Figure 6-25 The solenoid uses two windings. Both are energized to draw the plunger, then only the hold-in winding is used to hold the plunger in position.

The two windings of the solenoid are called the **pull-in** and the **hold-in** **windings.** Their names explain their functions.

solenoid was directly from the battery through the ignition switch. However, most of today's vehicles use a starter relay in conjunction with a solenoid. The relay is used to reduce the amount of current flow through the ignition switch and is usually controlled by the powertrain control module (PCM). This system will be discussed later in this chapter.

When the circuit is closed and current flows to the solenoid, current from the battery is directed to the **pull-in** and **hold-in windings** (Figure 6-25). Because it may require up to 50 amperes to create a magnetic force large enough to pull the plunger in, both windings are energized to create a combined magnetic field that pulls the plunger. Once the plunger is moved, the current required to hold the plunger is reduced. This allows the current that was used to pull the plunger in to be used to rotate the starter motor.

When the ignition switch is placed in the START position, voltage is applied to the S terminal of the solenoid (Figure 6-26). The hold-in winding has its own ground to the case of the solenoid. The pull-in winding's ground is through the starter motor. Current will flow through both windings to produce a strong magnetic field. When the plunger is moved into contact with the main battery and motor terminals, the pull-in winding is de-energized. The pull-in winding is not energized because the contact places battery voltage on both sides of the coil (Figure 6-27). The current that was directed through the pull-in winding is now sent to the motor.

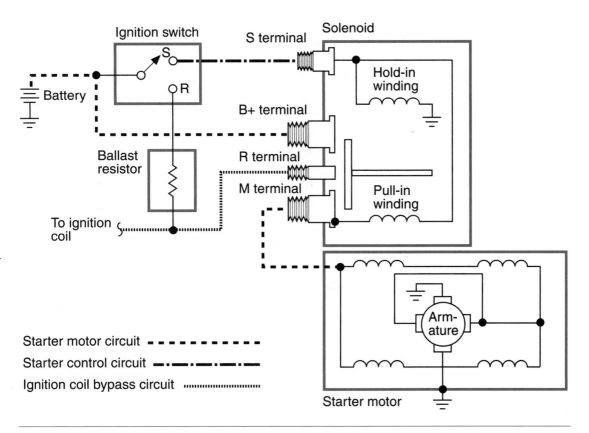

Figure 6-26 Schematic of solenoid-operated starter motor circuit.

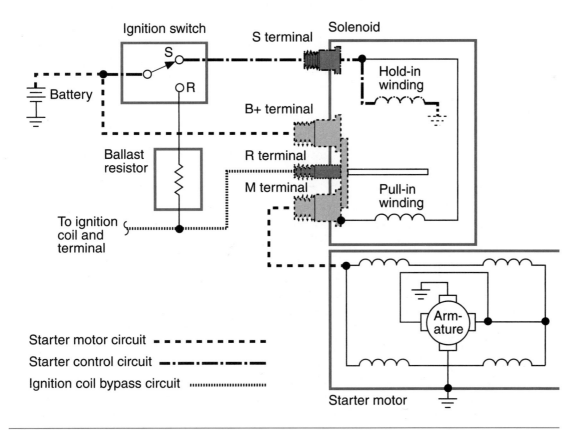

Figure 6-27 Once the contact disc closes the terminals, the hold-in winding is the only one that is energized.

Because the contact disc does not close the circuit from the battery to the starter motor until the plunger has moved the shift lever, the pinion gear is in full mesh with the flywheel before the armature starts to rotate.

After the engine is started, releasing the key to the RUN position opens the control circuit. Voltage no longer is supplied to the hold-in windings, and the return spring causes the plunger to return to its neutral position.

In Figures 6-26 and 6-27, an R terminal is illustrated. This terminal provides current to the ignition bypass circuit that is used to provide full battery voltage to the ignition coil while the engine is cranking. This circuit bypasses the ballast resistor. The bypass circuit is not used on most ignition systems today.

A common problem with the control circuit is that low system voltage or an open in the hold-in windings will cause an oscillating action to occur. The combination of the pull-in winding and the hold-in winding is sufficient to move the plunger. However, once the contacts are closed, there is insufficient magnetic force to hold the plunger in place. This condition is recognizable by a series of clicks when the ignition switch is turned to the START position. Before replacing the solenoid, check the battery condition; a low battery charge will cause the same symptom.

> **AUTHOR'S NOTE:** Some manufacturers use a starter relay in conjunction with a solenoid relay. The relay is used to reduce the amount of current flow through the ignition switch.

Many manafacturers call the remote solenoid the starter relay.

Remote Solenoids. Some manufacturers use a starter solenoid that is mounted near the battery on the fender well or radiator support (Figure 6-28). Unlike the starter-mounted solenoid, the remote solenoid does not move the pinion gear into mesh with the flywheel ring gear.

When the ignition switch is turned to the START position, current is supplied through the switch to the solenoid windings. The windings produce a magnetic field that pulls the moveable core into contact with the internal contacts of the battery and starter terminals (Figure 6-29). With the contacts closed, full battery current is supplied to the starter motor.

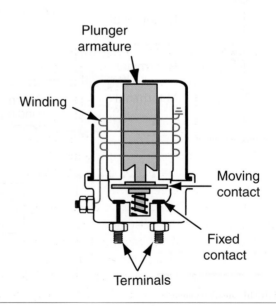

Figure 6-28 A remote starter solenoid, often referred to as the starter relay.

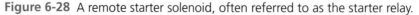

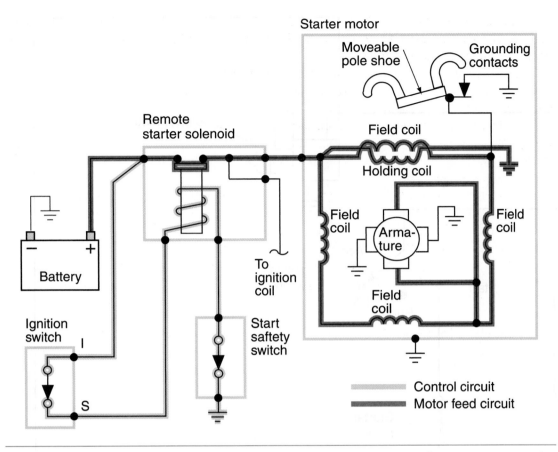

Figure 6-29 Current flow when the remote starter solenoid is energized.

A secondary function of the starter relay is to provide for an alternate path for current to the ignition coil during cranking. This is done by an internal connection that is energized by the relay core when it completes the circuit between the battery and the starter motor.

Starter Relay Controls

Most modern vehicles will use a starter relay in conjunction with a starter motor–mounted solenoid to control starter motor operation. The relay can be controlled through the ignition switch or by the powertrain control module (PCM).

In a system that uses the ignition switch to control the relay, the switch will usually be installed on the insulated side of the relay control circuit (Figure 6-30). When the ignition switch is turned to the START position, battery voltage is applied to the coil of the relay. Since the relay coil is grounded, the coil is energized and pulls the contacts closed. With the contacts closed, battery voltage is applied to the control side of the starter solenoid. The solenoid operates in the same manner as discussed previously.

In this type of system, a very small wire can be used through the steering column to the ignition switch. This reduces the size of the wiring harness.

In a PCM-controlled system, the PCM will monitor the ignition switch position to determine if the starter motor should be energized. System operation differs among manufacturers. However, in most systems, the PCM will control the starter relay coil ground circuit (Figure 6-31). Control by the PCM allows the manufacturer to install software commands such as **double start override,** which prevents the starter motor from being energized if the engine is already running, and **sentry key** within the PCM.

Sentry key is one of the terms used to describe a sophisticated antitheft system that prevents the engine from starting unless a special key is used.

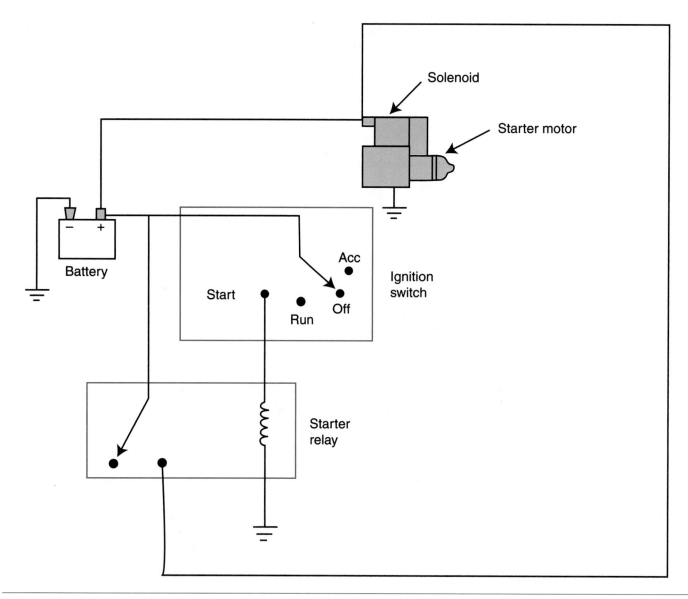

Figure 6-30 Starter control circuit using an insulated side relay to control current to the starter solenoid.

Ignition Switch

The ignition switch is the power distribution point for most of the vehicle's primary electrical systems (Figure 6-32). Most ignition switches have five positions:

1. **ACCESSORIES:** Supplies current to the vehicle's electrical accessory circuits. It will not supply current to the engine control circuits, starter control circuit, or the ignition system.

2. **LOCK:** Mechanically locks the steering wheel and transmission gear selector. All electrical contacts in the ignition switch are open. Most ignition switches must be in this position to insert or remove the key from the cylinder.

3. **OFF:** All circuits controlled by the ignition switch are opened. The steering wheel and transmission gear selector are unlocked.

4. **ON or RUN:** The switch provides current to the ignition, engine controls, and all other circuits controlled by the switch. Some systems will power a chime or light with the key in the ignition switch. Other systems power an antitheft system when the key is removed and turn it off when the key is inserted.

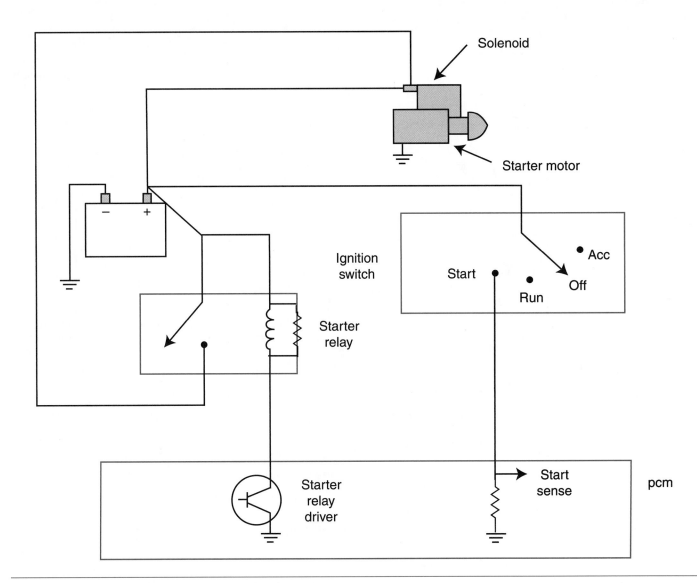

Figure 6-31 Typical PCM starter control circuit.

5. START: The switch provides current to the starter control circuit, ignition system, and engine control circuits.

The ignition switch is spring loaded in the START position. This momentary contact automatically moves the contacts to the RUN position when the driver releases the key. All other ignition switch positions are detent positions.

Starting Safety Switch

The neutral safety switch is used on vehicles equipped with automatic transmissions. It opens the starter control circuit when the transmission shift selector is in any position except PARK or NEUTRAL. The actual location of the neutral safety switch depends on the kind of transmission and the location of the shift lever. Some manufacturers place the switch in the transmission (Figure 6-33).

Vehicles equipped with automatic transmissions require a means of preventing the engine from starting while the transmission is in gear. Without this feature, the vehicle would lunge forward or backward once it was started, causing personal injury or property damage. The normally open neutral safety switch is connected in series in the starting system control circuit and is usually operated by the shift lever. When in the PARK or NEUTRAL position, the switch

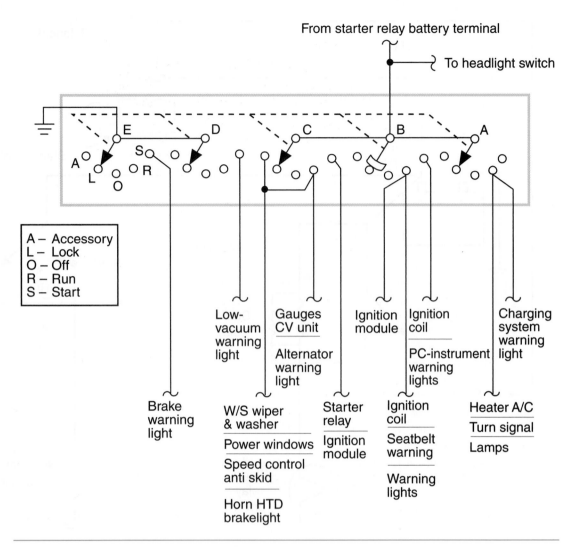

From starter relay battery terminal

To headlight switch

A – Accessory
L – Lock
O – Off
R – Run
S – Start

Brake warning light

Low-vacuum warning light

W/S wiper & washer

Power windows

Speed control anti skid

Horn HTD brakelight

Gauges CV unit

Alternator warning light

Starter relay

Ignition module

Ignition module

Ignition coil

Ignition coil

Seatbelt warning

Warning lights

PC-instrument warning lights

Charging system warning light

Heater A/C

Turn signal

Lamps

Figure 6-32 Ganged ignition switch.

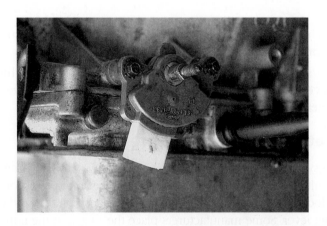

Figure 6-33 The neutral safety switch can be combined with the backup light switch and installed on the transmission case.

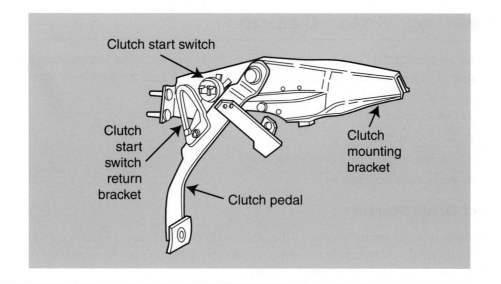

Figure 6-34 Most vehicles with a manual transmission use a clutch start switch to prevent the engine from starting unless the clutch pedal is pressed.

is closed, allowing current to flow to the starter circuit. If the transmission is in a gear position, the switch is opened and current cannot flow to the starter circuit.

Many vehicles equipped with manual transmissions use a similar type of safety switch. The **startclutch interlock switch** is usually operated by movement of the clutch pedal (Figure 6-34). When the clutch pedal is pushed downward, the switch closes and current can flow through the starter circuit. If the clutch pedal is left up, the switch is open and current cannot flow.

Some vehicles use a mechanical linkage that blocks movement of the ignition switch cylinder unless the transmission is in PARK or NEUTRAL (Figure 6-35).

AUTHOR'S NOTE: One-touch and remote starting systems will be discussed in Chapter 15.

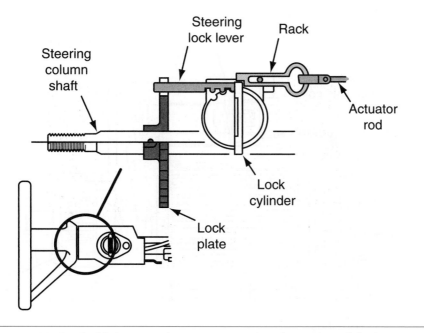

Figure 6-35 Mechanical linkage used to prevent starting the engine while the transmission is in gear.

Cranking Motor Designs

The most common type of starter motor used today incorporates the overrunning clutch starter drive instead of the old inertia-engagement bendix drive. There are four basic groups of starter motors:

1. Direct drive.
2. Gear reduction.
3. Positive-engagement (moveable pole).
4. Permanent magnet.

Direct Drive Starters

The direct drive starter motor can be either series-wound or compound motors.

A common type of starter motor is the solenoid-operated direct drive unit (Figure 6-36). Although there are construction differences between applications, the operating principles are the same for all solenoid-shifted starter motors.

When the ignition switch is placed in the START position, the control circuit energizes the pull-in and hold-in windings of the solenoid. The solenoid plunger moves and pivots the shift lever, which in turn locates the drive pinion gear into mesh with the engine flywheel.

When the solenoid plunger is moved all the way, the contact disc closes the circuit from the battery to the starter motor. Current now flows through the field coils and the armature. This develops the magnetic fields that cause the armature to rotate, thus turning the engine.

Gear Reduction Starters

Some gear reduction starter motors are compound motors.

Most gear reduction starters have the commutator and brushes located in the center of the motor.

Some manufacturers use a gear reduction starter to provide increased torque (Figure 6-37). The gear reduction starter differs from most other designs in that the armature does not drive the pinion gear directly. In this design, the armature drives a small gear that is in constant mesh with a larger gear. Depending on the application, the ratio between these two gears is between 2:1 and 3.5:1. The additional reduction allows for a small motor to turn at higher speeds and greater torque with less current draw.

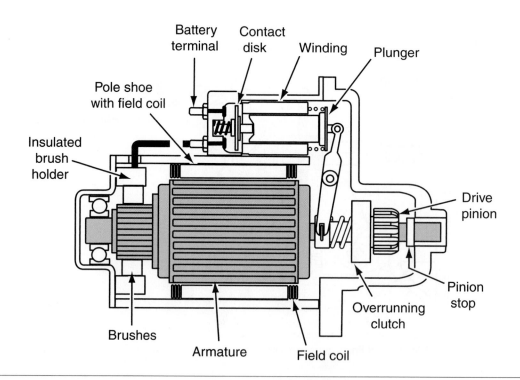

Figure 6-36 Solenoid-operated Delco MT series starter motor.

156

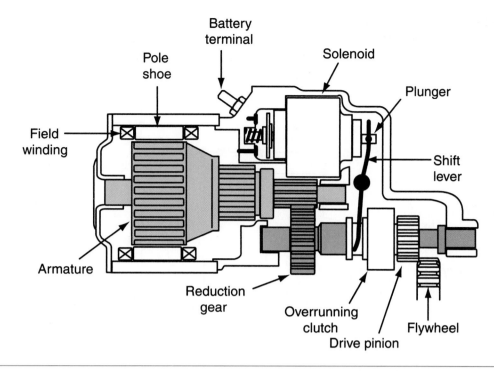

Figure 6-37 Gear reduction starter motor construction.

The solenoid operation is similar to that of the solenoid-shifted direct drive starter in that the solenoid moves the plunger, which engages the starter drive.

Positive-Engagement Starters

A commonly used starter on Ford applications in the past was the positive-engagement starter (Figure 6-38). Positive-engagement starters use the shunt coil windings of the starter motor to engage the starter drive. The high starting current is controlled by a starter solenoid mounted close to the battery. When the solenoid contacts are closed, current flows through a drive coil.

Positive-engagement starters are also called moveable-pole shoe starters.

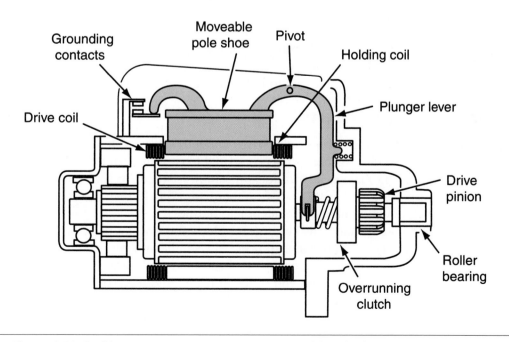

Figure 6-38 Positive-engagement starters use a moveable pole shoe.

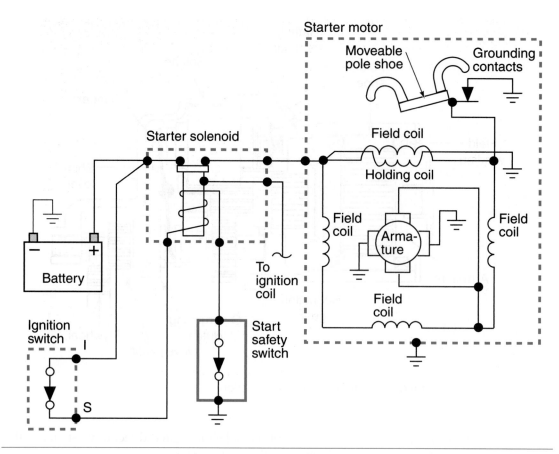

Figure 6-39 Schematic of positive-engagement starter.

The drive coil creates an electromagnetic field that attracts a moveable pole shoe. The moveable pole shoe is attached to the starter drive through the plunger lever. When the moveable pole shoe moves, the drive gear engages the engine flywheel.

As soon as the starter drive pinion gear contacts the ring gear, a contact arm on the pole shoe opens a set of normally closed grounding contacts (Figure 6-39). With the return to ground circuit opened, all the starter current flows through the remaining three field coils and through the brushes to the armature. The starter motor then begins to rotate. To prevent the starter drive from disengaging from the ring gear if battery voltage drops while cranking, the moveable pole shoe is held down by a holding coil. The holding coil is a smaller coil inside the main **drive coil** and is strong enough to hold the starter pinion gear engaged.

Permanent Magnet Starters

The **permanent magnet gear reduction (PMGR)** starter design provides for less weight, simpler construction, and less heat generation as compared to conventional field coil starters (Figure 6-40). The permanent magnet gear reduction starter uses four or six permanent magnet field assemblies in place of field coils. Because there are no field coils, current is delivered directly to the armature through the commutator and brushes.

The permanent magnet starter also uses gear reduction through a planetary gear set (Figure 6-41). The planetary geartrain transmits power between the armature and the pinion shaft. This allows the armature to rotate at greater speed and increased torque. The planetary gear assembly consists of a sun gear on the end of the armature and three planetary carrier gears inside a ring gear. The ring gear is held stationary. When the armature is rotated, the sun gear causes the carrier gears to rotate about the internal teeth of the ring gear. The planetary carrier

The **drive coil** is a hollowed field coil that is used to attract the moveable pole shoe.

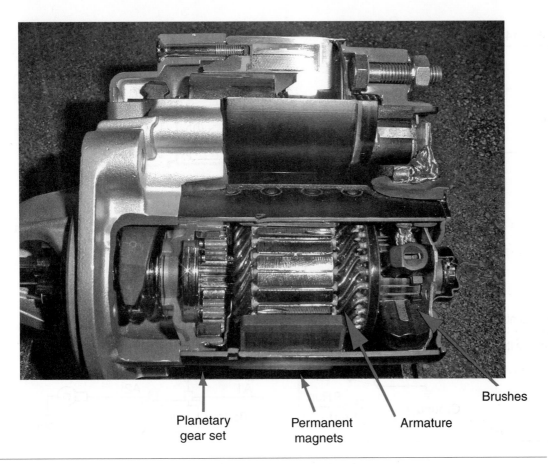

Brushes

Planetary
gear set

Permanent
magnets

Armature

Figure 6-40 The PMGR motor uses a planetary gear set and permanent magnets.

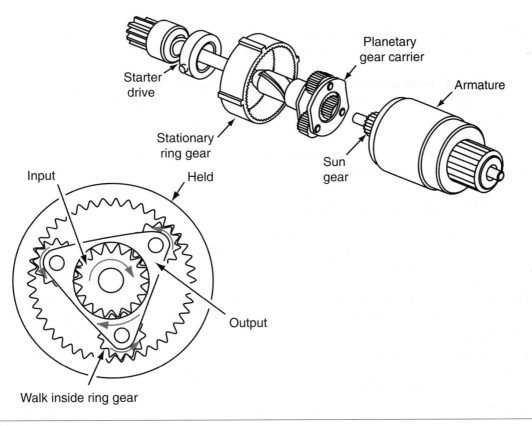

Starter
drive

Planetary
gear carrier

Armature

Stationary
ring gear

Sun
gear

Input

Held

Output

Walk inside ring gear

Figure 6-41 Planetary gear set.

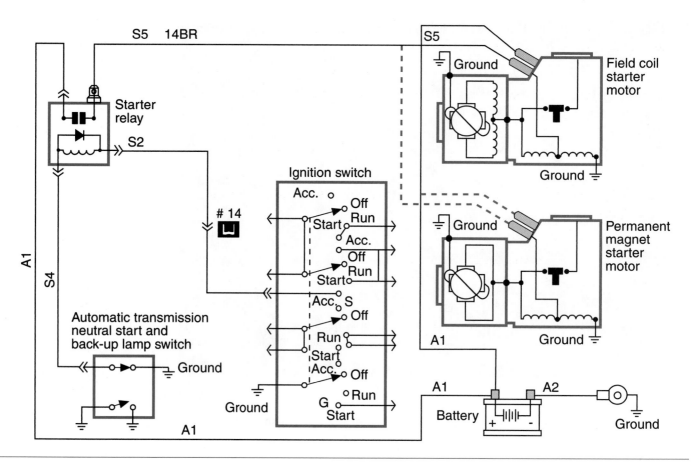

Figure 6-42 Comparison of the electrical circuits used in field coil and PMGR starters.

is attached to the output shaft. The gear reduction provided for by this gear arrangement is 4.5:1. By providing for this additional gear reduction, the demand for high current is lessened.

AUTHOR'S NOTE: The greatest amount of gear reduction from a planetary gear set is accomplished by holding the ring gear, inputting the sun gear, and outputting the carrier.

The electrical operation between the conventional field coil and PMGR starters remains basically the same (Figure 6-42).

Terms to Know

Armature

Brushes

Commutator

Compound motor

Counter
  electromotive
  force (CEMF)

Double-start
override

Summary

❏ The starting system is a combination of mechanical and electrical parts that work together to start the engine.

❏ The starting system components include the battery, cable and wires, the ignition switch, the starter solenoid or relay, the starter motor, the starter drive and flywheel ring gear, and the starting safety switch.

❏ The armature is the moveable component of the motor that consists of a conductor wound around a laminated iron core. It is used to create a magnetic field.

- ❏ Pole shoes are made of high–magnetic permeability material to help concentrate and direct the lines of force in the field assembly.

- ❏ The magnetic forces will cause the armature to turn in the direction of the weaker field.

- ❏ Within an electromagnetic style of starter motor, the inside windings are called the armature. The armature rotates within the stationary outside windings, called the field, which has windings coiled around pole shoes.

- ❏ The commutator is a series of conducting segments located around one end of the armature.

- ❏ A split-ring commutator is in contact with the ends of the armature loops. So, as the brushes pass over one section of the commutator to another, the current flow in the armature is reversed.

- ❏ Two basic winding patterns are used in the armature: lap winding and wave winding.

- ❏ The field coils are electromagnets constructed of wire coils wound around a pole shoe.

- ❏ When current flows through the field coils, strong stationary electromagnetic fields are created.

- ❏ In any DC motor, there are three methods of connecting the field coils to the armature: in series, in parallel (shunt), and a compound connection that uses both series and shunt coils.

- ❏ A starter drive includes a pinion gear set that meshes with the engine flywheel ring gear on the engine.

- ❏ To prevent damage to the pinion gear or the ring gear, the pinion gear must mesh with the ring gear before the starter motor rotates.

- ❏ The bendix drive depends on inertia to provide meshing of the drive pinion with the ring gear.

- ❏ The most common type of starter drive is the overrunning clutch. This is a roller-type clutch that transmits torque in one direction only and freewheels in the other direction.

- ❏ The starting system consists of two circuits called the starter control circuit and the motor feed circuit.

- ❏ The components of the control circuit include the starting portion of the ignition switch, the starting safety switch (if applicable), and the wire conductor to connect these components to the relay or solenoid.

- ❏ The motor feed circuit consists of heavy battery cables from the battery to the relay and the starter or directly to the solenoid if the starter is so equipped.

- ❏ There are four basic groups of starter motors: direct drive, gear reduction, positive engagement (moveable pole), and permanent magnet.

Terms to Know
(continued)

Drive coil

Eddy currents

Field coils

Hold-in windings

Laminated construction

Overrunning clutch

Permanent magnet gear reduction (PMGR)

Pole shoes

Pull-in windings

Ratio

Sentry key

Shunt

Start/clutch interlock switch

Starter drive

Static neutral point

Review Questions

Short-Answer Essays

1. What is the purpose of the starting system?
2. List and describe the purpose of the major components of the starting system.
3. Explain the principle of operation of the DC motor.
4. Describe the types of magnetic switches used in starting systems.
5. Describe the operation of the overrunning clutch drive.

6. Describe the differences between the positive-engagement and solenoid shift starter.

7. Explain the operating principles of the permanent magnet starter.

8. Describe the purpose and operation of the armature.

9. Describe the purpose and operation of the field coil.

10. Describe the operation of the two circuits of the starter system.

Fill in the Blanks

1. DC motors use the interaction of magnetic fields to convert the _____ energy into _____ energy.

2. The _____ is the moveable component of the motor, which consists of a conductor wound around a _____ iron core and is used to create a _____ field.

3. Pole shoes are made of high–magnetic _____ material to help concentrate and direct the _____ _____ _____ in the field assembly.

4. The starter motor electrical connection that permits all of the current that passes through the field coils to also pass through the armature is called the _____ motor.

5. _____ _____ _____ is voltage produced in the starter motor itself. This current acts against the supply voltage from the battery.

6. A starter motor that uses the characteristics of a series motor and a shunt motor is called a _____ motor.

7. The _____ _____ is the part of the starter motor that engages the armature to the engine flywheel ring gear.

8. The _____ _____ is a roller-type clutch that transmits torque in one direction only and freewheels in the other direction.

9. The two circuits of the starting system are called the _____ _____ circuit and the _____ _____ circuit.

10. There are two basic types of magnetic switches used in starter systems: the _____ and the _____ .

Multiple Choice

1. The armature:
   A. Is the stationary component of the starter that creates a magnetic field.
   B. Is the rotating component of the starter that creates a magnetic field.
   C. Carries electrical current to the commutator.
   D. Prevents the starter from engaging if the transmission is in gear.

2. What is the purpose of the commutator?
   A. To prevent the field windings from contacting the armature.
   B. To maintain constant electrical contact with the field windings.
   C. To reverse current flow through the armature.
   D. All of the above.

3. The field coils:
    A. Are made of wire wound around a nonmagnetic pole shoe.
    B. Are always shunt wound to the armature.
    C. Are always series wound with the armature.
    D. None of the above.

4. Which of the following describes the operation of the starter solenoid?
    A. An electromagnetic device that uses movement of a plunger to exert a pulling or holding force.
    B. Both the pull-in and hold-in windings are energized to engage the starter drive.
    C. When the starter drive plunger is moved, the pull-in winding is de-energized.
    D. All of the above.

5. The purpose of the overrunning clutch is to:
    A. Allow the ring gear to drive the armature.
    B. Prevent the field windings from rotating.
    C. Prevent the armature from rotating.
    D. Transmit torque to the ring gear but prevent the ring gear from turning the armature.

6. Permanent magnet starters are being discussed.
    Technician A says the permanent magnet starter uses four or six permanent magnet field assemblies in place of field coils.
    Technician B says the permanent magnet starter uses a planetary gear set.
    Who is correct?
    A. A only
    B. B only
    C. Both A and B
    D. Neither A nor B

7. Typical components of the control circuit of the starting system include:
    A. Ring gear.
    B. Magnetic switch.
    C. Pinion gear.
    D. All of the above.

8. The characteristic of the series-wound motor is:
    A. Current flows from the armature, to the brushes, then to the field windings.
    B. Current flows from the field windings, to the brushes, and to the armature.
    C. Current flows through shunts to the field windings and the armature.
    D. All of the above.

9. The gear reduction starter uses:
    A. A starter drive that is connected directly to the armature.
    B. A larger gear to drive a smaller gear that is attached to the starter drive.
    C. A smaller gear to drive a larger gear that is attached to the starter drive.
    D. A starter drive that is attached to the commutator ring.

10. A characteristic of permanent magnet starters is:
    A. The use of planetary gears.
    B. Current flows from the field windings, to the brushes, and to the armature.
    C. Connection directly to the armature.
    D. All of the above.

CHAPTER 7

Charging Systems

Upon completion and review of this chapter, you should be able to:

❏ Explain the purpose of the charging system.

❏ Identify the major components of the charging system.

❏ Explain the function of the major components of the AC generator.

❏ Describe the two styles of stators.

❏ Describe how AC current is rectified to DC current in the AC generator.

❏ Describe the three principle circuits used in the AC generator.

❏ Explain the relationship between regulator resistance and field current.

❏ Explain the relationship between field current and AC generator output.

❏ Identify the differences between A circuit, B circuit, and isolated circuit.

❏ Explain the operation of charge indicators, including lamps, electronic voltage monitors, ammeters, and voltmeters.

Introduction

The automotive storage battery is not capable of supplying the demands of the electrical system for an extended period of time. Every vehicle must be equipped with a means of replacing the current being drawn from the battery. A charging system is used to restore the electrical power to the battery that was used during engine starting. In addition, the charging system must be able to react quickly to high load demands required of the electrical system. It is the vehicle's charging system that generates the current to operate all of the electrical accessories while the engine is running.

Two basic types of charging systems have been used. The first was a DC generator, which was discontinued in the 1960s. Since that time the AC generator has been the predominant charging device. The DC generator and the AC generator both use similar operating principles.

The purpose of the charging system is to convert the mechanical energy of the engine into electrical energy to recharge the battery and run the electrical accessories. When the engine is first started, the battery supplies all the current required by the starting and ignition systems.

As the battery drain continues, and engine speed increases, the charging system is able to produce more voltage than the battery can deliver. When this occurs, the electrons from the charging device are able to flow in a reverse direction through the battery's positive terminal. The charging device is now supplying the electrical system's load requirements; the reserve electrons build up and recharge the battery (Figure 7-1).

If there is an increase in the electrical demand and a drop in the charging system's output equal to the voltage of the battery, the battery and charging system work together to supply the required current.

The entire charging system consists of the following components:

1. Battery.

2. Generator.

3. Drive belt.

4. Voltage regulator.

5. Charge indicator (lamp or gauge).

In an attempt to standardize terminology in the industry, the term *alternator* is being replaced with *generator*. Often an alternator is referred to as an AC generator.

164

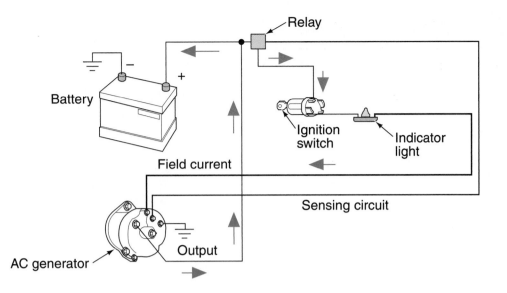

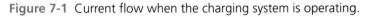

Figure 7-1 Current flow when the charging system is operating.

6. Ignition switch.

7. Cables and wiring harness.

8. Starter relay (some systems).

9. Fusible link (some systems).

Principle of Operation

All charging systems use the principle of electromagnetic induction to generate electrical power (Figure 7-2). Electromagnetic principle states that a voltage will be produced if motion between a conductor and a magnetic field occurs. The amount of voltage produced is affected by:

1. The speed at which the conductor passes through the magnetic field.

2. The strength of the magnetic field.

3. The number of conductors passing through the magnetic field.

The ignition switch is considered a part of the charging system because it closes the circuit that supplies current to the indicator lamp and stimulates the field coil.

Shop Manual
Chapter 7, page 249

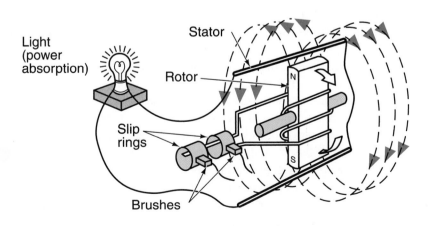

Figure 7-2 Simplified AC generator indicating electromagnetic induction.

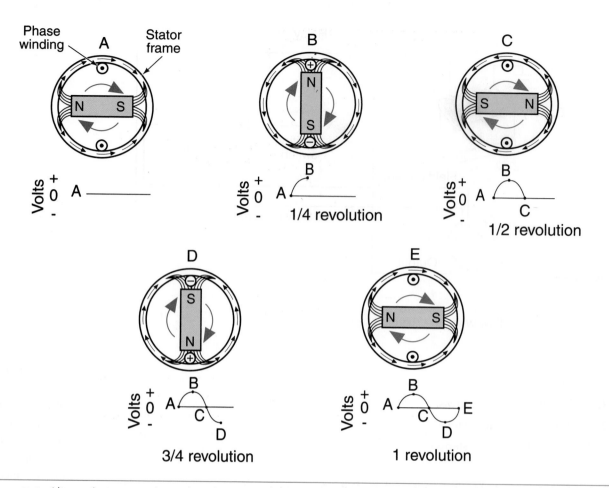

Figure 7-3 Alternating current is produced as the magnetic field is rotated.

To see how electromagnetic induction produces an AC voltage, refer to the illustration (Figure 7-3). When the conductor is parallel with the magnetic field, the conductor is not cut by any flux lines (Figure 7-3A). At this point in the revolution, zero voltage and current are being produced.

As the magnetic field is rotated 90 degrees, the magnetic field is at a right angle to the conductor (Figure 7-3B). At this point in the revolution, the maximum number of flux lines cut the conductor at the north pole. With the maximum amount of flux lines cutting the conductor, voltage is at its maximum positive value.

When the magnetic field is rotated an additional 90 degrees, the conductor returns to being parallel with the magnetic field (Figure 7-3C). Once again, no flux lines cut the conductor and voltage drops to zero.

An additional 90-degree revolution of the magnetic field results in the magnetic field being reversed at the top conductor (Figure 7-3D). At this point in the revolution, the maximum number of flux lines cuts the conductor at the south pole. Voltage is now at maximum negative value.

When the magnetic field completes one full revolution, it returns to a parallel position with the magnetic field. Voltage returns to zero. The sine wave is determined by the angle between the magnetic field and the conductor. It is based on the trigonometry sine function of angles. The sine wave shown (Figure 7-4) plots the voltage generated during one revolution.

It is the function of the drive belt to turn the magnetic field. Drive belt tension should be checked periodically to assure proper charging system operation. A loose belt can inhibit charging system efficiency, and a belt that is too tight can cause early bearing failure.

The sine wave produced by a single conductor during one revolution is called single-phase voltage.

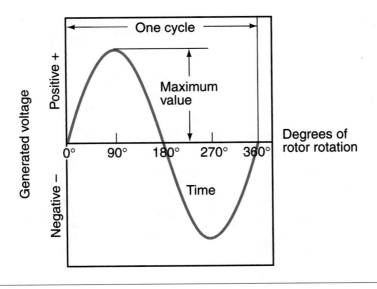

Figure 7-4 Sine wave produced in one revolution of the conductor or magnetic field.

AC Generators

AUTHOR'S NOTE: The first charging systems used a DC generator that had two field coils that created a magnetic field. Output voltage was generated in the wire loops of the armature as it rotated inside the magnetic field. Current sent to the battery was through the commutator and the generator's brushes.

The DC generator was unable to produce the sufficient amount of current required when the engine was operating at low speeds. With the addition of more electrical accessories and components, the AC (alternating current) generator, or alternator, replaced the DC generator. The main components of the AC generator are (Figure 7–5):

1. The rotor.
2. Brushes.
3. The stator.
4. The rectifier bridge.
5. The housing.
6. Cooling fan.

Rotors

The **rotor** creates the rotating magnetic field of the AC generator. It is the portion of the AC generator that is rotated by the drive belt. The rotor is constructed of many turns of copper wire around an iron core. There are metal plates bent over the windings at both ends of the rotor windings (Figure 7-6). The poles (metal plates) do not come into contact with each other, but they are interlaced. When current passes through the coil (1.5 to 3.0 amperes), a magnetic field is produced. The strength of the magnetic field is dependent on the amount of current flowing through the coil and the number of windings.

The poles will take on the polarity (north or south) of the side of the coil they touch. The right-hand rule will show whether a north or south pole magnet is created. When the rotor is

Shop Manual
Chapter 7, page 272

The current flow through the coil is referred to as field current.

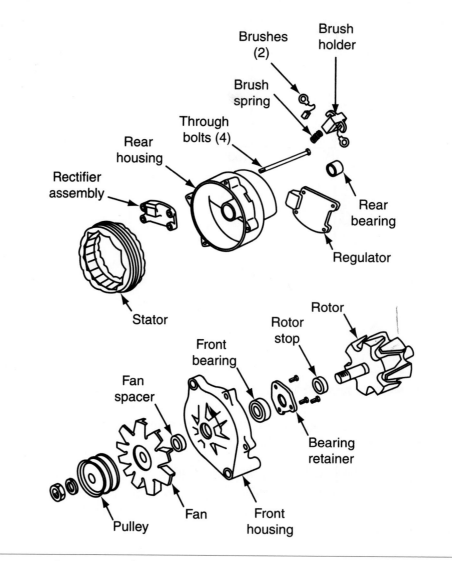

Figure 7-5 Components of an AC generator.

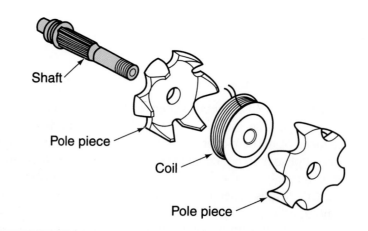

Figure 7-6 Components of a typical AC generator rotor.

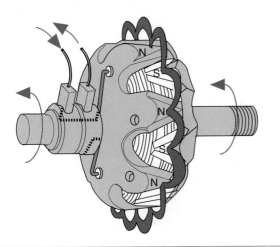

Figure 7-7 The north and south poles of a rotor's field alternate.

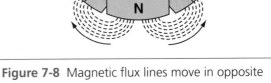

Figure 7-8 Magnetic flux lines move in opposite directions between the rotor poles.

assembled, the poles alternate from north to south around the rotor (Figure 7-7). As a result of this alternating arrangement of poles, the magnetic flux lines will move in opposite directions between adjacent poles (Figure 7-8). This arrangement provides for several alternating magnetic fields to intersect the stator as the rotor is turning. These individual magnetic fields produce a voltage by induction in the stationary stator windings.

The wires from the rotor coil are attached to two **slip rings** that are insulated from the rotor shaft. The slip rings function much like the armature commutator in the starter motor, except they are smooth. The insulated stationary carbon brush passes field current into a slip ring, then through the field coil, and back to the other slip ring. Current then passes through a grounded stationary brush (Figure 7-9) or to a voltage regulator.

Most rotors have twelve to fourteen poles.

Brushes

The field winding of the rotor receives current through a pair of brushes that ride against the slip rings. The brushes and slip rings provide a means of maintaining electrical continuity between stationary and rotating components. The brushes (Figure 7-10) ride the surface of the slip rings on the rotor and are held tight against the slip rings by spring tension provided by the brush holders. The brushes conduct only the field current (2 to 5 amperes). The low current that the brushes must carry contributes to their longer life.

Shop Manual
Chapter 7, page 271

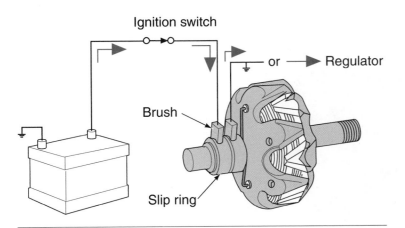

Figure 7-9 The slip rings and brushes provide a current path to the rotor coil.

Figure 7-10 Brushes are the stationary electrical contact to the rotor's slip rings.

Direct current from the battery is supplied to the rotating field through the field terminal and the insulated brush. The second brush may be the ground brush, which is attached to the AC generator housing or to a voltage regulator.

Stators

Shop Manual
Chapter 7, pages
263, 272

The **stator** contains three main sets of windings wrapped in slots around a laminated, circular iron frame (Figure 7-11). The stator is the stationary coil in which electricity is produced. Each of the three windings has the same number of coils as the rotor has pairs of north and south poles. The coils of each winding are evenly spaced around the core. The three sets of windings alternate and overlap as they pass through the core (Figure 7-12). The overlapping is needed to produce the required phase angles.

The rotor is fitted inside the stator (Figure 7-13). A small air gap (approximately 0.015 inch) is maintained between the rotor and the stator. This gap allows the rotor's magnetic field to energize all of the windings of the stator at the same time and to maximize the magnetic force.

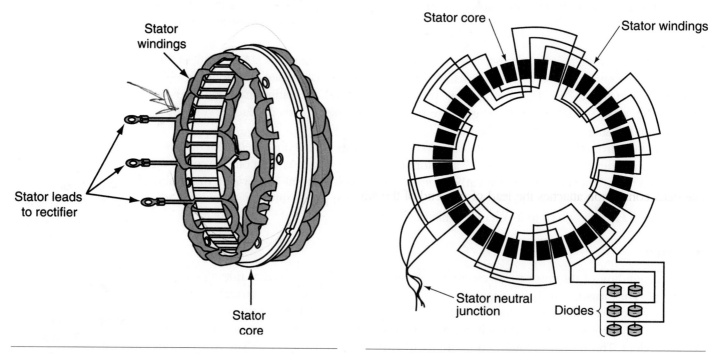

Figure 7-11 Components of a typical stator.

Figure 7-12 Overlapping stator windings produce the required phase angles.

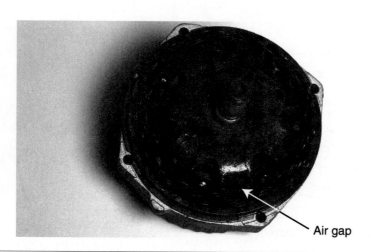

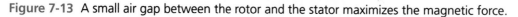

Figure 7-13 A small air gap between the rotor and the stator maximizes the magnetic force.

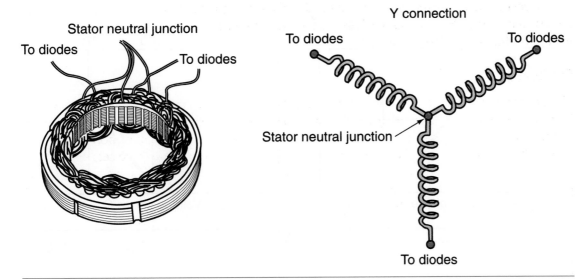

Figure 7-14 Wye-connected stator winding.

Each group of windings has two leads. The first lead is for the current entering the winding. The second lead is for current leaving. There are two basic means of connecting the leads. The first method is the **wye wound connection** (Figure 7-14). In the wye connection, one lead from each winding is connected to one common junction. From this junction, the other leads branch out in a Y pattern. A wye wound AC generator is usually found in applications that do not require high amperage output.

The second method of connecting the windings is called the **delta connection** (Figure 7-15). The delta connection attaches the lead of one end of the winding to the lead at the other end of the next winding. The delta connection is commonly used in applications that require high amperage output.

In a wye wound or delta wound stator winding, each group of windings occupies one third of the stator, or 120 degrees of the circle. As the rotor revolves in the stator, a voltage is produced in each loop of the stator at different phase angles. The resulting overlap of sine waves that is

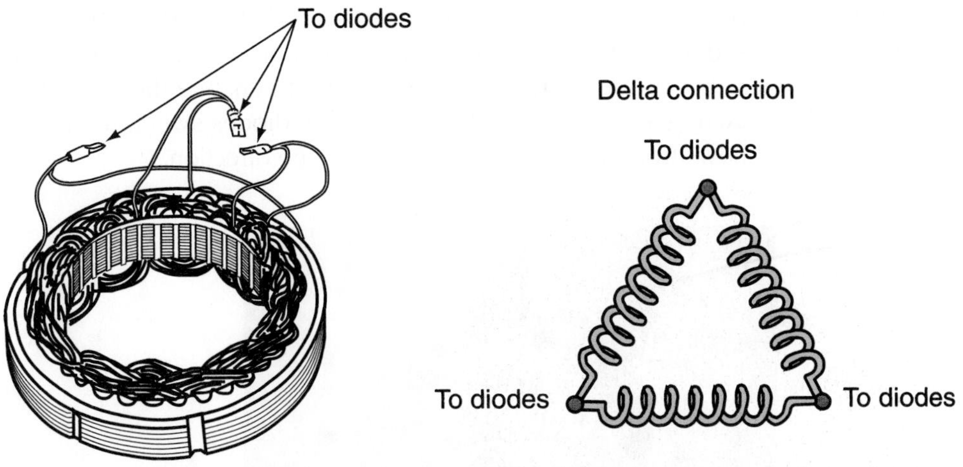

The common junction in the wye connected winding is called the stator neutral junction.

Figure 7-15 Delta-connected stator winding.

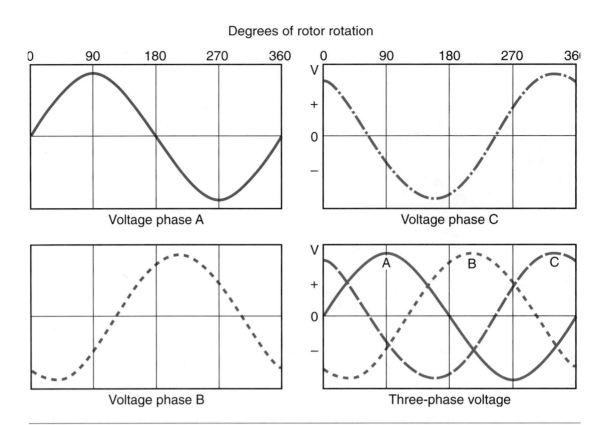

Voltage phase A

Voltage phase C

Voltage phase B

Three-phase voltage

Figure 7-16 The voltage produced in each stator winding is added together to create a three-phase voltage.

Shop Manual
Chapter 7, pages
263, 273

The rectifier bridge is also known as a rectifier stack.

produced is shown (Figure 7-16). Each of the sine waves is at a different phase of its cycle at any given time. As a result, the output from the stator is divided into three phases.

Diode Rectifier Bridge

The battery and the electrical system cannot accept or store AC voltage. For the vehicle's electrical system to be able to use the voltage and current generated in the AC generator, the AC current needs to be converted to DC current. This process is called **rectification.** A split-ring commutator cannot be used to rectify AC current to DC current because the stator is stationary in an AC generator. Instead, a **diode rectifier bridge** is used to change the current in an AC generator (Figure 7-17). Acting as a one-way check valve, the diodes switch the current flow back and forth so that it flows from the AC generator in only one direction.

Figure 7-17 General Motors' rectifier bridge.

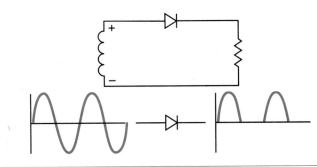

Figure 7-18 AC current rectified to a pulsating DC current after passing through a positive-biased diode. This is called half-wave rectification.

When AC current reverses itself, the diode blocks and no current flows. If AC current passes through a positively biased diode, the diode will block off the negative pulse. The result is the scope pattern shown in Figure 7-18. The AC current has been changed to a pulsing DC current. This process is called **half-wave rectification.**

An AC generator usually uses a pair of diodes for each stator winding, for a total of six diodes (Figure 7-19). Three of the diodes are positive biased and are mounted in a **heat sink** to dissipate the heat (Figure 7-20). The three remaining diodes are negative biased and are attached directly to the frame of the AC generator (Figure 7-21). By using a pair of diodes that are reverse-biased to each

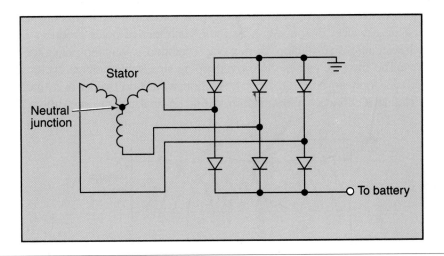

Figure 7-19 A simplified schematic of the AC generator windings connected to the diode rectifier bridge.

Figure 7-20 The positive-biased diodes are mounted into a heat sink to provide protection.

Figure 7-21 Negative-biased diodes pressed into the AC generator housing.

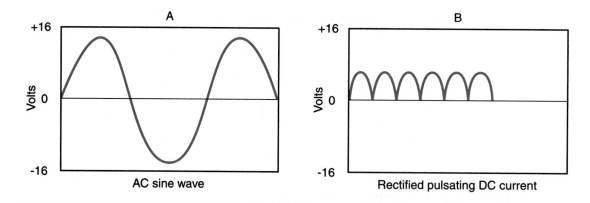

Figure 7-22 Full-wave rectification uses both sides of the AC sine wave to create a pulsating DC current.

other, rectification of both sides of the AC sine wave is achieved (Figure 7-22). The process of converting both sides of the sine wave to a DC voltage is called **full-wave rectification.** The negative-biased diodes allow for conducting current from the negative side of the AC sine wave and putting this current into the circuit. Diode rectification changes the negative current into positive output.

With each stator winding connected to a pair of diodes, the resultant waveform of the rectified voltage would be similar to that shown (Figure 7-23). With six peaks per revolution, the voltage will vary only slightly during each cycle.

The examples used so far have been for single-pole rotors in a three-winding stator. Most AC generators use either a twelve- or fourteen-pole rotor. Each pair of poles produces one complete sine wave in each winding per revolution. During one revolution, a fourteen-pole rotor will produce seven sine waves. The rotor generates three overlapping sine wave voltage cycles in the stator. The total output of a fourteen-pole rotor per revolution would be twenty-one sine wave cycles (Figure 7-24). With final rectification, the waveform would be similar to the one shown (Figure 7-25).

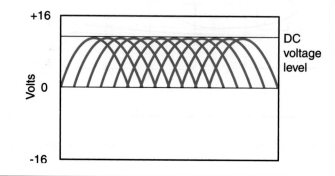

Figure 7-23 With three-phase rectification, the DC voltage level is uniform.

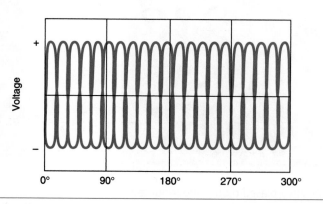

Figure 7-24 Sine wave cycle of a fourteen-pole rotor and three-phase stator.

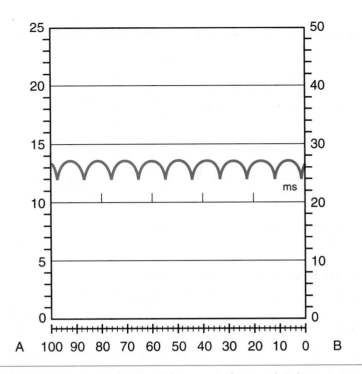

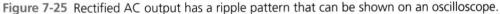

Figure 7-25 Rectified AC output has a ripple pattern that can be shown on an oscilloscope.

Full-wave rectification is desired because using only half-wave rectification wastes the other half of the AC current. Full-wave rectification of the stator output uses the total potential by redirecting the current from the stator windings so that all current is in one direction.

A wye wound stator with each winding connected to a pair of diodes is shown (Figure 7-26). Each pair of diodes has one negative and one positive diode. During rotor movement, two stator windings will be in series and the third winding will be neutral. As the rotor revolves, it will energize a different set of windings. Also, current flow through the windings is reversed as the rotor passes. Current in any direction through two windings in series will produce DC current.

The action that occurs when the delta wound stator is used is shown (Figure 7-27). Instead of two windings in series, the three windings of the delta stator are in parallel. This makes more current available because the parallel paths allow more current to flow through the diodes. Since the three outputs of the delta winding are in parallel, current flows from each winding continuously.

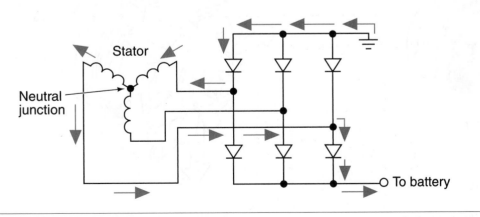

Figure 7-26 Current flow through a wye wound stator.

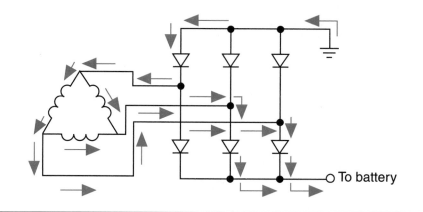

Figure 7-27 Current flow through a delta wound stator.

 AUTHOR'S NOTE: Not only do the diodes rectify stator output, but they also block battery drain back when the engine is not running.

AC Generator Housing and Cooling Fan

Shop Manual
Chapter 7, page 271

Most AC generator housings are a two-piece construction, made from cast aluminum (Figure 7-28). The two end frames provide support of the rotor and the stator. In addition, the end frames contain the diodes, regulator, heat sinks, terminals, and other components of the AC generator. The two end pieces are referred to as:

1. The drive end housing: This housing holds a bearing to support the front of the rotor shaft. The rotor shaft extends through the drive end housing and holds the drive pulley and cooling fan.

2. The slip ring end housing: This housing also holds a rotor shaft that supports a bearing. In addition, it contains the brushes and has all of the electrical terminals. If the AC generator has an integral regulator, it is also contained in this housing.

The cooling fan draws air into the housing through the openings at the rear of the housing. The air leaves through openings behind the cooling fan (Figure 7-29).

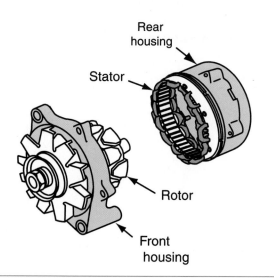

Figure 7-28 Typical two-piece AC generator housing.

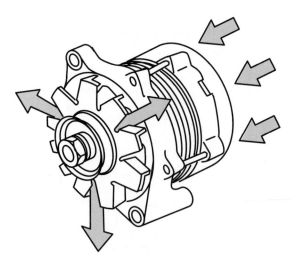

Figure 7-29 The cooling fan draws air in from the rear of the AC generator to keep the diodes cool.

AC Generator Circuits

There are three principal circuits used in an AC generator:

1. The charging circuit: Consists of the stator windings and rectifier circuits.
2. The excitation circuit: Consists of the rotor field coil and the electrical connections to the coil.
3. The preexcitation circuit: Supplies the initial current for the field coil that starts the buildup of the magnetic field.

For the AC generator to produce current, the field coil must develop a magnetic field. The AC generator creates its own field current in addition to its output current.

For excitation of the field to occur, the voltage induced in the stator rises to a point that it overcomes the forward voltage drop of at least two of the rectifier diodes. Before the **diode trio** can supply field current, the anode side of the diode must be at least 0.6 volt more positive than the cathode side (Figure 7-30). When the ignition switch is turned on, the warning lamp current acts as a small magnetizing current through the field (Figure 7-31). This current preexcites the field, reducing the speed required to start its own supply of field current.

Shop Manual
Chapter 7, page 274

The **diode trio** is used by some manufacturers to rectify current from the stator so that it can be used to create the magnetic field in the field coil of the rotor. This eliminates extra wiring.

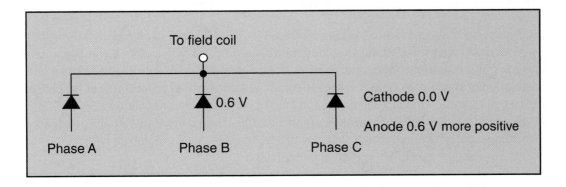

Figure 7-30 The diode trio connects the phase windings to the field. To conduct, there must be 0.6 V more positive on the anode side of the diodes.

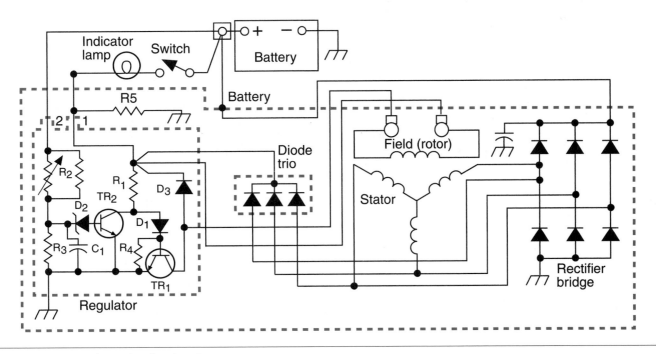

Figure 7-31 Schematic of a charging system.

AUTHOR'S NOTE: If the battery is completely discharged, the vehicle cannot be push started because there is no excitation of the field coil.

AC Generator Operation Overview

When the engine is running, the drive belt spins the rotor inside the stator windings. This magnetic field inside the rotor generates a voltage in the windings of the stator. Field current flowing through the slip rings to the rotor creates alternating north and south poles on the rotor.

The induced voltage in the stator is an alternating voltage because the magnetic fields are alternating. As the magnetic field begins to induce voltage in the stator's windings, the induced voltage starts to increase. The amount of voltage will peak when the magnetic field is the strongest. As the magnetic field begins to move away from the stator windings, the amount of voltage will start to decrease. Each of the three windings of the stator generates voltage, so the three combine to form a three-phase voltage output.

In the wye connection (Figure 7-32), output terminals (A, B, and C) apply voltage to the rectifier. Because only two stator windings apply voltage (because the third winding is always connected to diodes that are reverse-biased), the voltages come from points A to B, B to C, and C to A.

To determine the amount of voltage produced in the two stator windings, find the difference between the two points. For example, to find the voltage applied from points A and B, subtract the voltage at point B from the voltage at point A. If the voltage at point A is 8 volts positive and the voltage at point B is 8 volts negative, the difference is 16 volts. This procedure can be performed for each pair of stator windings at any point in time to get the sine wave patterns (Figure 7-33). The voltages in the windings are designated as Va, Vb, and Vc. Designations of Vab, Vbc, and Vca refer to the voltage difference in the two stator windings. In addition, the numbers refer to the diodes used for the voltages generated in each winding pair.

 AUTHOR'S NOTE: Alternating current is constantly changing, so this formula would have to be performed at several different times.

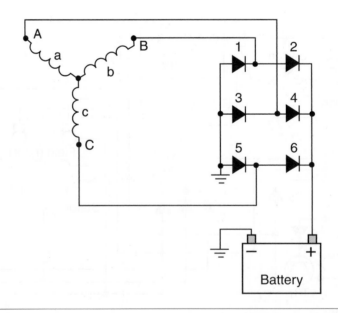

Figure 7-32 Wye-connected stator windings attached to the rectifier bridge.

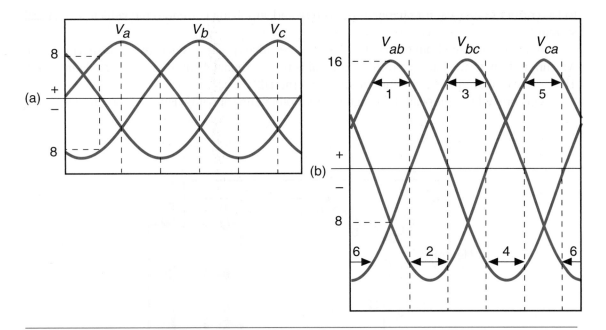

Figure 7-33 (A) Individual stator winding voltages; (B) voltages across the stator terminals A, B, and C.

The current induced in the stator passes through the diode rectifier bridge, consisting of three positive and three negative diodes. At this point, there are six possible paths for the current to follow. The path that is followed depends on the stator terminal voltages. If the voltage from points A and B is positive (point A is positive in respect to point B), current is supplied to the positive terminal of the battery from terminal A through diode 2 (Figure 7-34). The negative return path is through diode 3 to terminal B.

Both diodes 2 and 3 are forward-biased. The stator winding labeled C does not produce current because it is connected to diodes that are reverse-biased. The stator current is rectified

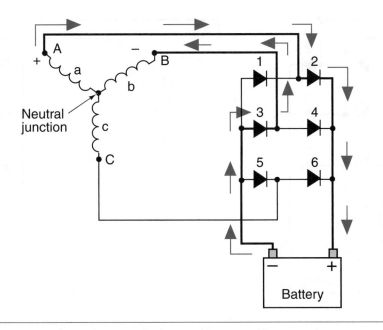

Figure 7-34 Current flow when terminals A and B are positive.

to DC current to be used for charging the battery and supplying current to the vehicle's electrical system.

When the voltage from terminals C and A is negative (point C is negative in respect to point A), current flow to the battery positive terminal is from terminal A through diode 2 (Figure 7-35). The negative return path is through diode 5 to terminal C.

This procedure is repeated through the four other current paths (Figures 7-36 through 7-39).

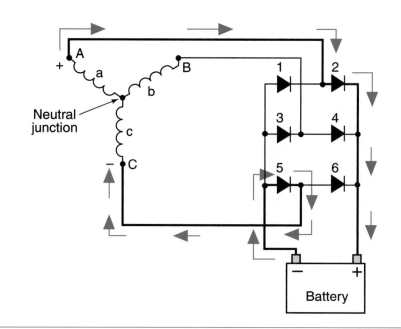

Figure 7-35 Current flow when terminals A and C are negative.

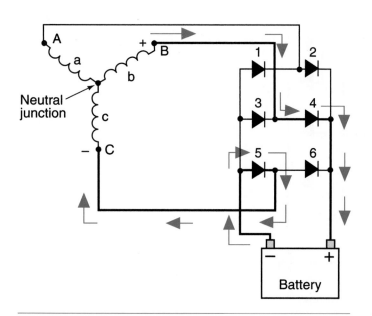

Figure 7-36 Current flow when terminals B and C are positive.

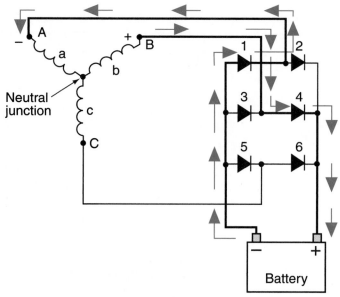

Figure 7-37 Current flow when terminals A and B are negative.

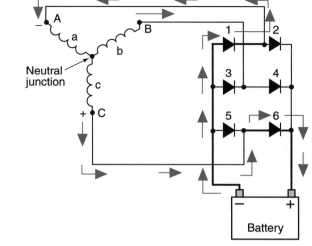

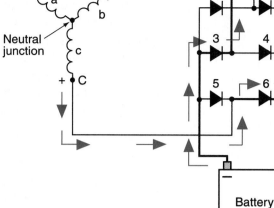

Figure 7-38 Current flow when terminals A and C are positive.

Figure 7-39 Current flow when terminals B and C are negative.

Regulation

The battery, and the rest of the electrical system, must be protected from excessive voltages. To prevent early battery and electrical system failure, regulation of the charging system voltage is very important. Also, the charging system must supply enough current to run the vehicle's electrical accessories when the engine is running.

AC generators do not require current limiters; because of their design, they limit their own current output. Current limit is the result of the constantly changing magnetic field because of the induced AC current. As the magnetic field changes, an opposing current is induced in the stator windings. This **inductive reactance** in the AC generator limits the maximum current that the AC generator can produce. Even though current (amperage) is limited by its operation, voltage is not. The AC generator is capable of producing as high as 250 volts, if it were not controlled.

Regulation of voltage is done by varying the amount of field current flowing through the rotor. The higher the field current, the higher the output voltage. Control of field current can be done either by regulating the resistance in series with the field coil or by turning the field circuit on and off (Figure 7-40). By controlling the amount of current in the field coil, control of the field

Shop Manual
Chapter 7, page 262

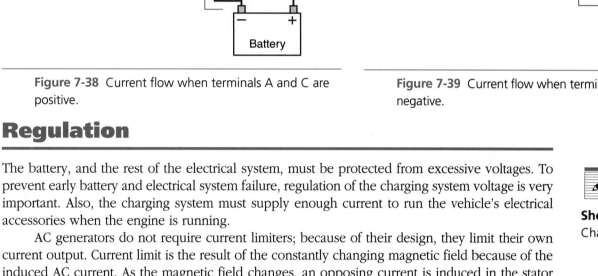

(a)

(b)

Figure 7-40 The regulator can control the field current by (A) controlling the resistance in series with the coil or (B) by switching the field on and off.

current and the AC generator output is obtained. To ensure a full battery charge, and operation of accessories, most voltage regulators are set for a system voltage between 13.5 and 14.5 volts.

The regulator must have system voltage as an input in order to regulate the output voltage. The input voltage to the AC generator is called **sensing voltage.** If sensing voltage is below the regulator setting, an increase in charging voltage output results by increasing field current. Higher sensing voltage will result in a decrease in field current and voltage output. A vehicle being driven with no accessories on and a fully charged battery will have a high sensing voltage. The regulator will reduce the charging voltage until it is at a level to run the ignition system while trickle charging the battery. If a heavy load is turned on (such as the headlights), the additional draw will cause a drop in the battery voltage. The regulator will sense this low system voltage and will increase current to the rotor. This will allow more current to the field windings. With the increase of field current, the magnetic field is stronger and AC generator voltage output is increased. When the load is turned off, the regulator senses the rise in system voltage and cuts back the amount of field current and ultimately AC generator voltage output.

Another input that affects regulation is temperature. Because ambient temperatures influence the rate of charge that a battery can accept, regulators are temperature compensated (Figure 7-41). Temperature compensation is required because the battery is more reluctant to accept a charge at lower ambient temperatures. The regulator will increase the system voltage until it is at a higher level so the battery will accept it.

Shop Manual
Chapter 7, page 254

The A circuit is called an external grounded field circuit.

Usually the B circuit regulator is mounted externally of the AC generator. The B circuit is an internally grounded circuit.

Field Circuits

To properly test and service the charging system, it is important to identify the field circuit being used. Automobile manufacturers use three basic types of field circuits. The first type is called the A circuit. It has the regulator on the ground side of the field coil. The B+ for the field coil is picked up from inside the AC generator (Figure 7-42). By placing the regulator on the ground side of the field coil, the regulator will allow the control of field current by varying the current flow to ground.

	Volts		
Temperature	Minimum	Maximum
20° F	14.3	15.3
80° F	13.8	14.4
140° F	13.3	14.0
Over 140° F	Less than 13.3	–

Figure 7-41 Chart indicating relationship between temperature and charge rate.

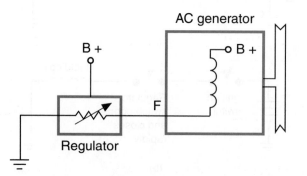

Figure 7-42 Simplified diagram of an A circuit field.

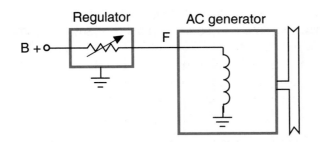

Figure 7-43 Simplified diagram of a B circuit field.

The second type of field circuit is called the B circuit. In this case, the voltage regulator controls the power side of the field circuit. Also, the field coil is grounded from inside the AC generator (Figure 7-43).

 AUTHOR'S NOTE: To remember these circuits: Think of "A" for "After" the field and "B" for "Before" the field.

The third type of field circuit is called the isolated field. The AC generator has two field wires attached to the outside of the case. The voltage regulator can be located on either the ground (A circuit) or on the B+ (B circuit) side (Figure 7-44).

Regardless of which type is used, the field circuit is designed to control the amount of voltage output by controlling the amount of current through the field windings. The relationship between the field current, rotor speed, and regulated voltage is illustrated in Figure 7-45. As rotor speed increases, field current must be decreased to maintain regulated voltage.

Electronic Regulators

The **electronic regulator** uses solid-state circuitry to perform the regulatory functions. Electronic regulators can be mounted either externally or internally of the AC generator. There are no moving parts, so it can cycle between 10 and 7,000 times per second. This quick cycling provides more accurate control of the field current through the rotor.

Pulse width modulation controls AC generator output by varying the amount of time the field coil is energized. For example, assume that a vehicle is equipped with a 100-ampere generator. If the electrical demand placed on the charging system requires 50 amperes of current, the

Isolated field
AC generators pick up B+ and ground externally.

Shop Manual
Chapter 7, page 262

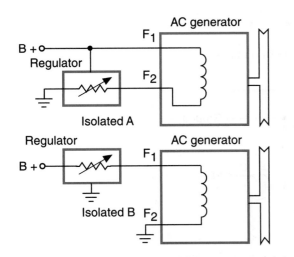

Figure 7-44 In the isolated circuit field AC generator, the regulator can be installed on either side of the field.

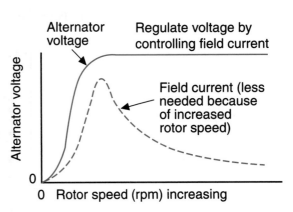

Figure 7-45 Graph showing the relationship between field current, rotor speed, and regulated voltage changes depending on electrical load.

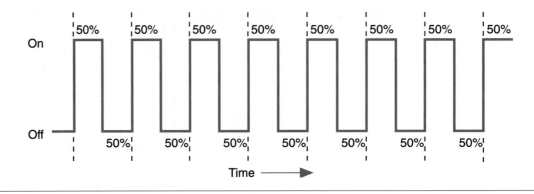

Figure 7-46 Pulse width modulation with 50% on time.

regulator would energize the field coil for 50% of the time (Figure 7-46). If the electrical system's demand was increased to 75 amperes, the regulator would energize the field coil 75% of the cycle time (Figure 7-47).

The electronic regulator uses a zener diode that blocks current flow until a specific voltage is obtained, at which point it allows the current to flow. An electronic regulator is shown (Figure 7-48).

Battery voltage is applied to the anode side of the zener diode as well as to the base of transistor number 1. No current will flow through the zener diode, since battery voltage is too low to push through the zener. However, as the AC generator produces voltage, the voltage at the anode will increase until it reaches the upper limit (14.5 volts) and is able to push through the zener diode. Current will now flow from the battery, through the resistor (R_1) to the anode, through the zener diode, through the resistor (R_2) and thermistor in parallel, and to ground. Since current is flowing, each resistance in the circuit will drop voltage. As a result, voltage to the base of transistor number

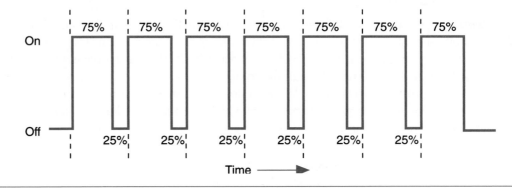

Figure 7-47 Pulse width modulation with 75% on time.

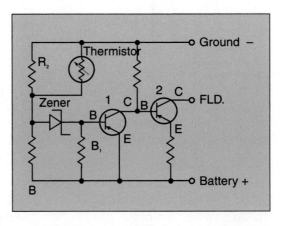

Figure 7-48 A simplified circuit diagram of an electronic regulator utilizing a zener diode.

1 will be less than the voltage applied to the emitter. Since transistor number 1 is a PNP transistor and the base voltage is less than the emitter voltage, transistor number 1 is turned on. The base of transistor number 2 will now have battery voltage applied to it. Since the voltage applied to the base of transistor number 2 is greater than that applied to its emitter, transistor number 2 is turned off. Transistor number 2 is in control of the field current and generator output.

The thermistor changes circuit resistance according to temperature. This provides for the temperature-related voltage change necessary to keep the battery charged in cold-weather conditions.

Many manufacturers are installing the voltage regulator inside the AC generator. This eliminates some of the wiring needed for external regulators. The diode trio rectifies AC current from the stator to DC current that is applied to the field windings (Figure 7-49).

Current flow with the engine off and the ignition switch in the RUN position is illustrated (Figure 7-50). Battery voltage is applied to the field through the common point above R_1. TR_1

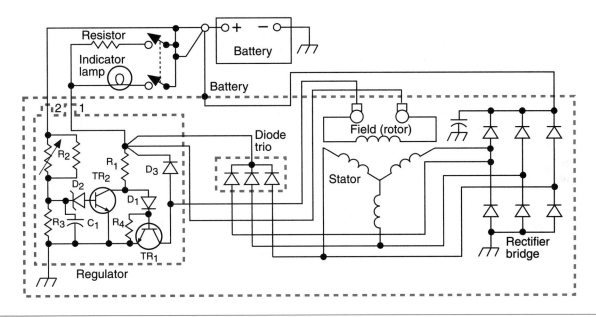

Figure 7-49 AC generator circuit diagram with internal regulator. This system uses a diode trio to rectify stator current to be applied to the field coil. The resistor above the indicator lamp is used to ensure current will flow through terminal 1 if the lamp burns out.

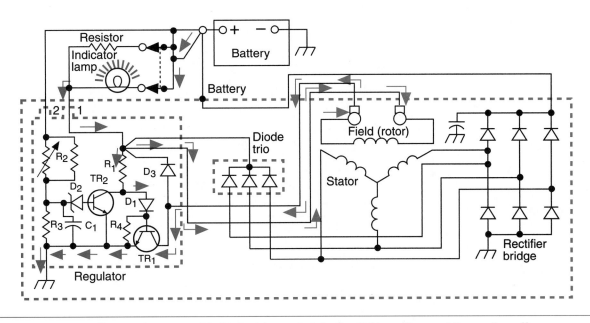

Figure 7-50 Current flow to the rotor with the ignition switch in the RUN position and the engine off.

conducts the field current coming from the field coil, producing a weak magnetic field. The indicator lamp lights because TR_1 directs current to ground and completes the lamp circuit.

Current flow with the engine running is illustrated (Figure 7-51). When the AC generator starts to produce voltage, the diode trio will conduct and battery voltage is available for the field and terminal 1 at the common connection. Placing voltage on both sides of the lamp gives the same voltage potential at each side; therefore, current doesn't flow and the lamp goes out.

Current flow as the voltage output is being regulated is illustrated (Figure 7-52). The sensing circuit from terminal 2 passes through a thermistor to the zener diode (D2). When the system voltage reaches the upper voltage limit of the zener diode, the zener diode conducts current

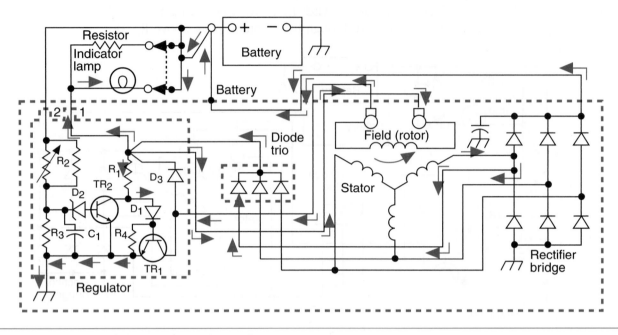

Figure 7-51 Current flow with the engine running and AC generator producing voltage.

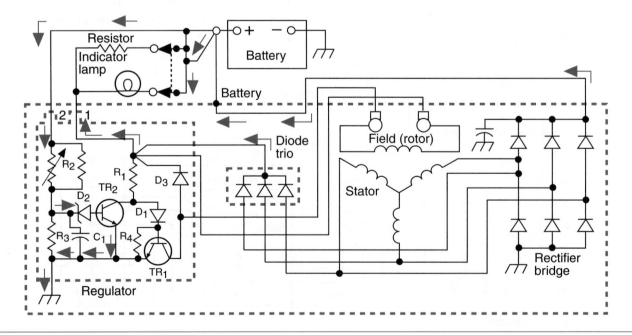

Figure 7-52 When the system voltage is high enough to allow the zener diode to conduct, TR_2 is turned on and TR_1 is shut off, which opens the field circuit.

to TR_2. When TR_2 is biased, it opens the field coil circuit and current stops flowing through the field coil. Regulation of this switching on and off is based on the sensing voltage received through terminal 2. With the circuit to the field coil opened, the sensing voltage decreases and the zener diode stops conducting. TR_2 is turned off and the circuit for the field coil is closed.

Computer-Controlled Regulation

On many vehicles after the mid-1980s, the regulator function has been incorporated into the powertrain control module (PCM). The operation is similar to the internal electronic regulator (Figure 7-53). Regulation of the field circuit is through the ground (A circuit).

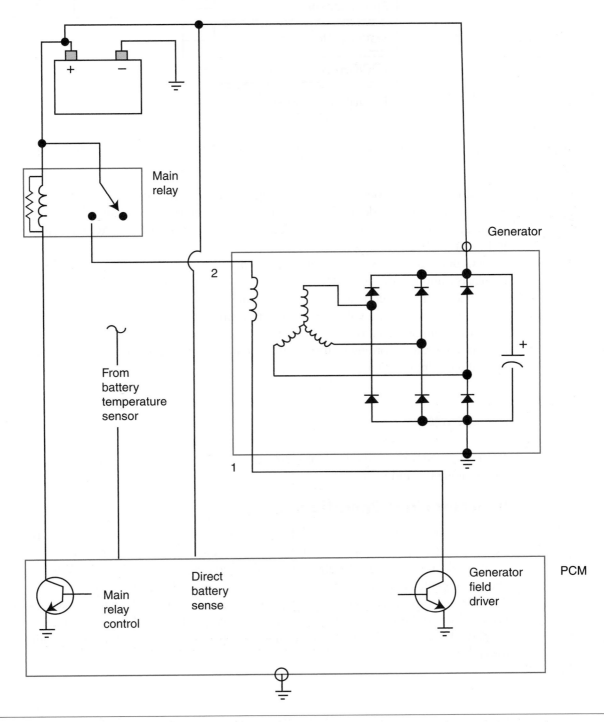

Figure 7-53 Computer-controlled voltage regulator circuit.

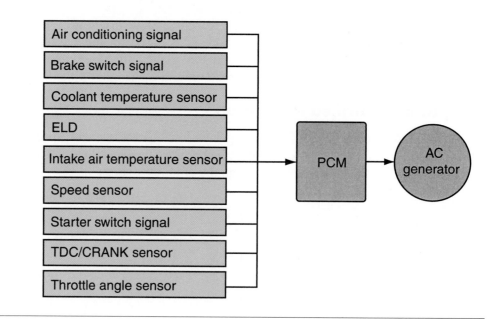

Figure 7-54 The PCM will use various inputs to regulate AC generator output.

The PCM's decisions, concerning voltage regulation, are based on battery voltage and battery temperature. When the desired output voltage is obtained (based on battery temperature), the PCM switches the transistor on or off as needed. This transistor grounds the AC generator's field to control output voltage.

General Motors' CS series generators may be connected directly to the PCM through terminals L and F at the generator. The voltage regulator portion of the PCM switches the field current on and off at a frequency of about 400 times per second. Varying the on and off time of the field current controls the voltage output of the generator.

The computer-controlled regulation system has the ability to precisely maintain and control the changing rate according to the electrical requirements, battery (or ambient) temperature, and several other inputs (Figure 7-54).

Charging Indicators

There are three basic methods of informing the driver of the charging system's condition: indicator lamps, ammeter, and voltmeter.

Indicator Light Operation

As discussed earlier, most indicator lamps operate on the basis of opposing voltages. If the AC generator output is less than battery voltage, there is an electrical potential difference in the lamp circuit and the lamp will light. If there is no stator output through the diode trio, then the lamp circuit is completed to ground through the rotor field and TR_1 (Figure 7-55).

On most systems, the warning lamp will be "proofed" when the ignition switch is in the RUN position before the engine starts. This indicates that the bulb and indicator circuit are operating properly. Proofing the bulb is accomplished because there is no stator output without the rotor turning.

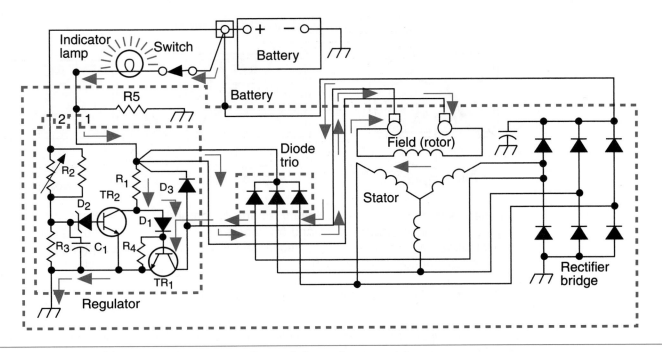

Figure 7-55 Electronic regulator with an indicator light on due to no AC generator output.

Ammeter Operation

In place of the indicator light, some manufacturers install an ammeter. The ammeter is wired in series between the AC generator and the battery (Figure 7-56). Most ammeters work on the principle of d'Arsonval movement.

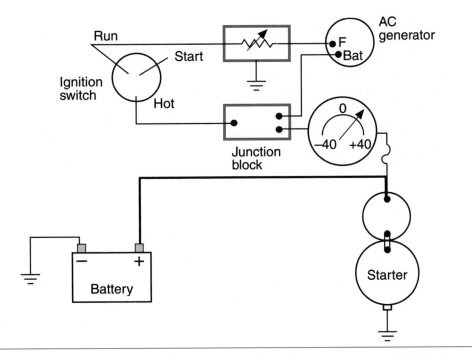

Figure 7-56 Ammeter connected in series to indicate charging system operation.

Ammeter conditions

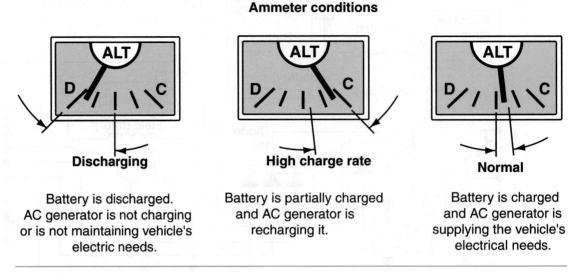

Discharging	High charge rate	Normal
Battery is discharged. AC generator is not charging or is not maintaining vehicle's electric needs.	Battery is partially charged and AC generator is recharging it.	Battery is charged and AC generator is supplying the vehicle's electrical needs.

Figure 7-57 Ammeter needle movement indicates charging conditions.

The movement of the ammeter needle under different charging conditions is illustrated (Figure 7-57). If the charging system is operating properly, the ammeter needle will remain within the normal range. If the charging system is not generating sufficient current, the needle will swing toward the discharge side of the gauge. When the charging system is recharging the battery, or is called on to supply high amounts of current, the needle deflects toward the charge side of the gauge.

It is normal for the gauge to read a high amount of current after initial engine startup. As the battery is recharged, the needle should move more toward the normal range.

Voltmeter Operation

Because the ammeter is a complicated gauge for most people to understand, many manufacturers use a voltmeter to indicate charging system operation. In early systems, the voltmeter is connected between the battery positive and negative terminals (Figure 7-58).

When the engine is started, it is normal for the voltmeter to indicate a reading between 13.2 and 15.2 volts. If the voltmeter indicates a voltage level that is below 13.2, it may mean that the

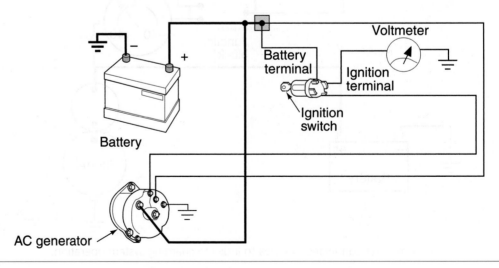

Figure 7-58 Voltmeter connected to the charging circuit to monitor operation.

battery is discharging. If the voltmeter indicates a voltage reading that is above 15.2 volts, the charging system is overcharging the battery. The battery and electrical circuits can be damaged as a result of higher-than-normal charging system output.

In most modern systems, the voltmeter is controlled either directly by the PCM or by information sent to the instrument cluster by the PCM. A dedicated circuit from the battery to the PCM allows the PCM to constantly monitor the battery voltage.

AC Generator Design Differences

Although all AC generators operate on the same principles, there are differences in the styles and construction.

General Motors 10SI Series

The 10SI series AC generator uses an internal voltage regulator that is mounted to the inside of the slip ring end frame (Figure 7-59). There are three terminals on the rear-end frame of the AC generators:

Shop Manual
Chapter 7, page 258

❏ **Terminal number 1:** Connects to the field through one brush and slip ring and to the output of the diode trio. In addition, this terminal is connected to a portion of the regulator and warning light circuitry.

❏ **Terminal number 2:** Connects to the regulator to supply battery voltage to a portion of the regulator circuitry that senses system voltage.

❏ **BAT terminal:** Connects to the output of the stator windings and supplies the battery with charging voltage.

General Motors CS Series

Beginning in 1986 and continuing through the 1999 model year, General Motors used the smaller CS series AC generator with an internal regulator. This generator uses a delta wound stator. The

Shop Manual
Chapter 7, page 259

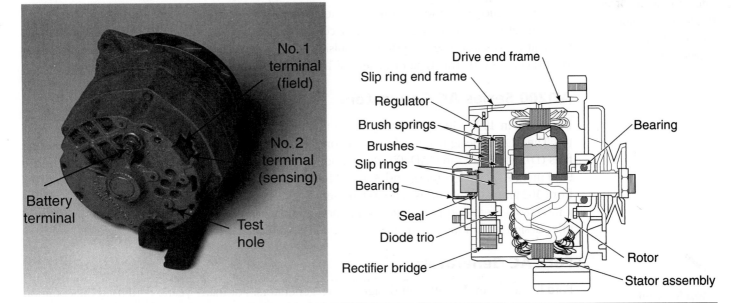

Figure 7-59 10SI AC generator.

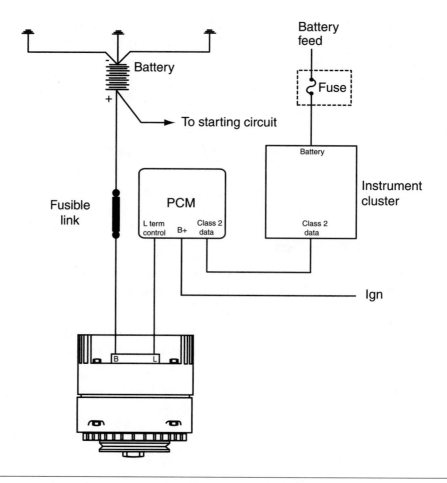

Figure 7-60 General Motors PCM-controlled charging system using high side pulse width control.

The series designation number refers to the diameter of the outer stator lamination in millimeters.

field current is supplied directly from the stator, thus eliminating the need for a diode trio. The generators in this series include the CS-121, CS-130, and CS-144, which represent the unit size in millimeters.

As mentioned earlier, recent CS series generators use computer control regulation of the AC generator. In addition to regulation control by varying the ground of the field windings, General Motors also uses a system of pulsing the voltage output to the field windings from the PCM (Figure 7-60). This type of generator has a constant field winding ground connection.

AD200 Series AC Generators

Beginning in 1999, General Motors began to change to a Delphi-designed AD200 series generator. The AD200 designation refers to second-generation (200), air-cooled (A) and dual internal fans (D). There are three AD200 series models being used, depending on unit diameters: AD230 (130 mm), AD237 (137 mm), and AD244 (144 mm). Amperage output of these alternators ranges from 102 amps to 150 amps. The AD200 series generator uses an offset-wound stator to achieve a more consistent output voltage. Some models also use a pulley with a built-in clutch. The rectifier design has an increased surface area and uses avalanche diodes.

Ford AC Generators

Ford has used several different designs of AC generators. For many years, Ford used the common rear- or side-terminal AC generator. The rear-terminal AC generators used two different types of rectifiers. One rectifier had a single plate that contained all six of the diodes (Figure 7-61).

Shop Manual
Chapter 7, pages 255, 259

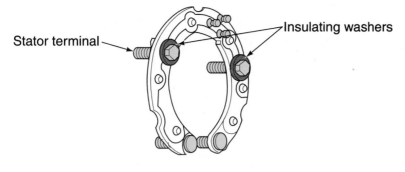

Rectifier with built-in diodes

Figure 7-61 Flat-type rectifier has single plate containing all six diodes.

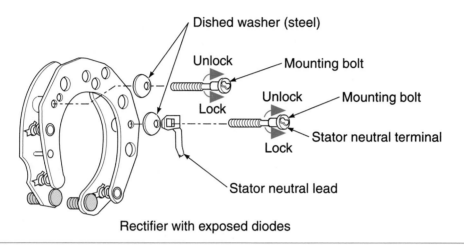

Rectifier with exposed diodes

Figure 7-62 Stacked-type rectifier with two plates.

The second type of rectifier used two plates that were stacked on top of each other (Figure 7-62). One plate contained the positive diodes and the other contained the negative diodes.

On the rear-terminal AC generators there were four terminals on the end frame:

❏ **BAT terminal:** Stator output connection to the battery.

❏ **FLD terminal:** Connection to one side of the field winding through the insulated brush and the slip ring.

❏ **STA terminal:** Connection to the neutral stator junction.

❏ **GRD terminal:** Connection for the ground wire from the regulator.

The Ford side terminal AC generator is larger and has higher output capacities. The same four terminals are used; however, they are arranged differently.

In 1984, Ford introduced an integral alternator regulator (IAR) AC generator. The regulator is mounted on the exterior of the rear-end frame, which simplifies testing and replacement of the regulator. The F and A terminals are used to test the charging system. Additional modifications include the brushes being a part of the regulator.

When the ignition switch is in the RUN position, voltage is sent to the I terminal of the regulator (Figure 7-63). Regulator terminal A senses system voltage. Field current is drawn through this terminal also.

Ford's **EVR** charging system uses a solid-state external regulator. This style is built with either rear or side terminals. The side-terminal design provides higher output by using delta wound stators.

The single-plate rectifier is called a flat rectifier.

The two-plate rectifier is called a stacked rectifier.

EVR stands for external voltage regulator.

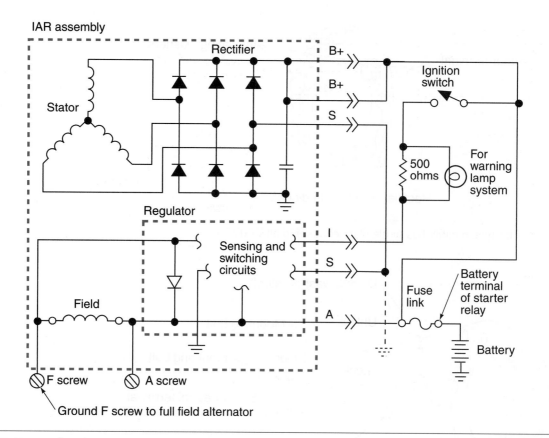

Figure 7-63 IAR charging system schematic.

DaimlerChrysler AC Generators

Early Chrysler AC generators used separate heat sinks for the positive and negative diodes. Both heat sinks are attached to the rear-end frame. Also, the brushes are attached to the exterior of the end frame. This allows for brush replacement without having to disassemble the AC generator (in fact, the brushes can usually be replaced without having to remove the AC generator from the vehicle).

The three terminals on the rear-end frame are connected as follows:

❏ **BAT terminal:** Connects the stator output to the battery to supply charging voltage.

❏ **FLD terminal:** There are two field terminals. Battery voltage is applied to one of the field terminals; the regulator connects to the second field terminal.

In 1985, Chrysler introduced a delta wound, dual-output, computer-controlled charging system (Figure 7-64). This system has some unique capabilities:

1. The system is capable of varying charging system output based on the ambient temperature and the system's voltage needs.

2. The computer monitors the charging system and is capable of self-diagnosis.

A BIT OF HISTORY

Chrysler equipped its vehicles with AC generators in the late 1950s, making it the first manufacturer to use an AC generator. Chrysler introduced the dual-output AC generator (40 or 90 amperes) with computer control in 1985.

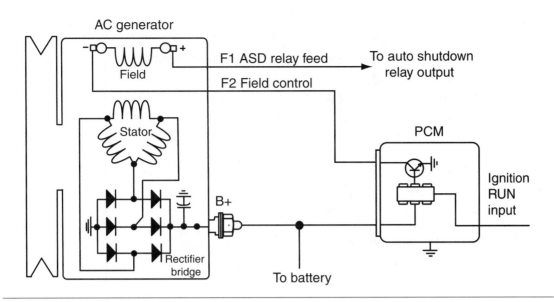

Figure 7-64 Computer-controlled, dual-output charging system.

When the ignition switch is in the RUN position, the PCM checks the ambient temperature and determines required field current. Based on inputs relating to temperature and system requirements, the PCM determines when current output adjustments are required.

In recent years, DaimlerChrysler used a Denso- or Melco-built AC generator that has a wye wound stator. This system also uses a PCM, to control voltage regulation by varying the field winding ground. Vehicles equipped with next-generation controllers (NGC) have high side control, similar to that of General Motors CS series discussed earlier.

Mitsubishi AC Generators

Some Mitsubishi AC generators use an internal voltage regulator (Figure 7-65). It also has two separate wye connected stator windings (Figure 7-66). Each of the stator windings has its own set of six diodes for rectification.

This AC generator also uses a diode trio to rectify stator voltage to be used in the field winding. The three terminals are connected as follows:

❏ **B terminal:** Connects the output of both stator windings to the battery, supplying charging voltage.

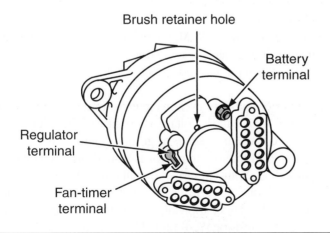

Figure 7-65 Mitsubishi AC generator terminals.

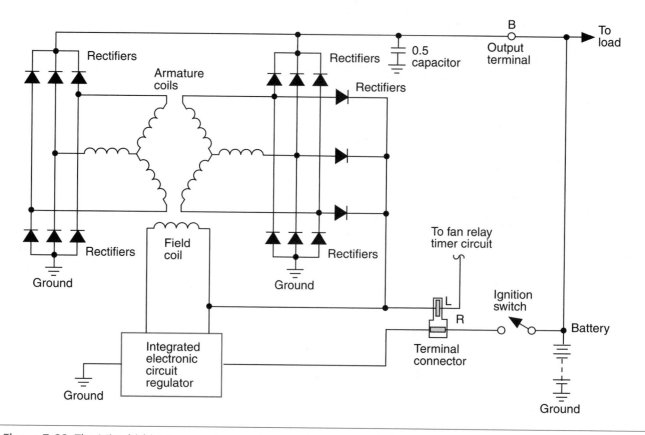

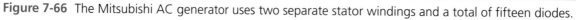

Figure 7-66 The Mitsubishi AC generator uses two separate stator windings and a total of fifteen diodes.

❑ **R terminal:** Supplies 12 volts to the regulator.

❑ **L terminal:** Connects to the output of the diode trio to provide rectified stator voltage to designated circuits.

Another method that Mitsubishi uses for charging control involves a single wye wound stator generator with an internal regulator that interfaces with the PCM (Figure 7-67). The PCM monitors the state of the field coil through terminal FR of the generator. The PCM sends 5 volts to the FR terminal. As the field coil is turned on and off by the internal regulator, the voltage will cycle between 5 and 0 volts. When the PCM senses 5 volts, it knows the field coil is turned off, and when the voltage drops close to 0 volts, it knows the field coil is turned on. This allows the PCM to control idle speed when the regulator is applying full field. In addition, the PCM can dampen the effects of full fielding during high electrical loads. This will prevent occurrences such as lights flicking bright and dim as the field coil is turned on and off.

When a full field condition is sensed by the PCM, it will modulate its internal transistor. This will then control the power transistor in the internal regulator. The PCM has a maximum authority of 14.4 volts. If the charging system output exceeds this level, the internal regulator turns the power transistor off.

To perform this function, the PCM will duty cycle its internal transistor, which turns the TR_1 in the generator on and off. In Figure 7-67, battery voltage is applied to the S terminal of the generator. This voltage goes through three resistors in series to ground (Figure 7-68). Each of the resistors has 2 ohms of resistance. If the PCM internal transistor is turned on, the voltage to the base of TR_1 is pulled low and TR_1 is turned off. With TR_1 turned off, all three of the resistors are involved in the circuit. Each resistor will drop 4 volts. Since R_1 drops 4 volts, 8 volts is applied to the zener diode. This is enough to blow through the diode, applying voltage to the base of TR_2 and turning it on. With TR_2 turned on, base voltage to transistor TR_3 is pulled low and transistor TR_3 will be turned off. With TR_3 turned off, the field coil is de-energized.

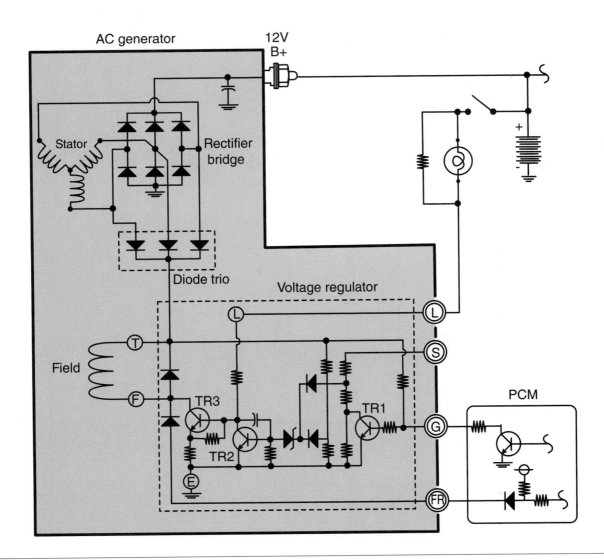

Figure 7-67 Schematic of Mitsubishi charging systems using both an internal voltage regulator and the PCM to control output.

If the internal transistor in the PCM is turned off, battery voltage will be applied to the base of TR_1 and turn it on. TR_1 will now supply an alternate path to ground, bypassing R_3. Now, only R_1 and R_2 are in series and each resistor will drop 6 volts. This means that 6 volts will be applied to the zener diode. This is not enough to blow through the zener; therefore, TR_2 will be off. With TR_2 off, battery voltage is applied to the base of transistor TR_3 and it will be turned on. Since transistor TR_3 is on, the field coil circuit to ground is complete and the coil is energized. The PCM internal transistor switches on and off several times a second to prevent the generator from going to full field too rapidly.

Brushless AC Generators

Some manufacturers have developed AC generators that do not require the use of brushes or slip rings. In these AC generators, the field winding and the stator winding are stationary (Figure 7-69). A screw terminal is used to make the electrical connection. The rotor contains the pole pieces and is fitted between the field winding and the stator winding.

The magnetic field is produced when current is applied to the field winding. The air gaps in the magnetic path contain a nonmetallic ring to divert the lines of force into the stator winding.

Brushless AC generators are normally used in heavy-duty trucks.

197

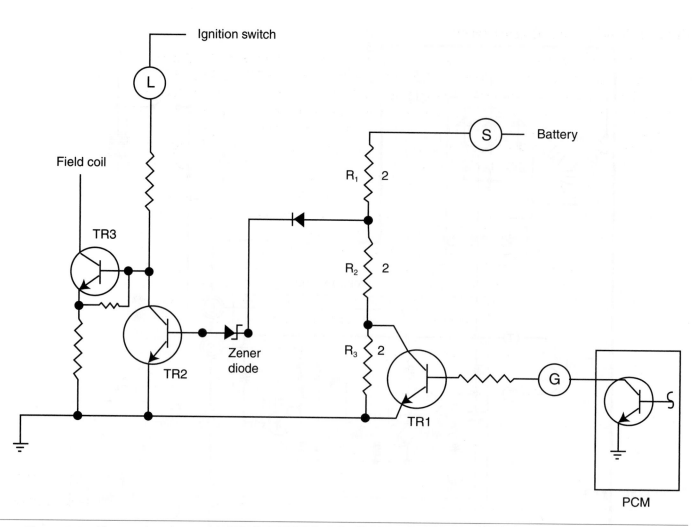

Figure 7-68 Three resistors in series on the sense circuit.

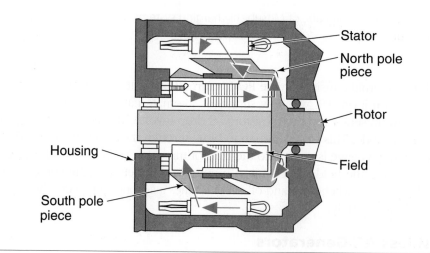

Figure 7-69 Brushless AC generator with stationary field and stator windings.

The pole pieces on the rotor concentrate the magnetic field into alternating north and south poles. When the rotor is spinning, the north and south poles alternate as they pass the stator winding. The moving magnetic field produces an electrical current in the stator winding. The alternating current is rectified in the same manner as in conventional AC generators.

Hybrid Design AC Generators

The hybrid AC generator design consists of a rotor assembly with both wire wound and permanent magnet sections. The stator is made up of two separate lamination stacks, and rectification is accomplished with traditional diodes. The advantage of this design is that output at idle is more than two times greater than that of conventional AC generators. Some hybrid AC generator manufacturers claim an idle output of 240 amps and a maximum output of over 300 amps.

Remember that the output of conventional AC generators is dependent upon the intensity of the magnetic field, the number of conductors passing through the magnetic field at any given time, the number of magnets, and the speed at which the lines of flux or the conductors are moving when the intercept occurs. Since the speed at which the magnetic poles move influences the amount of current output, then output will be the lowest at idle. Maximum output will not be achieved until higher engine speeds. It is during the times of low engine speeds that current demand is likely to be at its highest. To increase output, the hybrid rotor has permanent magnets located between the pole pieces of the rotor. The magnet flux from these permanent magnets goes into the pole piece, through the rotor shaft, and then back through the pole piece on the opposite side of the magnet. The permanent magnet fills the gap between the pole pieces, forcing more of the flux from the rotor into the stator windings. This results in an increase of the alternator's output.

Regulation uses a technique that is referred to as "boost-buck." At low speed and high electrical loads, the wire wound section is fully energized. This extra magnetic flux then boosts the output of the permanent magnet section. When the vehicle is being operated at a medium speed and with a medium electrical load, the field current is off. During this time only the permanent magnet section is producing the output. During high-speed, low-electrical-load conditions, the field current is reversed. This bucks the permanent magnet's field and maintains a constant output voltage.

Summary

- The purpose of the charging system is to convert the mechanical energy of the engine into electrical energy to recharge the battery and run the electrical accessories.

- The charging system consists of the battery, AC generator, drive belt, voltage regulator, charge indicator (lamp or gauge), ignition switch, cables and wiring harness, starter relay (some systems), and fusible links (some systems).

- All charging systems use the principle of electromagnetic induction to generate the electrical power.

- The electromagnetic principle states that a voltage will be produced if motion between a conductor and a magnetic field occurs. The amount of current produced is affected by the speed that the conductor passes through the magnetic field, the strength of the magnetic field, and the number of conductors passing through the magnetic field.

- The main components of the AC generator are the rotor, the brushes, the stator, the rectifier bridge, the housing, and the cooling fan.

- The rotor creates the rotating magnetic field of the AC generator. It is the portion of the AC generator that is rotated by the drive belt.

- The insulated stationary carbon brush passes field current into a slip ring, then through the field coil, and back to the other slip ring. Current then passes through to the grounded stationary brush or voltage regulator.

- The stator is the stationary coil in which current is produced.

- The stator contains three main sets of windings wrapped in slots around a laminated, circular iron frame.

Terms to Know

Delta connection

Diode rectifier
  bridge

Diode trio

Electronic
  regulator

EVR

Full-wave
  rectification

Half-wave
  rectification

Heat sink

Inductive reactance

Pulse width
  modulation

Rectification

Rotor

Sensing voltage

Slip rings

Stator

Wye wound
  connection

❑ The most common method of stator connection is called the wye connection. In the wye connection, one lead from each winding is connected to one common junction. From this junction, the other leads branch out in a Y pattern.

❑ Another method of stator connection is called the delta connection. The delta connection connects the lead of one end of the winding to the lead at the other end of the next winding.

❑ The diode rectifier bridge provides reasonably constant DC voltage to the vehicle's electrical system and battery. The diode rectifier bridge is used to change the current in an AC generator.

❑ The converting of AC current to DC current is called rectification.

❑ Most AC generator housings are two-piece construction, made from cast aluminum. The two end frames provide support of the rotor and the stator. In addition, the end frames contain the diodes, regulator, heat sinks, terminals, and other components.

❑ The cooling fan draws air into the AC generator through the openings at the rear of the housing. The air leaves through openings behind the cooling fan.

❑ The three principal circuits used in the AC generator are the charging circuit, which consists of the stator windings and rectifier circuits; the excitation circuit, which consists of the rotor field coil and the electrical connections to the coil; and the preexcitation circuit, which supplies the initial current for the field coil that starts the buildup of the magnetic field.

❑ The voltage regulator controls the output voltage of the AC generator, based on charging system demands, by controlling field current. The higher the field current, the higher the output voltage.

❑ The regulator must have system voltage as an input in order to regulate the output voltage. The input voltage to the regulator is called sensing voltage.

❑ Because ambient temperatures influence the rate of charge that a battery can accept, regulators are temperature compensated.

❑ The A circuit is called an external grounded field circuit and is always an electronic-type regulator. In the A circuit, the regulator is on the ground side of the field coil. The B+ for the field coil is picked up from inside the AC generator.

❑ Usually the B circuit regulator is mounted externally of the AC generator. The B circuit is an internally grounded circuit. In the B circuit, the voltage regulator controls the power side of the field circuit.

❑ Isolated field AC generators pick up B+ and ground externally. The AC generator has two field wires attached to the outside of the case. The voltage regulator can be located either on the ground (A circuit) or on the B+ (B circuit) side.

❑ There are two basic types of regulators: electromechanical and electronic. Also, on many newer model vehicles, the computer controls regulation.

❑ In the electromechanical regulator, the field relay applies voltage to the field coil. The voltage limiter is connected through the resistor network and determines whether the field will receive high, low, or no voltage.

❑ An electronic regulator uses solid-state circuitry to perform the regulatory functions. Electronic regulators can be mounted either externally or internally of the AC generator. Because there are no moving parts, this type of regulator can cycle between 10 and 7,000 times per second.

❑ The electronic regulator uses a zener diode that blocks current flow until a specific voltage is obtained, at which point it allows the current to flow.

❑ On many vehicles made after the mid-1980s, the regulator function has been incorporated into the vehicle's engine computer. Regulation of the field circuit is through the ground (A circuit).

❏ There are three basic methods of informing the driver of the charging system's condition: indicator lamps, ammeter, and voltmeter.

❏ Most indicator lamps operate on the basis of voltage drop. If the charging system output is less than battery voltage, there is an electrical potential difference in the lamp circuit and the lamp will light.

❏ The ammeter measures charging and discharging current in amperes. The ammeter is wired in series between the AC generator and the battery.

❏ The voltmeter is usually connected between the battery's positive and negative terminals.

❏ The hybrid AC generator design consists of a rotor assembly with both wire wound and permanent magnet sections. The permanent magnets are located between the pole pieces of the rotor.

Review Questions

Short-Answer Essays

1. List the major components of the charging system.

2. List and explain the function of the major components of the AC generator.

3. How does the regulator control the charging system's output?

4. What is the relationship between field current and AC generator output?

5. Identify the differences between A, B, and isolated circuits.

6. Explain the operation of charge indicator lamps.

7. Describe the two styles of stators.

8. What is the difference between half-wave and full-wave rectification?

9. Describe how AC voltage is rectified to DC voltage in the AC generator.

10. What is the purpose of the charging system?

Fill in the Blanks

1. The charging system converts the _____ energy of the engine into _____ energy to recharge the battery and run the electrical accessories.

2. All charging systems use the principle of _____ _____ to generate the electrical power.

3. The _____ creates the rotating magnetic field of the AC generator.

4. _____ are electrically conductive sliding contacts, usually made of copper and carbon.

5. In the _____ connection stator, one lead from each winding is connected to one common junction.

6. The _____ _____ controls the output voltage of the AC generator, based on charging system demands, by controlling _____ current.

7. In an electronic regulator, _____ _____ _____ controls AC generator output by varying the amount of time the field coil is energized.

8. In most electronic regulators that use an indicator lamp, if there is no _____ _____ , then the lamp circuit is completed to ground.

9. Full-wave rectification in the AC generator requires _____ pair of diodes.

10. The _____ is the stationary coil that produces current in the AC generator.

Multiple Choice

1. The magnetic field current of the AC generator is carried in the:
   - **A.** Rotor.
   - **B.** Diode trio.
   - **C.** Rectifier bridge.
   - **D.** Stator.

2. The voltage induced in one conductor by one revolution of the rotor is called:
   - **A.** Three-phase.
   - **B.** Half-wave.
   - **C.** Single-phase.
   - **D.** Full-wave.

3. Rectification is being discussed.
   Technician A says the AC generator uses a segmented commutator to rectify AC current.
   Technician B says the DC generator uses a pair of diodes to rectify AC current.
   Who is correct?
   - **A.** A only
   - **B.** B only
   - **C.** Both A and B
   - **D.** Neither A nor B

4. Rotor construction is being discussed.
   Technician A says the poles will take on the polarity of the side of the coil that they touch.
   Technician B says the magnetic flux lines will move in opposite directions between adjacent poles.
   Who is correct?
   - **A.** A only
   - **B.** B only
   - **C.** Both A and B
   - **D.** Neither A nor B

5. The amount of voltage output of the AC generator is related to:
   - **A.** Field strength.
   - **B.** Stator speed.
   - **C.** Number of rotor segments.
   - **D.** All of the above.

6. The delta wound stator:
   - **A.** Shares a common connection point.
   - **B.** Has each winding connected in series.
   - **C.** Does not require rectification.
   - **D.** None of the above.

7. Indicator lamp operation is being discussed.
   Technician A says in a system with an electronic regulator, the lamp will light if there is no stator output through the diode trio.
   Technician B says when there is stator output, the lamp circuit has voltage applied to both sides and the lamp will not light.
   Who is correct?
   - **A.** A only
   - **B.** B only
   - **C.** Both A and B
   - **D.** Neither A nor B

8. AC generator differences are being discussed.
   Technician A says the Mitsubishi AC generator uses two separate wye connected stator windings.
   Technician B says Ford rear-terminal AC generators use two different types of rectifiers.
   Who is correct?
   - **A.** A only
   - **B.** B only
   - **C.** Both A and B
   - **D.** Neither A nor B

9. *Technician A* says only two stator windings apply voltage because the third winding is always connected to diodes that are reverse-biased.
   Technician B says AC generators that use half-wave rectification are the most efficient.
   Who is correct?
   - **A.** A only
   - **B.** B only
   - **C.** Both A and B
   - **D.** Neither A nor B

10. Charging system regulation is being discussed.
    Technician A says the regulation of voltage is done by varying the amount of field current flowing through the rotor.
    Technician B says control of field current can be done either by regulating the resistance in series with the field coil or by turning the field circuit on and off.
    Who is correct?
    - **A.** A only
    - **B.** B only
    - **C.** Both A and B
    - **D.** Neither A nor B

Lighting Circuits

Upon completion and review of this chapter, you should be able to:

❑ Describe the operation and construction of automotive lamps.

❑ Describe the differences between conventional sealed-beam, halogen, and composite headlight lamps.

❑ Describe the operation and controlled circuits of the headlight switch.

❑ Describe the operation of the dimmer switch.

❑ Explain the operation of the most common styles of concealed headlight systems.

❑ Describe the operation of the various exterior light systems, including parking, tail, brake, turn, side, clearance, and hazard warning lights.

❑ Explain the operating principles of the turn signal and hazard light flashers.

❑ Describe the operation of the various interior light systems, including courtesy and instrument panel lights.

Introduction

Today's technician is required to understand the operation and purpose of the various lighting circuits on the vehicle. The addition of computers and their many sensors and actuators (some that interlink to the lighting circuits) make it impossible for technicians to just bypass part of the circuit and rewire the system to their own standards. If a lighting circuit is not operating properly, the safety of the driver, the passengers, people in other vehicles, and pedestrians are in jeopardy. When today's technician performs repairs on the lighting systems, the repairs must meet at least two requirements: They must assure vehicle safety and meet all applicable laws.

The lighting circuits of today's vehicles can consist of more than 50 light bulbs and hundreds of feet of wiring. Incorporated within these circuits are circuit protectors, relays, switches, lamps, and connectors. In addition, more sophisticated lighting systems use computers and sensors. The lighting circuits consist of an array of interior and exterior lights, including headlights, taillights, parking lights, stoplights, marker lights, dash instrument lights, courtesy lights, and so on.

The lighting circuits are largely regulated by federal laws, so the systems are similar between the various manufacturers. However, there are variations. Before attempting to do any repairs on an unfamiliar circuit, the technician should always refer to the manufacturer's service manuals. This chapter provides information about the types of lamps used, describes the headlight circuit, discusses different types of concealed headlight systems, and explores the various exterior and interior light circuits individually.

Lamps

A **lamp** generates light through a process of changing energy forms called **incandescence.** The lamp produces light as a result of current flow through a filament. The filament is enclosed within a glass envelope and is a type of resistance wire that is generally made from tungsten. As current flows through the tungsten filament, it gets very hot (Figure 8-1). The changing of electrical energy to heat energy in the resistive wire filament is so intense that the filament starts to glow and emits light. The lamp must have a vacuum surrounding the filament to prevent it from burning so hot that the filament burns in two. The glass envelope that encloses the filament maintains the presence of vacuum. When the lamp is manufactured, all the air is removed and the glass envelope seals out the air. If air is allowed to enter the lamp, the oxygen would cause the filament to oxidize and burn up.

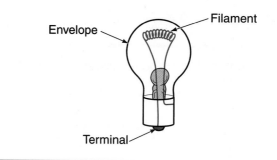

Figure 8-1 A single-filament bulb.

Many lamps are designed to execute more than one function. A **double-filament lamp** has two filaments so the bulb can perform more than one function (Figure 8-2). It can be used in the stop light circuit, taillight circuit, and turn signal circuit combined.

It is important that any burned-out lamp be replaced with the correct lamp. The technician can determine what lamp to use by checking the lamp's standard trade number (Figures 8-3).

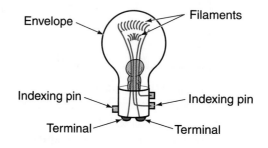

Figure 8-2 A double-filament lamp.

TYPICAL AUTOMOTIVE LIGHT BULBS

Trade Number	Design Volts	Design Amperes	Watts: P = A x V
37	14.0	0.09	1.26
37E	14.0	0.09	1.3
51	7.5	0.22	1.7
53	14.4	0.12	1.7
55	7.0	0.41	2.9
57	14.0	0.24	3.4
57X	14.0	0.24	3.4
63	7.0	0.63	4.4
67	13.5	0.59	8.0
68	13.5	0.59	8.0
70	14.0	0.15	2.1
73	14.0	0.08	1.1
74	14.0	0.10	1.4
81	6.5	1.02	6.6
88	13.0	0.58	7.5
89	13.0	0.58	7.5
90	13.0	0.58	7.5
93	12.8	1.04	13.3
94	12.8	1.04	13.3
158	14.0	0.24	3.4
161	14.0	0.19	2.7
168	14.0	0.35	4.9

Figure 8-3 Chart of typical automotive light bulbs.

TYPICAL AUTOMOTIVE LIGHT BULBS (continued)

Trade Number	Design Volts	Design Amperes	Watts: P = A x V
192	13.0	0.33	4.3
194	14.0	0.27	3.8
194E-1	14.0	0.27	3.8
194NA	14.0	0.27	3.8
209	6.5	1.78	11.6
211-2	12.8	0.97	12.4
212-2	13.5	0.74	10.0
214-2	13.5	0.52	7.0
561	12.8	0.97	12.4
562	13.5	0.74	10.0
563	13.5	0.52	7.0
631	14.0	0.63	8.8
880	12.8	2.10	27.0
881	12.8	2.10	27.0
906	13.0	0.69	8.97
912	12.8	1.00	12.8
921	12.8	1.40	17.92
1003	12.8	0.94	12.0
1004	12.8	0.94	12.0
1034	12.8	1.80/0.59	23.0/7.6
1073	12.8	1.80	23.0
1076	12.8	1.80	23.0
1129	6.4	2.63	16.8
1133	6.2	3.91	24.2
1141	12.8	1.44	18.4
1142	12.8	1.44	18.4
1154	6.4	2.63/.75	16.8/4.5
1156	12.8	2.10	26.9
1157	12.8	2.10/0.59	26.9/7.6
1157A	12.8	2.10/0.59	26.9/7.6
1157NA	12.8	2.10/0.59	26.9/7.6
1176	12.8	1.34/0.59	17.2/7.6
1195	12.5	3.00	37.5
1196	12.5	3.00	37.5
1445	14.4	0.135	1.9
1816	13.0	0.33	4.3
1889	14.0	0.27	3.8
1891	14.0	0.24	3.4
1892	14.4	0.12	1.7
1893	14.0	0.33	4.6
1895	14.0	0.27	3.8
2057	12.8	2.10/0.48	26.9/6.1
2057NA	12.8	2.10/0.48	26.9/6.1

Figure 8-3 (continued)

Trade Number	Design Volts	Design Amperes	Watts: P = A x V
3022	5.0	0.021	0.105
3057	12.8-14.0	2.1/0.48	26.9/6.72
3122	5.0	0.125	0.625
3156	12.8	2.10	26.9
3157	12.8-14.0	2.1/0.59	26.9/8.26
3457	12.8-14.0	2.23/0.59	28.5/8.26
4157	12.8-14.0	2.23/0.59	28.5/8.26
6411	12.0	0.833	10.0
6418	12.0	0.417	5.0
7225	1.5	0.01	0.015
7440	12.0	1.75	21.0
7443	12.0	1.75/0.417	21.0/5.0
7507	12.0	1.75	21.0
9005SX	12.8	5.10	65.0
9006SX	12.8	4.30	55.0
9040	12.8	3.10	40.0
9055	12.8	4.30	55.0
9145	12.8	3.50	45.0
P25-1	13.5	1.86	25.1
P25-2	13.5	1.86	25.1
R19/5	13.5	0.37	
R19/10	13.5	0.74	
W10/3	13.5	0.25	

Figure 8-3 (*continued*)

Headlights

There are four basic types of headlights used on automobiles today: (1) standard sealed beam, (2) halogen sealed beam, (3) composite, and (4) high-intensity discharge (HID).

Sealed-Beam Headlights

Shop Manual
Chapter 8, page 288

From 1939 to about 1975, the headlights used on vehicles remained virtually unchanged. During this time, the headlight was a round lamp. The introduction of the rectangular headlight in 1975 enabled the vehicle manufacturers to lower the hood line of their vehicles. Both the round and rectangular headlights were **sealed-beam** construction (Figure 8-4). The sealed-beam headlight is a self-contained glass unit made up of a filament, an inner reflector, and an outer glass lens. The standard sealed-beam headlight does not surround the filament with its own glass envelope (bulb). The glass lens is fused to the parabolic reflector, which is sprayed with **vaporized aluminum** that gives a reflecting surface that is comparable to silver. The inside of the lamp is filled with argon gas. All oxygen must be removed from the standard sealed-beam headlight to prevent the filament from becoming oxidized. The reflector intensifies the light that the filament produces, and the lens directs the light to form the required light beam pattern.

The lens is designed to produce a broad, flat beam. The light from the reflector is passed through concave **prisms** in the glass lens (Figure 8-5). Lens prisms redirect the light beam and create a broad, flat beam. The illustration (Figure 8-6) shows the horizontal spreading and the vertical control of the light beam to prevent upward glaring.

By placing the filament in different locations on the reflector, the direction of the light beam is controlled (Figure 8-7). In a dual-filament lamp, the lower filament is used for the high beam and the upper filament is used for the low beam.

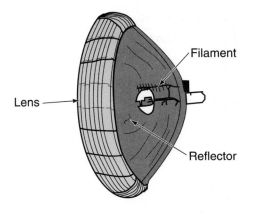

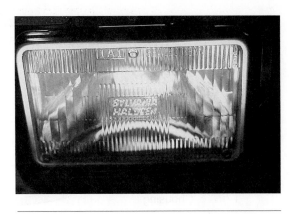

Figure 8-4 Sealed-beam headlight construction.

Figure 8-5 The lens uses prisms to redirect the light.

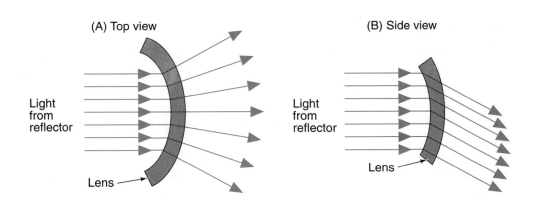

Figure 8-6 The prism directs the beam into (A) a flat horizontal pattern and (B) downward.

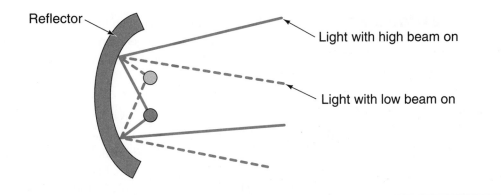

Figure 8-7 Filament placement controls the projection of the light beam.

A BIT OF HISTORY

Improved sealed-beam headlamps were introduced in 1955.

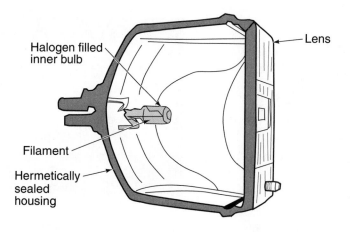

Figure 8-8 A halogen sealed-beam headlight with iodine vapor bulb.

Halogen Headlights

The **halogen** lamp most commonly used in automotive applications consists of a small bulb filled with iodine vapor. The bulb is made of a high-temperature-resistant quartz that surrounds a tungsten filament. This inner bulb is installed in a sealed glass housing (Figure 8-8). With the halogen added to the bulb, the tungsten filament is capable of withstanding higher temperatures than that of conventional sealed-beam lamps. The halogen lamp can withstand higher temperatures and thus is able to burn brighter.

In a conventional sealed-beam headlight, the heating of the filament causes atoms of tungsten to be released from the surface of the filament. These released atoms deposit on the glass envelope and create black spots that affect the light output of the lamp. In a halogen lamp, the iodine vapor causes the released tungsten atoms to be redeposited onto the filament. This virtually eliminates any black spots. It also allows for increased high-beam output of 25% over conventional lamps and for longer bulb life.

AUTHOR'S NOTE: Because the filament is contained in its own bulb, cracking or breaking of the lens does not prevent halogen headlight operation. As long as the filament envelope has not been broken, the filament will continue to operate. However, a broken lens will result in poor light quality and should be replaced.

Composite Headlights

By using the **composite headlight** system, vehicle manufacturers are able to produce any style of headlight lens they desire (Figure 8-9). This improves the aerodynamics, fuel economy, and styling of the vehicle. Composite headlight systems use a replaceable halogen bulb.

Many manufacturers vent the composite headlight housing because of the increased amount of heat these bulbs develop. Because the housings are vented, condensation may develop inside the lens assembly. This condensation is not harmful to the bulb and does not affect headlight operation. When the headlights are turned on, the heat generated from the halogen bulbs will dissipate the condensation quickly. Ford uses integrated nonvented composite headlights. On these vehicles, condensation is not considered normal. The assembly should be replaced.

HID Headlights

High-intensity discharge (HID) headlamps are the latest headlight development. HID headlights use an inert gas to amplify the light produced by arcing across two electrodes. These headlamps (Figure 8-10) put out three times more light and twice the light spread on the road than conventional halogen headlamps. They also use about two-thirds less power to operate and will last

Halogen is the term used to identify a group of chemically related nonmetallic elements. These elements include chlorine, fluorine, and iodine.

Shop Manual
Chapter 8, page 290

Many of today's vehicles have a halogen headlight system. This system is called **composite headlights.**

Shop Manual
Chapter 8, page 292

two to three times longer. HID lamps produce light in both ultraviolet and visible wavelengths. This advantage allows highway signs and other reflective materials to glow. This type of lamp first appeared on select models from BMW in 1993, Ford in 1995, and Porsche in 1996.

The HID lamp consists of an outer bulb made of cerium-doped quartz that houses the inner bulb (arc tube). The inner bulb is made of fused quartz and contains two tungsten electrodes. It also is filled with xenon gas, mercury, and metal halides (salts).

The HID lamp does not rely on a glowing filament for light. Instead it uses a high-voltage arcing bridge across the air gap between the electrodes. The xenon gas amplifies the light intensity given off by the arcing. The HID system requires the use of an ignitor and ballast to provide the electrical energy required to arc the electrodes (Figure 8-11). The ignitor is usually built into the base of the HID bulb and will provide the initial 15,000 to 25,000 volts to jump the gap. Once the gap has been jumped and the gas warms, then the ballast will provide the required voltage to maintain current flow across the gap. The ballast must deliver 35 watts to the lamp when the voltage across the lamp is between 70 and 110 volts.

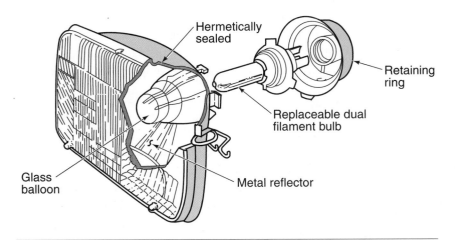

Figure 8-9 A composite headlight system with a replaceable halogen bulb.

Figure 8-10 A Lincoln equipped with HID headlamps; note the reduced size of the headlamp assemblies.

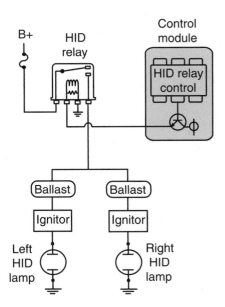

Figure 8-11 HID headlight schematic showing the use of a ballast and ignitor.

Figure 8-12 7 lens HID headlamp.

The great light output of these lamps allows the headlamp assembly to be smaller and lighter. These advantages allow designers more flexibility in body designs as they attempt to make their vehicles more aerodynamic and efficient. Infiniti Q45 models are equipped with a seven-lens HID system (Figure 8-12).

Shop Manual
Chapter 8, page 301

Headlight Switches

The headlight switch may be located either on the dash by the instrument panel or on the steering column (Figure 8-13). It controls most of the vehicle's lighting systems. The most common style of headlight switch is the three-position type with OFF, PARK, and HEADLIGHT positions. The headlight switch will generally receive direct battery voltage to two terminals of the switch. This allows the light circuits to be operated without having the ignition switch in the RUN or ACC (accessory) position.

When the headlight switch is in the OFF position, the open contacts prevent battery voltage from continuing to the lamps (Figure 8-14). When the switch is in the PARK position, battery voltage that is present at terminal 5 is able to be applied through the closed contacts to the side marker, taillight, license plate, and instrument cluster lights (Figure 8-15). This circuit is usually protected by a 15- to 20-ampere fuse that is separate from the headlight circuit.

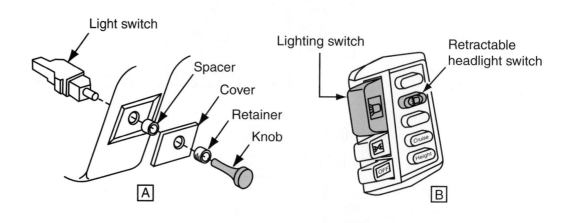

Figure 8-13 (A) Instrument panel–mounted headlight switch. (B) Steering column–mounted headlight switch.

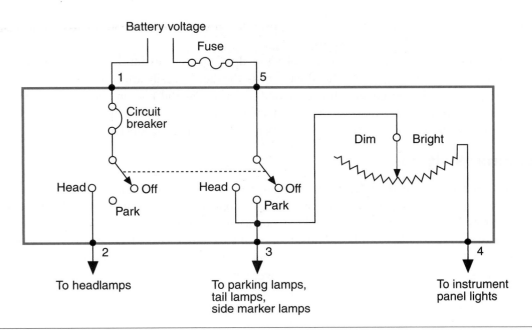

Figure 8-14 Operation with the headlight switch in the OFF position.

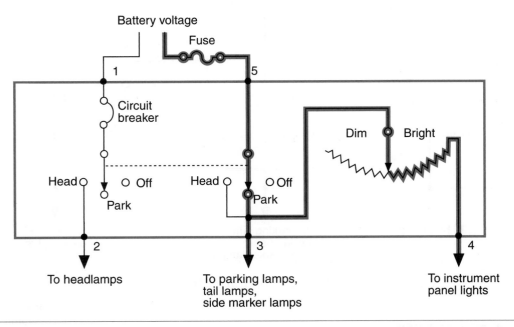

Figure 8-15 Operation with the headlight switch in the PARK position.

The Model T Ford used a headlight system that had a replaceable bulb. The owner's manual warned against touching the reflector except with a soft cloth.

When the switch is located in the HEADLIGHT position, battery voltage that is present at terminal 1 is able to be applied through the circuit breaker and the closed contacts to light the headlights. Battery voltage from terminal 5 continues to light the lights that were on in the PARK position (Figure 8-16). The circuit breaker is used to prevent temporary overloads to the system from totally disabling the headlights.

The rheostat is a variable resistor that the driver uses to control the instrument cluster illumination lamp brightness. As the driver turns the light switch knob, the resistance in the rheostat is changed. The greater the resistance, the dimmer the instrument panel illumination lights glow. In vehicles that have the headlight switch located in the steering column, the rheostat may be a separate unit located on the dash near the instrument panel (Figure 8-17).

Shop Manual
Chapter 8, page 304

Dimmer Switches

The **dimmer switch** provides the means for the driver to select either high- or low-beam operation, and to switch between the two. The dimmer switch is connected in series within the headlight circuit and controls the current path for high and low beams. In the past, the most common location of the dimmer switch was on the floor board next to the left kick panel. The driver operates this switch by pressing on it with a foot. Positioning the switch on the floor board made the switch subject to damage because of rust, dirt, and so forth. Most newer vehicles locate the dimmer switch on the steering column to prevent early failure and to increase driver accessibility (Figure 8-18). This switch is activated by the driver pulling the stock switch (turn signal lever) rearward.

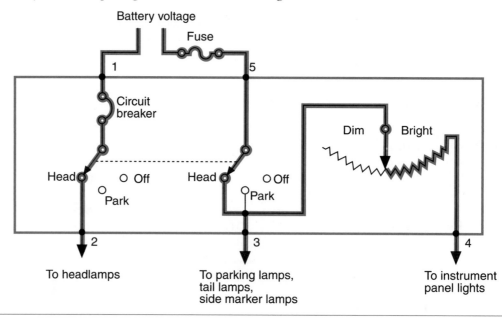

Figure 8-16 Operation with the headlight switch in the HEADLIGHT position.

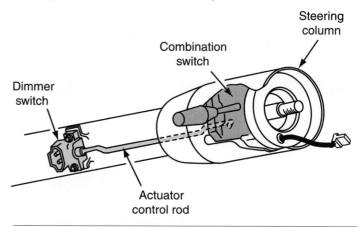

Figure 8-17 A headlight switch mounted on the steering column with a separate control on the dash to control the brightness of the instrument panel lights.

Figure 8-18 A steering column–mounted dimmer switch.

Foot-operated dimmer switches became standard equipment in 1923.

Headlight Circuits

The complete headlight circuit consists of the headlight switch, dimmer switch, high-beam indicator, and the headlights. When the headlight switch is pulled to the HEADLIGHT position, current flows to the dimmer switch through the closed contacts (Figure 8-19). If the dimmer switch is in the LOW position, current flows through the low-beam filament of the headlights. When the

Shop Manual
Chapter 8, page 298

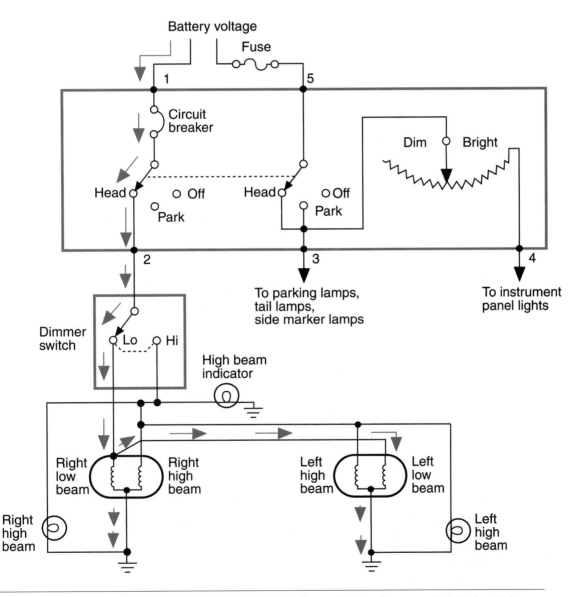

Figure 8-19 A headlight circuit indicating current flow with the dimmer switch in the LO beam position.

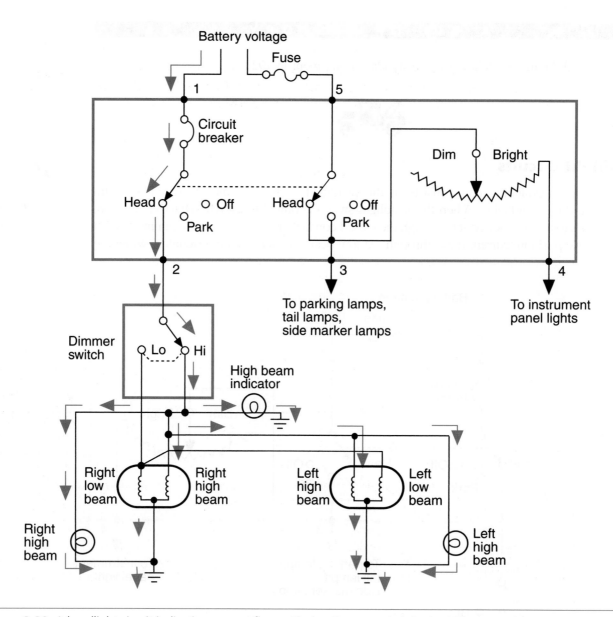

Figure 8-20 A headlight circuit indicating current flow with the dimmer switch in the HI beam position.

dimmer switch is placed on the HIGH position, current flows through the high-beam filaments of the headlights (Figure 8-20).

The headlight circuits just discussed are designed with insulated side switches and grounded bulbs. In this system, battery voltage is present to the headlight switch. The switch must be closed in order for current to flow through the filaments and to ground. The circuit is complete because the headlights are grounded to the vehicle body or chassis. Many import manufacturers use a system design that has insulated bulbs and ground side switches. In this system, when the headlight switch is located in the HEADLIGHT position, the contacts are closed to complete the circuit path to ground. The headlight switch is located after the headlight lamps in the circuit. Battery voltage is applied directly to the headlights when the relays are closed. But the headlights will not light until the switch completes the ground side of the relay circuits. In this system, both the headlight and dimmer switches complete the circuits to ground.

No matter if the headlights use insulated side switches or ground side switches, each system is wired in parallel. This prevents total headlight failure if one filament burns out.

Concealed Headlights

Shop Manual
Chapter 8, page 300

A vehicle equipped with a **concealed headlight** system hides the lamps behind doors when the headlights are turned off. When the headlight switch is turned to the HEADLIGHT position, the headlight doors open (Figure 8-21). Early systems used vacuum-controlled doors. Today most systems use electric motors.

Electrically controlled systems can use either a torsion bar and a single motor to open both headlight doors, or a separate motor for each headlight door. Most systems will use limit switches to stop current flow when the doors are full up or full down. These switches generally operate from a cam on the reaction motor (Figure 8-22). Only one limit switch can be closed at a time. When the door is full up, the opening limit switch opens and the closing limit switch closes. When the door is full down, the closing limit switch is open and the opening limit switch closes. This prepares the reaction motor for the next time that the system is activated or deactivated.

The electrically operated concealed headlight system provides a provision for manually opening the doors in the event of a system failure (Figure 8-23).

Figure 8-24 illustrates a system that incorporates an integrated chip (IC). Each motor has its own relay and limiting switches. When the limit switches are in the A-B position, the doors are full open. When the switches are in the A-C position, the doors are full closed.

Figure 8-21 Concealed headlights enhance the vehicle's styling and aerodynamics.

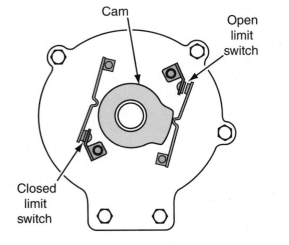

Figure 8-22 Most limit switches operate off of a cam on the motor.

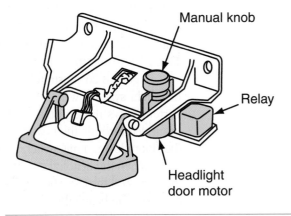

Figure 8-23 An electrically controlled concealed headlight system with a manual control valve.

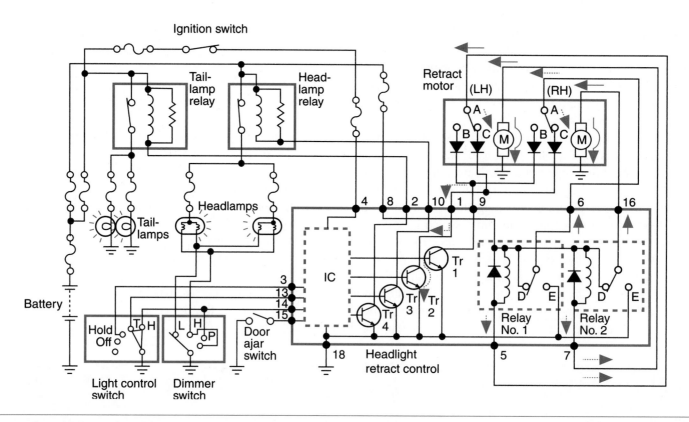

Figure 8-24 Pop-up headlight system wiring schematic.

When the headlights are turned on, terminal 14 is grounded through the headlight switch (Figure 8-24). This signals the IC to bias the leg of transistor TR2 for 10 seconds. When TR2 is biased, it closes the circuits for relays 1 and 2, through position A-C of the limit switches and terminal 18. With the relay coils energized, battery voltage is supplied to the motors and they operate until the limit switches close to A-B or the 10-second timer turns off. TR3 and TR4 are also biased to complete the circuit for the headlight and taillight relays. When the relays are energized, the circuit is completed to the headlights and taillights, and they illuminate.

The motors used in this system provide for 360-degree rotation. The first 180 degrees of rotation opens the doors and an additional 180 degrees closes them. The timer circuit is used to protect the motors. The diodes in the motor relays are called clamping diodes. These protect the IC from induced voltage when the relay coils collapse.

When the headlights are turned off, ground through terminal 14 is opened by the headlight switch (Figure 8-25). This signals the IC to bias transistor TR1 for 10 seconds. TR1 closes the circuit to ground for the relay circuit through the A-B positions of the limit switches and terminal 18. With the relays energized, battery voltage is sent to the motors. Also the IC ceases to bias TR3 and TR4. As a result, the relays are de-energized and the lamps turn off.

Figure 8-26 illustrates another method used to operate the electric motors of a concealed headlight system. When the ignition switch is in the RUN position but the headlight switch is off, current flows through the ignition switch to the relay. The relay is energized because the coil is grounded through the headlight filaments. With the coil energized, the relay points close. However, the door closing limit switch is open. This results in a de-energized door closing field winding.

When the headlight switch is turned to the HEADLIGHT position, current continues to flow to the relay coil through the ignition switch. However, current is also sent to the other side of the relay coil from the headlight switch. Voltage is equal on both sides of the relay coil, so there is no voltage potential and the coil is de-energized. The relay contact points close to the door opening field winding. With the door opening limit switch closed, the motor operates until the limit switch is opened.

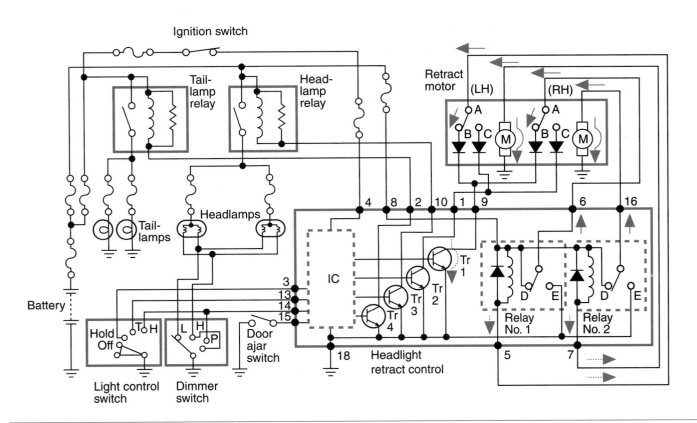

Figure 8-25 Operation when the headlight switch is turned to the HEADLIGHT position on Toyota's pop-up system.

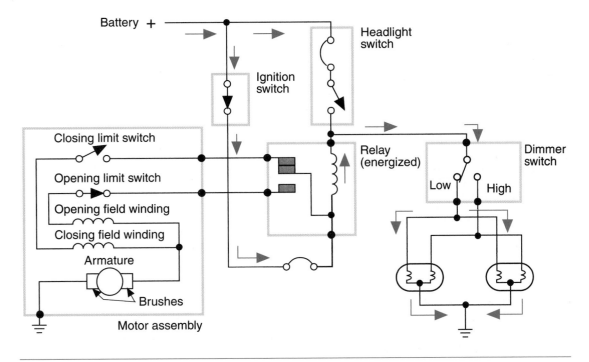

Figure 8-26 Current flow with the headlight switch OFF and the headlight doors closed.

When the headlights are turned off again, voltage is applied to only one side of the coil through the ignition switch. Ground is provided through the headlight filaments. This energizes the relay, closing the contact point to the door closing winding. Since the door closing limit switch is closed, the motor operates until the limit switch opens again.

Flash to Pass

Many steering column–mounted dimmer switches have an additional feature called "flash to pass." This circuit illuminates the high-beam headlights even with the headlight switch in the OFF or PARK position (Figure 8-27). In this illustration, battery voltage is supplied to terminal B1 of the headlight switch and on to the dimmer switch. Battery voltage is available to the dimmer switch through this wire in both the OFF and PARK positions of the headlight switch. When the driver activates the flash-to-pass feature, the contacts in the dimmer switch complete the circuit to the high-beam filaments.

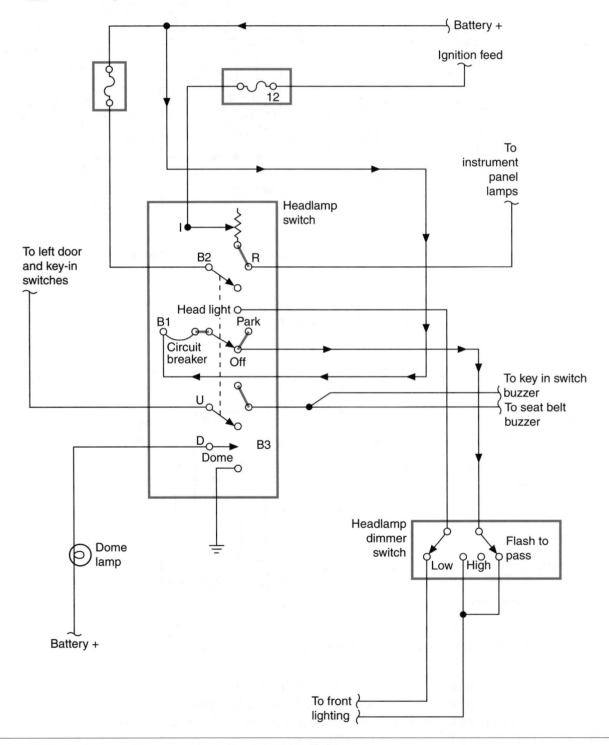

Figure 8-27 Flash-to-pass feature added to the headlight circuit.

Exterior Lights

When the headlight switch is placed in the PARK or HEADLIGHT position, the front parking lights, taillights, side marker lights, and rear license plate light are all turned on. The front parking lights usually use dual-filament bulbs. The other filament is used for the turn signals and hazard lights.

Most taillight assemblies include the brake, parking, rear turn signal, and rear hazard lights. The center high mounted stop light (CHMSL), back-up lights, and license plate lights can be included as part of the taillight circuit design. Depending on the manufacturer, the taillight assembly can be wired to use single-filament or dual-filament bulbs. When single-filament bulbs are used, the taillight assembly is wired as a three-bulb circuit. A three-bulb circuit uses one bulb each for the tail, brake, and turn signal lights on each side of the vehicle. When dual-filament bulbs are used, the system is wired as a two-bulb circuit. Each bulb can perform more than one function.

Taillight Assemblies

The headlight switch controls parking lights and taillights. They can be turned on without having to turn on the headlights. Usually, the first detent on the headlight switch is provided for this function. Figure 8-28 illustrates a parking light and taillight circuit. This circuit is controlled by the headlight switch. Thus the lights can be operated with the ignition switch in the OFF position.

Shop Manual
Chapter 8, page 308

Parking lights can also be referred to as running lights.

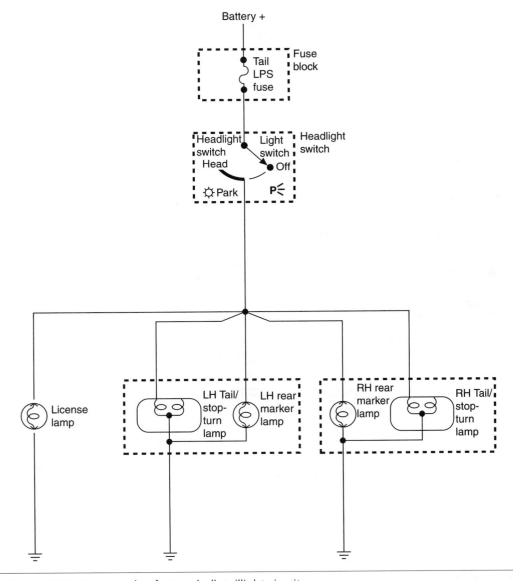

Figure 8-28 An example of a two-bulb taillight circuit.

Taillights on both sides of the car didn't appear until 1929.

In a three-bulb taillight system, the brake lights are controlled directly by the brake light switch. In most applications, the brake light switch is attached to the brake pedal. When the brakes are applied, the pedal moves down and the switch plunger closes the contact points and lights the brake lights (Figure 8-29). On some vehicles, the brake light switch may be a pressure-sensitive switch located in the brake master cylinder. When the brakes are applied, the pressure developed in the master cylinder closes the switch to light the lamps.

The brake light switch receives direct battery voltage through a fuse, which allows the brake lights to operate when the ignition switch is in the OFF position. Once the switch is closed, voltage is applied to the brake lights. The brake lights on both sides of the vehicle are wired in parallel. The bulb is grounded to complete the circuit.

Most brake light systems use dual-filament bulbs that perform multifunctions. Usually, the filament of the dual-filament bulb, which is also used by the turn signal and hazard lights, is used for the brake lights (the high-intensity filament). In this type of circuit, the brake lights are wired through the turn signal and hazard switches (Figure 8-30). If neither turn signal is on, the current is sent to both of the brake lights (Figure 8-31). If the left turn signal is on, current for the right brake light is sent to the lamp through the turn signal switch terminal 5. The left brake light does not receive any voltage from the brake switch because the turn signal switch opens that circuit (Figure 8-32). The left-rear lamp will flash as the turn signal flasher provides pulsed voltage into switch terminal 3 and out terminal 8.

Shop Manual
Chapter 8, page 310

Studies indicate that locating a brake light at eye level significantly reduces rear-end collisions.

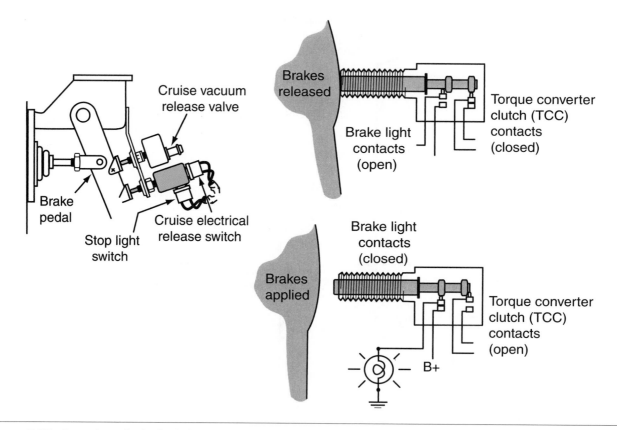

Figure 8-29 Operation of a brake light switch.

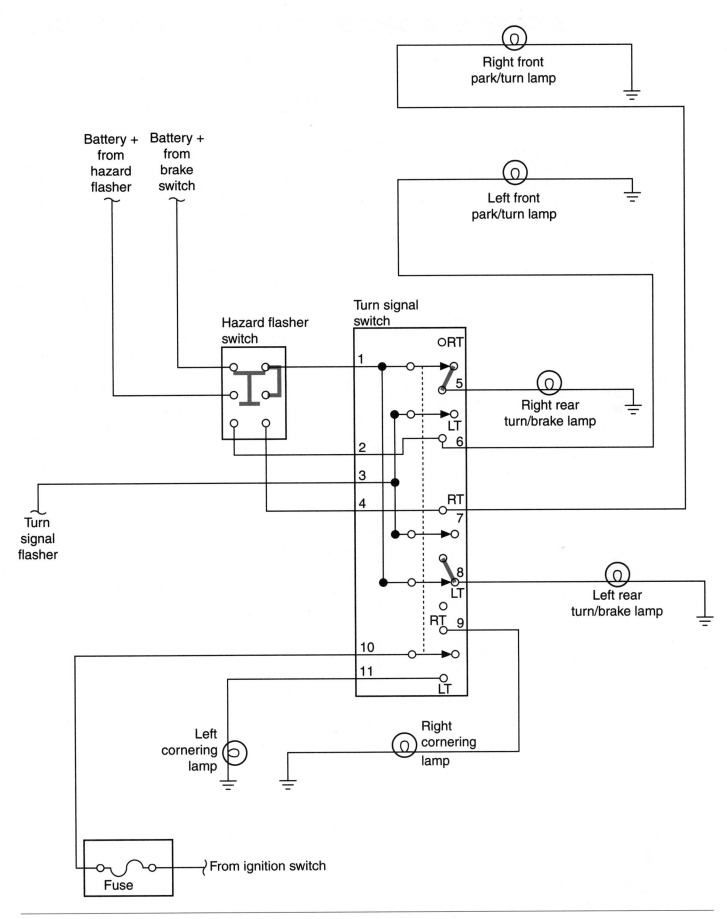

Battery +
from
hazard
flasher

Battery +
from
brake
switch

Hazard flasher
switch

Turn signal
switch

Right front
park/turn lamp

Left front
park/turn lamp

1

ORT

5

Right rear
turn/brake lamp

LT

2

6

Turn
signal
flasher

3

4

RT

7

8
LT

RT

9

10

11
LT

Left rear
turn/brake lamp

Left
cornering
lamp

Right
cornering
lamp

Fuse

From ignition switch

Figure 8-30 A turn signal switch used in a two-bulb taillight circuit.

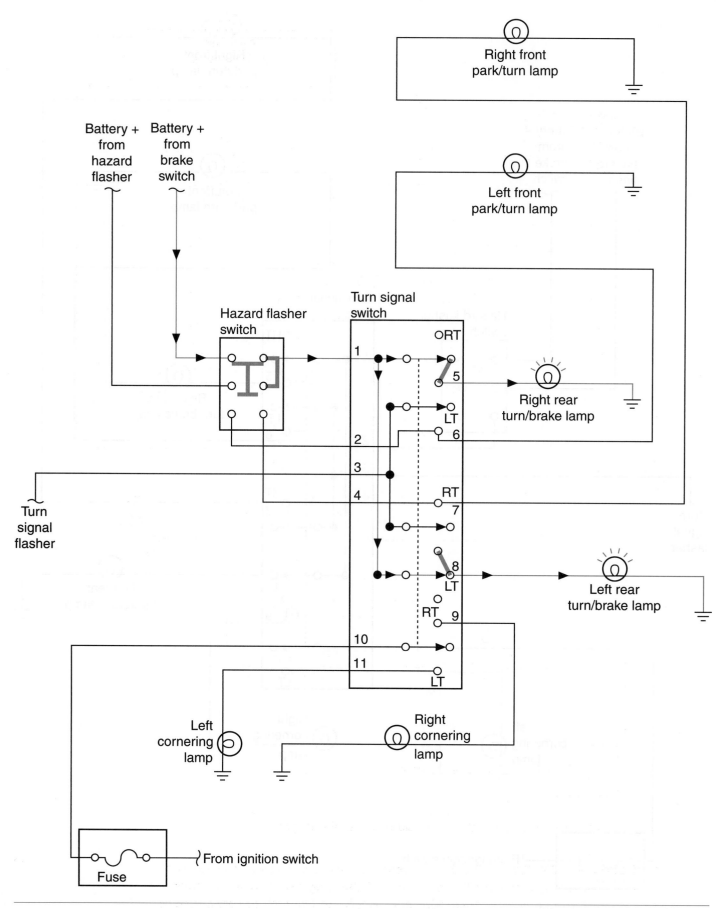

Figure 8-31 Brake light operation with the turn signals in the neutral position.

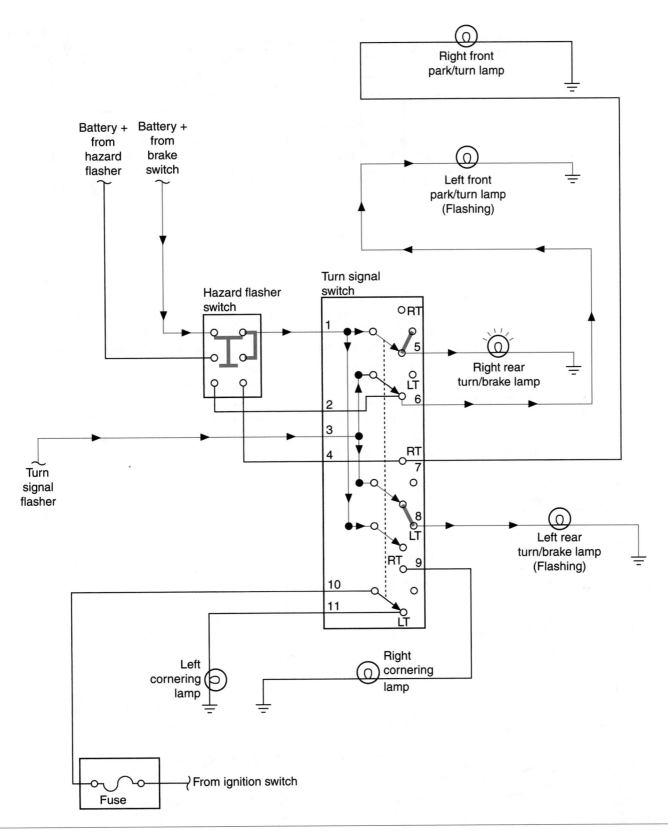

Figure 8-32 Brake light operation with the turn signal in the left-turn position.

AUTHOR'S NOTE: Because the turn signal switches used in a two-bulb system also control a portion of the operation of the brake lights, they have a complex system of contact points. The technician must remember that many brake light problems are caused by worn contact points in the turn signal switch.

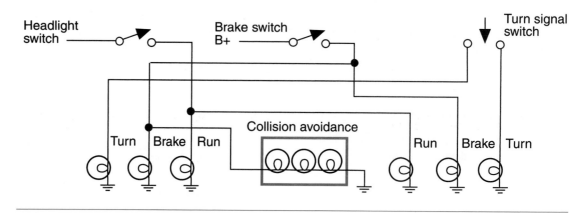

Figure 8-33 Wiring of a CHMSL in a three-bulb circuit.

All brake lights must be red and, starting in 1986, the vehicle must have a center high mounted stop lamp (CHMSL). This lamp must be located on the center line of the vehicle and no lower than 3 inches below the bottom of the rear window (6 inches on convertibles). In a three-bulb system, wiring for the CHMSL is in parallel to the brake lights (Figure 8-33).

In a two-bulb circuit, the CHMSL can be wired in one of two common methods. The first method is to connect to the brake light circuit between the brake light switch and the turn signal switch (Figure 8-34). This method is simple to perform. However, it increases the number of conductors needed in the harness.

The most common method used by manufacturers is to install diodes in the conductors that are connected between the left- and right-side bulbs (Figure 8-35). If the brakes are applied with the turn signal switch in the neutral position, the diodes will allow voltage to flow to the CHMSL.

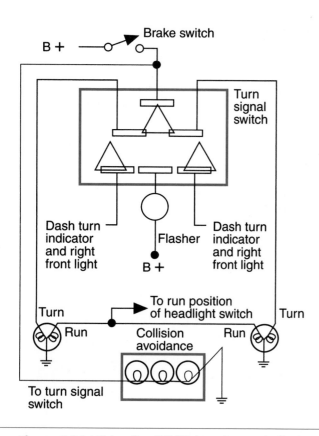

Figure 8-34 Wiring the CHMSL into the two-bulb circuit between the brake light switch and the turn signal.

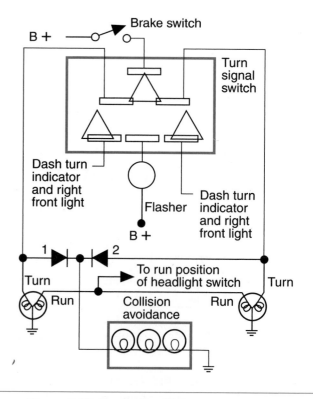

Figure 8-35 Two-bulb taillight circuit incorporating a CHMSL into the brake light system.

If the turn signal switch is placed in the left-turn position, the left light must receive a pulsating voltage from the flasher. However, the steady voltage being applied to the right brake light would cause the left light to burn steady if the diode was not used. Diode 1 will block the voltage from the right lamp, preventing it from reaching the left light. Diode 2 will allow this voltage from the right brake light circuit to be applied to the CHMSL.

A BIT OF HISTORY

In 1921, turn signals were made standard equipment by Leland Lincoln. This marque later joined Ford Motor Company. Leland Lincolns were built by Henry Leland, who was the originator of Cadillac. Early turn signals were not like those used today; many were steel arms with reflective material on them. These arms pivoted out on the side of the car as it was turning. This style continued for many years until Buick introduced electric turn signals to the public in 1939.

Turn Signals

Turn signals are used to indicate the driver's intention to change direction or lanes. The driver will actuate a turn signal switch that is usually located in the steering column (Figure 8-36). Figures 8-37 through 8-39 illustrate the operation of the turn signal circuit. In the neutral position, the contacts are opened, preventing current flow (Figure 8-37). When the driver moves the turn signal lever to indicate a left turn, the turn signal switch closes the contacts to direct voltage to the front and rear lights on the left side of the vehicle (Figure 8-38).

Shop Manual
Chapter 8, page 310

When the turn signal switch is moved to indicate a right turn, the contacts are moved to direct voltage to the front and rear turn signal lights on the right side of the vehicle (Figure 8-39).

A **flasher** is used to open and close the turn signal circuit at a set rate. With the contacts closed, power flows from the flasher through the turn signal switch to the lamps. The flasher (Figure 8-40) consists of a set of normally closed contacts, a bimetallic strip, and a coil heating element

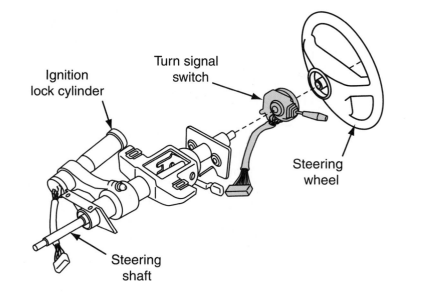

Figure 8-36 Typical turn signal switch location.

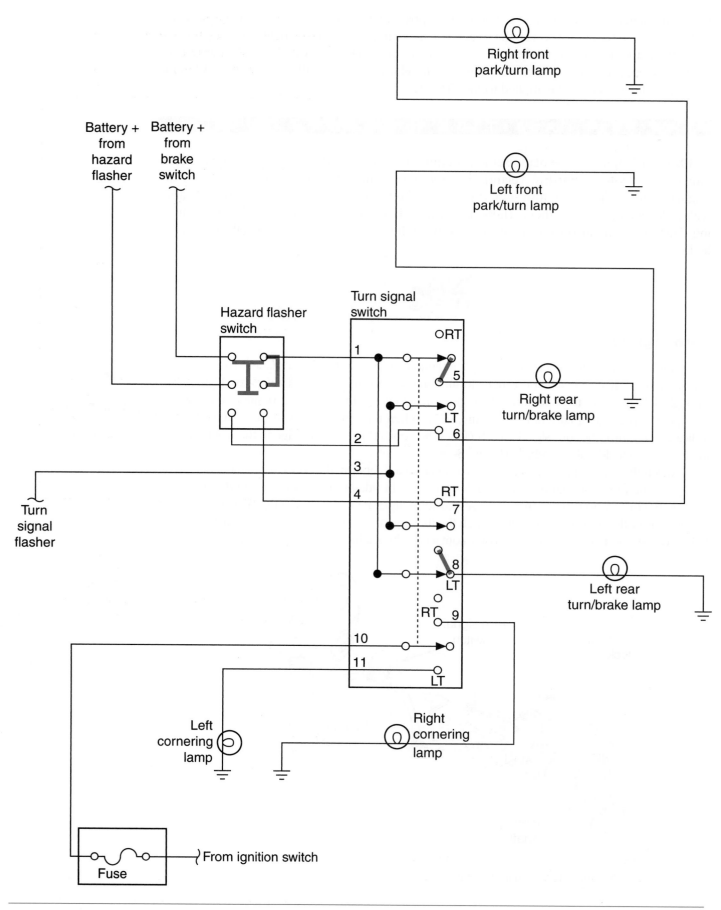

Figure 8-37 Turn signal circuit with switch in neutral position.

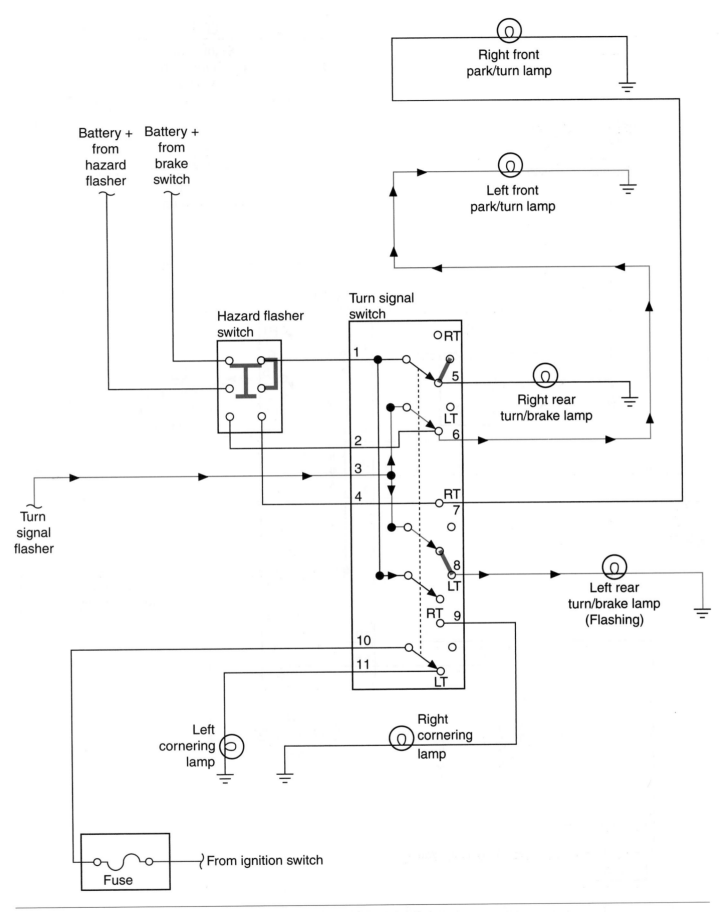

Figure 8-38 Turn signal circuit with the switch in the left-turn position.

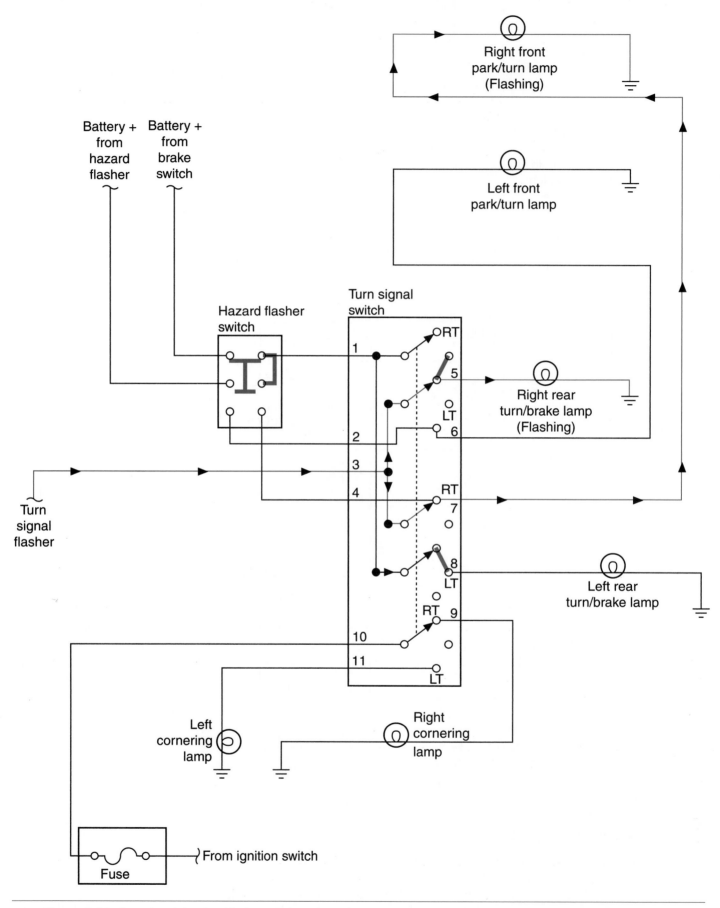

Figure 8-39 Turn signal circuit with the switch in the right-turn position.

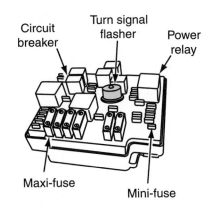

Figure 8-40 A turn signal flasher located in the fuse box.

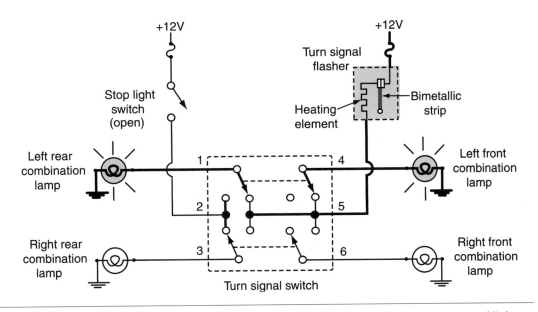

Figure 8-41 The flasher uses a bimetallic strip and a heating coil to flash the turn signal lights.

(Figure 8-41). These three components are wired in series. As current flows through the heater element, it increases in temperature, which heats the bimetallic strip. The strip then bends and opens the contact points. Once the points are open, current flow stops. The bimetallic strip cools and the contacts close again. With current flowing again, the process is repeated. Because the flasher is in series with the turn signal switch, this action causes the turn signal lights to turn on and off.

The hazard warning system is part of the turn signal system. It has been included on all vehicles sold in North America since 1967. All four turn signal lamps flash when the hazard warning switch is turned on. Depending on the manufacturer, a separate flasher can be used for the hazard lights than the one used for the turn signal lights. The operation of the hazard flasher is identical to that of the turn signal. The only difference is that the hazard flasher is capable of carrying the additional current drawn by all four turn signals. And, it receives its power source directly from the battery. Figure 8-42 shows the current flow through the hazard warning system.

Neon Third Brake Lights

In 1995, Ford Motor Company began equipping some models with rear high-mount brake lights that use neon lights. These lights are more energy efficient and turn on more quickly than the regular lights. Behind the third brake light lens is a single neon bulb. Since neon bulbs have no filament, the neon bulb should last much longer than a regular bulb.

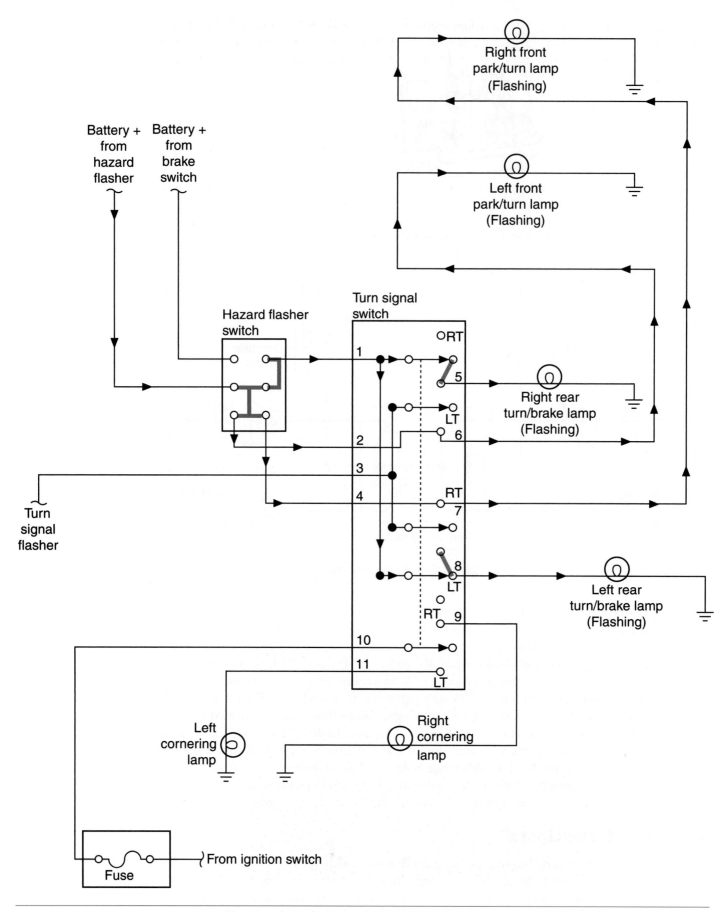

Figure 8-42 Current flow when the hazard warning system is activated.

The neon bulbs turn on within 3 milliseconds after being activated. Halogen lamps require 300 milliseconds. The importance of this quickness is that it gives the driver behind the vehicle an earlier warning to stop. This early warning can give the approaching driver 19 more feet (6 meters) for stopping when driving at 60 miles per hour (96 kmh).

LED Exterior Lighting

Many car manufacturers use LED lighting technology for several exterior lighting functions. The most common use of LEDs is in the CHMSL. Other uses include taillight assemblies, side marker lights, and turn-indicating outside mirrors. LEDs used in rear-lighting applications (especially the CHMSL) provide one means of increasing traffic safety. Driver reaction times in response to the brake light function is shorter for CHMSLs equipped with conventional incandescent bulbs than for those equipped with LEDs. This is due to the shorter LED illumination time of about 30 ms. Another advantage of LED lighting is the extended life of the LED compared to a bulb.

An example of the use of LED technology in rear-lighting systems occurs in the Cadillac STS. Each tail lamp assembly has thirty points of illumination by using two vertical boards, each board consisting of fifteen LEDs. The CHMSL is approximately 1/2 inch (12 mm) thick and consists of seventy-eight points of illumination.

Cornering Lights

Cornering lights are lamps that illuminate when the turn signals are activated. They burn steady when the turn signal switch is in a turn position to provide additional illumination of the road in the direction of the turn (Figure 8-43). Vehicles equipped with cornering lights have an additional set of contacts in the turn signal switch. These contacts operate the cornering light circuit only. The contacts can receive voltage from either the ignition switch or the headlight switch. If the ignition switch provides the power, the cornering lights will be activated any time the turn signals are used (Figure 8-44). If the contacts receive the voltage from the headlight switch, the cornering lights do not operate unless the headlight switch is in the PARK or HEADLIGHT position.

Fog Lights

For increased safety when driving in snow, sleet, heavy rains, and heavy fog conditions, some vehicles are equipped with fog lights. Fog lights can also be installed as an after-market accessory. Headlights will reflect off heavy, dense fog and cause a white haze that can reduce visibility. Fog lights emit a specialized beam to penetrate through the snow, rain, or fog, providing the driver with a better and safer field of vision.

Figure 8-43 Cornering lights are used to provide additional illumination during turns.

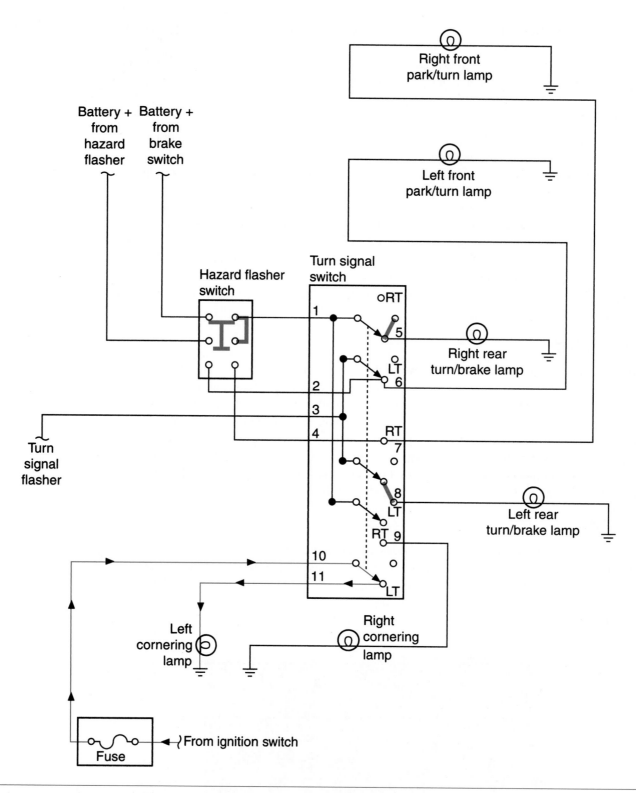

Figure 8-44 Cornering light circuit powered through the ignition switch.

Fog lights are installed on each side of the vehicle, generally low on the front fascia. Due to their mounting location, fog lights illuminate below the normal line of sight. This minimizes the amount of reflected light, to help the driver see better.

Common fog light circuits use a relay (Figure 8-45). Also, most fog light circuits are wired so the fog lights will come on only if the headlight switch is in the PARK or Low beam positions.

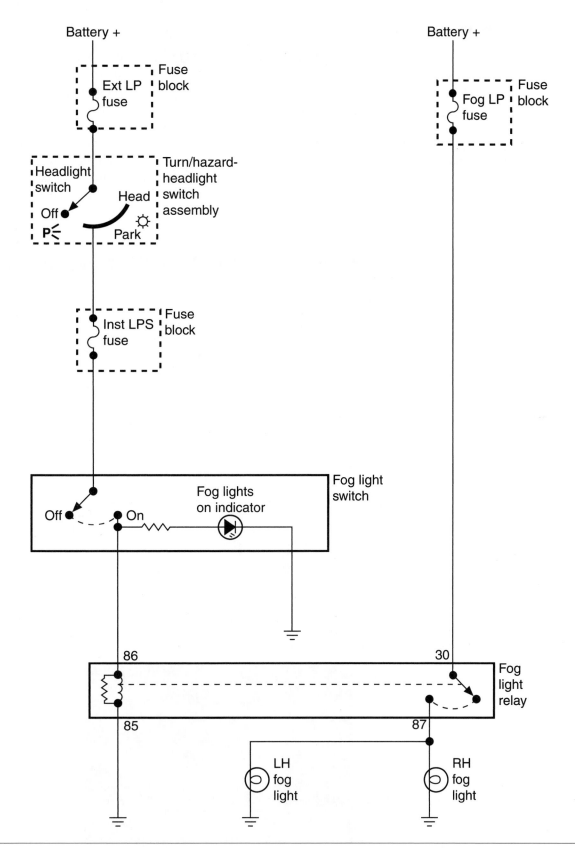

Figure 8-45 Typical fog light circuit.

Back-up Lights

All vehicles sold in North America after 1971 are required to have back-up lights. Back-up lights illuminate the road behind the vehicle and warn other drivers and pedestrians of the driver's intention to back up. Figure 8-46 illustrates a back-up light circuit. Power is supplied through the ignition switch when it is in the RUN position. When the driver shifts the transmission into reverse, the back-up light switch contacts are closed and the circuit is completed.

Many vehicles equipped with automatic transmissions incorporate the back-up light switch into the neutral safety switch. Most manual transmissions are equipped with a separate switch. Either style of switch can be located on the steering column, on the floor console, or on the transmission (Figure 8-47). Depending on the type of switch used, there may be a means of adjusting the switch to assure that the lights are not on when the vehicle is in a forward gear selection.

A BIT OF HISTORY

The 1921 Wills-St. Claire was the first car to display a backup lamp.

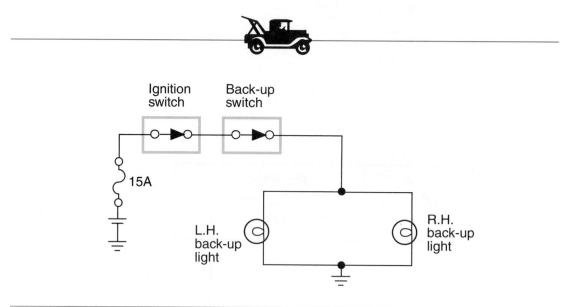

Figure 8-46 Back-up light circuit.

Figure 8-47 A combination back-up and neutral safety switch installed on an automatic transmission.

Side Marker Lights

Side marker lights are installed on all vehicles sold in North America since 1969. These lamps permit the vehicle to be seen when entering a roadway from the side. This also provides a means for other drivers to determine vehicle length. The front side marker light lens must be amber and the rear lens must be red. Vehicles that use wrap-around headlight and taillight assemblies also use this lens for the side marker lights (Figure 8-48). Vehicles that surpass certain length and height limits are also required to have clearance lights that face both to the front and rear of the vehicle.

The common method of wiring the side marker lights is in parallel with the parking lights. Wired in this manner, the side marker lights would illuminate only when the headlight switch was in the PARK or HEADLIGHT position.

Many vehicle manufacturers use a method of wiring in the side marker lights so that they flash when the turn signals are activated. The side marker light is wired across the parking light and turn signal light (Figure 8-49). If the parking lights are on, voltage is applied to the side marker light from the parking light circuit. Ground for the side marker light is provided through the turn signal filament. Because of the large voltage drop across the side marker lamp, the turn signal bulb will barely illuminate. In this condition, the side marker light stays on constantly (Figure 8-50).

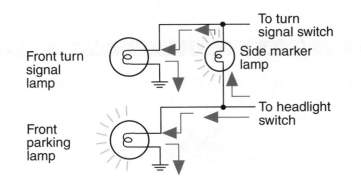

Figure 8-48 Wrap-around headlights serve as side marker lights.

Figure 8-49 A side marker light wired across two circuits.

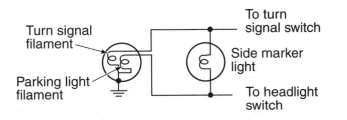

Figure 8-50 Current flow to the side marker light with the parking light on.

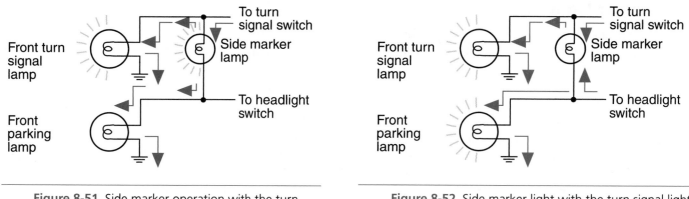

Figure 8-51 Side marker operation with the turn signal switch activated.

Figure 8-52 Side marker light with the turn signal light and parking light on.

If the parking lights are off and the turn signal is activated, the side marker light receives its voltage source from the turn signal circuit. Ground for the side marker light is provided through the parking light filament. The voltage drop over the side marker light is so high that the parking light will not illuminate. The side marker light will flash with the turn signal light (Figure 8-51).

If the turn signal is activated while the parking lights are illuminated, the side marker light will flash alternately with the turn signal light. When both the turn signal light and the parking light are on, there is equal voltage applied to both sides of the side marker light. There is no voltage potential across the bulb, so the light does not illuminate (Figure 8-52). The turn signal light turns off as a result of the flasher opening. Then the turn signal light filament provides a ground path and the side marker light comes on. The side marker light will stay on until the flasher contacts close to turn on the turn signal light again.

Interior Lights

Shop Manual
Chapter 8, page 314

Interior lighting includes courtesy lights, map lights, and instrument panel lights.

Courtesy Lights

Courtesy lights illuminate the vehicle's interior when the doors are open. Courtesy lights operate from the headlight and door switches and receive their power source directly from a fused battery connection. The switches can be either ground switch circuit (Figure 8-53) or insulated switch circuit design (Figure 8-54). In the insulated switch circuit, the switch is used as the power relay to the lights. In the grounded switch circuit, the switch controls the grounding portion of the circuit for the lights. The courtesy lights may also be activated by the headlight switch. When the headlight switch knob is turned to the extreme counterclockwise position, the contacts in the switch close and complete the circuit.

A B I T O F H I S T O R Y

In 1913, the Spaulding touring car had such luxuries as four seats with folding backs, air mattresses, and electric reading lamps.

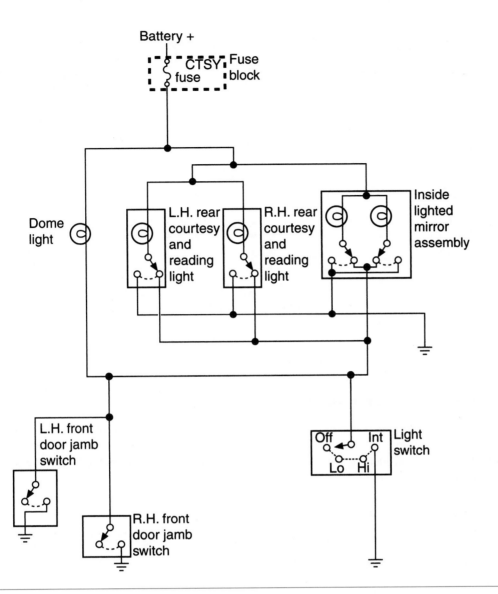

Figure 8-53 Courtesy lights using ground side switches.

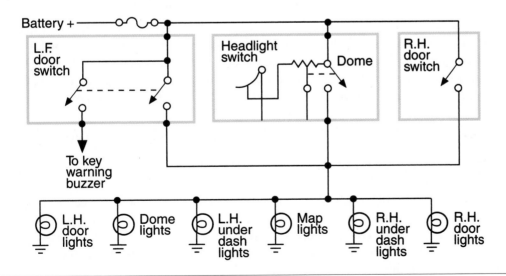

Figure 8-54 Courtesy lights using insulated side switches.

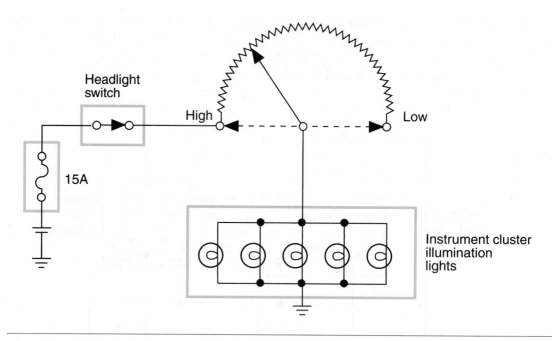

Figure 8-55 A rheostat controls the brightness of the instrument panel lights.

Reading and Map Lights

Individual switches and controls to allow passengers in the vehicle to turn on individual lights are incorporated within most courtesy light systems (refer to Figure 8-53). The system shown has individual two-position switches that allow the passenger to turn on a light. When the switch is pressed, it completes the circuit to ground for that light only.

Shop Manual
Chapter 8, page 317

Instrument Cluster and Panel Lights

Consider the following three types of lighting circuits within the instrument cluster:

1. **Warning lights** alert the driver to potentially dangerous conditions such as brake failure or low oil pressure.

2. **Indicator lights** include turn signal indicators.

3. **Illumination lights** provide indirect lighting to illuminate the instrument gauges, speedometer, heater controls, clock, ashtray, radio, and other controls.

The power source for the instrument panel lights is provided through the headlight switch. The contacts are closed when the headlight switch is located in the PARK or HEADLIGHT position. The current must flow through a variable resistor (rheostat) that is either a part of the headlight switch or a separate dial on the dash. The resistance of the rheostat is varied by turning the knob. By varying the resistance, changes in the current flow to the lamps control the brightness of the lights (Figure 8-55).

Lighting System Complexity

Today's vehicles have a sophisticated lighting system and electrical interconnections. It is possible to have problems with lights and accessories that cause them to operate when they are not supposed to. This is through a condition called **feedback.** Feedback occurs when electricity seeks a path of lower resistance. This alternate path operates another component than the one intended. If there is an open in the circuit, electricity will seek another path to follow. This may cause any lights or accessories in that path to turn on.

Examples of feedback and how it may cause undesired operation are illustrated in Figures 8-56 through 8-63. The illustration (Figure 8-56) shows a system that has the dome light, taillight, and brake light circuits on one fuse; the cigarette lighter circuit has its own fuse. The two fuses are located in the main fuse block and share a common bus bar on the power side.

If the dome light fuse blew and the headlight switch was in the PARK or HEADLIGHT position, the courtesy lights, dome light, taillight, parking lights, and instrument lights would all be very dimly lit (Figure 8-57). Current would flow through the cigarette lighter fuse to the courtesy light and on to the door light switch. Current will then continue through the dome light to the headlight switch. Because the headlight switch is now closed, the instrument panel lights are also in the circuit. The lights are dim because all the bulbs are now connected in series.

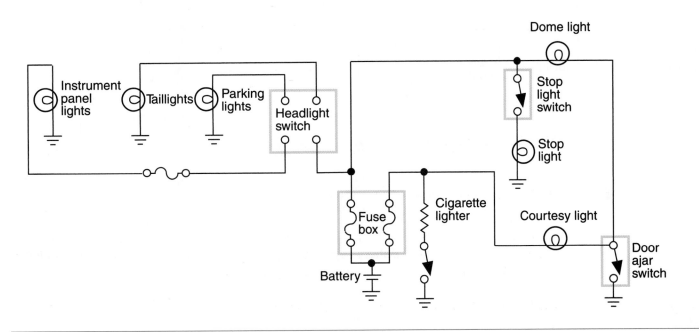

Figure 8-56 A normally operating light circuit.

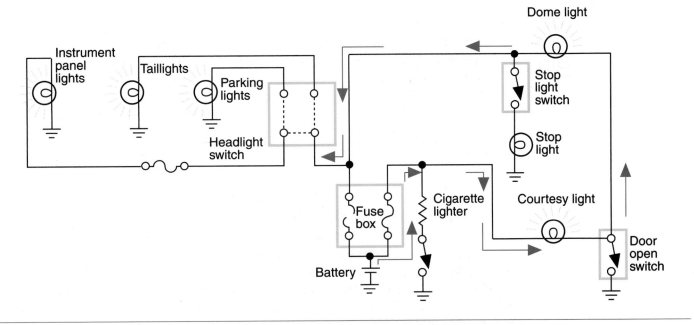

Figure 8-57 An open (blown) dome light fuse can cause feedback into other circuits when the headlights are turned on.

If the dome light fuse is blown and the headlight switch is in the OFF position, all lights will turn off. However, if the door is opened, the courtesy lights will come on but the dome light will not (Figure 8-58).

With the same blown fuse and the brake light switch closed, the dome light, courtesy light, and brake light will all illuminate dimly because the loads are in series (Figure 8-59).

In this example, if the dome light and courtesy lights come on dimly when the cigarette lighter is pushed in, the problem can be caused by a blown cigarette lighter fuse (Figure 8-60). With the cigarette lighter pushed in, a path to ground is completed. The lights and cigarette lighter are

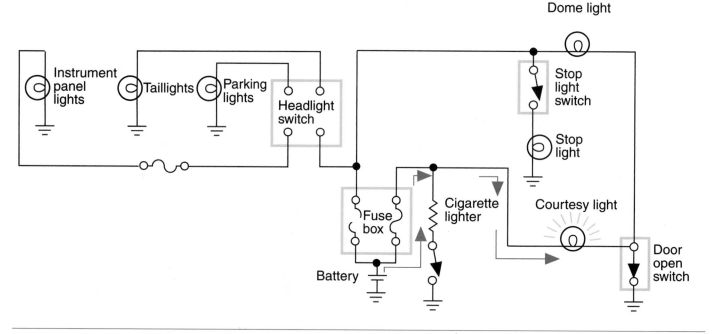

Figure 8-58 Feedback with a blown dome light fuse and the door switch closed.

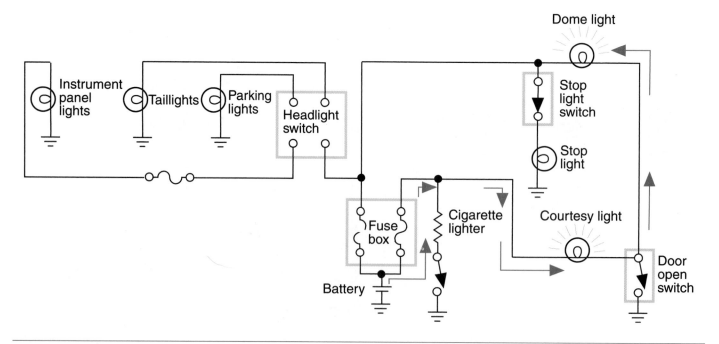

Figure 8-59 Feedback when brake light switch is closed.

now in series; thus the lights are dim and there is not enough current to heat and release the cigarette lighter. If the cigarette lighter was left in this position, the battery would eventually drain down.

A blown cigarette lighter fuse will also cause the dome light to get brighter when the doors are open, and the courtesy lights will go out (Figure 8-61). Also, if the lighter is pushed in and the brake light switch is closed, the dome and courtesy lights will go out (Figure 8-62).

Feedback can also be the result of a conductive corrosion that is developed at a connection. If the corrosion allows for current flow from one conductor to an adjacent conductor in the con-

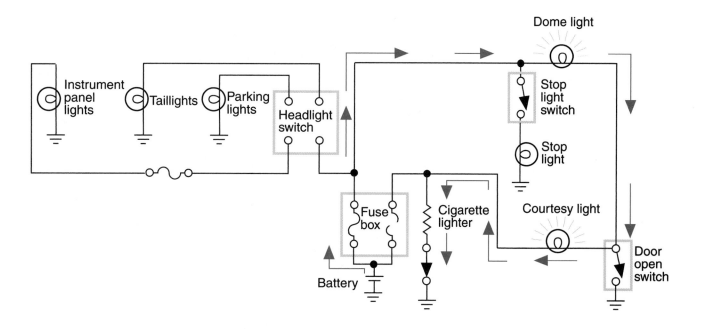

Figure 8-60 Feedback as a result of the cigarette lighter fuse being blown.

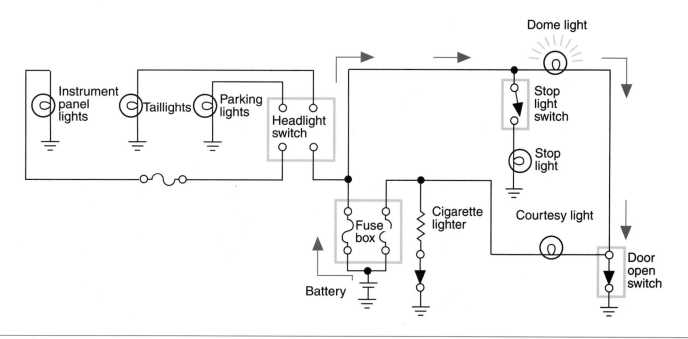

Figure 8-61 Opening the door will make the courtesy light go out and the dome light get brighter.

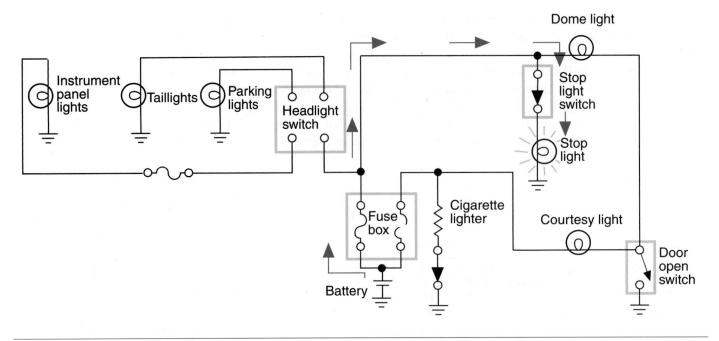

Figure 8-62 Dome and courtesy lights go out when the brakes are applied.

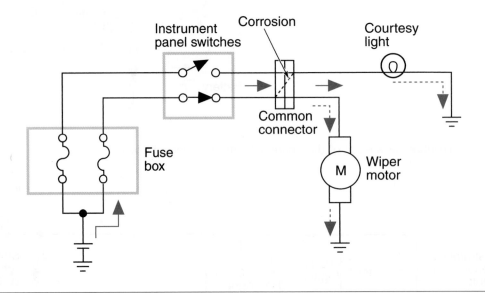

Figure 8-63 A corroded common connection can cause feedback.

nection, the other circuit will also be activated. The illustration (Figure 8-63) shows how corrosion in a common connector can cause the dome light to illuminate when the wiper motor is turned on. Because the wiper motor has a greater resistance than the light bulb, more voltage will flow through the bulb than through the motor. The bulb will light brightly, but the motor will turn very slowly or not at all. The same effect will result if the courtesy light switch is turned on with the motor switch off.

AUTHOR'S NOTE: Dual-filament light bulb sockets are subject to corrosion that can cause feedback. Also, dual-filament bulbs can have a filament burn out and attach to the other filament and result in feedback. The most common indicator of this problem is parking lights that illuminate when the brake pedal is applied.

Summary

❑ The most commonly thought of light circuit is the headlights. But there are many lighting systems in the vehicle.

❑ Different types of lamps are used to provide illumination for the systems. The lamp may be either a single-filament bulb that performs a single function, or a double-filament bulb that performs several functions.

❑ The headlight lamps can be one of four designs: standard sealed beam, halogen sealed beam, composite, or high-intensity discharge (HID).

❑ The headlight filament is located on a reflector that intensifies the light, which is then directed through the lens. The lens is designed to change the circular light pattern into a broad, flat light beam. Placement of the filament in the reflector provides for low- and high-beam light patterns.

❑ Some manufacturers use concealed headlights to improve the aerodynamics of the vehicle. The concealed headlight doors can operate from vacuum or by electrically controlled motors. Some systems incorporate the use of IC chips into the concealed headlight door control.

❑ In addition to the headlight system, the lighting systems include:

Stop lights.
Turn signals.
Hazard lights.
Parking lights.
Taillights.
Back-up lights.
Side marker lights.
Courtesy lights.
Instrument panel lights.

❑ The headlight switch can be used as the control of many of these lighting systems. Most headlight switches have a circuit breaker that is an integral part of the switch. The circuit breaker provides protection of the headlight system without totally disabling the headlight operation if a circuit overload is present.

❑ A rheostat is used in conjunction with the headlight switch to control the brightness of the instrument panel illumination lights.

Terms to Know

Composite headlight
Concealed headlight
Cornering lights
Courtesy lights
Dimmer switch
Double-filament lamp
Feedback
Flasher
Halogen
High-intensity discharge (HID)
Incandescence
Lamp
Prisms
Sealed-beam
Vaporized aluminum

Review Questions

Short-Answer Essays

1. What lighting systems are controlled by the headlight switch?

2. How is the brightness of the instrument cluster lamps controlled?

3. What is CHMSL?

4. What is the purpose of the lamp filament?

5. What three lighting circuits are incorporated within the instrument cluster?

6. List and describe the four types of headlight lamps used.

7. Describe the influence that the turn signal switch has on the operation of the brake lights in a two-bulb taillight assembly.

8. What is the purpose of the diodes on some CHMSL circuits?

9. How are most cornering light circuits wired to allow the cornering light to be on steady when the turn signal switch is activated?

10. Describe what the term *feedback* means and how it can affect the operation of the electrical system.

Fill in the Blanks

1. When today's technician performs repairs on the lighting systems, the repairs must meet at least two requirements: They must assure vehicle _____ and meet all _____ _____.

2. A _____ is a device that produces light as a result of current flow through a _____.

3. _____ _____ redirect the light beam and create a broad, flat beam.

4. The _____ _____ controls most of the vehicle's lighting systems.

5. The _____ _____ provides the means for the driver to select either high- or low-beam operation.

6. The complete headlight circuit consists of the _____ _____, _____ _____, _____ _____ _____, and the _____.

7. The headlight doors of a concealed system can be controlled by either _____ _____ or by _____.

8. On vehicles equipped with cornering lights, the turn signal switch has an additional set of _____ that operate the cornering light circuit only.

9. Most limit switches operate off of a _____ on the motor.

10. In most vehicles equipped with automatic transmissions, the back-up light switch is part of the _____ _____ _____. Most manual transmissions are equipped with a _____ switch.

Multiple Choice

1. In the two-bulb taillight assembly:
   A. The brake lights use the high-intensity filament of the taillight bulb.
   B. The current to the brake lights flows through the turn signal switch.
   C. All of the above.
   D. None of the above.

2. In a composite headlight:
   A. The bulb is replaceable.
   B. A cracked lens will prevent lamp operation.
   C. All of the above
   D. None of the above.

3. Current to the brake light switch usually comes from the:
   A. Ignition switch feed.
   B. Headlight switch feed.
   C. Turn signal switch feed.
   D. Direct battery feed.

4. Which of the following statements about the turn signal circuit is NOT correct?
   A. The dimmer switch is not a part of the circuit.
   B. Most flashers use a bimetallic strip to open and close the circuit.
   C. The turn signals operate only when the neutral safety switch is open.
   D. The turn signal switch provides the means for lighting brake lights when the turn signals are operating.

5. The turn signal system is being discussed.
   Technician A says there is a separate flasher unit used for the left and the right turn signals.
   Technician B says on most vehicles, the flasher unit is located in the lamp socket.
   Who is correct?
   A. A only
   B. B only
   C. Both A and B
   D. Neither A nor B

6. The concealed headlight system is being discussed.
   Technician A says the system can use vacuum to operate the doors.
   Technician B says the system can use electric motors to operate the doors.
   Who is correct?
   A. A only
   B. B only
   C. Both A and B
   D. Neither A nor B

7. The CHMSL circuit is being discussed.
   Technician A says the diodes are used to assure proper turn signal operation.
   Technician B says the diodes are used to prevent radio static when the brake light is activated.
   Who is correct?
   A. A only
   B. B only
   C. Both A and B
   D. Neither A nor B

8. Which statement about the cornering light system is NOT correct?
   A. The cornering lights use an additional set of contacts in the turn signal switch.
   B. The cornering lights can receive voltage from the ignition switch.
   C. The cornering lights can receive voltage from the headlight switch.
   D. The cornering lights can operate only if the vehicle speed sensor input indicates speeds over 3 mph (4.8 kph).

9. *Technician A* says the side marker lights can be wired to flash with the turn signals.
   Technician B says wrap-around lenses can be used for side marker lights.
   Who is correct?
   A. A only
   B. B only
   C. Both A and B
   D. Neither A nor B

10. Which of the following best describes feedback?
    A. Feedback is normal during the operation of the electrical system.
    B. Feedback can be cause circuits to operate in series.
    C. Feedback can cause circuits to operate in parallel.
    D. All of the above.

Electrical Accessories

Upon completion and review of this chapter, you should be able to:

❏ Describe the operation and function of the horn circuit.

❏ Explain and detail the operation of standard two- and three-speed wiper motors, both permanent magnet and electromagnetic field designs.

❏ Describe the operation of intermittent wipers.

❏ Explain how depressed-park wipers operate.

❏ Explain the operation of windshield washer pump systems.

❏ Describe the operation and methods used to control blower fan motor speeds.

❏ Describe the operation of electric defoggers.

❏ Describe operational principles of power mirrors.

❏ Explain the principles of operation for power windows, power seats, and power locks.

Introduction

Electrical accessories provide additional safety and comfort. There are many electrical accessories that can be installed into today's vehicles. This chapter explains the principles of operation for some of the most common electrical accessories. Systems not discussed here are similar in concept.

This chapter explores the operation of safety accessories, such as the horn, windshield wipers, and windshield washers. Comfort accessories explored in this chapter include the blower motor, electric defoggers, power mirrors, power windows, power seats, and power door locks.

Shop Manual
Chapter 9, page 335

A horn is a device that produces an audible warning signal.

Horns

The automotive electrical horn operates on an electromagnetic principle that vibrates a **diaphragm** to produce a warning signal. The diaphragm is a thin, flexible, circular plate that is held around its outer edge by the horn housing, allowing the middle to flex. Most electrical horns consist of an electromagnet, a moveable armature, a diaphragm, and a set of normally closed contact points (Figure 9-1). The contact points are wired in series with the field coil. One of the points is attached to the armature. When current flows through the field coil, a magnetic field is developed that

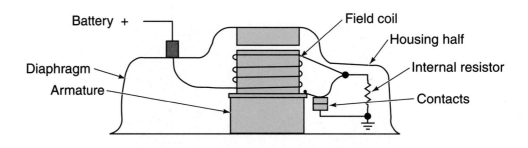

Figure 9-1 Horn construction. The internal resistor allows a weak magnetic field to remain after the contacts open, reducing the amount of time required to rebuild the field when the contacts close again.

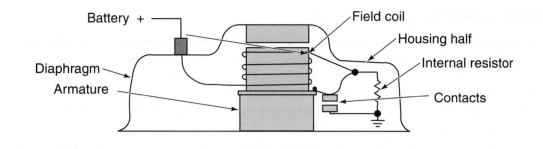

Figure 9-2 When the armature pulls the diaphragm up, the contacts open.

attracts the moveable armature (Figure 9-2). The diaphragm is attached to and moves with the armature. Movement of the armature results in the contact points opening, which breaks the circuit. The diaphragm is released and returns to its normal position. The contact points close again and the cycle repeats. This vibration of the diaphragm is repeated several times per second.

As the diaphragm vibrates, it causes a column of air in the horn to vibrate. The vibration of the column of air produces the sound.

Most vehicles are equipped with two horns that are wired in parallel with each other and in series with the switch. One of the horns will have a slightly lower pitch than the other. The design and shape of the horn determines the frequency and tone of the sound (Figure 9-3). Pitch is controlled by the number of times the diaphragm vibrates per second. The faster the vibration, the higher the pitch. The horn pitch can be adjusted by changing the spring tension of the armature. This alters the magnetic pull on the armature and changes the rate of vibrations. The adjustment is made on the outside of the horn (Figure 9-4).

When two horns are used on a vehicle, one is called the low-note horn and the other is called the high-note horn.

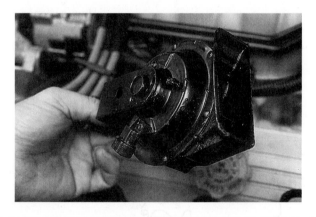

Figure 9-3 Horn design affects sound quality.

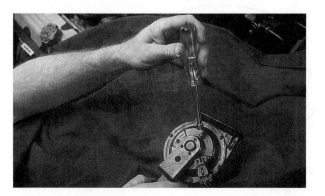

Figure 9-4 Horn pitch adjustment screw.

Horn Switches

Horn switches are installed either in the steering wheel or as a part of the multifunction switch. Most horn switches are normally open switches.

The steering wheel–mounted horn switch can be a single button in the middle of the steering wheel (Figure 9-5). Another design is to have multiple buttons in the horn pad (Figure 9-6). Switches mounted on the steering wheel require the use of a slip ring (Figure 9-7). The slip ring has contacts that provide continuity for the horn control in all steering wheel positions. The contacts consist of a circular contact in the steering wheel that slides against a spring-loaded contact in the steering column. Most vehicle manufacturers now use a **clockspring** (Figure 9-8) to provide continuity between the steering wheel components—horn switch, cruise control switches, air bag, and so on—and the steering column wiring harness. A clockspring is a winding of electrical

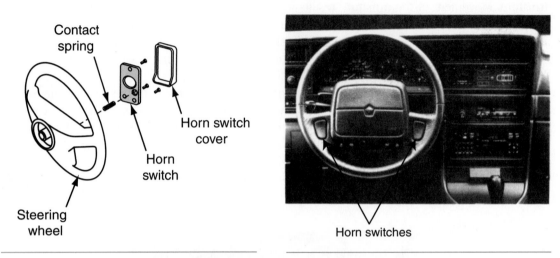

Figure 9-5 Single-button-type horn pad.

Figure 9-6 Multiple-button horn pad.

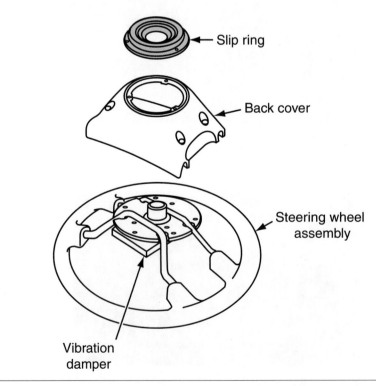

Figure 9-7 Slip ring contact provides horn continuity in all steering wheel positions.

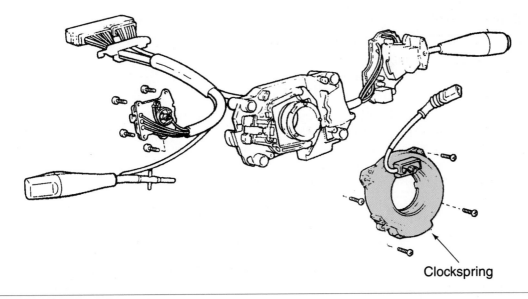

Clockspring

Figure 9-8 Most vehicle manufacturers now use a clockspring instead of sliding contacts.

conducting tape enclosed within a plastic housing. The clockspring maintains continuity between the steering wheel switches, the air bag, and the wiring harness in all steering wheel positions. The clockspring provides a more reliable connection than the sliding contacts.

Horn switches that are a part of the multifunction switch usually operate by a push button on the end of the lever.

Horn Circuits

There are two methods of circuit control: with or without a relay. If the horn circuit does not use a relay, the horns must be of low current design because the horn switch carries the total current. Depressing the horn switch completes the circuit from the battery to the horns (Figure 9-9).

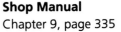

Shop Manual
Chapter 9, page 335

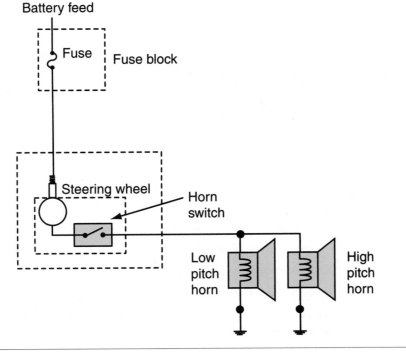

Figure 9-9 Insulated side switch without relay.

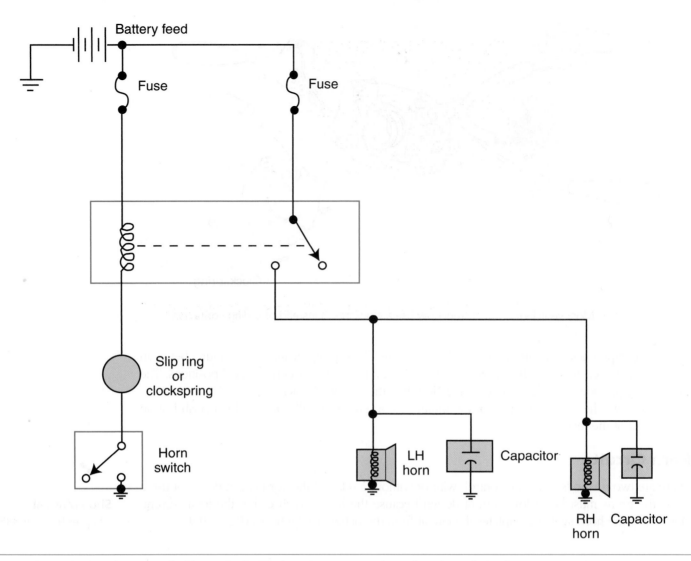

Figure 9-10 Relay-controlled horn circuit. The horn button completes the relay coil circuit.

The most common type of circuit control is to use a relay (Figure 9-10). Most circuits have battery voltage present to the lower contact plate of the horn switch. When the switch is depressed, the contacts close and complete the circuit to ground. Only low current is required to operate the relay coil, so the horn switch does not have to carry the heavy current requirements of the horns. When the horn switch is closed, it energizes the relay core. The core attracts the relay armature, which closes the contacts and completes the horn circuit. Current flows from the battery to the grounded horns.

Windshield Wipers

Shop Manual
Chapter 9, page 339

There are two types of windshield wiper systems being used today:

1. A standard two- or three-speed system.

2. A two- or three-speed system with an intermittent feature.

On most models, the same motor is used for either the standard or intermittent system. Both can have a **depressed-park** feature. In systems equipped with depressed park, the blades drop down below the lower windshield molding to hide them.

Figure 9-11 Headlight wipers.

A single-speed rear window wiper and washer is used on many vehicles. In addition, headlight wipers are installed on many luxury vehicles that operate in union with the windshield wipers (Figure 9-11). The operation of these accessories is the same as the windshield wipers.

A BIT OF HISTORY

Windshield wipers were introduced at the 1916 National Auto Show by several manufacturers. These were hand operated. Vacuum-operated wipers were standard equipment by 1923. Electric wipers became common in the 1950s.

Wiper Motors

Windshield wipers are mechanical arms that sweep back and forth across the windshield to remove water, snow, or dirt. Most windshield wiper motors use permanent magnet fields. Motor speed is controlled by the placement of the brushes on the commutator. Three brushes are used: common, high speed, and low speed. The common brush carries current whenever the motor is operating. The low-speed and high-speed brushes are placed in different locations, based on motor design:

1. The high-speed and common brushes oppose each other, with the low-speed brush offset (Figure 9-12).

2. The low-speed and common brushes oppose each other with the high-speed brush offset or centered between them (Figure 9-13). This is the most common brush arrangement.

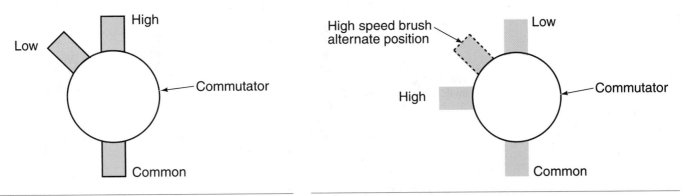

Figure 9-12 One style of brush arrangement has the high-speed brush opposite the common brush.

Figure 9-13 The most common brush arrangement is to place the low-speed brush opposite the common brush.

The placement of the brushes determines how many armature windings are connected in the circuit. There are fewer armature windings connected between the common and high-speed brushes. When battery voltage is applied to fewer windings, there is less magnetism in the armature and a lower counterelectromotive force (CEMF). With less CEMF in the armature, there is greater armature current. The greater armature current results in higher speeds.

There are more armature windings connected in the circuit between the common and low-speed brushes. With more windings, the magnetic field in the armature is increased. This results in greater CEMF. The increased CEMF reduces the amount of current in the armature and slows the speed of the motor.

Some two-speed, and all three-speed wiper motors use two electromagnetic field windings instead of permanent magnets (Figure 9-14). The two field coils are wound in opposite directions so that their magnetic fields will oppose each other. The series field is wired in series

Most wiper circuits use a circuit breaker to prevent temporary overloads from totally disabling the windshield wipers due to a blown fuse.

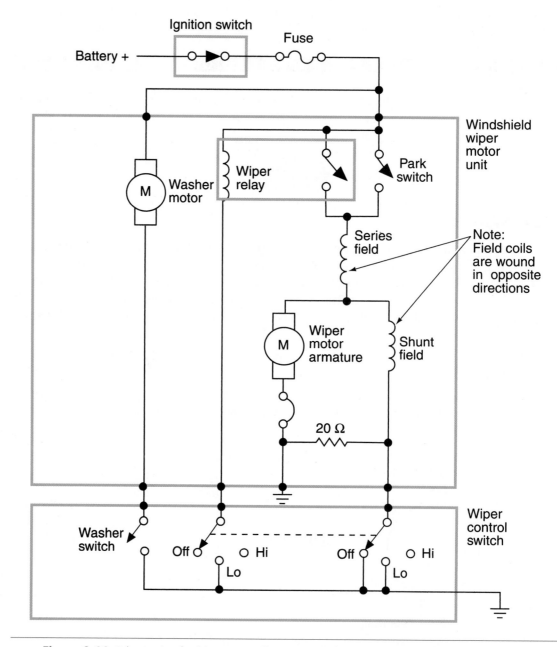

Figure 9-14 Schematic of wiper system that uses an electromagnetic field motor.

with the motor brushes and commutator. The shunt field forms a separate circuit branch off the series field to ground. The strength of the total magnetic field will determine the speed of the motor.

A BIT OF HISTORY

The first windshield wipers were hand operated. Most of the early powered wiper systems were operated by engine vacuum. Many times the wipers would not operate when the vehicle was driven up a hill because of the loss of engine vacuum under those conditions. In 1968, federal law mandated that all vehicles must be equipped with a two-speed wiper and washer system. American Motors Corporation continued to use a two-speed vacuum motor until the 1970s; all other manufacturers converted to electric motors.

Permanent Magnet Wiper Circuits

A set of **park contacts** are incorporated into the motor assembly and operate off a cam or latch arm on the motor gear (Figure 9-15). Park contacts are located inside the motor assembly, and supply current to the motor after the wiper control switch has been turned to the PARK position. This allows the motor to continue operating until the wipers have reached their PARK position. The park switch changes position with each revolution of the motor. The switch remains in the RUN position for approximately 90% of the revolution. It is in the PARK position for the remaining 10% of the revolution. This does not affect the operation of the motor until the wiper control switch is placed in the PARK position.

Shop Manual
Chapter 9, page 339

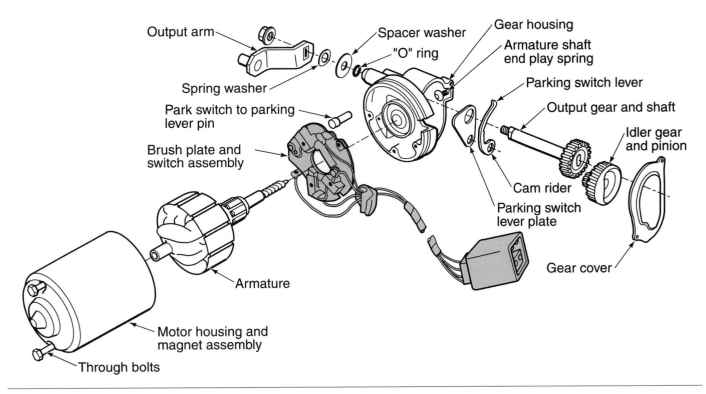

Figure 9-15 Park switch components.

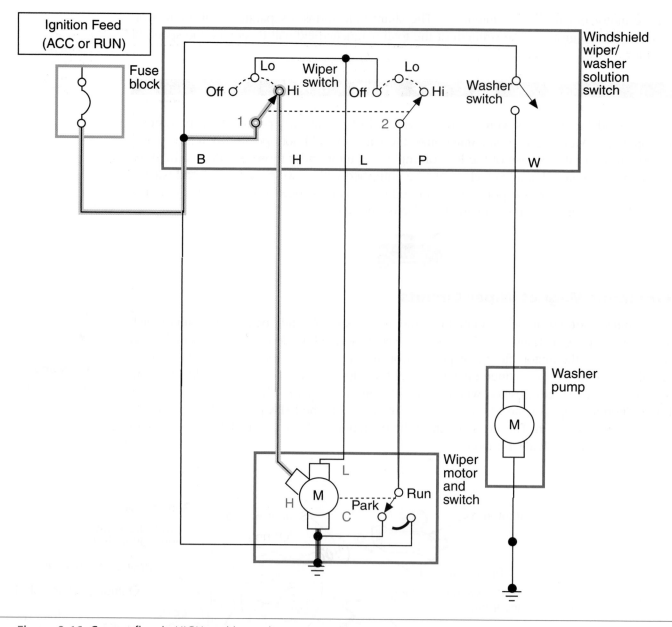

Figure 9-16 Current flow in HIGH position, using a permanent magnet motor.

When the wiper control switch is in the HIGH-SPEED position, battery voltage is supplied through wiper 1 to the high-speed brush (Figure 9-16). Wiper 2 moves with wiper 1 but does not complete any circuits. Current flows through the low-speed brush, the armature, and the common brush to ground. Because the ground connection is before the park switch, the park switch position has no effect on motor operation.

When the switch is placed in the LOW-SPEED position, battery voltage is supplied through wiper 1 to the low-speed brush (Figure 9-17). Wiper 2 also moves to the LOW position, but does not complete any circuits. Current flows through the armature, the low-speed brush, and the common brush to ground. Park switch position has no effect on motor operation.

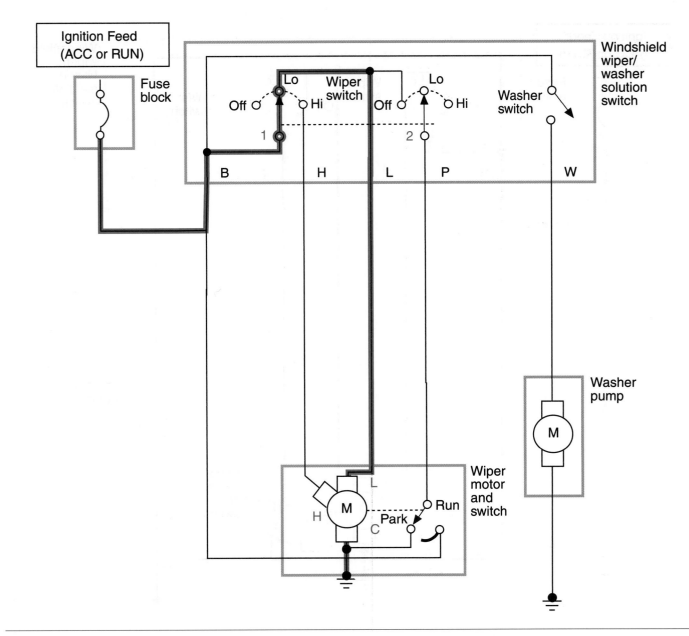

Figure 9-17 Current flow in LOW position.

When the switch is returned to the OFF position, wiper 1 opens (Figure 9-18). Battery voltage is applied to the park switch. Wiper 2 is closed to allow current to flow to the low-speed brush, through the armature, and to ground. When the wiper blades are in their lowest position, the park switch is moved to the PARK position. This opens the circuit to the low-speed brush and the motor shuts off.

Electromagnetic Field Wiper Circuits

Electromagnetic field motors use two fields that are wired in opposite directions: the series field and the shunt field. Resistors are used in series with one of the fields to control the strength of the total magnetic field. The wiper control switch directs the current flow through the resistors to obtain the desired motor speed. Circuit operation varies between two- and three-speed systems.

Shop Manual
Chapter 9, page 339

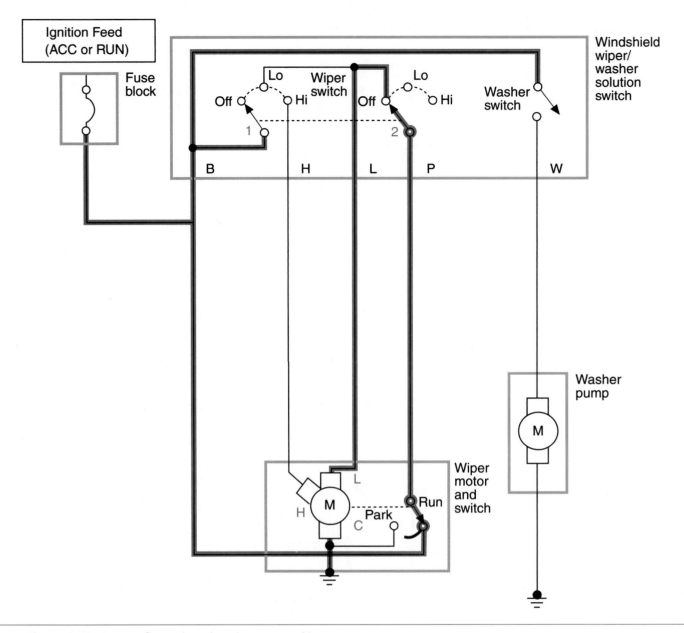

Figure 9-18 Current flow when the wipers are parking.

Two-Speed Motors. The ground side switch will determine the current path. One path is directly to ground after the field coil, the other is through a 20-ohm resistor (Figure 9-19).

With the switch in the OFF position, switch wiper 1 breaks the circuit through the relay to ground. Because the relay is not energized, current is not supplied to the motor.

When the switch is placed in the LOW-SPEED position, wiper 1 completes the relay coil circuit to ground (Figure 9-20). The energized relay closes the contacts and applies current to the motor, through the series field and shunt field coils. Wiper 2 provides the ground path for the shunt field coil. Because there is less resistance to ground through wiper 2, the 20-ohm resistor is bypassed. With no resistance in the shunt field coil, the shunt field is very strong and bucks the magnetic field of the series field. The result is slow motor operation.

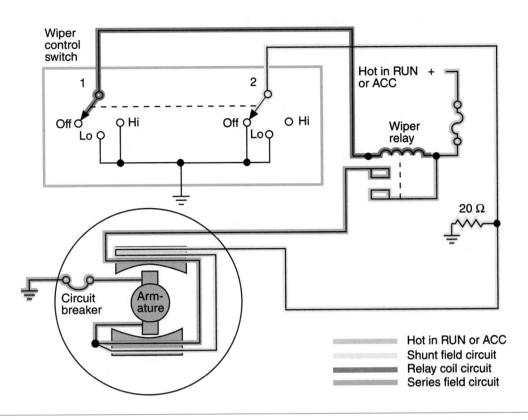

Figure 9-19 Simplified diagram of an electromagnetic field, two-speed wiper system.

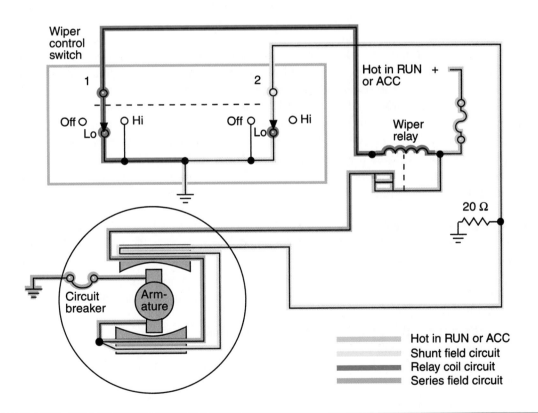

Figure 9-20 Current flow in the LOW-SPEED position. Wiper 2 provides the ground path, bypassing the 20-ohm resistor.

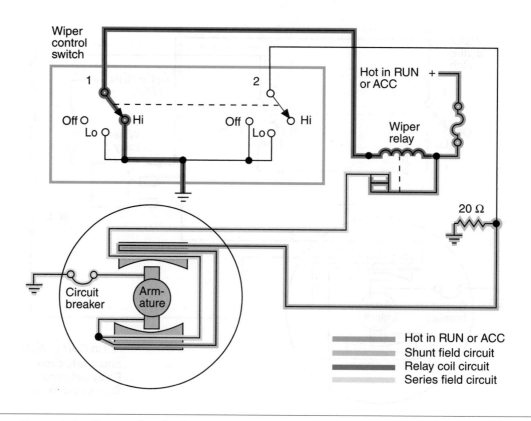

Wiper control switch

1
2

Off O
Lo O
Hi

Off O
Lo O
Hi

Circuit breaker

Arm-ature

Hot in RUN or ACC +

Wiper relay

20 Ω

Hot in RUN or ACC
Shunt field circuit
Relay coil circuit
Series field circuit

Figure 9-21 Current flow in the HIGH-SPEED position. Wiper 2 opens, causing the current to flow through the resistor.

When the switch is located in the HIGH-SPEED position, wiper 1 completes the relay coil circuit to ground. This closes the relay contacts to the series field and shunt field coils. Wiper 2 opens the circuit to ground and current must now pass through the 20-ohm resistor to ground (Figure 9-21). The resistor reduces the current flow and strength of the shunt field. With less resistance from the shunt field, the series field is able to turn the motor at an increased speed.

When the switch is returned to the OFF position, wiper 1 opens the relay coil circuit. The relay is de-energized, but the contact points to the series and shunt field coils are manually held closed by the park switch. The park switch is closed until the wipers are in their lowest position. Wiper 2 closes the circuit path to ground. As long as the park switch is closed, current flows through the series field, shunt field, and wiper 2 to ground (Figure 9-22). Once the wipers are in their lowest point of travel, the park switch opens and the motor turns off.

Three-Speed Motors. The three-speed motor system offers a low-, medium-, and high-speed selection. The wiper control switch position determines what resistors, if any, will be connected to the circuit of one of the fields (Figure 9-23).

When the wiper control switch is placed in the LOW-SPEED position, both field coils receive equal current (Figure 9-24). Both field coils have the same amount of current flow, so the total magnetic field is weak and the motor speed is slow.

When the switch is placed in the MEDIUM-SPEED position, the current flows through a resistor before flowing to the shunt field (Figure 9-25). The resistor weakens the strength of the shunt coil but strengthens the total magnetic field of the motor. The speed is increased over that of the LOW-SPEED position.

Shop Manual
Chapter 9, page 339

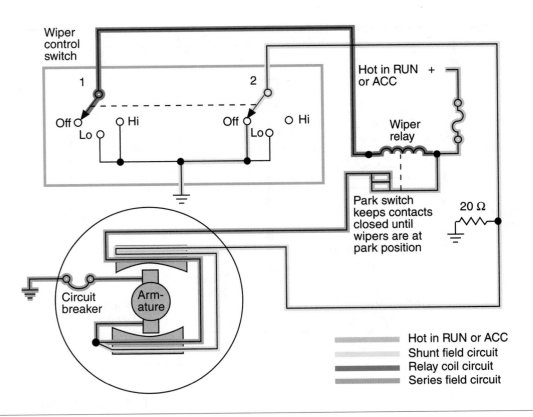

Figure 9-22 Current flow when the wipers are parking.

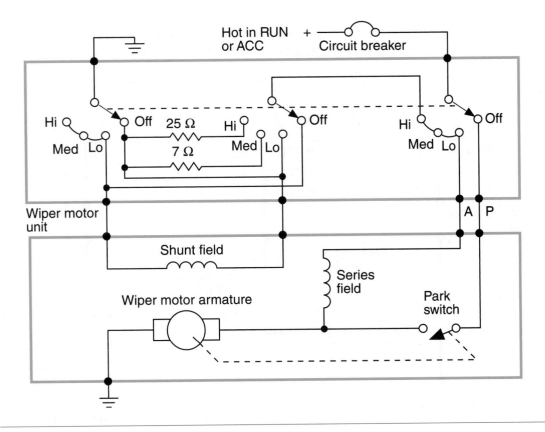

Figure 9-23 Three-speed wiper motor schematic.

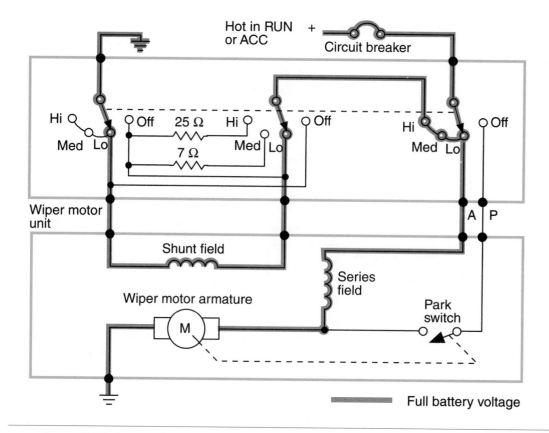

Figure 9-24 Current flow in LOW position.

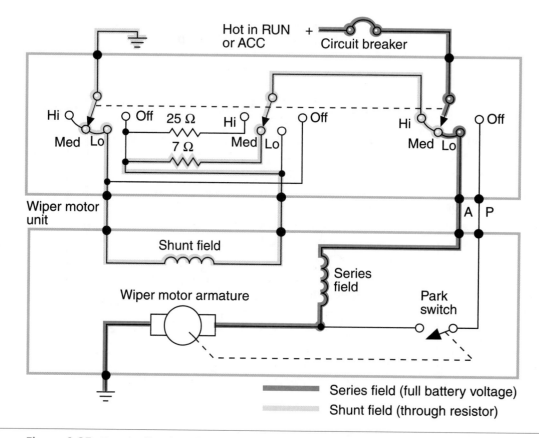

Figure 9-25 Current flow in MEDIUM position.

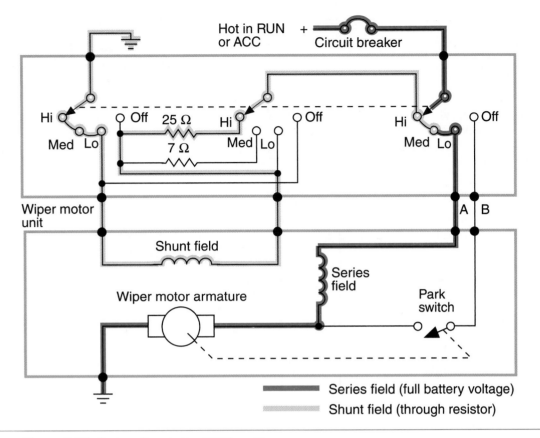

Figure 9-26 Current flow in the HIGH position.

When the switch is placed in the HIGH-SPEED position, a resistor of greater value is connected into the shunt field circuit (Figure 9-26). The resistor weakens the magnetic field of the shunt coil, allowing a stronger total field to rotate the motor at a high speed.

Relay-Controlled, Two-Speed Wiper Systems

To reduce the size of the wires through the steering column harness, a wiper system may use relays to supply current to the wiper motor (Figure 9-27). With the ignition switch in the RUN or ACC position, battery voltage is supplied to the wiper motor and to the coils of the ON/OFF and HI/LO relays. If the wiper switch is placed in the LOW speed position, the coil of the ON/OFF relay is energized (Figure 9-28). The contacts of the ON/OFF relay now provide ground for the wiper motor. Since the Hi/Lo relay coil is not energized, current flows through the motor's low-speed brush to ground.

When the wiper switch is placed in the HIGH-SPEED position, the switch provides ground for both the wiper ON/OFF and the wiper Hi/Lo relay coils (Figure 9-29). The wiper Hi/Lo relay is a circuit diverter and now provides ground for the wiper motor's high-speed brush through the contacts of the wiper ON/OFF relay (which remains energized).

This system may also incorporate an intermittent wipe module that provides a parallel path to ground for the ON/OFF relay coil. The relay is energized long enough to activate the wiper motor. Once the wipers move, the park switch closes and the motor continues to run until the wipers are in the park location. At this time, the park switch opens and the motor stops. After the delay period has expired, the process is repeated.

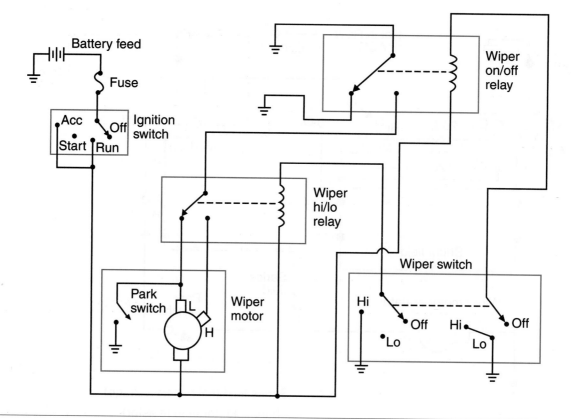

Figure 9-27 Relay-controlled wiper system.

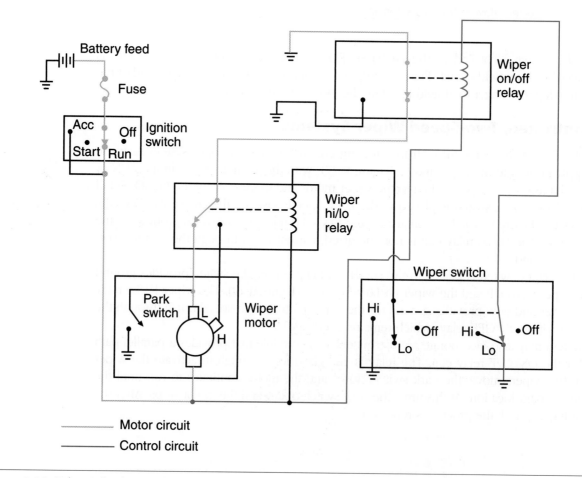

Motor circuit

Control circuit

Figure 9-28 Relay activation to obtain low-speed operation.

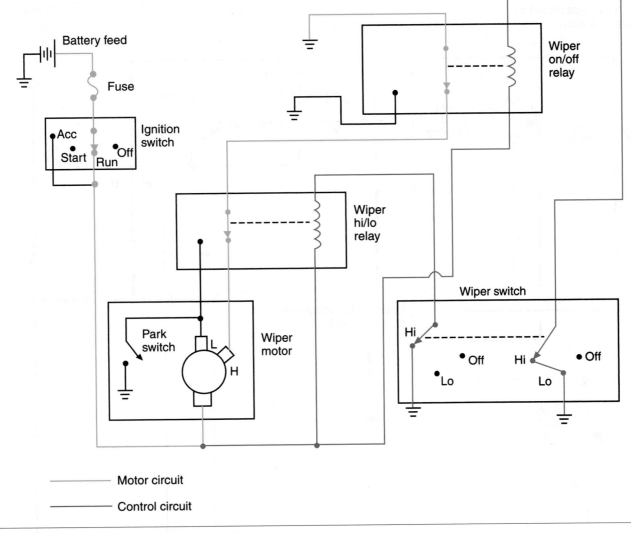

Figure 9-29 Relay activation to obtain high-speed operation.

Intermittent Wipers

Many wiper systems offer an intermittent mode that provides a variable interval between wiper sweeps. Many of these systems use a module located in the steering column (Figure 9-30). If the intermittent wiper mode is initiated when the wipers are in their parked position, the park switch is in the ground position. Current is sent to the solid-state module to the "timer activate" terminal

Shop Manual
Chapter 9, page 345

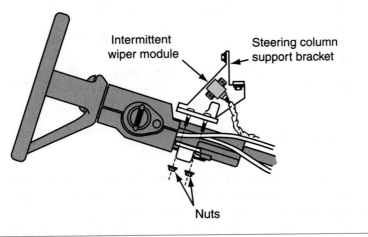

Figure 9-30 Intermittent wiper module.

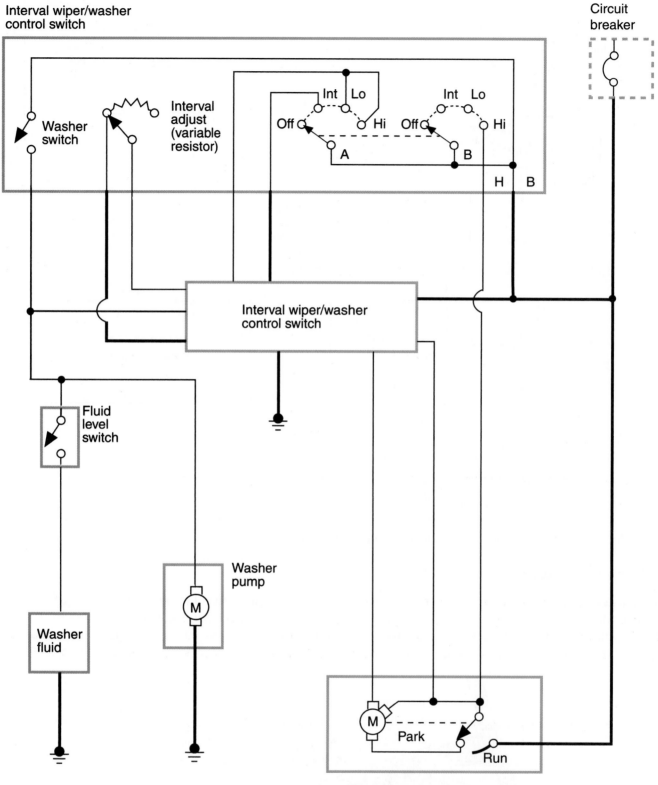

Figure 9-31 Intermittent wiper system schematic.

(Figure 9-31). The internal timer unit "triggers" the electronic switch to close the circuit for the governor relay, which then closes the circuit to the low-speed brush (Figure 9-32). The wiper will operate until the park switch swings back to the PARK position.

The delay between wiper sweeps is determined by the amount of resistance the driver puts into the potentiometer control. By rotating the intermittent control knob, the resistance value is

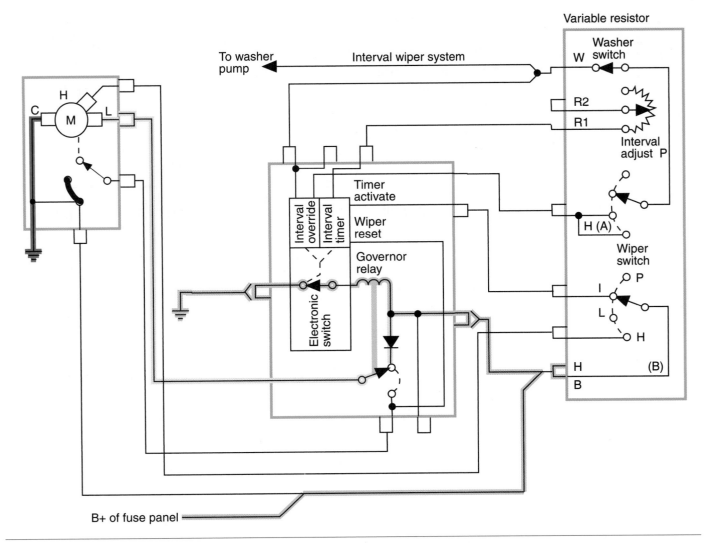

Figure 9-32 Current flow when intermittent wiper mode is initiated.

altered. The module contains a capacitor that is charged through the potentiometer. Once the capacitor is saturated, the electronic switch is "triggered" to send current to the wiper motor. The capacitor discharge is long enough to start the wiper operation, and the park switch is returned to the RUN position. The wiper will continue to run until one sweep is completed and the park switch opens. The amount of time between sweeps is based on the length of time required to saturate the capacitor. As more resistance is added to the potentiometer, it takes longer to saturate the capacitor.

AUTHOR'S NOTE: Many manufacturers are incorporating this function into the body computer. Also, some manufacturers are equipping their vehicles with speed sensitive wiper systems. The delay between wiper sweeps is determined by the speed of the vehicle.

Depressed-Park Wiper Systems

Systems that have a depressed-park feature use a second set of contacts with the park switch. These contacts are used to reverse the rotation of the motor for about 15 degrees after the wipers have reached the normal PARK position. The circuitry of the depressed circuit is different from that of standard wiper motors.

The operation of a depressed-park wiper system in the LOW-SPEED position is shown (Figure 9-33). Current flows through the number 3 wiper to the common brush. Ground is provided through the low-speed brush and switch wiper 2.

When the switch is placed in the OFF position, current is supplied through the park switch wiper B and switch wiper 3 (Figure 9-34). Ground is supplied through the low-speed brush and switch wiper 1, then to park switch wiper A.

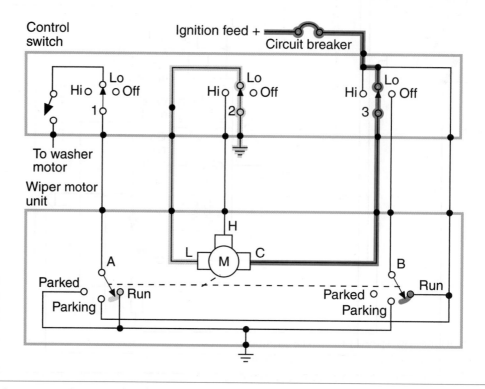

Figure 9-33 Depressed-park wiper system in LOW-SPEED position.

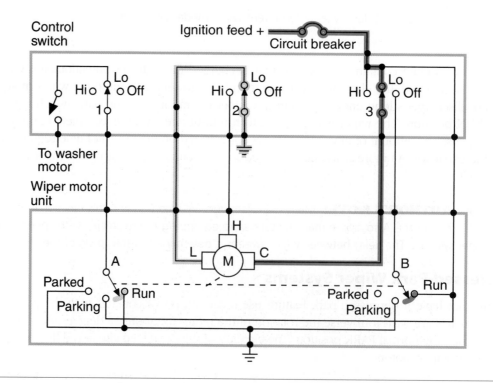

Figure 9-34 Current flow when the switch is turned to OFF after operation (parking).

266

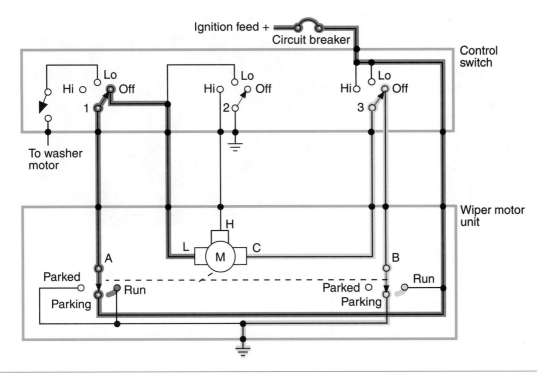

Figure 9-35 Current flow when wipers are parking into the depressed position.

When the wipers reach their PARK position, the park switch swings to the PARKING position (Figure 9-35). Current flow is through the park switch wiper A, to switch wiper 1. Wiper 1 directs the current to the low-speed brush. The ground path is through the common brush, switch wiper 3, and park switch wiper B. This reversed current flow is continued until the wipers reach the depressed-park position, when park switch wiper A swings to the PARKED position.

 AUTHOR'S NOTE: Note the difference in operation between the depressed-park wiper system and the one shown in Figure 9-16.

Washer Pumps

Windshield washers spray a washer fluid solution onto the windshield and work in conjunction with the wiper blades to clean the windshield of dirt. Some vehicles that have composite headlights incorporate a headlight washing system along with the windshield washer (Figure 9-36).

Shop Manual
Chapter 9, page 348

The washer system includes plastic or rubber hoses to direct fluid flow to the nozzles and produce the spray pattern.

Figure 9-36 Headlight washer system may operate with the windshield washer or have a separate switch.

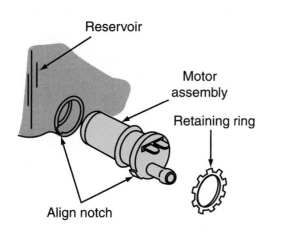

Figure 9-37 Washer motor installed into the reservoir.

Figure 9-38 General Motors' pulse-type washer system incorporates the washer motor into the wiper motor.

Most systems have the washer pump motor installed into the reservoir (Figure 9-37). General Motors uses a pulse-type washer pump that operates off the wiper motor (Figure 9-38).

The system is activated by holding the washer switch (Figure 9-39). If the wiper/washer system also has an intermittent control module, a signal is sent to the module when the washer switch is activated (Figure 9-40). An override circuit in the module operates the wipers on low speed for a programmed length of time. The wipers either will return to the parked position or will operate in intermittent mode, depending on system design.

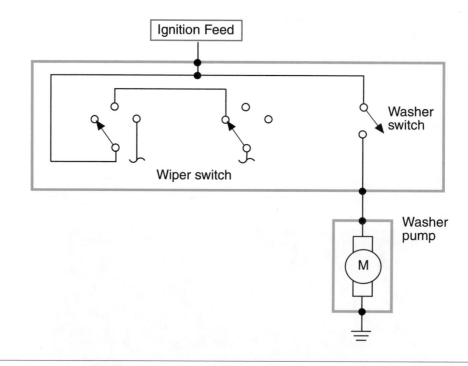

Figure 9-39 Windshield washer motor circuit.

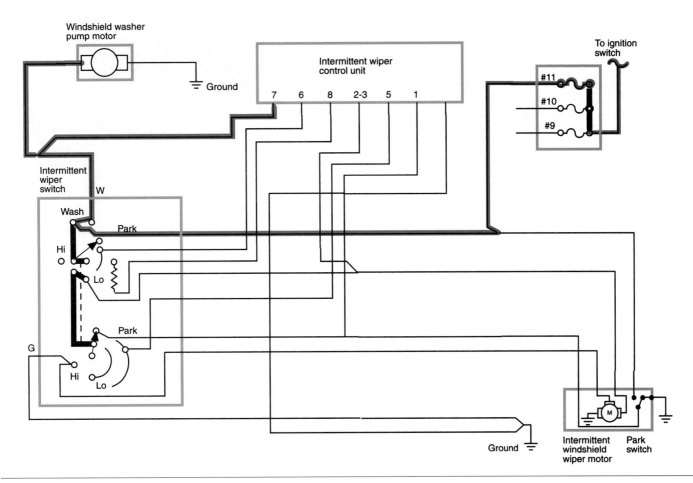

Figure 9-40 Input signal alerts the module that the washers are activated.

Blower Motor Circuit

The blower motor is used to move air inside the vehicle for air conditioning, heating, defrosting, and ventilation. The motor is usually a permanent magnet, single-speed motor and is located in the heater housing assembly (Figure 9-41). A blower motor switch mounted on the dash controls

Shop Manual
Chapter 9, page 349

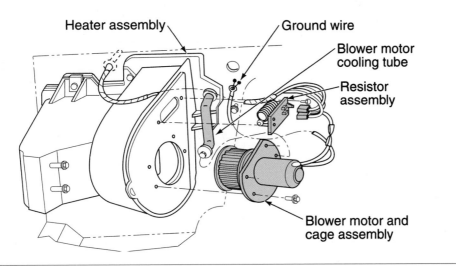

Figure 9-41 The blower motor is usually installed into the heater assembly. Mode doors control if vent, heater, or A/C-cooled air is blown by the motor cage.

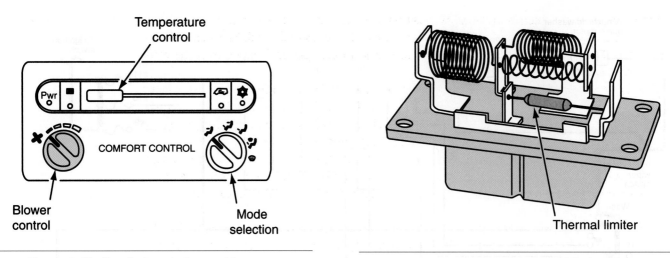

Figure 9-42 Comfort control assembly.

Figure 9-43 Fan motor resistor block.

the fan speed (Figure 9-42). The switch position directs current flow to a **resistor block** that is wired in series between the switch and the motor (Figure 9-43) and consists of two or three helically wound wire resistors wired in series.

The blower motor circuit includes the control assembly, blower switch, resistor block, and blower motor (Figure 9-44). This system uses an insulated side switch and a grounded motor. Battery voltage is applied to the control head when the ignition switch is in the RUN or ACC positions. The current can flow from the control head to the blower switch and resistor block in any control head position except OFF.

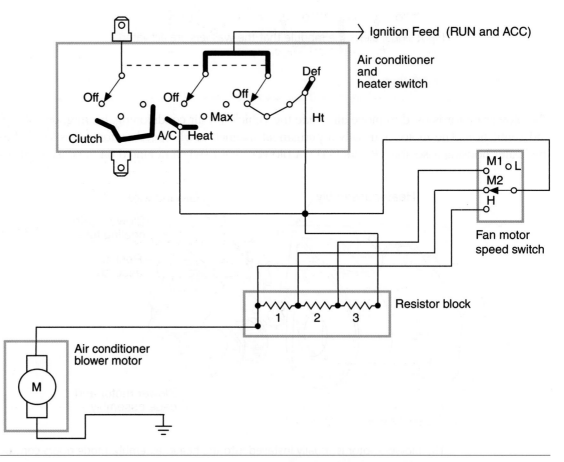

Figure 9-44 Blower motor circuit.

When the blower switch is in the LOW position, the blower switch wiper opens the circuit. Current can only flow to the resistor block directly through the control head. The current must pass through all the resistors before reaching the motor. With the voltage dropped over the resistors, the motor speed is slowed (Figure 9-45).

When the blower switch is placed in the MED 1, MED 2, or HIGH position, the current flows through the blower switch to the resistor block. Depending on the speed selection, the current must pass through one, two, or none of the resistors (Figure 9-46). With more applied voltage to the motor, the fan speed is increased as the amount of resistance decreases.

Current through the circuit will remain constant; varying the amount of resistance changes the voltage applied to the motor. Because the motor is a single-speed motor, it obtains its fastest rotational speed with full battery voltage. The resistors drop the amount of voltage to the motor, resulting in slower speeds.

Some manufacturers use ground side switching with an insulated motor (Figure 9-47). The switch completes the circuit to ground. Depending on wiper position, current flow is directed through the resistor block. The operating principles are identical to that of the insulated switch already discussed.

 AUTHOR'S NOTE: Many of today's vehicles are using the body computer to control fan speed by pulse width modulation.

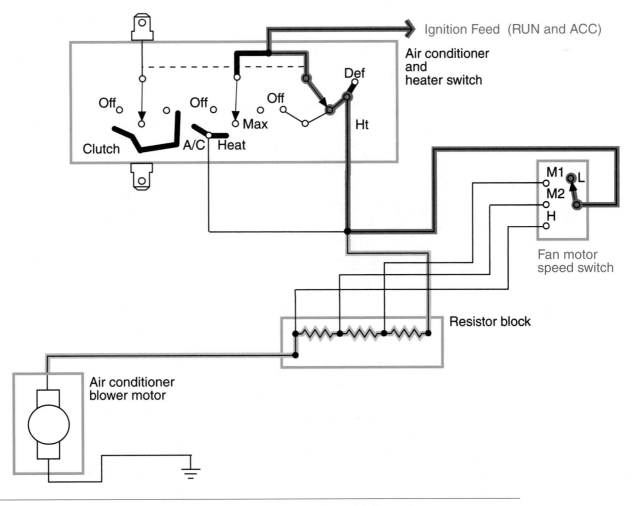

Figure 9-45 Current flow in the LOW position.

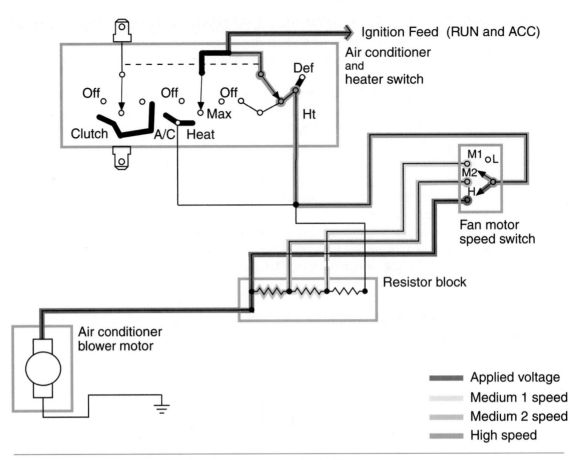

Figure 9-46 Current flow in the different speed selections.

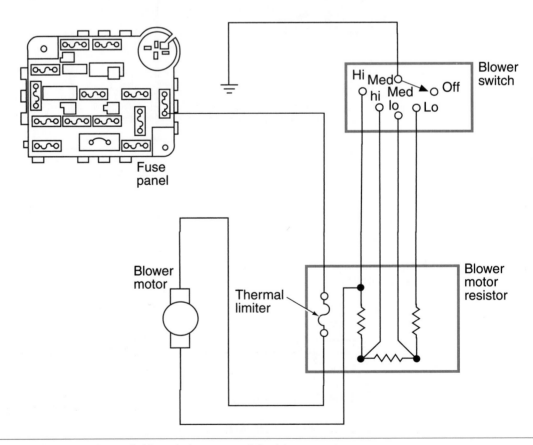

Figure 9-47 Ground side switch to control the blower motor system.

Automotive electric heaters were introduced at the 1917 National Auto Show. Hot water in-car heaters were introduced in 1926.

Electric Defoggers

Shop Manual
Chapter 9, page 352

When electrons are forced to flow through a resistance, heat is generated. Rear window electric defoggers use this principle of controlled resistance to heat the glass. Electric defoggers heat the rear window to remove ice and/or condensation. Some vehicles use the same circuit to heat the outside driver's side mirror. The resistance is through a **grid** that is baked on the inside of the glass (Figure 9-48). The rear window defogger grid is a series of horizontal, ceramic silver-compounded lines baked into the surface of the window. The terminals are soldered to the vertical bus bars. One terminal supplies the current from the switch; the other provides the ground (Figure 9-49).

Most systems incorporate a timer circuit to control the relay (Figure 9-50). The timer is used due to the high amount of current required to operate the system (approximately 30 amperes). If this drain were allowed to continue for extended periods of time, battery and charging system failure could result. Because of the high current draw, most vehicles equipped with a rear window defogger use a high output AC generator.

Figure 9-48 Rear window defogger grid.

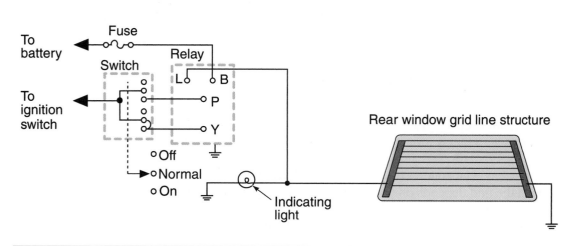

Figure 9-49 Rear window defogger circuit schematic.

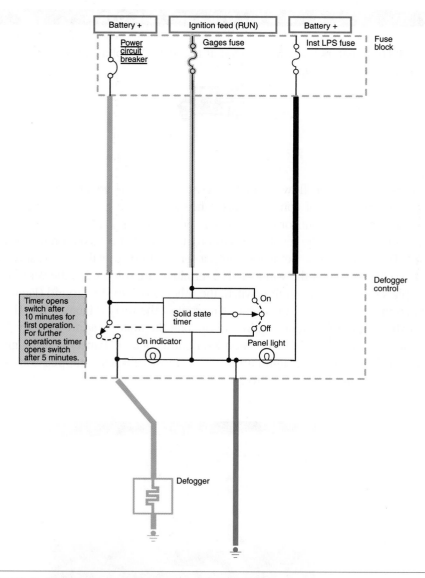

Figure 9-50 Defogger circuit using a solid-state timer.

The ON indicator can be either a bulb or light-emitting diode (LED).

Many manufacturers refer to the electric rear window defogger as an electric backlight (EBL).

The control switch may be a three-position, spring-loaded switch that returns to the center position after making momentary contact to the ON or OFF terminals. Activation of the switch energizes the electronic timing circuit, which energizes the relay coil. With the relay contacts closed, direct battery voltage is sent to the heater grid (Figure 9-51). At the same time, voltage is applied to the ON indicator. The timer is activated for 10 minutes. At the completion of the timed cycle, the relay is de-energized and the circuit to the grid and indicator light is broken. If the switch is activated again, the timer will energize the relay for 5 minutes.

The timer sequence can be aborted by moving the switch to the OFF position or by turning off the ignition switch. If the ignition switch is turned off while the timer circuit is activated, the rear window defogger switch will have to be returned to the ON position to activate the system again.

Ambient temperatures have an effect on electrical resistance; thus the amount of current flow through the grid depends on the temperature of the grid. As the ambient temperature decreases, the resistance value of the grid also decreases. A decrease in resistance increases the current flow and results in quick warming of the window. The defogger system tends to be self-regulated to match the requirements for defogging.

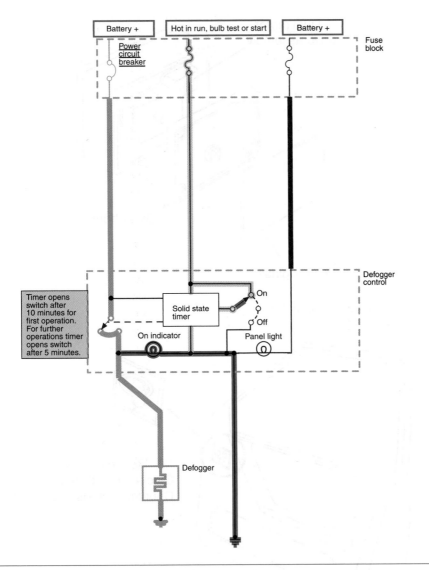

Figure 9-51 Current flow through the defogger circuit.

Power Mirrors

Electrically controlled power mirrors allow the driver to position the outside mirrors by use of a switch (Figure 9-52). The mirror assembly will use built-in, dual drive, reversible permanent magnet (PM) motors (Figure 9-53).

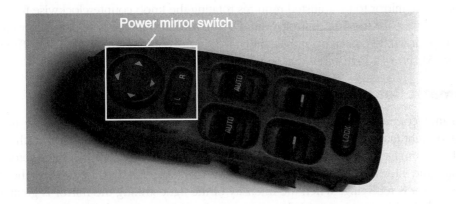

Figure 9-52 Control switch for power mirrors.

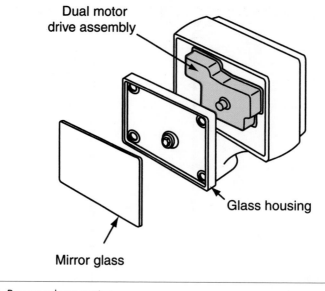

Figure 9-53 Power mirror motor.

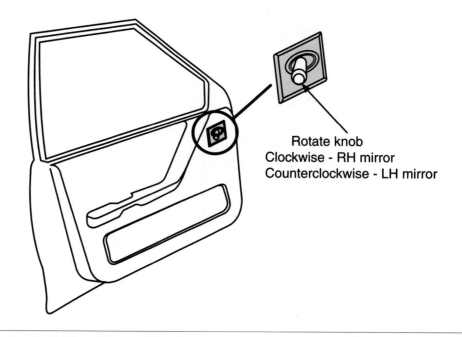

Figure 9-54 Mirror selection from the control switch.

A single switch for controlling both the left and right side mirrors is used. On many systems, selection of the mirror to be adjusted requires rotating the knob counterclockwise for the left mirror and clockwise for the right mirror (Figure 9-54). After the mirror is selected, movement of the joystick (up, down, left, or right) moves the mirror in the corresponding direction. The illustration (Figure 9-55) shows a logic table for the mirror switch and motors.

Automatic Rear View Mirror

Some manufacturers have developed interior rear view mirrors that automatically tilt when the intensity of light that strikes the mirror is sufficient enough to cause discomfort to the driver.

The system has two photocells mounted in the mirror housing. One of the photocells is used to measure the intensity of light inside the vehicle. The second is used to measure the intensity of light the mirror is receiving. When the intensity of the light striking the mirror is greater than that of ambient light, by a predetermined amount, a solenoid is activated that tilts the mirror.

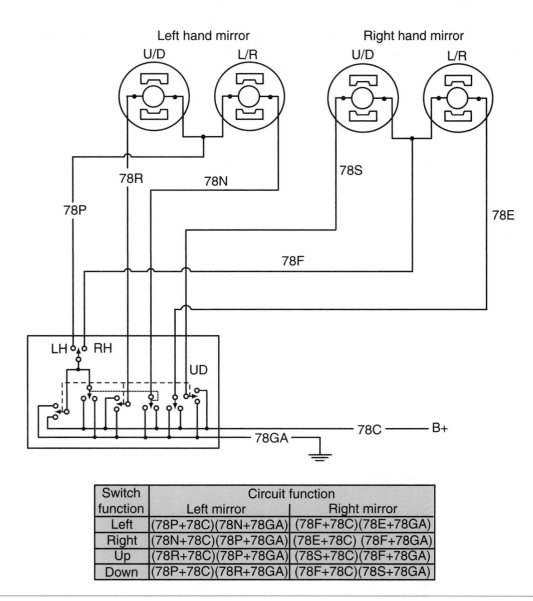

Switch	Circuit function	
function	Left mirror	Right mirror
Left	(78P+78C)(78N+78GA)	(78F+78C)(78E+78GA)
Right	(78N+78C)(78P+78GA)	(78E+78C) (78F+78GA)
Up	(78R+78C)(78P+78GA)	(78S+78C)(78F+78GA)
Down	(78P+78C)(78R+78GA)	(78F+78C)(78S+78GA)

Figure 9-55 Power mirror logic table.

Electrochromic Mirrors

Electrochromic mirrors automatically adjust the amount of reflectance based on the intensity of glare (Figure 9-56). The electrochromic mirror uses forward- and rearward-facing photo sensors and a solid-state chip. Based on light intensity differences, the chip applies a small voltage to the silicon layer. As voltage is applied, the molecules of the layer rotate and redirect the light beams. Thus the

Figure 9-56 Electrochromic mirror operation (A) Daytime; (B) Mild glare; (C) High glare.

mirror reflection appears dimmer. If the glare is heavy, the mirror darkens to about 6% reflectivity. The electrochromic mirror has the advantage that it provides a comfort zone where the mirror will provide 20% to 30% reflectivity. When no glare is present, the mirror changes to the daytime reflectivity rating of up to 85%. The reduction of the glare by darkening of the mirror does not impair visibility.

Electrochromic mirrors can be installed as the outside mirror and/or inside mirrors. The mirror is constructed of a thin layer of electrochromic material that is placed between two plates of conductive glass. There are two photocell sensors that measure light intensity in front and in back of the mirror. During night driving, the headlight beam striking the mirror causes the mirror to gradually become darker as the light intensity increases. The darker mirror absorbs the glare. Sensitivity of the mirror can be adjusted by the driver through a three-position switch (Figure 9-57).

The MIN position is used for city driving.

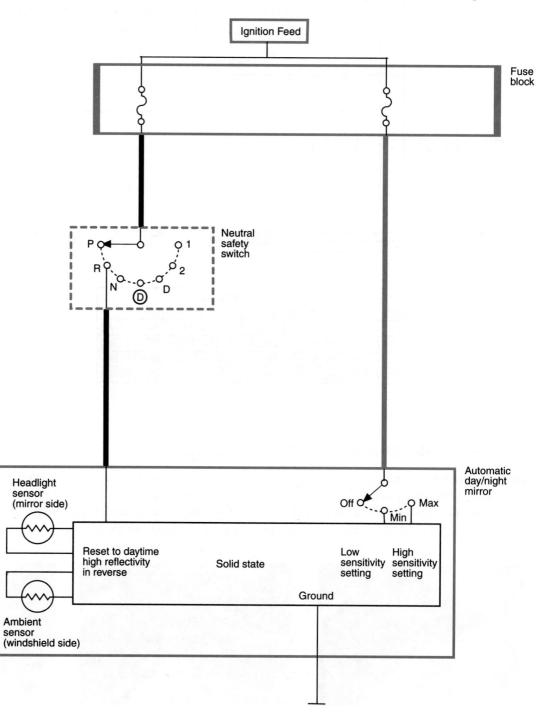

Figure 9-57 Automatic electrochromic day/night mirror diagram.

When the ignition switch is placed in the RUN position, battery voltage is applied to the three-position switch. If the switch is in the MIN position, battery voltage is applied to the solid-state unit and sets the sensitivity to a low level. The MAX setting causes the mirror to darken more at a lower glare level. When the transmission is placed in reverse, thereset circuit is activated. This returns the mirror to daytime setting for clearer viewing to back up.

Power Windows

Shop Manual
Chapter 9, page 355

Many luxury vehicles replace the conventional window crank with electric motors that operate the side windows. In addition, most station wagon models are equipped with electric rear tailgate windows. The motor used in the power window system is a reversible PM or two-field winding motor.

The power window system usually consists of the following components:

1. Master control switch.

2. Individual control switches.

3. Individual window drive motors.

4. Lock-out or disable switch.

Another design is to use rack-and-pinion gears. The rack is a flexible strip of gear teeth with one end attached to the window (Figure 9-58).

A B I T O F H I S T O R Y

Power windows were introduced in 1939.

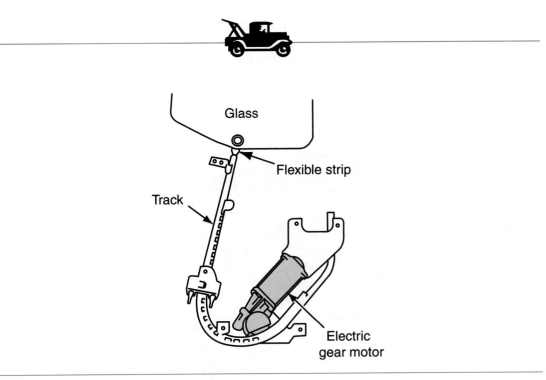

Figure 9-58 Rack and pinion–style power window motor and regulator.

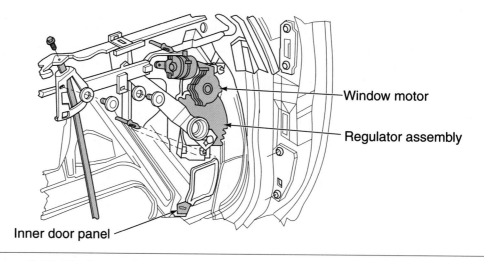

Figure 9-59 Window regulator.

A **window regulator** converts the rotary motion of the motor into the vertical movement of the window. The motor operates the window regulator either through a cable or directly. On direct drive motors, the motor pinion gear meshes with gear teeth on the regulator called the **sector gear** (Figure 9-59). As the window is lowered, the spiral spring is wound. The spring unwinds as the window is raised, to assist in raising the window. The spring reduces the amount of current that would be required to raise the window by the motor itself.

The master control switch provides the overall control of the system (Figure 9-60). Power to the individual switches is provided through the master switch. The master switch may also have a safety lock switch to prevent operation of the windows by the individual switches. When the safety switch is activated, it opens the circuit to the other switches and control is only by the master switch. As an additional safety feature, some systems prevent operation of the individual switches unless the ignition switch is in the RUN or ACC position (Figure 9-61).

Wiring circuits depend on motor design. Most PM-type motors are insulated, with ground provided through the master switch (Figure 9-62). When the master control switch is placed in the UP position, current flow is from the battery, through the master switch wiper, through the individual

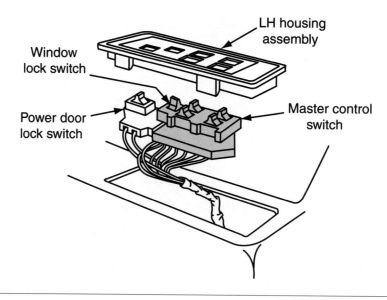

Figure 9-60 Power window master control switch.

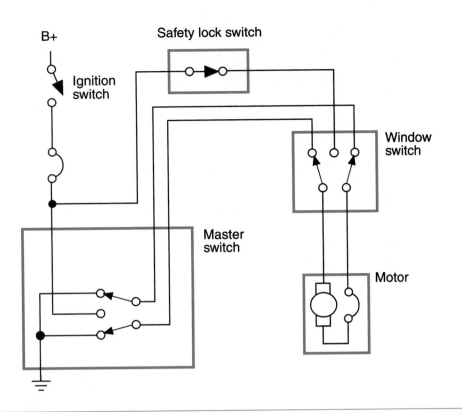

Figure 9-61 Simple power window wiring diagram.

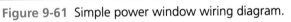

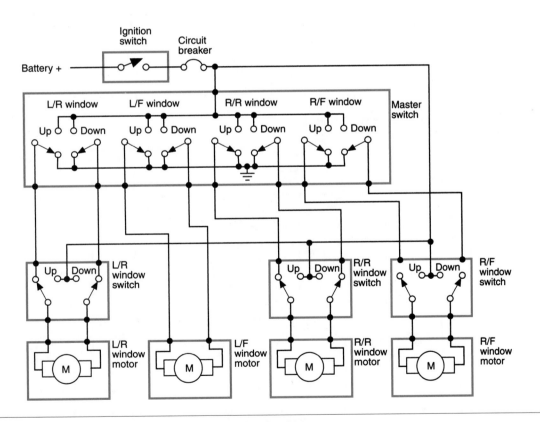

Figure 9-62 Typical power window circuit using PM motors.

switch wiper, to the top brush of the motor. Ground is through the bottom brush and circuit breaker to the individual switch wiper, to the master switch wiper and ground (Figure 9-63).

When the window is raised from the individual switch, battery voltage is supplied directly to the switch and wiper from the ignition switch. The ground path is through the master control switch (Figure 9-64).

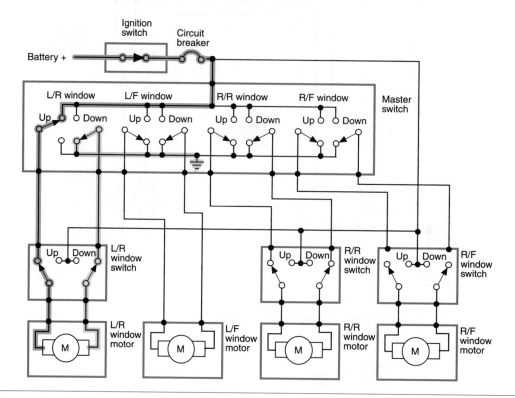

Figure 9-63 Current flow with the L/R window master control switch in the UP position.

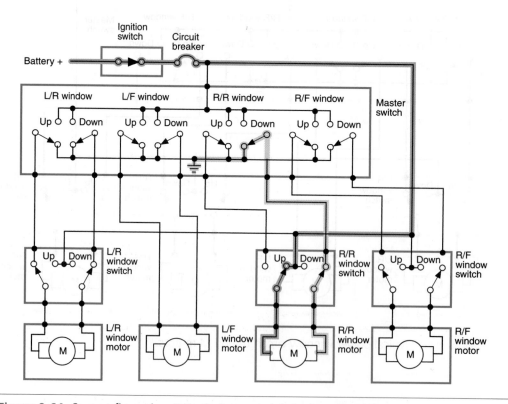

Figure 9-64 Current flow when R/R window switch is in the UP position.

When the window is lowered from the master control switch, the current path is reversed (Figure 9-65). In the illustration shown (Figure 9-66), current flows through the individual switch to lower the window.

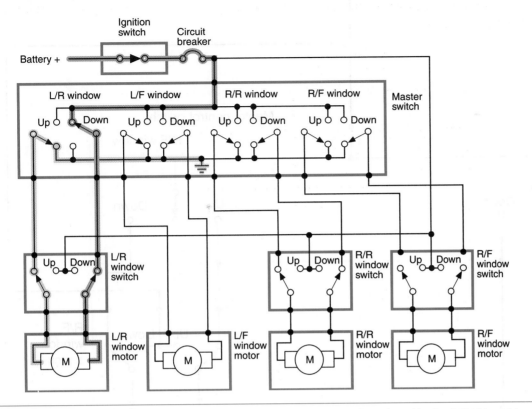

Figure 9-65 Current flow when L/R window master control switch is placed in the DOWN position.

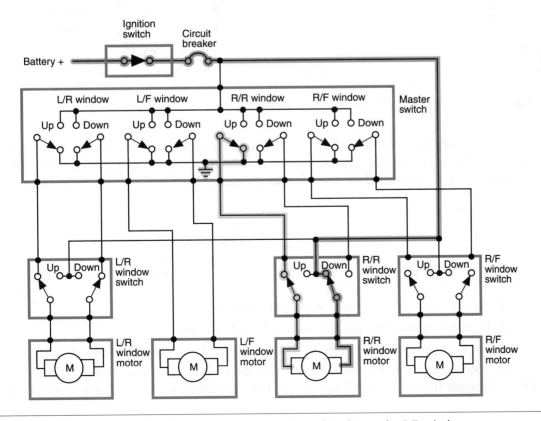

Figure 9-66 Current flow through the individual switch to lower the R/R window.

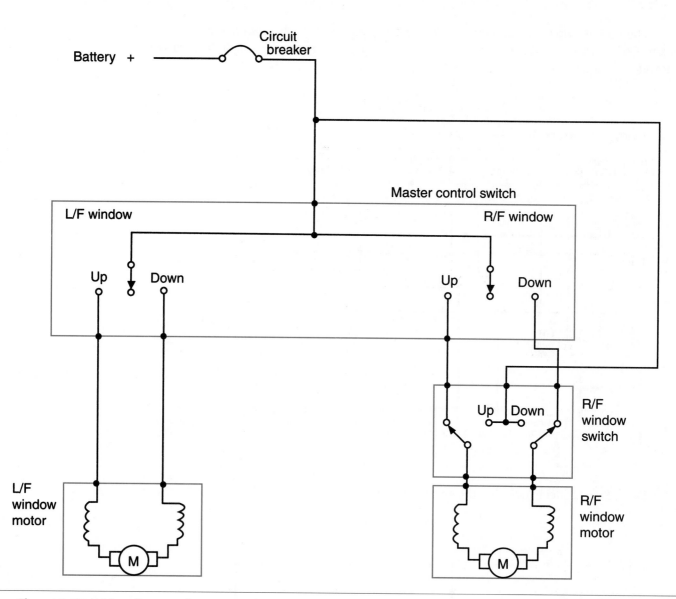

Figure 9-67 Wiring diagram of power window circuit using two-field coil motors.

Some manufacturers use a two-field coil motor that is grounded with insulated side switches. The two field coils are wired in opposite directions and only one coil is energized at a time. Direction of the motor is determined by which coil is activated (Figure 9-67).

Shop Manual
Chapter 9, page 357

Some seat back latches use a solenoid to lock the seat unless the door is open. The solenoid is controlled by the door jamb switch.

Power Seats

The power seat system is classified by the number of ways in which the seat is moved. The most common classifications are:

1. Two-way: Moves the seat forward and backward.
2. Four-way: Moves the seat forward, backward, up, and down.
3. Six-way: Moves the seat forward, backward, up, down, front tilt, and rear tilt.

All modern six-way power seats use a reversible, permanent magnet, three-armature motor called a **trimotor** (Figure 9-68). The motor may transfer rotation to a rack-and-pinion or to a worm

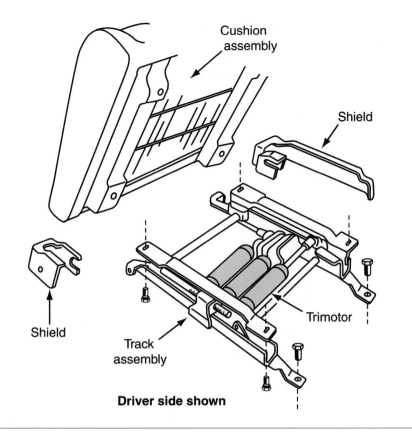

Driver side shown

Figure 9-68 Trimotor power seat installation.

gear drive transmission. A typical control switch consists of a four-position knob and a set of two-position switches (Figure 9-69). The four-position knob controls the forward, rearward, up, and down movement of the seat. The separate two-position switches are used to control the front tilt and rear tilt of the seat.

 AUTHOR'S NOTE: Early General Motors six-way power seats used a single motor. Solenoids were used to connect the motor to one of three transmissions that would move the seat.

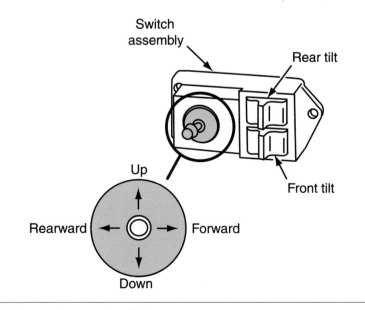

Figure 9-69 Power seat control switch.

Current direction through the motor determines the rotation direction of the motor. The switch wipers control the direction of current flow (Figure 9-70). If the driver pushes the four-way switch into the down position, the entire seat lowers (Figure 9-71). Switch wipers 3 and 4 are swung to the left and battery voltage is sent through wiper 4 to wipers 6 and 8. These wipers direct the current to the front and rear height motors. The ground circuit is provided through wipers 5 and 7, to wiper 3 and ground.

Some manufacturers equip their seats with adjustable support mats that shape the seat to fit the driver (Figure 9-72). The lumbar support mat provides the driver with additional comfort by supporting the back curvature. Some systems use air that is pumped into the mats; others use a motor to roll the lumbar support.

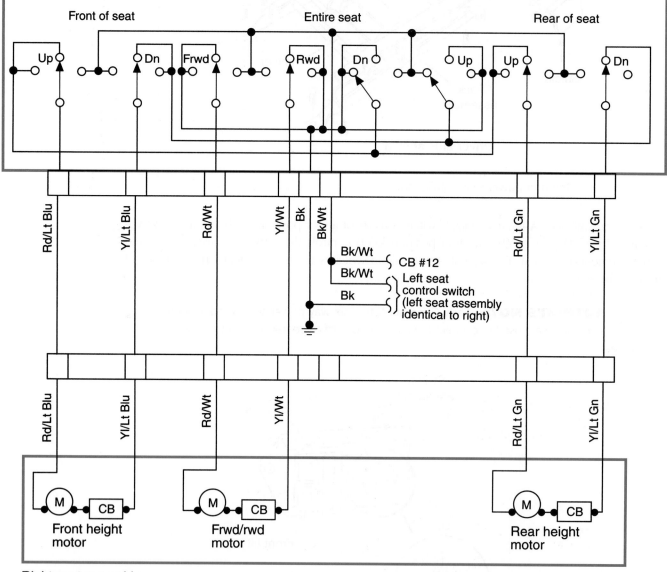

Figure 9-70 Six-way power seat circuit diagram.

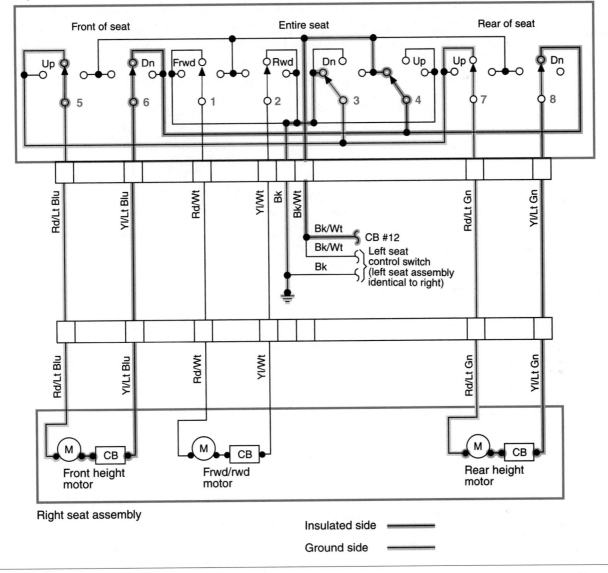

Figure 9-71 Current flow in the seat LOWER position.

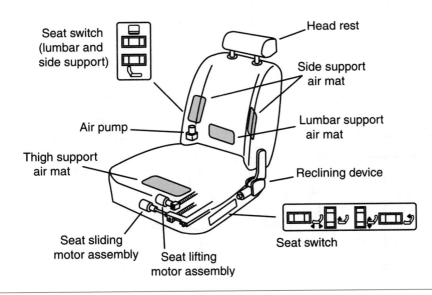

Figure 9-72 Adjustable seat cushions increase driver comfort and safety.

Power Door Locks

Electric power locks use either a solenoid or a permanent magnet reversible motor. Due to the high current demands of solenoids, most modern vehicles use PM motors (Figure 9-73). Depending on circuit design, the system may incorporate a relay (Figure 9-74). The relay has two coils and two sets of contacts to control current direction. In this system, the door lock switch energizes one

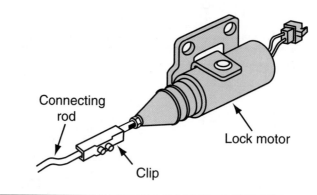

Figure 9-73 PM power door lock motor.

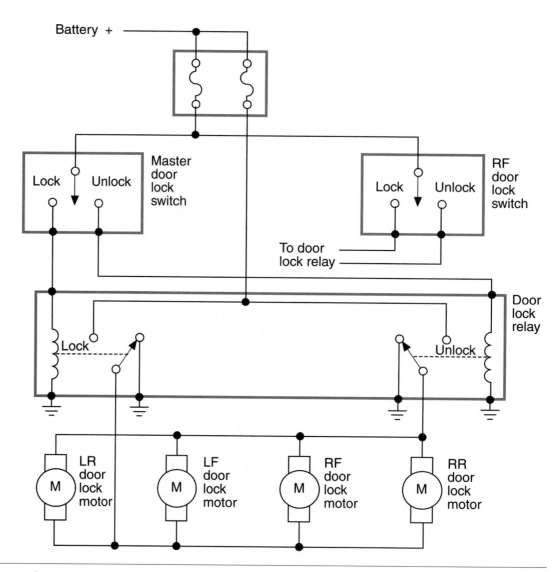

Figure 9-74 Power door lock circuit using a relay.

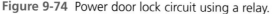

of the door lock relay coils to send battery voltage to the motor. If the door lock switch is placed in the LOCK position, current flow is that shown (Figure 9-75). In the illustration (Figure 9-76), current flows when the door lock switch is placed in the UNLOCK position.

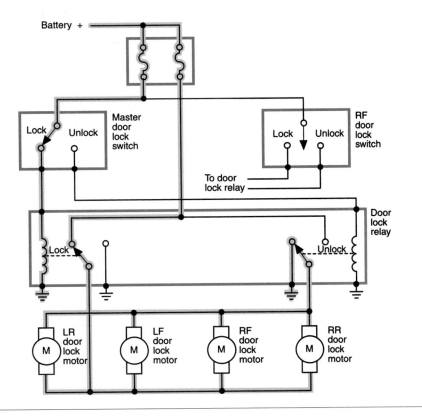

Figure 9-75 Current flow in the LOCK position.

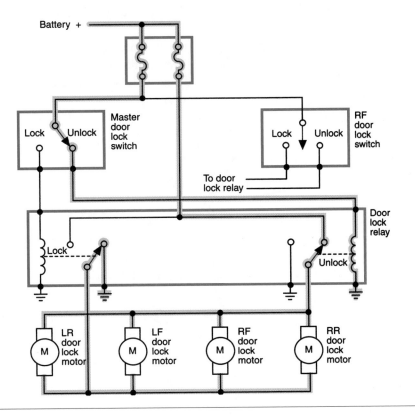

Figure 9-76 Current flow in the UNLOCK position.

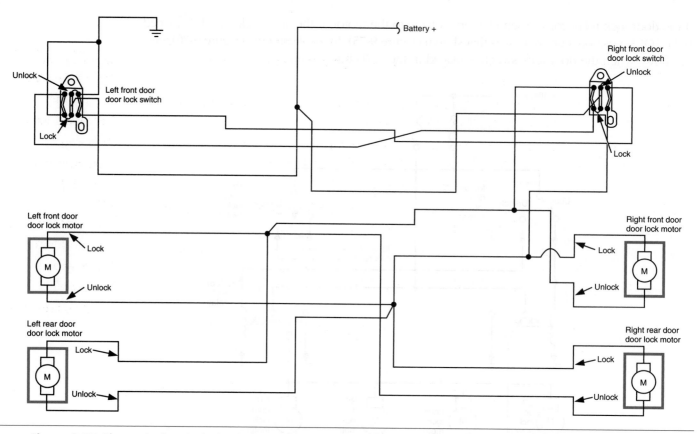

Figure 9-77 Electronic door lock circuit diagram.

A system that does not use relays is shown (Figure 9-77). The switch provides control of current flow in the same manner as power seats or windows.

Many vehicles are equipped with automatic door locks that are activated when the gear shift lever is placed in the DRIVE position. The doors unlock when the selector is returned to the PARK position.

Some manufacturers use a **child safety latch** in the door lock system that prevents the door from being opened from the inside, regardless of the position of the door lock knob. The child safety latch is activated by a switch designed into the latch bellcrank (Figure 9-78). By placing the latch in the deactived mode, the door operates as normal.

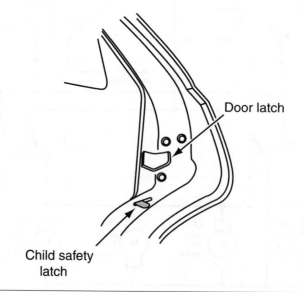

Figure 9-78 Child safety latch.

Summary

❏ Automotive electrical horns operate on an electromagnetic principle that vibrates a diaphragm to produce a warning signal.

❏ Horn switches are either installed in the steering wheel or as a part of the multifunction switch. Most horn switches are normally open switches.

❏ Horn switches that are mounted on the steering wheel require the use of sliding contacts. The contacts provide continuity for the horn control in all steering wheel positions.

❏ The most common type of horn circuit control is to use a relay.

❏ Most two-speed windshield wiper motors use permanent magnet fields whereby the motor speed is controlled by the placement of the brushes on the commutator.

❏ Some two-speed and all three-speed wiper motors use two electromagnetic field windings: series field and shunt field. The two field coils are wound in opposite directions so that their magnetic fields will oppose each other. The strength of the total magnetic field will determine at what speed the motor will operate.

❏ Park contacts are located inside the wiper motor assembly and supply current to the motor after the switch has been turned to the PARK position. This allows the motor to continue operating until the wipers have reached the PARK position.

❏ Intermittent wiper mode provides a variable interval between wiper sweeps and is controlled by a solid-state module.

❏ Systems that have a depressed-park feature use a second set of contacts with the park switch, which are used to reverse the rotation of the motor for about 15 degrees after the wipers have reached the normal PARK position.

❏ Blower fan motors use a resistor block that consists of two or three helically wound wire resistors that are connected in series to control fan speed.

❏ The blower motor circuit includes the control assembly, blower switch, resistor block, and blower motor.

❏ Electric defoggers heat the rear window by means of a resistor grid.

❏ Electric defoggers may incorporate a timer circuit to prevent the high current required to operate the system from damaging the battery or charging system.

❏ The electrically controlled mirror allows the driver to position the outside mirrors by use of a switch that controls dual-drive, reversible PM motors.

❏ Power windows, seats, and door locks usually use reversible PM motors, whereby motor rotational direction is determined by the direction of current flow through the switch wipers.

Terms to Know
Child safety latch
Clockspring
Depressed-park
Diaphragm
Electrochromic mirrors
Grid
Park contacts
Resistor block
Sector gear
Trimotor
Window regulator

Review Questions

Short-Answer Essays

1. Describe the operation of a relay-controlled horn circuit.

2. Explain how brush placement determines the speed of a two-speed, permanent magnet motor.

3. How do wiper motor systems that use a three-speed, electromagnetic motor control wiper speed?

4. What is the purpose of the capacitor in some intermittent wiper systems?

5. Describe the method used to control blower fan motor speeds.

6. Describe the operation of electric defoggers.

7. Briefly explain the principles of operation for power windows.

8. Describe the operation of a six-way, trimotor power seat.

9. Explain how depressed-park wipers operate.

10. What is the advantage of PM motors over solenoids in the power lock system?

Fill in the Blanks

1. Electrical accessories provide for additional _____ and _____.

2. The _____ is a thin, flexible, circular plate that is held around its outer edge by the horn housing, allowing the middle to flex.

3. Horn switches that are mounted on the steering wheel require the use of _____ _____ to provide continuity in all steering wheel positions.

4. In systems equipped with _____ _____, the blades drop down below the lower windshield molding to hide them.

5. PM windshield wiper motors use _____ brushes.

6. The operational speed of electromagnetic field winding motors used in wiper systems depends on the strength of the _____ _____ _____.

7. Most blower motor fan speeds are controlled through a _____ _____ that is wired in series to the fan motor.

8. Electric defoggers operate on the principle that when electrons are forced to flow through a _____, heat is generated.

9. A window _____ converts the rotary motion of the motor into the vertical movement of the window.

10. The three-armature motor is called a _____.

Multiple Choice

1. Sound from the horn is generated by:
   A. Heat causing the diaphragm to vibrate.
   B. Vibrating a column of air.
   C. Pulse-width modulating of the horn relay.
   D. All of the above.

2. The horn circuit is being discussed.
   Technician A says if the circuit does not use a relay, the horn switch carries the total current requirements of the horns.
   Technician B says most systems that use a relay have battery voltage present to the lower contact plate of the horn switch and the switch closes the path for the relay coil.
   Who is correct?
   A. A only   C. Both A and B
   B. B only   D. Neither A nor B

3. In a permanent magnet wiper motor:
   A. The placement of the brushes on the commutator controls motor speed.
   B. Two field coils wound in opposite directions control motor speed.
   C. Series and shunt fields control motor speed.
   D. None of the above.

4. Which statement about permanent magnet wiper motors is true?
   A. The more armature windings between the high-speed and common brushes results in less magnetism in the armature and lower CEMF.
   B. The lower the CEMF in the armature, the greater the armature current.
   C. The fewer windings between the low-speed and common brushes results in increased magnetism and higher CEMF.
   D. All of the above.

5. All of the following statements about electromagnetic wiper motors are true EXCEPT:
   A. The two field coils are wired in opposite directions.
   B. The magnetic fields oppose each other.
   C. The two field coils are wired in parallel.
   D. The strength of the total magnetic field determines motor speed.

6. In a conventional intermittent wiper system:
   A. The capacitor operates the motor directly until it is fully discharged.
   B. The capacitor's time to recharge is controlled by the resistance through a potentiometer.
   C. The capacitor is used to park the wipers.
   D. All of the above.

7. The blower motor circuit is being discussed.
   Technician A says the motor is usually a permanent magnet, single-speed motor.
   Technician B says the switch position directs current flow to one of the three different brushes in the motor.
   Who is correct?
   A. A only
   B. B only
   C. Both A and B
   D. Neither A nor B

8. *Technician A* says a resistor block that is wired in series between the switch and the motor controls blower fan speed.
   Technician B says a thermal limiter acts as a circuit breaker to protect the circuit if the resistor block gets too hot.
   Who is correct?
   A. A only
   B. B only
   C. Both A and B
   D. Neither A nor B

9. The electric rear window defogger is being discussed.
   Technician A says the grid is a series of controlled voltage amplifiers.
   Technician B says the system may incorporate a timer circuit to protect the vehicle's electrical system.
   Who is correct?
   A. A only
   B. B only
   C. Both A and B
   D. Neither A nor B

10. *Technician A* says the master control switch for power windows provides the overall control of the system.
    Technician B says current direction through the power seat motor determines the rotation direction of the motor.
    Who is correct?
    A. A only
    B. B only
    C. Both A and B
    D. Neither A nor B

Introduction to the Body Computer

Upon completion and review of this chapter, you should be able to:

❏ Describe the principle of analog and digital voltage signals.

❏ Explain the principle of computer communications.

❏ Describe the basics of logic gate operation.

❏ Describe the basic function of the central processing unit (CPU).

❏ Explain the basic method by which the CPU is able to make determinations.

❏ List and describe the differences in memory types.

❏ List and describe the functions of the various sensors the computer uses.

❏ List and describe the operation of output actuators.

Introduction

A computer is an electronic device that stores and processes data. It is also capable of controlling other devices. This chapter introduces the basic theory and operation of the digital computer used to control many of the vehicle's accessories. The use of computers on automobiles has expanded to include control and operation of several functions, including climate control, lighting circuits, cruise control, antilock braking, electronic suspension systems, and electronic shift transmissions. Some of these are functions of what is known as a body computer module (BCM). Some body computer–controlled systems include direction lights, rear window defoggers, illuminated entry, intermittent wipers, and other systems once thought of as basic.

AUTHOR'S NOTE: When computer controls were first installed on the automobile, the aura of mystery surrounding these computers was so great that some technicians were afraid to work on them. Most technicians coming into the field today have grown up around computers and the fear is not as great. Regardless of your comfort level with computers, knowledge is key to understanding their function. Although it is not necessary to understand all of the concepts of computer operation in order to service the systems they control, knowledge of the digital computer will help you feel more comfortable when working on these systems.

A computer processes the physical conditions that represent information (data). The operation of the computer is divided into four basic functions:

1. *Input:* A voltage signal sent from an input device. This device can be a sensor or a switch activated by the driver or technician.

2. *Processing:* The computer uses the input information and compares it to programmed instructions. The logic circuits process the input signals into output demands.

3. *Storage:* The program instructions are stored in an electronic memory. Some of the input signals are also stored for later processing.

4. *Output:* After the computer has processed the sensor input and checked its programmed instructions, it will put out control commands to various output devices. These output devices may be the instrument panel display or a system actuator. The output of one computer can also be used as an input to another computer.

Understanding these four functions will help today's technician organize the troubleshooting process. When a system is tested, the technician will be attempting to isolate the problem to one of these functions.

Analog and Digital Principles

Remembering the basics of electricity, voltage does not flow through a conductor; current flows and voltage is the pressure that "pushes" the current. However, voltage can be used as a signal; for example, difference in voltage levels, frequency of change, or switching from positive to negative values can be used as a signal.

The computer is capable of reading only voltage signals. A **program** is a set of instructions the computer must follow to achieve desired results. The program used by the computer is "burned" into integrated circuit (IC) chips using a series of numbers. These numbers represent various combinations of voltages that the computer can understand. The voltage signals to the computer can be either analog or digital. Many of the inputs from the sensors are analog variables. For example, ambient temperature sensors do not change abruptly. The temperature varies in infinite steps from low to high. The same is true for several other inputs such as engine speed, vehicle speed, fuel flow, and so on.

Compared to an analog voltage representation, digital voltage patterns are square-shaped because the transition from one voltage level to another is very abrupt (Figure 10-1). A digital signal is produced by an on/off or high/low voltage. The simplest generator of a digital signal is a switch (Figure 10-2). If 5 volts are applied to the circuit, the voltage sensor will read 5 volts (a high voltage value) when the switch is open. Closing the switch will result in the voltage sensor reading close to 0 volts. This measuring of voltage drops sends a digital signal to the computer. The voltage values are represented by a series of digits, which create a **binary code.** Binary code is represented by the numbers 1 and 0. Any number and word can be translated into a combination of binary 1's and 0's.

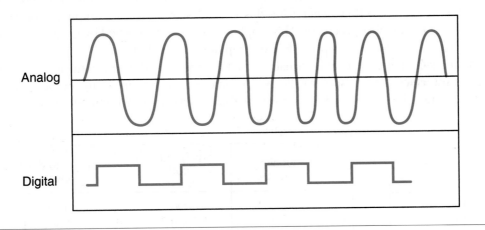

Figure 10-1 Analog voltage signals are constantly variable. Digital voltage patterns are either on or off. Digital signals are referred to as a square sine wave.

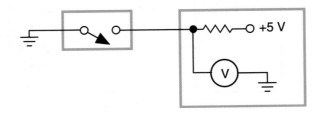

Figure 10-2 Simplified voltage sensing circuit that indicates if the switch is opened or closed.

Binary Numbers

A transistor that operates as a relay is the basis of the digital computer. As the input signal switches from off to on, the transistor output switches from cutoff to saturation. The on and off output signals represent the binary digits 1 and 0.

The computer converts the digital signal into binary code by translating voltages above a given value to 1 and voltages below a given value to 0. As shown (Figure 10-3), when the switch is open and 5 volts are sensed, the voltage value is translated into a 1 (high voltage). When the switch is closed, lower voltage is sensed and the voltage value is translated into a 0. Each 1 or 0 represents one bit of information.

In the binary system, whole numbers are grouped from right to left. Because the system uses only two digits, the first portion must equal a 1 or a 0. To write the value of 2, the second position must be used. In binary, the value of 2 would be represented by 10 (one two and zero ones). To continue, a 3 would be represented by 11 (one two and one one). Figure 10-4 illustrates the conversion of binary numbers to digital base ten numbers. If a thermistor is sensing 150 degrees, the binary code would be 10010110. If the temperature increases to 151 degrees, the binary code changes to 10010111.

The computer contains a crystal oscillator or **clock circuit** that delivers a constant time pulse. The clock is a crystal that electrically vibrates when subjected to current at certain voltage levels. As a result, the chip produces a very regular series of voltage pulses. The clock maintains an orderly flow of information through the computer circuits by transmitting one **bit** of binary code for each pulse (Figure 10-5). In this manner, the computer is capable of distinguishing between the binary codes such as 101 and 1001.

> A **bit** is a 0 or a 1 of the binary code. Eight bits is called a byte.

Signal Conditioning and Conversion

The input and/or output signals may require conditioning in order to be used. This conditioning may include amplification and/or signal conversion.

Some input sensors produce a very low voltage signal of less than 1 volt. This signal also has an extremely low current flow. Therefore, this type of signal must be amplified, or increased, before it is sent to the microprocessor. This amplification is accomplished by the amplification circuit in the input conditioning chip inside the computer (Figure 10-6).

For the computer to receive information from the sensor and give commands to actuators, it requires an **interface.** The computer will have two interface circuits: input and output. An interface is used to protect the computer from excessive voltage levels and to translate input and output

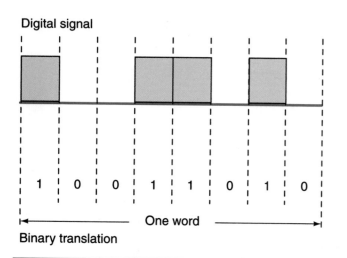

Figure 10-3 Each binary 1 and 0 is one bit of information. Eight bits equal one byte.

Decimal number	Binary number code 8 4 2 1	Binary to decimal conversion
0	0000	= 0 + 0 = 0
1	0001	= 0 + 1 = 1
2	0010	= 2 + 0 = 2
3	0011	= 2 + 1 = 3
4	0100	= 4 + 0 = 4
5	0101	= 4 + 1 = 5
6	0110	= 4 + 2 = 6
7	0111	= 4 + 2 + 1 = 7
8	1000	= 8 + 0 = 8
9	1001	= 8 + 1 = 9

Figure 10-4 Binary number code conversion to base ten numbers.

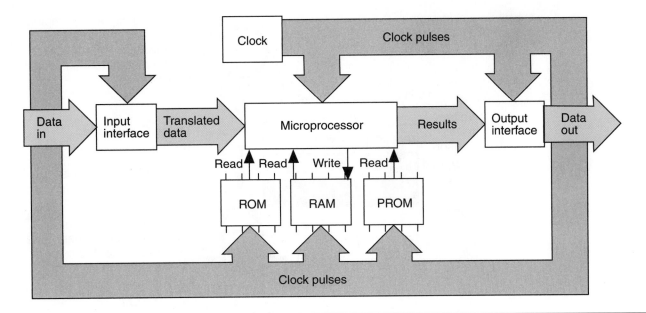

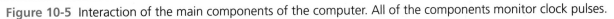

Figure 10-5 Interaction of the main components of the computer. All of the components monitor clock pulses.

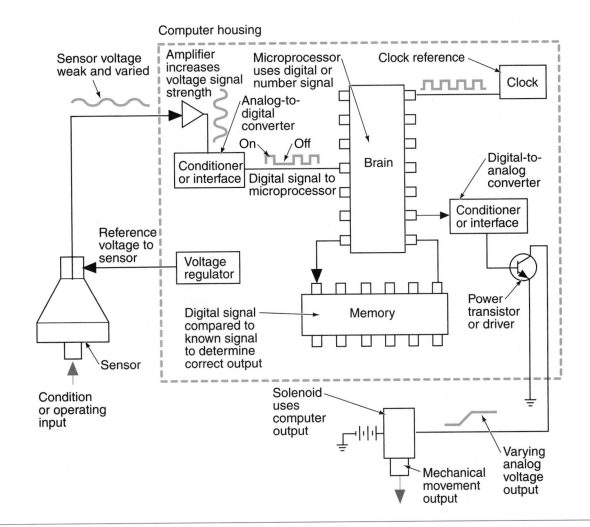

Figure 10-6 Amplification and interface circuits in the computer. The amplification circuit boosts the voltage and conditions it. The interface converts analog inputs into digital signals. The digital-to-analog converter changes the output from digital to analog.

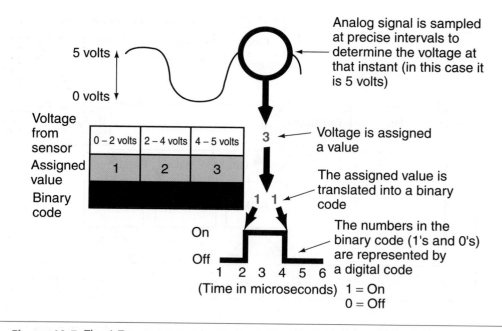

Figure 10-7 The A/D converter assigns a numeric value to input voltages and changes this numeric value to a binary code.

signals. The digital computer cannot accept analog signals from the sensors and requires an input interface to convert the analog signal to digital. The analog to digital (A/D) converter continually scans the analog input signals at regular intervals. For example, if the A/D converter scans the TPS signal and finds the signal at 5 volts, the A/D converter assigns a numeric value to this specific voltage. Then the A/D converter changes this numeric value to a binary code (Figure 10-7).

Also, some of the controlled actuators may require an analog signal. In this instance, an output digital to analog (D/A) converter is used.

Central Processing Unit

The terms *microprocessor* and *central processing unit* are basically interchangeable.

The **central processing unit (CPU)** is the brain of the computer. The CPU is constructed of thousands of transistors that are placed on a small chip. The CPU brings information into and out of the computer's memory. The input information is processed in the CPU and checked against the program in memory. The CPU also checks memory for any other information regarding programmed parameters. The information obtained by the CPU can be altered according to the program instructions. The program may have the CPU apply logic decisions to the information. Once all calculations are made, the CPU will deliver commands to make the required corrections or adjustments to the operation of the controlled system.

The program guides the microprocessor in decision making. For example, the program may inform the microprocessor when sensor information should be retrieved and then tell the microprocessor how to interpret this information. Finally, the program guides the microprocessor regarding the activation of output control devices such as relays and solenoids. The various memories contain the programs and other vehicle data that the microprocessor refers to as it performs calculations. As the microprocessor performs calculations and makes decisions, it works with the memories by either reading or writing information to them.

The CPU has several main components (Figure 10-8). The registers used include the accumulator, the data counter, the program counter, and the instruction register. The control unit implements the instructions located in the instruction register. The arithmetic logic unit (ALU) performs the arithmetic and logic functions.

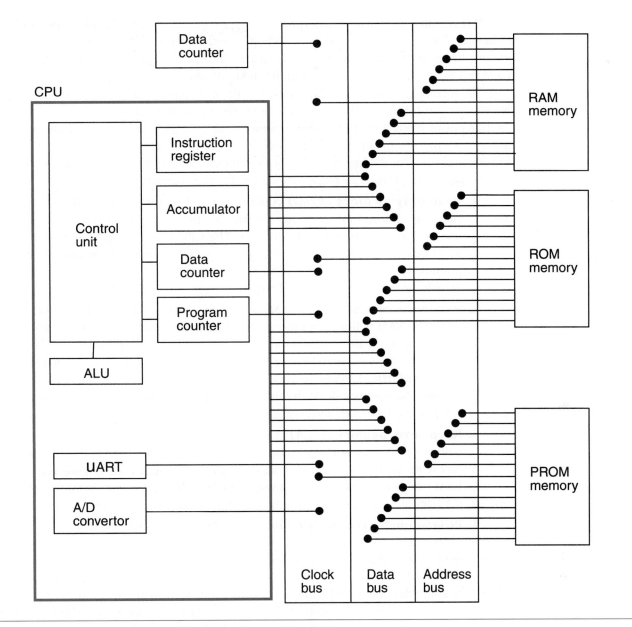

Figure 10-8 Main components of the computer and the CPU.

Computer Memory

The computer requires a means of storing both permanent and temporary memory. The memories contain many different locations. These locations can be compared to file folders in a filing cabinet, with each location containing one piece of information. An address is assigned to each memory location. This address may be compared to the lettering or numbering arrangement on file folders. Each address is written in a binary code, and these codes are numbered sequentially beginning with 0.

While the engine is running, the engine computer receives a large quantity of information from a number of sensors. The computer may not be able to process all this information immediately. In some instances, the computer may receive sensor inputs that the computer requires to make a number of decisions. In these cases, the microprocessor writes information into memory by specifying a memory address and sending information to this address.

The terms *ROM*, *RAM*, and *PROM* are used fairly consistently in the computer industry. However, the names vary between automobile manufacturers.

When stored information is required, the microprocessor specifies the stored information address and requests the information. When stored information is requested from a specific address, the memory sends a copy of this information to the microprocessor. However, the original stored information is still retained in the memory address.

The memories store information regarding the ideal air-fuel ratios for various operating conditions. The sensors inform the computer about the engine and vehicle operating conditions. The microprocessor reads the ideal air-fuel ratio information from memory and compares this information with the sensor inputs. After this comparison, the microprocessor makes the necessary decision and operates the injectors to provide the exact air-fuel ratio the engine requires.

Several types of memory chips may be used in the computer:

1. **Read only memory (ROM)** contains a fixed pattern of 1's and 0's that represent permanent stored information. This information is used to instruct the computer on what to do in response to input data. The CPU reads the information contained in ROM but it cannot write to it or change it. ROM is permanent memory that is programmed in. This memory is not lost when power to the computer is lost. ROM contains formulas, calibrations, and so on.

2. **Random access memory (RAM)** is constructed from flip-flop circuits formed into the chip. The RAM will store temporary information that can be read from or written to by the CPU. RAM stores information that is waiting to be acted upon and it stores output signals that are waiting to be sent to an output device. RAM can be designed as **volatile** or **nonvolatile.** In volatile RAM, the data will be retained as long as current flows through the memory. RAM that is connected to the battery through the ignition switch will lose its data when the switch is turned off (see number 7, nonvolatile RAM).

3. **Keep alive memory (KAM)** is a version of RAM. KAM is connected directly to the battery through circuit protection devices. For example, the microprocessor can read and write information to and from the KAM and erase KAM information. However, the KAM retains information when the ignition switch is turned off. KAM will be lost when the battery is disconnected, the battery drains too low, or the circuit opens.

4. **Programmable read only memory (PROM)** contains specific data that pertains to the exact vehicle in which the computer is installed. This information may be used to inform the CPU of the accessories that are equipped on the vehicle. The information stored in the PROM is the basis for all computer logic. The information in PROM is used to define or adjust the operating perimeters held in ROM. In many instances, the computer is interchangeable between models of the same manufacturers; however, the PROM is not. Consequently, the PROM may be replaceable and plug into the computer (Figure 10-9).

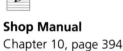

Shop Manual
Chapter 10, page 394

Figure 10-9 Assortment of PROM chips. Many manufacturers design PROMs that are removable by use of a special tool.

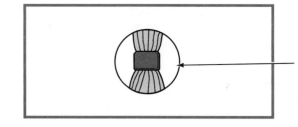

The Erasable PROM has a window such as this that the microcircuitry can be viewed through. This is normally covered by a piece of Mylar™-type material so that the information in it will not be erased by exposing it to ultraviolet light rays.

Figure 10-10 EPROM memory is erased when the ultraviolet ray contacts the microcircuitry.

5. **Erasable PROM (EPROM)** is similar to PROM except that its contents can be erased to allow new data to be installed. A piece of Mylar tape covers a window. If the tape is removed, the microcircuit is exposed to ultraviolet light that erases its memory (Figure 10-10).

6. **Electrically erasable PROM (EEPROM)** allows changing the information electrically one bit at a time. Some manufacturers use this type of memory to store information concerning mileage, vehicle identification number, and options. The flash EEPROM may be reprogrammed through the data link connector (DLC) using the manufacturer's specified diagnostic equipment.

7. **Nonvolatile RAM (NVRAM)** is a combination of RAM and EEPROM in the same chip. During normal operation, data is written to and read from the RAM portion of the chip. If the power is removed from the chip, or at programmed timed intervals, the data is transferred from RAM to the EEPROM portion of the chip. When the power is restored to the chip, the EEPROM will write the data back to the RAM.

Adaptive Strategy

If a computer has adaptive strategy capabilities, the computer can actually learn from past experience. For example, the normal voltage input range from an ambient temperature sensor may be 0.6 volt to 4.5 volts. If the sensor sends a 0.4-volt signal to the computer, the microprocessor interprets this signal as an indication of component wear and stores this altered calibration in the KAM. The microprocessor now refers to this new calibration during calculations and normal system performance is maintained. If a sensor output is erratic or considerably out of range, the computer may ignore this input. When a computer has adaptive strategy, a short learning period is necessary under the following conditions:

1. After the battery has been disconnected.
2. When a computer system component has been replaced or disconnected.
3. On a new vehicle.

Information Processing

The air charge temperature (ACT) sensor input will be used as an example of how the computer processes information. If the air temperature is low, the air is denser and contains more oxygen per cubic foot. Warmer air is less dense and therefore contains less oxygen per cubic foot. The cold, dense air requires more fuel compared to the warmer air that is less dense. The microprocessor must supply the correct amount of fuel in relation to air temperature and density.

An ACT sensor is positioned in the intake manifold where it senses air temperature. This sensor contains a resistive element that has an increased resistance when the sensor is cold.

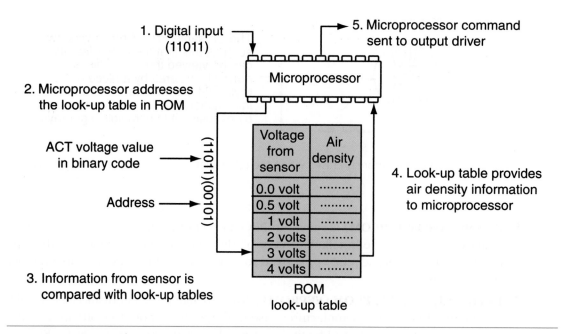

Figure 10-11 The microprocessor addresses the look-up tables in the ROM, retrieves air density information, and issues commands to the output devices.

Conversely, the ACT sensor resistance decreases as the sensor temperature increases. When the ACT sensor is cold, it sends a high analog voltage signal to the computer, and the A/D converter changes this signal to a digital signal.

When the microprocessor receives this ACT signal, it addresses the tables in the ROM. The look-up tables list air density for every air temperature. When the ACT sensor voltage signal is very high, the look-up table indicates very dense air. This dense air information is relayed to the microprocessor, and the microprocessor operates the output drivers and injectors to supply the exact amount of fuel the engine requires (Figure 10-11).

Logic Gates

Logic gates are the thousands of field effect transistors (FETs) incorporated into the computer circuitry. These circuits are called logic gates because they act as gates to output voltage signals depending on different combinations of input signals. The FETs use the incoming voltage patterns to determine the pattern of pulses leaving the gate. The following are some of the most common logic gates and their operations. The symbols represent functions and not electronic construction:

1. **NOT gate:** A NOT gate simply reverses binary 1's to 0's and vice versa (Figure 10-12). A high input results in a low output and a low input results in a high output.

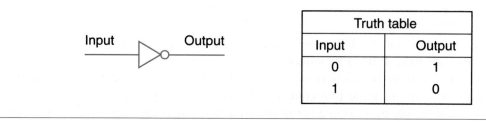

Truth table	
Input	Output
0	1
1	0

Figure 10-12 The NOT gate symbol and truth table. The NOT gate inverts the input signal.

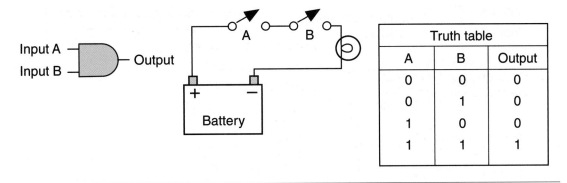

Truth table		
A	B	Output
0	0	0
0	1	0
1	0	0
1	1	1

Figure 10-13 The AND gate symbol and truth table. The AND gate operation is similar to that of switches in series.

2. **AND gate:** The AND gate will have at least two inputs and one output. The operation of the AND gate is similar to two switches in series to a load (Figure 10-13). The only way the light will turn on is if switches A *and* B are closed. The output of the gate will be high only if both inputs are high. Before current can be present at the output of the gate, current must be present at the base of both transistors (Figure 10-14).

3. **OR gate:** The OR gate operates similarly to two switches that are wired in parallel to a light (Figure 10-15). If switch A *or* B is closed, the light will turn on. A high signal to either input will result in a high output.

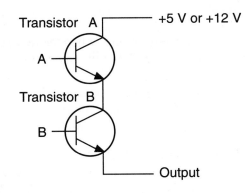

Figure 10-14 The AND gate circuit.

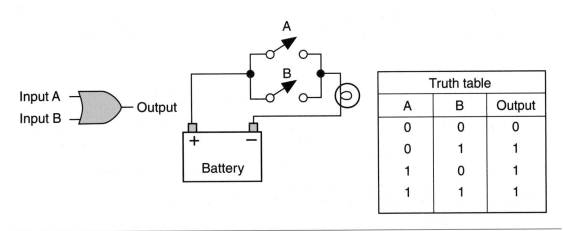

Truth table		
A	B	Output
0	0	0
0	1	1
1	0	1
1	1	1

Figure 10-15 OR gate symbol and truth table. The OR gate is similar to parallel switches.

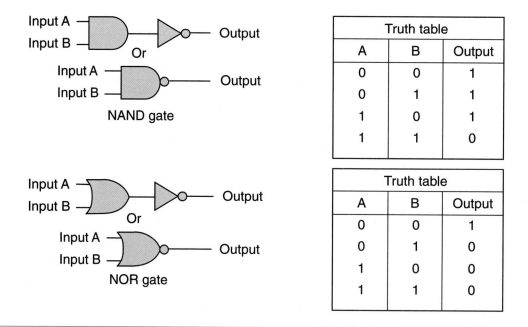

Truth table		
A	B	Output
0	0	1
0	1	1
1	0	1
1	1	0

Truth table		
A	B	Output
0	0	1
0	1	0
1	0	0
1	1	0

Figure 10-16 Symbols and truth tables for NAND and NOR gates. The small circle represents an inverted output on any logic gate symbol.

4. **NAND** and **NOR gates:** A NOT gate placed behind an OR or AND gate inverts the output signal (Figure 10-16).

5. **Exclusive-OR (XOR) gate:** A combination of gates that will produce a high output signal only if the inputs are different (Figure 10-17).

These different gates are combined to perform the processing function. The following are some of the most common combinations:

1. **Decoder circuit:** A combination of AND gates used to provide a certain output based on a given combination of inputs (Figure 10-18). When the correct bit pattern is received by the decoder, it will produce the high voltage signal to activate the relay coil.

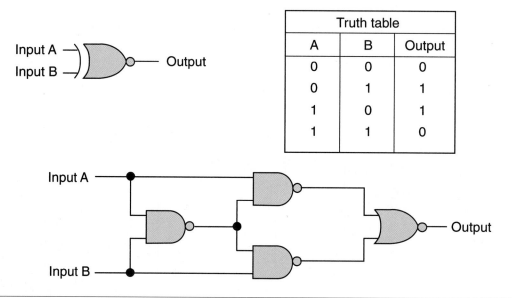

Truth table		
A	B	Output
0	0	0
0	1	1
1	0	1
1	1	0

Figure 10-17 XOR gate symbol and truth table. A XOR gate is a combination of NAND and NOR gates.

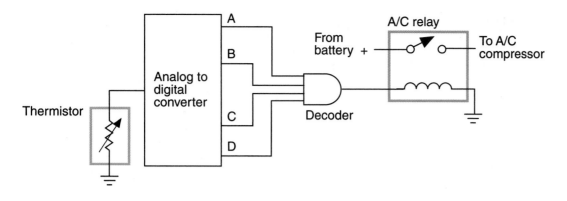

Figure 10-18 Simplified temperature sensing circuit that will turn on the air conditioning compressor when inside temperatures reach a predetermined value.

2. **Multiplexer (MUX):** The basic computer is not capable of looking at all of the inputs at the same time. A multiplexer is used to examine one of many inputs depending on a programmed priority rating (Figure 10-19). This process is called **sequential sampling.** This means the computer will deal with all of the sensors and actuators one at a time.

3. **Demultiplexer (DEMUX):** Operates similar to the MUX except that it controls the order of the outputs (Figure 10-20).

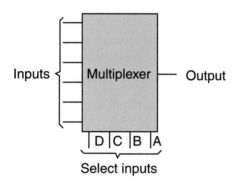

Figure 10-19 Selection at inputs D, C, B, A will determine which data input will be processed.

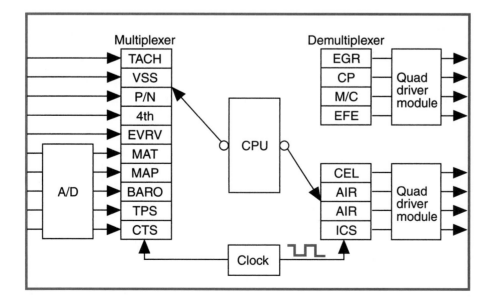

Figure 10-20 Block diagram representation of the MUX and DEMUX circuit.

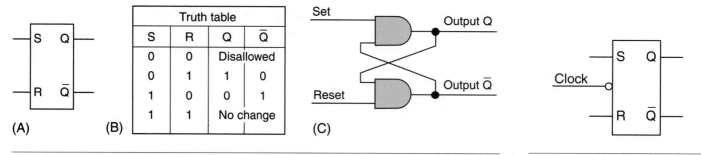

Figure 10-21 (A) RS flip-flop symbol. (B) Truth table. (C) Logic diagram. Variations of the circuit may include NOT gates at the inputs; if used, the truth table outputs would be reversed.

Figure 10-22 Clocked RS flip-flop symbol.

4. **RS and clocked RS flip-flop circuits:** Logic circuits that remember previous inputs and do not change their outputs until they receive new input signals. The illustration (Figure 10-21) shows a basic RS flip-flop circuit. The clocked flip-flop circuit has an inverted clock signal as an input so that circuit operations occur in the proper order (Figure 10-22). Flip-flop circuits are called **sequential logic circuits** because the output is determined by the sequence of inputs. A given input affects the output produced by the next input.

5. **Driver circuits:** A *driver* is a term used to describe a transistor device that controls the current in the output circuit. Drivers are controlled by a computer to operate such things as fuel injectors, ignition coils, and many other high-current circuits. The high currents handled by a driver are not really that high; they are just more than what is typically handled by a transistor. Several types of driver circuits are used on automobiles, such as Quad, Discrete, Peak and Hold, and Saturated Switch driver circuits.

6. **Registers:** A register is a combination of flip-flops that transfer bits from one to another every time a clock pulse occurs (Figure 10-23). It is used in the computer to temporarily store information.

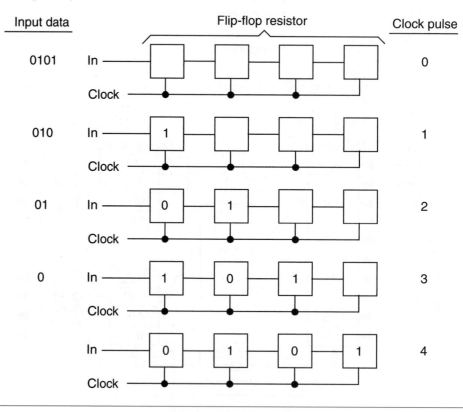

Figure 10-23 It takes four clock pulses to load four bits into the register.

7. Accumulators: Registers designed to store the results of logic operations that can become inputs to other modules.

Inputs

As discussed earlier, the CPU receives inputs that it checks with programmed values. Depending on the input, the computer will control the actuator(s) until the programmed results are obtained (Figure 10-24). The inputs can come from other computers, the driver, the technician, or through a variety of sensors.

Driver input signals are usually provided by momentarily applying a ground through a switch. The computer receives this signal and performs the desired function. For example, if the driver wishes to reset the trip odometer on a digital instrument panel, he would push the reset switch. This switch will provide a momentary ground that the computer receives as an input and sets the trip odometer to zero.

Switches can be used as an input for any operation that only requires a yes-no, or on-off, condition. Other inputs include those supplied by means of a sensor and those signals returned to the computer in the form of feedback.

Sensors

There are many different designs of **sensors.** Sensors convert some measurement of vehicle operation into an electrical signal. Some sensors are nothing more than a switch that completes the circuit. Others are complex chemical reaction devices that generate their own voltage under different conditions. Repeatability, accuracy, operating range, and **linearity** are all requirements of a sensor.

Thermistors. A **thermistor** is used to sense engine coolant or ambient temperatures. It is a solid-state variable resistor made from a semiconductor material that changes resistance in relation to temperature changes. By monitoring the thermistor's resistance value, the computer is capable of observing very small changes in temperature. The computer sends a reference

Shop Manual
Chapter 10, page 389

Shop Manual
Chapter 10, page 389

Linearity refers to the sensor signal being as constantly proportional to the measured value as possible. It is an expression of the sensor's accuracy.

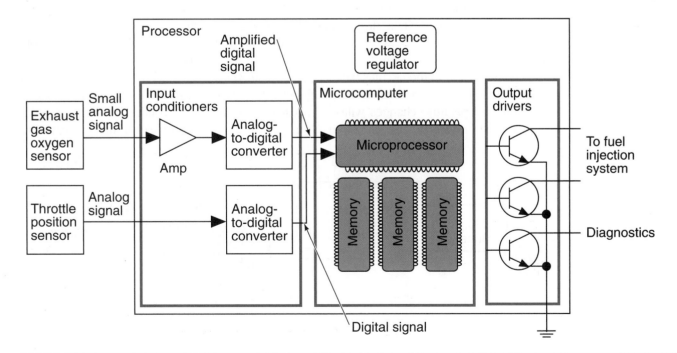

Figure 10-24 The input signals are processed in the microprocessor. The microprocessor directs the output drivers to activate actuators as instructed by the program.

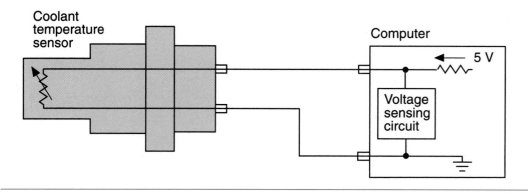

Figure 10-25 A thermistor is used to measure temperature. The sensing unit measures the resistance change and translates the data into temperature values.

voltage to the thermistor (usually 5 volts) through a fixed resistor. As the current flows through the thermistor resistance to ground, a voltage sensing circuit measures the voltage after the fixed resistor (Figure 10-25). The voltage dropped over the fixed resistor will change as the resistance of the thermistor changes. Using its programmed values, the computer is able to translate the voltage drop into a temperature value.

There are two types of thermistors: **negative temperature coefficient (NTC) thermistors** and **positive temperature coefficient (PTC) thermistors.** NTC thermistors reduce their resistance as the temperature increases, while PTC thermistors increase their resistance as the temperature increases. The NTC is the most commonly used.

An MAF is used to
measure the amount
of air flowing into
the engine.

Wheatstone Bridges. The illustration (Figure 10-26) shows the construction of the **Wheatstone bridge.** The Wheatstone bridge is a series-parallel arrangement of resistors between an input terminal and ground. A Wheatstone bridge is nothing more than two simple series circuits connected in parallel across a power supply. Usually three of the resistors are kept at exactly the same value and the fourth is the sensing resistor. When all four resistors have the same value, the bridge is balanced and the voltage sensor will indicate a value of 0 volts. The output from the amplifier acts as a voltmeter. Remember, since a voltmeter measures electrical pressure between two points, it will display this value. For example, if the reference voltage is 5 volts and the resistors have the same value, then the voltage drop over each resistor is 2.5 volts. Since the voltmeter is measuring the potential on the line between R_S and R_1 and R_2 and R_3, it will read 0 volts because both of these lines have 2.5 volts on them. If there is a change in the resistance value of the sense resistor, a change will occur in the circuit's balance. The sensing circuit will receive a voltage reading that is proportional to the amount of resistance change. If the Wheatstone bridge is used to measure temperature, temperature changes will be indicated as a change in voltage by the sensing circuit. Wheatstone bridges are also used to measure pressure (piezoresistive) and mechanical strain.

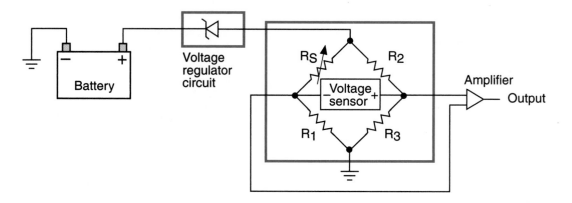

Figure 10-26 Wheatstone bridge.

A common use of a Wheatstone bridge is the hot wire sensor in a mass air flow (MAF) sensor. The sensor consists of a hot wire circuit, a cold wire circuit, and an electronic signal processing area. The hot and cold wire circuits form the Wheatstone bridge. The cold wire circuit is made of a fixed resistor and a thermistor. The amount of voltage dropped across the two resistors is determined by the temperature of the thermistor.

The hot wire circuit is made up of a fixed resistor and a variable resistance heat element (hot wire). The heat element generates heat in proportion to the amount of current flowing through it. This heat, in turn, changes its resistance. As air flows past the hot wire, it moves a small amount of heat from the element. This cooling of the element causes the voltage drop across it to change. The voltage drop across the hot wire is compared to the voltage drop across the fixed resistor in the cold wire circuit and air flow is determined.

Piezoelectric Devices. Piezoelectric devices are used to measure fluid and air pressures by generating their own voltages. The most commonly found piezoelectric device is the engine knock sensor. The knock sensor measures engine knock, or vibration, and converts the vibration into the sensor generates a voltage signal.

The word *piezoelectric* comes from the Greek word *piezo*, which means pressure.

The sensor is a voltage generator and has a resistor connected in series with it. The resistor protects the sensor from excessive current flow in case the circuit becomes shorted. The voltage generator is a thin ceramic disc attached to a metal diaphragm. When engine knock occurs, the vibration of the noise puts pressure on the diaphragm. This puts pressure on the piezoelectric crystals in the ceramic disc. The disc generates a voltage that is proportional to the amount of pressure. The voltage generated ranges from zero to one or more volts. Each time the engine knocks, the sensor generates a voltage spike.

Piezoresistive Devices. In construction, a piezoresistive device is similar to a piezoelectric one. However, they operate differently. The sensor in a piezoresistive device acts like a variable resistor. Its resistance value changes as the pressure applied to the crystal changes. A voltage regulator supplies a constant voltage to the sensor. Since the amount of voltage that the sensor drops will change with the change in resistance, the control module can determine the amount of pressure on the crystals by measuring the voltage drop across the sensor. Piezoresistive sensors are commonly used as gauge sending units.

Potentiometers. A potentiometer is a voltage divider that provides a variable DC voltage readig to the computer. The potentiometer usually consists of a wire wound resistor with a moveable center wiper (Figure 10-27). A constant voltage value (usually 5 volts) is applied to terminal A. If the wiper (which is connected to the shaft or moveable component of the unit that is being monitored) is located close to this terminal, there will be low voltage drop represented by high voltage

Shop Manual
Chapter 10, page 389

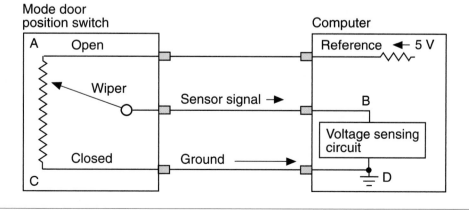

Figure 10-27 A potentiometer sensor circuit measures the amount of voltage drop to determine position.

signal back to the computer through terminal B. As the wiper is moved toward the C terminal, the sensor signal voltage to terminal B decreases. The computer interprets the different voltage value into different shaft positions. The potentiometer can measure linear or rotary movement. As the wiper is moved across the resistor, the position of the unit can be tracked by the computer.

Since applied voltage must flow through the entire resistance, temperature and other factors do not create false or inaccurate sensor signals to the computer. A rheostat is not as accurate and its use is limited in computer systems.

Magnetic Pulse Generators. **Magnetic pulse generators** are commonly used to send data to the computer about the speed of the monitored component. They use the principle of magnetic induction to produce a voltage signal. This data provides information about vehicle speed and individual wheel speed. Magnetic pulse generators are also used in some ignition systems. The signals from the speed sensors are used for computer-driven instrumentation, cruise control, antilock braking, speed sensitive steering, and automatic ride control systems. The magnetic pulse generator is also used to inform the computer of the position of a monitored component. This is common in engine controls where the computer needs to know the position of the crankshaft in relation to rotational degrees.

The components of the pulse generator are:

1. A **timing disc** that is attached to the rotating shaft or cable. The number of teeth on the timing disc is determined by the manufacturer and depends on application. The teeth will cause a voltage generation that is constant per revolution of the shaft. For example, a vehicle speed sensor may be designed to deliver 4,000 pulses per mile. The number of pulses per mile remains constant regardless of speed. The computer calculates how fast the vehicle is going based on the frequency of the signal.

2. A **pickup coil** consists of a permanent magnet that is wound around by fine wire.

An air gap is maintained between the timing disc and the pickup coil. As the timing disc rotates in front of the pickup coil, the generator sends an A/C signal (Figure 10-28). As a tooth on the timing disc aligns with the core of the pickup coil, it repels the magnetic field. The magnetic field is forced to flow through the coil and pickup core (Figure 10-29). Since the magnetic field is not expanding, a voltage of zero is induced in the pickup coil. As the tooth passes the core, the magnetic field is able to expand (Figure 10-30). The expanding magnetic field cuts across the windings of the pickup coil. This movement of the magnetic field induces a voltage in the windings. This action is repeated every time a tooth passes the core. The moving lines of magnetic force cut across the coil windings and induce a voltage signal.

When a tooth approaches the core, a positive current is produced as the magnetic field begins to concentrate around the coil (Figure 10-31). The voltage will continue to climb as long as the magnetic field is expanding. As the tooth approaches the magnet, the magnetic field gets smaller, causing the induced voltage to drop off. When the tooth and core align, there is no more expansion or contraction of the magnetic field (thus no movement) and the voltage drops to zero (Figure 10-32). When the tooth passes the core, the magnetic field expands and a negative current is produced (Figure 10-33). The resulting pulse signal is amplified, digitalized, and sent to the microprocessor.

Shop Manual
Chapter 10, page 390

The magnetic pulse generator is also called a permanent magnet (PM) generator.

The **timing disc** is known as an armature, reluctor, trigger wheel, pulse wheel, or timing core. It is used to conduct lines of magnetic force.

The **pickup coil** is also known as a stator, sensor, or pole piece. It remains stationary while the timing disc rotates in front of it. The changes of magnetic lines of force generate a small voltage signal in the coil.

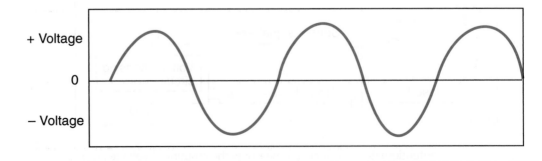

Figure 10-28 Pulse signal sine wave.

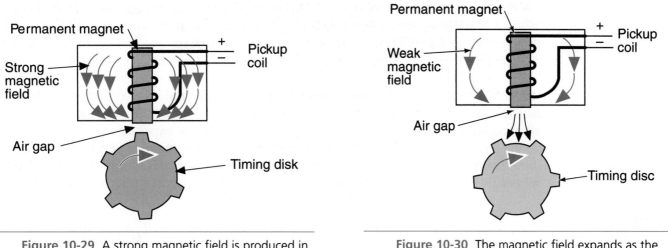

Figure 10-29 A strong magnetic field is produced in the pickup coil as the teeth align with the core.

Figure 10-30 The magnetic field expands as the teeth pass the core.

Figure 10-31 A positive voltage swing is produced as the tooth approaches the core.

Figure 10-32 When the tooth aligns with the core, there is no magnetic movement and no voltage.

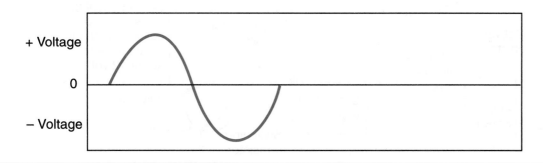

Figure 10-33 A negative waveform is created as the tooth passes the core.

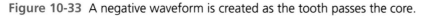

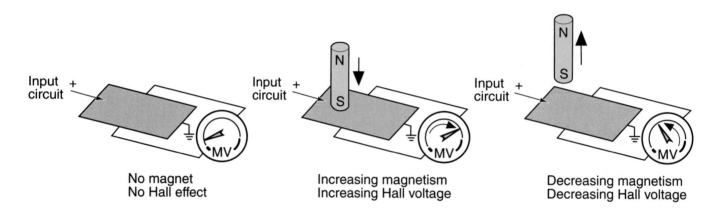

AUTHOR'S NOTE: The magnetic pulse (PM) generator operates on basic magnetic principles. Remember that a voltage can only be induced when a magnetic field is moved across a conductor. The magnetic field is provided by the pickup unit, and the rotating timing disc provides the movement of the magnetic field needed to induce voltage.

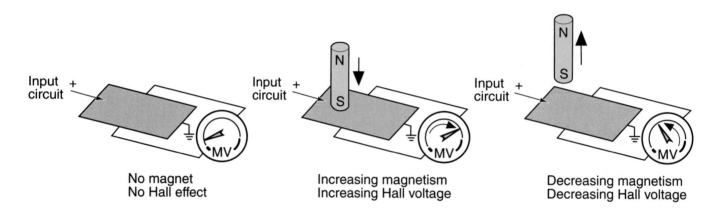

Shop Manual

Chapter 10, page 392

Hall-Effect Switches. The **Hall-effect switch** performs the same functions as the magnetic pulse generator. The Hall-effect switch operates on the principle that if a current is allowed to flow through thin conducting material that is exposed to a magnetic field, another voltage is produced (Figure 10-34). The switch contains a permanent magnet, a thin semiconductor layer made of gallium arsenate crystal (Hall layer), and a **shutter wheel** (Figure 10-35). The Hall layer has a negative and a positive terminal connected to

Figure 10-34 Hall-effect principles of voltage induction.

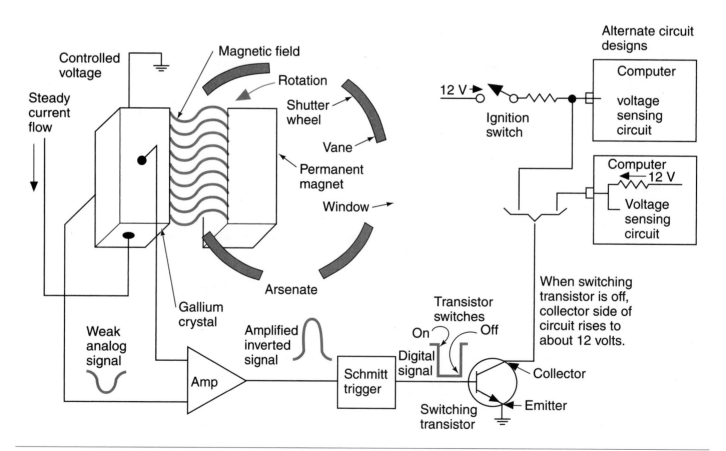

Figure 10-35 Typical circuit of a Hall-effect switch.

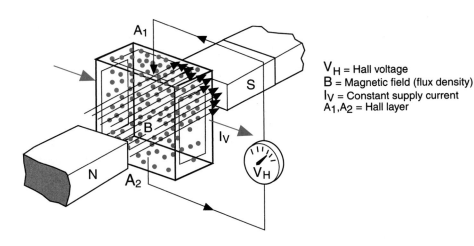

V$_H$ = Hall voltage
B = Magnetic field (flux density)
I$_V$ = Constant supply current
A$_1$,A$_2$ = Hall layer

Figure 10-36 The magnetic field causes the electrons from the supply current to gather at the Hall layer negative terminal. This creates a voltage potential.

it. Two additional terminals located on either side of the Hall layer are used for the output circuit. The shutter wheel consists of a series of alternating windows and vanes. It creates a magnetic shunt that changes the strength of the magnetic field from the permanent magnet.

The permanent magnet is located directly across from the Hall layer so that its lines of flux will bisect at right angles to the current flow. The permanent magnet is mounted so that a small air gap is between it and the Hall layer.

A steady current is applied to the crystal of the Hall layer. This produces a signal voltage that is perpendicular to the direction of current flow and magnetic flux. The signal voltage produced is a result of the effect the magnetic field has on the electrons. When the magnetic field bisects the supply current flow, the electrons are deflected toward the Hall layer negative terminal (Figure 10-36). This results in a weak voltage potential being produced in the Hall switch.

A shutter wheel is attached to a rotational component. As the wheel rotates, the shutters (vanes) will pass in this air gap. When a shutter vane enters the gap, it intercepts the magnetic field and shields the Hall layer from its lines of force. The electrons in the supply current are no longer disrupted and return to a normal state. This results in low voltage potential in the signal circuit of the Hall switch.

The signal voltage leaves the Hall layer as a weak analog signal. To be used by the computer, the signal must be conditioned. It is first amplified because it is too weak to produce a desirable result. The signal is also inverted so that a low input signal is converted into a high output signal. It is then sent through a **Schmitt trigger** where it is digitized and conditioned into a clean square wave signal. The signal is finally sent to a switching transistor. The computer senses the turning on and off of the switching transistor to determine the frequency of the signals and calculates speed.

A BIT OF HISTORY

The American physicist, Edwin Herbert Hall (1855–1938), discovered the principle of the Hall-effect switch in 1879.

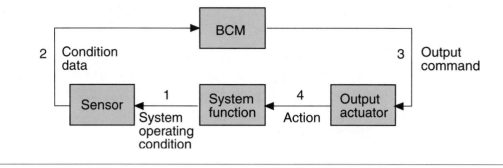

Figure 10-37 Principle of feedback signals.

Feedback Signals

If the computer sends a command signal to open a blend door in an automatic climate control system, a **feedback** signal may be sent back from the actuator to inform the computer the task was performed. The feedback signal will confirm both the door position and actuator operation (Figure 10-37). Another form of feedback is for the computer to monitor voltage as a switch, relay, or other actuator is activated. Changing states of the actuator will result in a predictable change in the computer's voltage sensing circuit. The computer may set a diagnostic code if it does not receive the correct feedback signal.

Oxygen Sensors. One of the most commonly used feedback sensors is the O_2 sensor. Although this is a principal input for engine control systems, its basic operation needs to be described here. The O_2 sensor is mounted in the exhaust gas stream and provides the PCM with a measurement of the oxygen in the engine's exhaust. The sensor is constructed of a zirconium dioxide ceramic thimble covered with a thin layer of platinum.

When the thimble is filled with oxygen-rich outside air and the outer surface of the thimble is exposed to oxygen-depleted exhaust gases, a chemical reaction in the sensor produces a voltage. The generation of voltage is similar to the same activity that takes place in a battery, except at much lower voltages. The voltage output varies with the level of oxygen present in the exhaust. As oxygen in the exhaust decreases, the voltage output increases. Likewise, as the oxygen level in the exhaust increases, the output voltage decreases.

For more information on the operation of the oxygen sensors, refer to *Today's Technician: Automotive Fuels and Emissions*.

An oxygen sensor is commonly referred to as an O_2 sensor.

Outputs

Once the computer's programming instructs that a correction or adjustment must be made in the controlled system, an output signal is sent to an actuator. This involves translating the electronic signals into mechanical motion.

An output driver is used within the computer to control the actuators. The circuit driver usually applies the ground circuit of the actuator. The ground can be applied steadily if the actuator must be activated for a selected amount of time. For example, if the BCM inputs indicate that the automatic door locks are to be activated, the actuator is energized steadily until the locks are latched. Then the ground is removed.

Other systems require the actuator to be turned either on and off very rapidly or for a set amount of cycles per second. It is duty cycled if it is turned on and off a set amount of cycles per second. Most duty cycled actuators cycle ten times per second. To complete a cycle it must go from off to on to off again. If the cycle rate is ten times per second, one actuator cycle is completed in one tenth of a second. If the actuator is turned on for 30% of each tenth of a second and off for 70%, it is referred to as a 30% duty cycle (Figure 10-38).

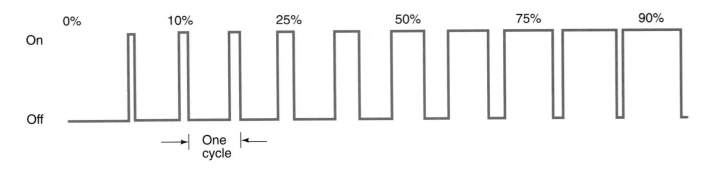

Figure 10-38 Duty cycle is the percentage of on-time per cycle. Duty cycle can be changed; however, total cycle times remain constant.

If the actuator is cycled on and off very rapidly, the pulse width can be varied to provide the programmed results. For example, the computer program will select an illumination level of the digital instrument panel based on the intensity of the ambient light in the vehicle. The illumination level is achieved through pulse width modulation of the lights. If the lights need to be bright, the pulse width is increased, which increases the length of on-time. As the light intensity needs to be reduced, the pulse width is decreased (Figure 10-39).

Actuators

Most computer-controlled **actuators** are electromechanical devices that convert the output commands from the computer into mechanical action. These actuators are used to open and close switches, control vacuum flow to other components, and operate doors or valves, depending on the requirements of the system.

    Although they do not fall into the strict definition of an actuator, the BCM can also control lights, gauges, and driver circuits.

Shop Manual
Chapter 10, page 387

Relays. A relay allows control of a high current draw circuit by a very low current draw circuit. The computer usually controls the relay by providing the ground for the relay coil (Figure 10-40). The use of relays protects the computer by keeping the high current from passing through it. For example, the motors used for power door locks require a high current draw to operate them. Instead of having the computer operate the motor directly, it will energize the relay. With the relay energized, a direct circuit from the battery to the motor is completed (Figure 10-41).

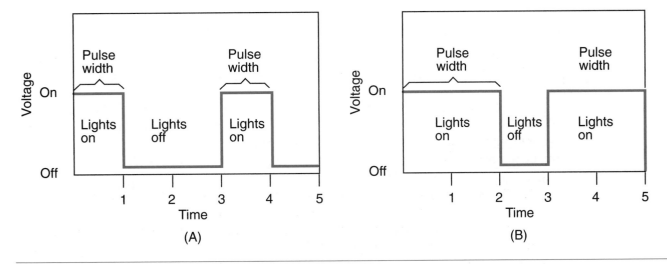

Figure 10-39 Pulse width is the duration of on-time. (A) Pulse-width modulation to achieve dimmer dash lights. (B) Pulse-width modulation to achieve brighter dash illumination.

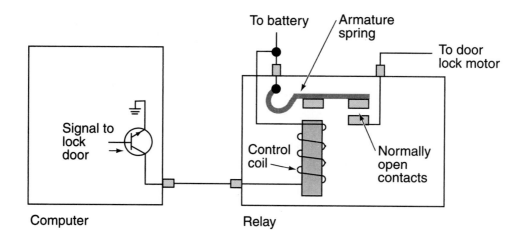

Figure 10-40 The computer's output driver applies the ground for the relay coil.

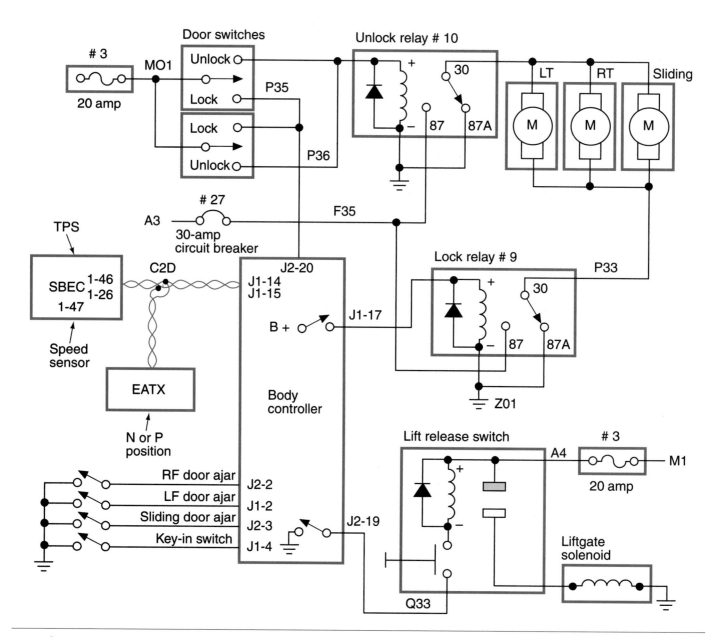

Figure 10-41 The computer controls the operation of the door lock motors by controlling the relays.

Solenoids. Computer control of the solenoid is usually provided by applying the ground through the output driver. A solenoid is commonly used as an actuator because it operates well under duty-cycling conditions.

One of the most common uses of the solenoid is to control vacuum to other components. Many automatic climate control systems use vacuum motors to move the blend doors. The computer can control the operation of the doors by controlling the solenoid.

Motors. Many computer-controlled systems use a **stepper motor** to move the controlled device to whatever location is desired. A stepper motor contains a permanent magnet armature with two, four, or more field coils (Figure 10-42). By applying voltage pulses to selected coils of the motor, the armature will turn a specific number of degrees. When the same voltage pulses are applied to the opposite coils, the armature will rotate the same number of degrees in the opposite direction.

Some applications require the use of a permanent magnet field **servomotor** (Figure 10-43). A servomotor produces rotation of less than a full turn. A feedback mechanism is used to position itself to the exact degree of rotation required. The polarity of the voltage applied to the armature windings determines the direction the motor rotates. The computer can apply a continuous voltage to the armature until the desired result is obtained.

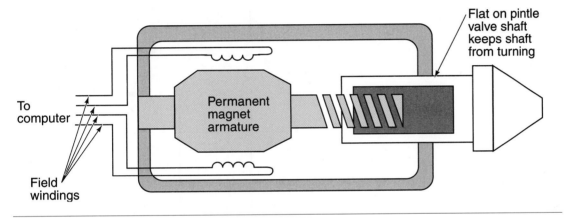

Figure 10-42 Typical stepper motor.

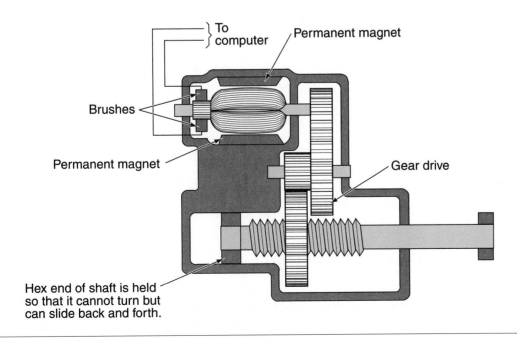

Figure 10-43 Reversible permanent magnet motor.

High Side and Low Side Drivers

Usually the computer will control an actuator by the use of **low side drivers.** These drivers will complete the path to ground through an FET transistor to control the output device. The computer may monitor the voltage on this circuit to determine if the actuator operates when commanded (Figure 10-44). Monitoring of the system can be done either by measuring voltage on the circuit or by measuring the current draw of the circuit.

Many newer vehicles are now using **high side drivers,** which control the output device by varying the positive (12-volt) side. High side drivers consist of a Metal Oxide Field Effect Transistor (MOSFET) that is controlled by a bipolar transistor. The bipolar transistor is controlled by the microprocessor. The advantage of the high side driver is that it can provide quick-response self-diagnostics for shorts, opens, and thermal conditions. It also reduces vehicle wiring.

High side driver diagnostic capabilities include the ability to determine a short circuit or open circuit condition. The high side driver will take the place of a fuse in the event of a short circuit condition. When it senses a high current condition, it will turn off the power flow and then

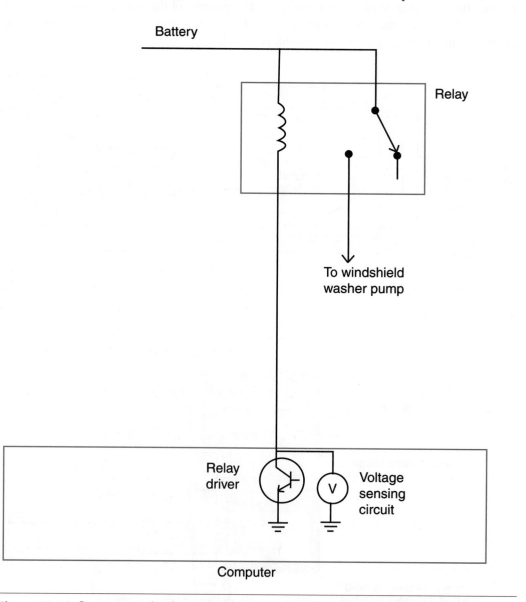

Figure 10-44 Computers using low-side drivers may be able to monitor the circuit for proper operation. When the relay coil is not energized, the sense circuit should see a high voltage (12 V). When the relay is turned on, the voltage should go low (0 V).

store a DTC in memory. The driver will automatically reset once the short circuit condition is removed. In addition, the high side driver monitors its temperature. The driver reports the junction temperature to the microprocessor. If a slow acting resistive short occurs in the circuit, the temperature will begin to climb. Once the temperature reaches 300°F (150°C) the driver will turn off and set a DTC.

The high side driver is also capable of detecting an open circuit, even if the system is turned off. This is done by reading a feedback voltage to the microprocessor when the driver is off. A 5-volt, 50 µA current is fed through the circuit, which also has a resistor wired in parallel. Low voltage (less than 2.25 volts) will indicate a normal circuit. If the voltage is high (above 2.25 volts), a high resistance or open circuit is detected. If the open circuit is detected, a DTC is set.

Summary

❏ A computer is an electronic device that stores and processes data and is capable of operating other devices.

❏ The operation of the computer is divided into four basic functions: input, processing, storage, and output.

❏ Binary numbers are represented by the numbers 1 and 0. A transistor that operates as a relay is the basis of the digital computer. As the input signal switches from off to on, the transistor output switches from cutoff to saturation. The on and off output signals represent the binary digits 1 and 0.

❏ Logic gates are the thousands of field effect transistors that are incorporated into the computer circuitry. The FETs use the incoming voltage patterns to determine the pattern of pulses that leave the gate. The most common logic gates are NOT, AND, OR, NAND, NOR, and XOR gates.

❏ There are several types of memory chips used in the body computer; ROM, RAM, and PROM are the most common types.

❏ ROM (read only memory) contains a fixed pattern of 1's and 0's representing permanent stored information used to instruct the computer on what to do in response to input data.

❏ RAM (random access memory) will store temporary information that can be read from or written to by the CPU.

❏ PROM (programmable read only memory) contains specific data that pertains to the exact vehicle in which the computer is installed.

❏ EPROM (Erasable PROM) is similar to PROM except its contents can be erased to allow new data to be installed.

❏ EEPROM (Electrically Erasable PROM) allows changing the information electrically one bit at a time.

❏ NVRAM (Nonvolatile RAM) is a combination of RAM and EEPROM into the same chip.

❏ Inputs provide the computer with system operation information or driver requests.

❏ Driver input signals are usually provided by momentarily applying a ground through a switch.

❏ Switches can be used as an input for any operation that only requires a yes-no, or on-off, condition.

❏ Sensors convert some measurement of vehicle operation into an electrical signal. There are many different designs of sensors: thermistors, Wheatstone bridge, potentiometers, magnetic pulse generator, and Hall-effect switches.

Terms to Know
Actuators
Binary code
Bit
Central processing unit (CPU)
Clock circuit
Feedback
Hall-effect switch
High side drivers
Interface
Linearity
Logic gates
Low side drivers
Magnetic pulse generator
Negative temperature coefficient (NTC) thermistors
Nonvolatile
Pickup coil
Positive temperature coefficient (PTC) thermistors
Program
Schmitt trigger
Sensors
Sequential logic circuits

❑ A thermistor is a solid-state variable resistor made from a semiconductor material that changes resistance in relation to temperature changes. Negative temperature coefficient (NTC) thermistors reduce their resistance as the temperature increases. Positive temperature coefficient (PTC) thermistors increase their resistance as the temperature increases.

❑ The Wheatstone bridge is a series-parallel arrangement of resistors between an input terminal and ground. Usually, three of the resistors are kept at exactly the same value while the fourth is the sensing resistor. When all four resistors have the same value, the bridge is balanced and the voltage sensor will indicate a value of 0 volts. If there is a change in the resistance value of the sense resistor, a change will occur in the circuit's balance. The sensing circuit will receive a voltage reading that is proportional to the amount of resistance change.

❑ A potentiometer is a variable resistor that usually consists of a wire wound resistor with a moveable center wiper.

❑ Magnetic pulse generators use the principle of magnetic induction to produce a voltage signal and are commonly used to send data concerning the speed of the monitored component to the computer.

❑ Hall-effect switches operate on the principle that if a current is allowed to flow through thin conducting material that is exposed to a magnetic field, another voltage is produced.

❑ Actuators are devices that perform the actual work commanded by the computer. They can be in the form of a motor, relay, switch, or solenoid.

❑ A servomotor produces rotation of less than a full turn. A feedback mechanism is used to position itself to the exact degree of rotation required.

❑ A stepper motor contains a permanent magnet armature with two, four, or more field coils. It is used to move the controlled device to whatever location is desired by applying voltage pulses to selected coils of the motor.

Review Questions

Short-Answer Essays

1. What is binary code?
2. Describe the basics of NOT, AND, and OR logic gate operation.
3. List and describe the four basic functions of the computer.
4. What is the difference between ROM, RAM, and PROM?
5. What is the difference between NTC and PTC thermistors?
6. How does the Hall-effect switch generate a voltage signal?
7. Describe the basic function of a stepper motor.
8. What is meant by feedback as it relates to computer control?
9. What is the difference between duty cycle and pulse width?
10. What are the purposes of the interface?

Fill in the Blanks

1. In binary code, the number 4 is represented by _____.

2. The _____ is a crystal that electrically vibrates when subjected to current at certain voltage levels.

3. _____ are registers designed to store the results of logic operations.

4. The _____ _____ _____ is the brain of the computer.

5. _____ contains specific data that pertains to the exact vehicle in which the computer is installed.

6. _____ convert some measurement of vehicle operation into an electrical signal.

7. Negative temperature coefficient (NTC) thermistors _____ their resistance as the temperature increases.

8. _____ switches operate on the principle that if a current is allowed to flow through thin conducting material exposed to a magnetic field, another voltage is produced.

9. Magnetic pulse generators use the principle of _____ _____ to produce a voltage signal.

10. _____ means that data concerning the effects of the computer's commands are fed back to the computer as an input signal.

Multiple Choice

1. *Technician A* says during the processing function the computer uses input information and compares it to programmed instructions.
   Technician B says during the output function the computer will put out control commands to various output devices.
   Who is correct?
   A. A only
   B. B only
   C. Both A and B
   D. Neither A nor B

2. Which of the following is correct?
   A. Analog signals are either high-low, on-off, or yes-no.
   B. Digital signals are infinitely variable within a defined range.
   C. All of the above.
   D. None of the above.

3. Logic gates are being discussed.
   Technician A says NOT gate operation is similar to that of two switches in series to a load.
   Technician B says an AND gate simply reverses binary 1's to 0's and vice versa.
   Who is correct?
   A. A only
   B. B only
   C. Both A and B
   D. Neither A nor B

4. All of the following statements about computer memory are true, EXCEPT:
   A. RAM stores temporary information that can be written to and read by the CPU.
   B. ROM can only be read by the CPU.
   C. All PROM memory is flashable.
   D. Volatile memory is erased when voltage is removed.

5. Nonvolatile memory is retained if removed from its power source.
   A. True
   B. False

6. *Technician A* says EPROM memory is erased if the tape is removed and the microcircuit is exposed to ultraviolet light.
   Technician B says electrostatic discharge will destroy the memory chip.
   Who is correct?
   A. A only
   B. B only
   C. Both A and B
   D. Neither A nor B

7. *Technician A* says negative temperature coefficient thermistors reduce their resistance as the temperature decreases.
   Technician B says positive temperature coefficient thermistors increase their resistance as the temperature increases.
   Who is correct?
   A. A only
   B. B only
   C. Both A and B
   D. Neither A nor B

8. *Technician A* says magnetic pulse generators are commonly used to send data to the computer concerning the speed of the monitored component.
   Technician B says an on-off switch sends a digital signal to the computer.
   Who is correct?
   A. A only
   B. B only
   C. Both A and B
   D. Neither A nor B

9. Speed sensors are being discussed.
   Technician A says the timing disc is stationary and the pickup coil rotates in front of it.
   Technician B says the number of pulses produced per mile increases as rotational speed increases.
   Who is correct?
   A. A only
   B. B only
   C. Both A and B
   D. Neither A nor B

10. *Technician A* says a Hall-effect switch uses a steady supply current to generate a signal.
    Technician B says a Hall-effect switch consists of a permanent magnet wound with a wire coil.
    Who is correct?
    A. A only
    B. B only
    C. Both A and B
    D. Neither A nor B

Vehicle Communication Networks

Upon completion and review of this chapter, you should be able to:

❏ Explain the principle of multiplexing.

❏ Describe the different OBD II multiplexing communication protocols.

❏ Explain the different classes of communications.

❏ Explain the operation of a class A multiplexing system.

❏ Explain the operation of a class B multiplexing system.

❏ Explain the operation of the Controller Area Network (CAN) bus system.

❏ Describe the purpose and operation of different supplemental data bus networks.

❏ Explain the operation of the Local Interconnect Network (LIN) data bus system.

❏ Describe the operation of the Media Oriented System Transport (MOST) data bus using fiber optics.

❏ Explain the operation of wireless networks using Bluetooth technology.

Introduction

In the past, if an accessory was added to the vehicle that required input information from sensors, either additional sensors had to also be added or the new accessory would have to be spliced into an existing sensor circuit. Either way, the cost of production was increased due to added components and wiring. For example, some early vehicles were equipped with as many as three engine coolant temperature sensors. One sensor was used by the powertrain control module (PCM) for fuel and ignition strategies, the other was used by the cooling fan module to operate the radiator fans at the correct speed based on temperature, and the third was used by the instrument cluster for temperature gauge operation.

Today vehicle manufacturers will use **multiplexing (MUX)** systems to enable different control modules to share information. Multiplexing provides the ability to use a single circuit to distribute and share data between several control modules throughout the vehicle. Because the data is transmitted through a single circuit, bulky wiring harnesses are eliminated. A MUX wiring system uses bus data links that connect each module and allow for the transporting of data from one module to another. Each module can transmit and receive digital codes over the bus data links. Each computer connected to the data bus is called a **node.** The signal sent from a sensor can go to any of the modules and can be used by the other modules. Before multiplexing, if information from the same sensing device was needed by several controllers, a wire from each controller needed to be connected in parallel to that sensor. If the sensor signal was analog, the controllers needed an analog to digital (A/D) convertor to be able to "read" the sensor information. By using multiplexing, the need for separate conductors from the sensor to each module is eliminated and the number of drivers in the controllers is reduced.

As discussed in Chapter 10, binary code is sent by digital signals to the nodes. The nodes use this code to communicate messages, both internally and with other controllers. A chip is used to prevent the digital codes from overlapping by allowing only one code to be transmitted at a time. Each digital message is preceded by an identification code that establishes its priority. If two modules attempt to send a message at the same time, the message with the higher priority code is transmitted first.

The major difference between a multiplexed system and a nonmultiplexed system is the way data is gathered and processed. In nonmultiplexed systems, the signal from a sensor is sent as an analog signal through a dedicated wire to the computer or computers. At the computer, the signal is changed from an analog to a digital signal. Because each sensor requires its own dedicated signal wire, the number of wires required to feed data from all of the sensors and transmit control signals to all of the output devices is great.

In a MUX system, the signal is sent to a computer where it is converted from analog to digital if needed. Since the computer or control module of any system can process only one input at a time, it calls for input signals as it needs them. By timing the transmission of data from the sensors to the control module, a single data circuit can be used. Between each transmission of data to the control module, the sensor is electronically disconnected from the control module.

Multiplexing Communication Protocols

A **protocol** is a language computers use to communicate with one another over the data bus. Protocols may differ in baud rate and in the method of delivery. For example, some manufacturers use pulse width modulation while others use variable pulse width. In addition, there may be differences in the voltage levels that equal a 1 or a 0.

The Society of Automotive Engineers (SAE) has defined different classes of protocols according to their **baud rate** (speed of communication):

❏ Class A—Generic Universal **Asynchronous** Receiver/Transmitter (UART) low-speed protocol that has a baud rate of up to 10 Kb/s (10,000 bits per second). Asynchronous protocol means that the communication between nodes is done only when needed instead of continuously.

❏ Class B—Medium-speed protocol that has a baud rate between 10 Kb/s and 125 Kb/s.

❏ Class C—A high-speed protocol with a baud rate between 125 Kb/s and 1,000 Kb/s, used for functions that require real-time control.

Newer automotive protocols fall under on-board diagnostics, second generation (OBD II) requirements. The SAE has adopted the OBD II protocol as the class B standard protocol (Figure 11-1).

ISO 9141-2 (K-line)
ISO 14230-4 (Keyword protocol (Kwp) 2000)
J 1850 10.4 Kb/s Variable pulse width
J 1850 41.6 Kb/s Pulse width modulated
J 2284/ISO 15765-4 Controller area network (CAN)

Figure 11-1 OBD II communication protocols.

ISO 9141-2 Protocol

The International Standards Organization (ISO) protocol known as the **ISO 9141-2** is a class B system with a baud rate of 10.4 Kb/s. ISO 9141-2 is not a network protocol since it can only be used for diagnostic purposes. ISO 9141 standardizes a protocol to be used between the nodes on the data bus and an OBD II standardized scan tool (as per SAE J1978 standards) for diagnostic purposes. This system is a two-wire system (Figure 11-2). One wire is called the **K-line** and is used for transmitting data from the module to the scan tool. The scan tool provides the bias voltage onto this circuit and the module pulls the voltage low to transmit its data. The other wire is called the **L-line** and is used by the module to receive data from the scan tool. The module provides the bias onto this circuit and the scan tool pulls the voltage low to communicate.

An adoption of the ISO 9141-2 protocol is the **ISO-K** bus that allows for bidirectional communication on a single wire. Vehicles that use the ISO-K bus require that the scan tool provide the bias voltage to power up the system. The scan tool provides up to 12 volts onto the circuit, and the data is transferred when the voltage is pulled low to create a digital signal (Figure 11-3). The actual voltage seen on the circuit can be a little less than 12 volts, based on the number of modules in the circuit.

ISO 14230-4 Protocol

The **ISO 14230-4** protocol uses a single-wire bidirectional data line to communicate between the scan tool and the nodes. This system is used by many European manufacturers. This data bus is only used for diagnostics and maintains the ISO 9141 protocol with a baud rate of 10.4 Kb/s. The operation of receive and transmit is similar to the UART system since communication requires a **master module.** The master module controls the transportation of messages by polling all of the **slave modules** and then waiting for the response. The communication occurs as the voltage on the wire is pulled low at a fixed pulse width. When there is no communication occurring, the voltage on the line will be 5 volts.

Shop Manual
Chapter 11, pages 410, 412

Biasing refers to the voltage supplied to the bus.

ISO 14230-4 is also called Keyword Protocol 2000 or KWP2000.

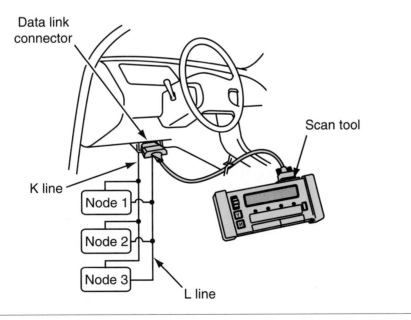

Figure 11-2 The two-wire ISO 9141-2 data bus used for diagnostic purposes.

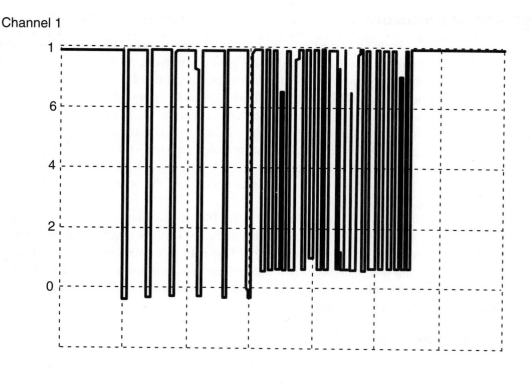

Channel 1

Figure 11-3 ISO K data bus transmission voltages.

Shop Manual
Chapter 11, page 416

J1850 Protocol

The **J1850** bus system is the class B standard for OBD II. The J1850 standard allows for two different versions based on baud rate. The first supports a baud rate of 41.6 Kb/s. Ford uses this protocol and calls it the Standard Corporate Protocol (SCP). This system uses a pulse width modulated (PWM) signal that is transferred using a twisted pair of wires. The second protocol supports a baud rate of 10.4 Kb/s average. General Motors and DaimlerChrysler have adopted this protocol. General Motors calls their system Class 2 and DaimlerChrysler calls theirs Programmable Communication Interface (PCI). These systems use a variable pulse width (VPW) data bus with a single wire. This system will be discussed in great detail later in this chapter.

Shop Manual
Chapter 11, page 418

J2284/ISO 15765-4 Protocol

The principal class C protocol is J2284. This system is referred to as **Controller Area Network (CAN).** The CAN network system can support baud rates up to 1 Mb/s and is designed for real-time control of specific systems. Robert Bosch developed CAN in the early 1980s, and it has been a very popular bus system in Europe. Until recently most CAN bus–equipped vehicles used the CAN bus for communications between modules only and not for diagnostics with a scan tool. Scan tool diagnostics are usually performed over the ISO-K bus.

New U.S. regulations are requiring the use of the CAN bus system under industry standard J2284. This will be used as the new protocol for communications with a scan tool on U.S.-sold vehicles. Although the CAN bus system has been used since the 1980s, J2284 makes it unique in that for the first time it will be used for diagnostics. Manufacturers will be required to use CAN for 2008 model year vehicles. This system will be discussed in greater detail later in this chapter.

Multiplexing Systems

The following are some examples of how data bus messages are transmitted. Although protocols are in place, manufacturers have some freedom to design the system they wish to use. The following examples will explain the common methods that are employed.

Class A Data Bus Network

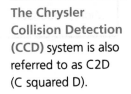

One of the earliest multiplexing systems was developed by Chrysler in 1988 and used through the 2003 model year. This system is called **Chrysler Collision Detection (CCD).** The term *collision* refers to the collision of data occurring simultaneously. This bus circuit uses two wires. The advantages of the CCD system include:

Shop Manual
Chapter 11, page 413

1. Reduction of wires.
2. Reduction of drivers required in the computers.
3. Reduced load across the sensors.
4. Enhanced diagnostics.

The CCD system uses a twisted pair of wires to transmit the data in digital form. One of the wires is called the **bus (+)** and the other is **bus (−)**. Negative voltages are not used. The (+) and (−) indicate that one wire is more positive than the other when the bus is sending the dominant bit "0." All modules that are connected to the CCD bus system have a special CCD chip installed (Figure 11-4). In most vehicles (but not all), the body control module (BCM) will

The Chrysler Collision Detection (CCD) system is also referred to as C2D (C squared D).

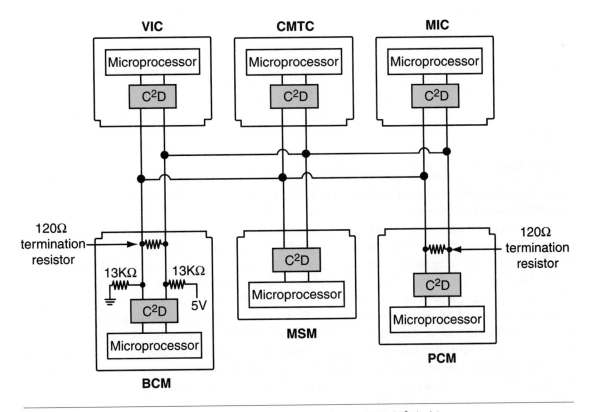

Figure 11-4 Each module on the CCD bus system has a CCD (C²D) chip.

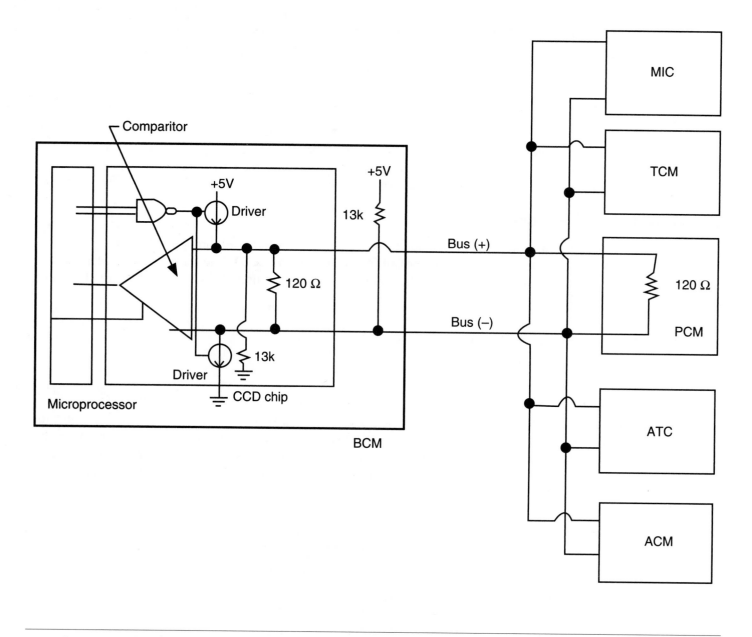

Figure 11-5 CCD bus circuit.

provide the bias voltage to power the bus circuits. Since the BCM powers the system, its internal components are illustrated (Figure 11-5). The other modules will operate the same as the BCM to send messages.

The bias voltage on the bus (+) and bus (−) circuits is approximately 2.50 volts when the system is idle (no data transmission occurring). This is accomplished through a regulated 5-volt circuit and a series of resistors. The regulated 5 volts sends current through a 13K-ohm resistor to the bus (−) circuit (Figure 11-6). The current is then sent through the two 120-ohm resistors that are wired in parallel and to the bus (+) circuit. Finally, the current is sent to ground through a second 13K-ohm resistor. A simplified schematic of this biasing circuit is shown along with the normal voltage drops that occur as a result of the resistors (Figure 11-7). The two 120-ohm resistors are referred to as **termination resistors.** Termination resistors are used to control induced voltages. Since voltage is dropped over resistors, the induced voltage is terminated. One termination resistor is internal to the BCM while the other is located in the powertrain control module (PCM).

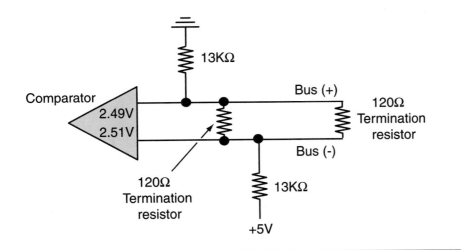

Figure 11-6 The bus is supplied 2.5 volts through the use of pull-up and pull-down resistors.

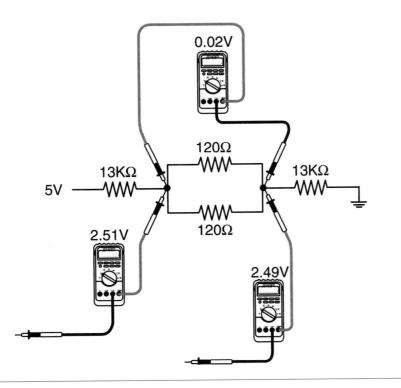

Figure 11-7 Simplified bus bias circuit for clarification.

With the bus circuits at the proper voltage levels, communication can occur. The comparator in the CCD chip acts as a voltmeter. If the positive lead of the voltmeter is connected to the bus (+) circuit and the negative lead is connected to the bus (–) circuit, the voltmeter will read the voltage difference between the circuits. At idle, the difference is 0.02 volts. When a module needs to send a message, the microprocessor will use the NAND gate to turn on and off the two drivers at the same time (Figure 11-8). The driver to the bus (+) circuit provides alternate 5 volts to the bus (+) circuit. The comparitor will measure this voltage. At the same time, the driver to the bus (–) circuit provides an alternate ground path for the original 5 volts. This alternate ground bypasses the termination resistors and the second 13K-ohm resistor. Since the first 13K-ohm resistor is now the only one in the circuit, all of the voltage is dropped over it and the comparator will see low voltage on the bus (–) circuit.

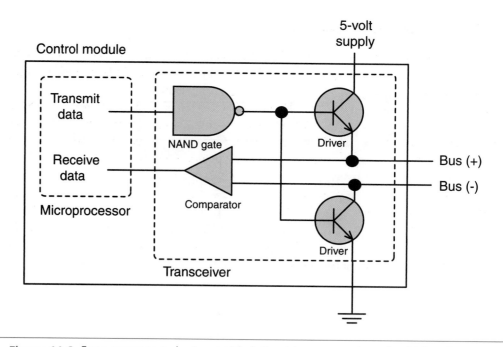

Figure 11-8 For a message to be transmitted, the drivers are activated, which pulls up bias on bus (+) and pulls down bias on bus (–).

Since the drivers are turned on and off at a rate of 7812.5 times per second, the voltage will not go to a full 5 volts on bus (+) nor to 0 volts on bus (–). However, bus (+) voltage is pulled higher *toward* 5 volts and bus (–) is pulled lower *toward* 0 volts (Figure 11-9). Once the comparator measures a voltage difference greater than 0.060 volts, the computers will recognize a bit value change. When the bus circuit is idle (0.02 voltage difference), the bit value is a 1. Once the voltage difference increases, the bit value is changed to a 0.

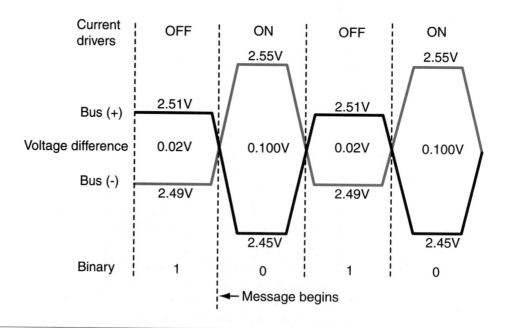

Figure 11-9 As the drivers are activated, the voltage difference between bus (+) and bus (–) increases over their voltage values at idle. The difference in voltage determines if a binary 1 or 0 is being transmitted.

Class B Data Bus Network

The J1850 protocol regulates the class B data bus network. As an example of this system, we will look at a VPW 10.4 Kb/s system that DaimlerChrysler uses. This system is similar to General Motors, class 2 bus.

Beginning in the 1998 model year, DaimlerChrysler began to phase out the CCD bus system and replace it with a new **Programmable Communication Interface (PCI)** bus system. Since this system is similar to that of other manufacturers, it will be used for discussion purposes.

The PCI system is a single-wire, bidirectional communication bus. Each module on the bus system supplies its own bias voltage and has its own termination resistors (Figure 11-10). Like the CCD system, the modules of the PCI system are connected in parallel. As a message is sent, a variable pulse-width modulation (VPWM) voltage between 0 and 7.75 volts is used to represent the 1 and 0 bits (Figure 11-11). The voltage signal is not a clean digital signal. Rather, the voltage traces appear to be trapezoidal in shape because the voltage is slowly ramped up and down to prevent magnetic induction.

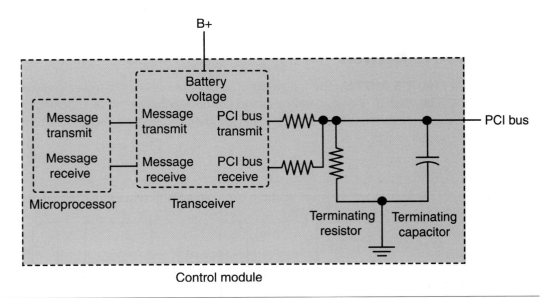

Figure 11-10 Bias and termination are supplied by each module on the PCI bus system.

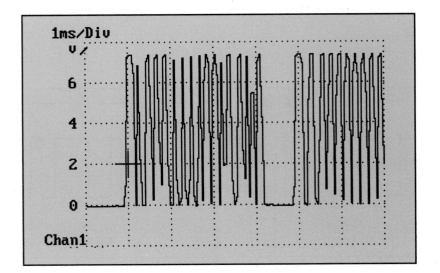

Figure 11-11 Lab scope trace of PCI bus voltages.

AUTHOR'S NOTE: The reference to "slowly ramped up and down" is relative. In the PCI bus, an average of 10,400 bits are transmitted per second. So, relatively speaking, the voltage is slow to go to 7.75 volts and slow to return to 0 volts.

The length of time the voltage is high or low determines if the bit value is a 1 or a 0 (Figure 11-12). The typical PCI bus message will have the following elements (Figure 11-13):

❏ Header—One to three bytes of information concerning the type, length, priority, target module, and sending module.

❏ Data byte(s)—The message that is being sent. This can be up to 8 bytes in length.

❏ Cyclic Redundancy Check (CRC) byte—Detects if the message has been corrupted or if there are any other errors.

❏ In-Frame Response (IFR) byte(s)—If the sending module requires an acknowledgment or an immediate response from the target module, this request will be received with the message. The IFR is the target module sending the requested information to the original sending module.

Figure 11-14 illustrates the type of information that is sent over the PCI bus system.

Controller Area Network

Shop Manual
Chapter 11, page 418

AUTHOR'S NOTE: The following is an example of the controller area network (CAN) bus system. Again, baud rates and design vary between manufacturers. This is provided as a common method used. By understanding this system, you should be able to grasp any system design.

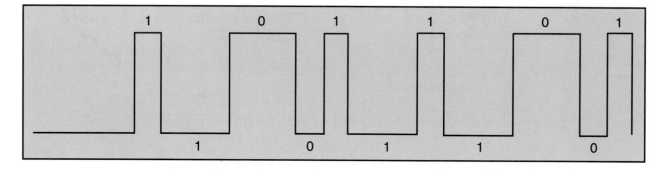

Figure 11-12 The VPWM determines the bit value.

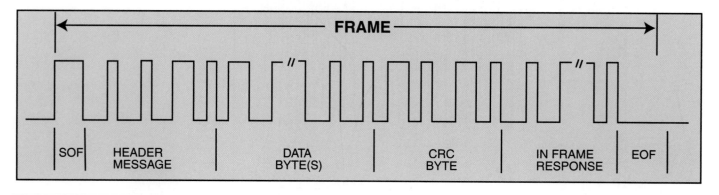

Figure 11-13 Components of a typical PCI bus message.

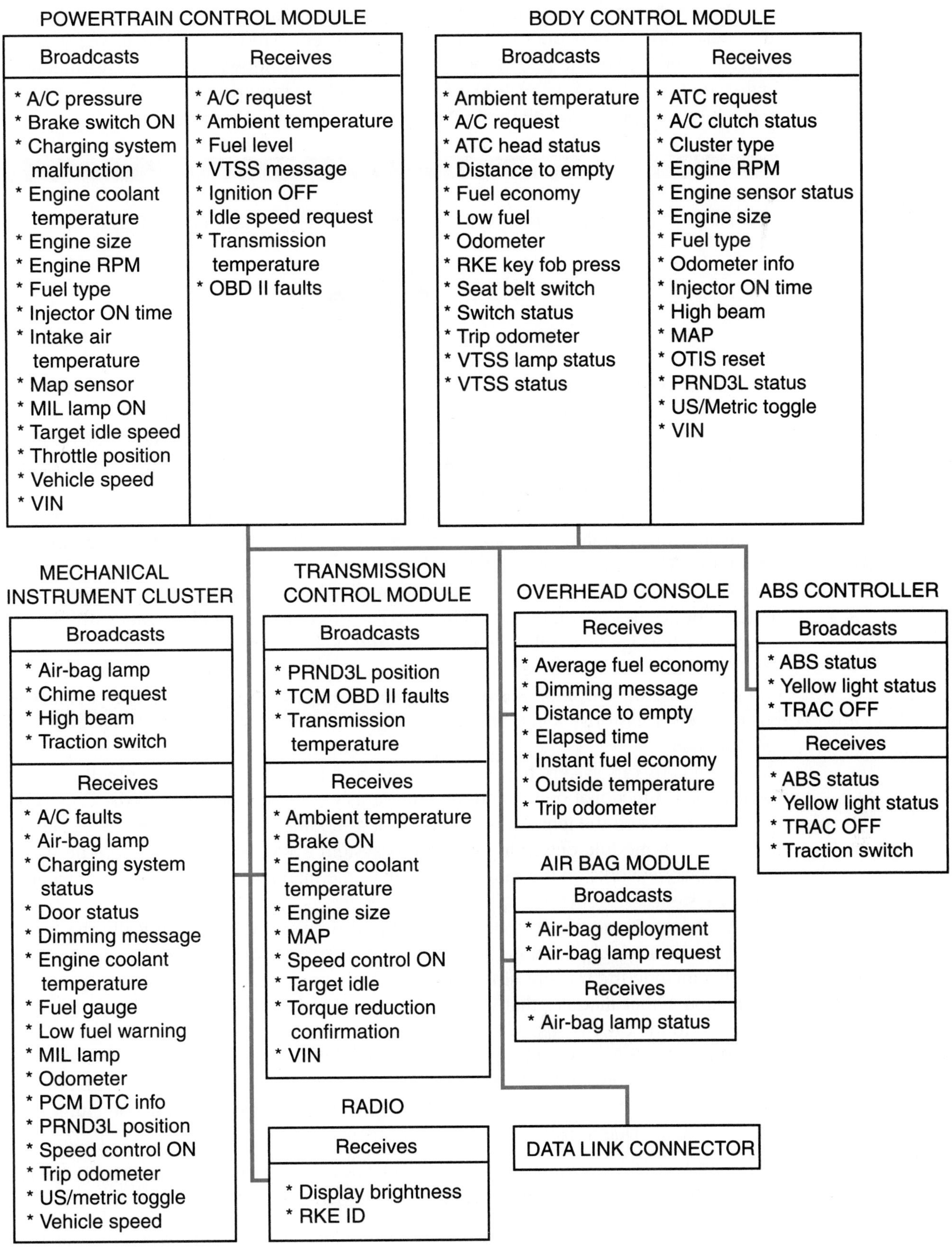

POWERTRAIN CONTROL MODULE

Broadcasts	Receives
* A/C pressure * Brake switch ON * Charging system malfunction * Engine coolant temperature * Engine size * Engine RPM * Fuel type * Injector ON time * Intake air temperature * Map sensor * MIL lamp ON * Target idle speed * Throttle position * Vehicle speed * VIN	* A/C request * Ambient temperature * Fuel level * VTSS message * Ignition OFF * Idle speed request * Transmission temperature * OBD II faults

BODY CONTROL MODULE

Broadcasts	Receives
* Ambient temperature * A/C request * ATC head status * Distance to empty * Fuel economy * Low fuel * Odometer * RKE key fob press * Seat belt switch * Switch status * Trip odometer * VTSS lamp status * VTSS status	* ATC request * A/C clutch status * Cluster type * Engine RPM * Engine sensor status * Engine size * Fuel type * Odometer info * Injector ON time * High beam * MAP * OTIS reset * PRND3L status * US/Metric toggle * VIN

MECHANICAL INSTRUMENT CLUSTER

Broadcasts
* Air-bag lamp * Chime request * High beam * Traction switch

Receives
* A/C faults * Air-bag lamp * Charging system status * Door status * Dimming message * Engine coolant temperature * Fuel gauge * Low fuel warning * MIL lamp * Odometer * PCM DTC info * PRND3L position * Speed control ON * Trip odometer * US/metric toggle * Vehicle speed

TRANSMISSION CONTROL MODULE

Broadcasts
* PRND3L position * TCM OBD II faults * Transmission temperature

Receives
* Ambient temperature * Brake ON * Engine coolant temperature * Engine size * MAP * Speed control ON * Target idle * Torque reduction confirmation * VIN

RADIO

Receives
* Display brightness * RKE ID

OVERHEAD CONSOLE

Receives
* Average fuel economy * Dimming message * Distance to empty * Elapsed time * Instant fuel economy * Outside temperature * Trip odometer

AIR BAG MODULE

Broadcasts
* Air-bag deployment * Air-bag lamp request

Receives
* Air-bag lamp status

DATA LINK CONNECTOR

ABS CONTROLLER

Broadcasts
* ABS status * Yellow light status * TRAC OFF

Receives
* ABS status * Yellow light status * TRAC OFF * Traction switch

Figure 11-14 Chart of messages received and broadcast by each module on the PCI bus.

Most vehicles that are following the J2284 protocol integrate either two or three controller area network bus networks that operate at different baud rates. The lower-speed bus is typically used for vehicle body functions such as seat, window, radio, and instrumentation. A high-speed bus is used for real-time functions such as required for engine management and antilock brake operation. The third CAN bus would be used for diagnostics.

The CAN bus system uses terminology such as CAN B and CAN C. The letters *B* and *C* distinguish the speed of the bus. CAN B is a medium-speed bus with a speed of up to 125,000 bits per second. The CAN C bus has a speed of 500,000 bits per second. A vehicle can be equipped with both of these bus networks. In addition, a new diagnostics CAN C bus is used to connect the scan tool. Diagnostic CAN C (which can be called by many different names) has a speed of 500,000 bits per second.

The CAN bus circuit consists of a pair of twisted wires. The transfer of digital data is done by simultaneously pulling the voltage on one circuit high, and pulling the voltage on the other circuit low. The wires for the CAN bus system are twisted to reduce electromagnetic interference (EMI). This requires 33 to 50 twists per meter. To maintain the twist, the bus wire pair is in adjacent cavities at connectors. Wires are routed to avoid parallel paths with high-current sources, such as ignition coil drivers, motors, and high-current PWM circuits.

On a CAN bus system, each module provides its own bias. Because of this, communication between groups of modules is still possible if an open occurs in the bus circuit. The CAN bus transceiver has drivers internal to the transceiver chip to supply the voltage and ground to the bus circuit.

Each CAN bus system has its advantages and limitations. For example, the high-speed CAN C bus is functional only when the ignition is on. On the other hand, the CAN B bus can remain active when the ignition is turned off, if a module requires it to be active. The requirements of each module determine which bus system it will be connected to. The use of two separate bus networks on the same vehicle gives the manufacturer the optimum characteristics of each system.

Vehicle systems that exchange data at real time use CAN C. Typically these modules would be the antilock brake module and the powertrain control module. The manufacturer may also include the transmission control module and other modules that require real-time information. Other modules that may need to transfer data with the ignition turned off will be connected to the CAN B bus.

Most CAN B bus systems are very fault tolerant and can operate with one of their conductors shorted to ground or both of their conductors shorted together. Provided there is an electrical potential between one of the CAN B circuits and chassis ground, communication may still be possible. Due to its high speed, CAN C is not fault tolerant.

YOU SHOULD KNOW: Since the CAN B bus can be designed to have a baud rate between 10.4 Kb/s and 125 Kb/s, fault tolerance diminishes as the baud rate increases.

The CAN C bus becomes active when the ignition is turned on. When the bus is biased, the voltage is approximately 2.5 volts. When both CAN C (+) and CAN C (−) are equal, the bus is recessive and the bit "1" is transmitted. When CAN C (+) is pulled high and CAN C (−) is

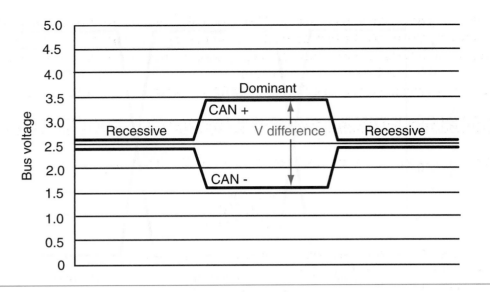

Figure 11-15 Voltages on the CAN C bus.

pulled low, the bit "0" is transmitted. When the bit "0" is transmitted, the bus is considered dominant (Figure 11-15). To be dominant, the voltage difference between CAN C (+) and CAN C (−) must be at least 1.5 volts and not more than 3.0 volts. To be recessive, the voltage difference between the two circuits must not be more than 50 mv.

The optimum CAN C bus termination is 60 ohms. Two CAN C modules will provide 120 ohms of termination each. Since the modules are wired in parallel, total resistance is 60 ohms. The two modules that provide termination are typically located the farthest apart from each other. The terminating modules have two 60-ohm resistors that are connected in series to equal the 120 ohms. Common to both resistors are the connections to the center tap and ultimately through a capacitor to ground. This center tap may also be connected to the transceiver (Figure 11-16). The other ends of the resistors are connected to CAN (+) and to CAN (−).

The CAN B bus can be active whether the ignition is on or off. When CAN B (+) is approximately 0 to 0.2 volts and CAN B (−) is 4.8 to 5 volts, the bus is idle or recessive. In this state, the logic is "1." When CAN B (+) is pulled between 3.6 and 5 volts and CAN B (−) is pulled low to

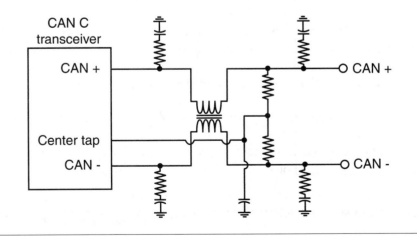

Figure 11-16 Termination resistance of a CAN C module.

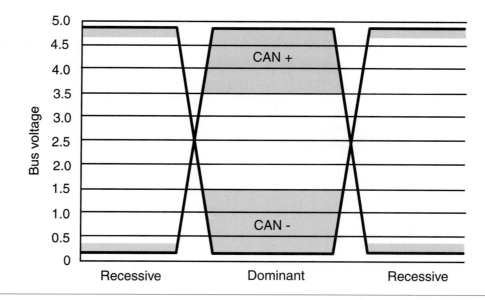

Figure 11-17 Typical CAN B bus voltages.

1.4 to 0 volts, the bus is considered dominant and the logic is "0" (Figure 11-17). When CAN B (+) is approximately 0 volts and CAN B (−) is near battery voltage, the bus is asleep.

Each module on the CAN B bus supplies its own termination resistance. Total bus termination resistance is determined by the number of modules installed on the vehicle. Internal to CAN B modules are two termination resistors. The resistors connect CAN B (+) and CAN B (−) to their respective transceiver termination pins (Figure 11-18). To provide termination and bias, the transceiver internally connects the CAN B (+) resistor to ground and the CAN B (−) resistor to a 5-volt source. When the CAN B bus goes into sleep mode, the termination pin connected to CAN B (−) switches from 5 volts to battery voltage by the transceiver.

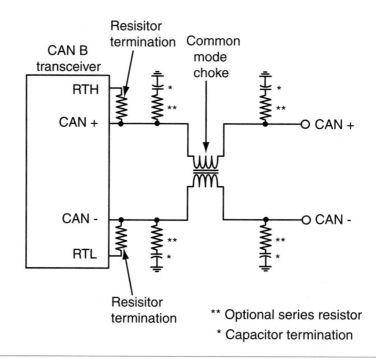

Figure 11-18 CAN B bus module termination resistance.

Some of the network messages are defined by the following:

❏ Cyclic: A message launched on a periodic schedule. An example is the ignition on status broadcasts on the CAN B bus every 100 ms.

❏ Spontaneous: An application-driven message.

❏ Cyclic and Change: A message launched on a periodic schedule as long as the signal D is not changing. The message is relaunched whenever the signal changes.

❏ By Active Function (BAF): A message that is only transmitted at a specific rate when the message does not equal a default value.

A **central gateway (CGW)** module is used where all three CAN bus networks (CAN B, CAN C, and diagnostic CAN C) connect together. Similar to a router in a computer network, this module allows data exchange between the different busses. The CGW can take a message on one bus and transfer that message to the other bus without changing the message. If several messages are being sent simultaneously, some of the messages will be captured in a buffer and sent out based on priority. The CGW also monitors the CAN network for failures and can log a network DTC (U code) if it detects a malfunction.

The CGW is also the gateway to the CAN network for the scan tool. The scan tool is connected to the gateway using its own CAN bus circuit known as diagnostic CAN C. Because CAN C is used for diagnostics, data can be exchanged with the scan tool at a real-time rate (500 Kb/s). As a result, a scan tool that is compatible with the CAN bus system is required for vehicle diagnosis.

Since a variety of generic scan tools may be connected to the vehicle, mandated regulations prohibit scan tools from containing any termination resistance. Due to the speed of the CAN C bus, termination resistors that are used need to be closely matched. If termination resistance resided in the tool, the variance in the tool termination could affect the operation of the diagnostic CAN C bus. For this reason, the entire termination for the diagnostic CAN C resides in the CGW. The configuration of the resistors is similar to a dominant CAN C module except two 30-ohm resistors are connected in series.

 AUTHOR'S NOTE: The most aggressive company advancing CAN bus networking into the automobile design is Intel.

Supplemental Data Bus Networks

Since no one data bus can handle all of the requirements of computer-controlled operations on today's vehicles, **supplemental bus networks** are also used. These are bus networks that are on the vehicle in addition to the main bus network. For example, a vehicle may have the CAN bus network and a second bus network to handle specified requirements. This section discusses the common bus networks that may also be on the vehicle.

Local Interconnect Network Data Bus

The **local interconnect network (LIN)** bus is a UART single master module, multiple slave module, low-speed network. The term *local interconnect* refers to all of the modules in the LIN network being located within a limited area. The LIN master module is connected to the CAN bus and controls the data transfer speed. The master module translates data between the slave module and the CAN bus (Figure 11-19). Diagnosis of the salve modules is performed through the master module. The termination resistance of the master module is 1K ohms.

The LIN bus system can support up to 15 slave modules. Slave modules use 30K-ohm termination resistors. The slave modules can be actual control modules or sensors and actuators.

Shop Manual
Chapter 11, page 422

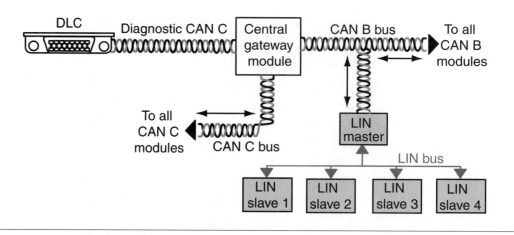

Figure 11-19 The LIN master communicates messages from the slaves onto the CAN bus.

Smart sensors are capable of sending digital messages on the LIN bus. The intelligent actuators receive commands in digital signals on the LIN bus from the master module. Only one pin of the master module is required to monitor several sensors and actuators.

The data transmission is variable between 1 Kb/s and 20 Kb/s over a single wire. The specific baud rate is programmed into each module. When a steady 12 volts is applied to the circuit, there is no message and the recessive bit is sent (bit 1). To transmit a dominant bit (bit "0"), the circuit is pulled low by a transceiver in the module that is transmitting the message (Figure 11-20). The master module or the slave module can send messages.

Media Oriented System Transport Data Bus

The **Media Oriented System Transport (MOST)** cooperation data bus system is based on standards established by cooperative efforts between automobile manufacturers, suppliers, and software programmers. The result is a data system that is specifically designed for the transmission of audio and video data. The MOST bus uses **fiber optics** to transmit data at a speed of to 25 megabits per second (25 Mb/s). The fiber optics use light waves to transmit the data without the effects of EMI or RFI. In the past, video and audio were transmitted as analog signals. With the MOST system using a fiber optics data bus, the data communications is digital.

Modules on the MOST data bus use an LED, photodiode, and a MOST transceiver to communicate with light signals (Figure 11-21). The LED and photodiode are part of the fiber optic transceiver. The photodiode changes light signals into voltage that is then transmitted to the MOST transceiver. The LED is used to convert voltage signals from the MOST transceiver into light signals.

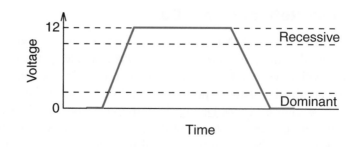

Figure 11-20 Voltages of the LIN bus.

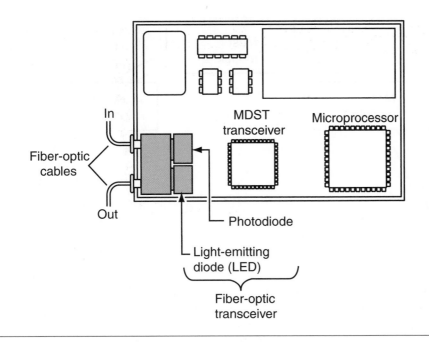

Figure 11-21 Typical MOST data system controller components.

The conversion of light to voltage signals in the photodiode is by subjecting the P-N junction of the photodiode with light. When the junction is penetrated with light, the energy is converted to free electrons and holes. The electrons and holes pass through the junction in direct proportion to the amount of light. The photodiode is connected in series with a resistor (Figure 11-22). As the voltage through the photodiode increases, the voltage drop across the resistor also increases. Since the voltage drop changes with light intensity, the light signals are changed to voltage signals.

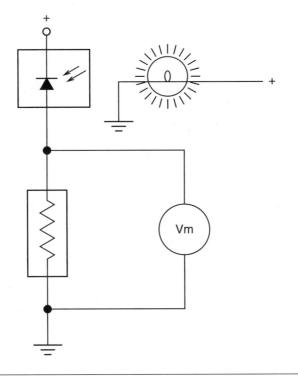

Figure 11-22 The voltage drop over the resistor changes in relation to the amount of light applied to the photodiode.

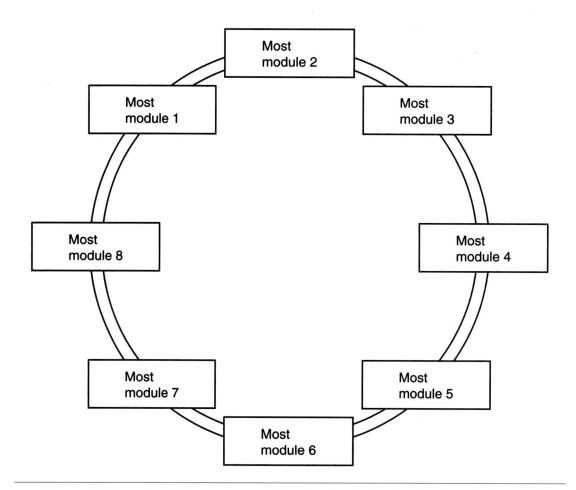

Figure 11-23 The MOST data system transfers data in a single direction through the use of a ring configuration.

The microprocessor commands the MOST transceiver to send messages to the fiber optic transceiver as voltage signals. Also, the MOST transceiver sends voltage signals from the fiber optic transceiver to the microprocessor.

The modules are connected in a ring fashion by fiber-optic cable (Figure 11-23). Messages are sent in one direction only. The message is usually started by the master module, but not always. The master module sends the message onto the data bus with a duty cycle frequency of 44.1 kHz. This frequency corresponds to the frequency of digital audio and video equipment. A message that is originated by a module is sent to the next module in the ring. That module then sends it to the next module, and this continues until the originating module receives its own message. At this time the ring is closed and the message is no longer passed on. If a module receives a message that it does not need, the message is sent through the MOST transceiver and back to the fiber-optic transceiver without being transmitted to the microprocessor. If the MOST bus is powered down (asleep), it can be awakened by the ignition switch input or an input from a module. When an input is received, the master module will send a "wake up" message to all of the modules in the ring.

The fiber-optic cable is constructed with several layers (Figure 11-24). The core consists of polymethyl methacrylate (PMMA). Light travels through the core of the cable based on the principle of **total reflection.** Total reflection is achieved when a light wave strikes a layer that is between a dense and a thin material. The core of the cable is coated with an optically transparent

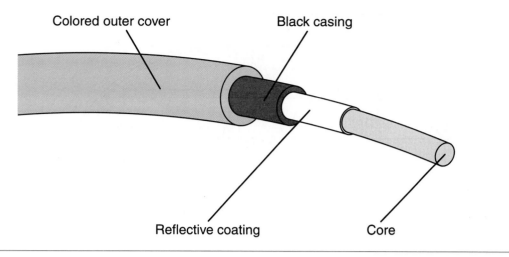

Figure 11-24 Fiber-optic cable construction.

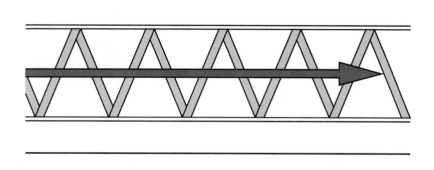

Figure 11-25 Light waves traveling through a straight section of the fiber-optic cable.

reflective coating. The core makes up the dense material, and the coating is the thin material. The casing of the cable is made from polyamide and protects the core from outside light. An outer cover is colored so the cable can easily be identified, but it also protects the cable from damage and high temperatures.

If the cable is laid out straight, some of the light travel through the core will be in a straight line. However, most of the light waves travel in a zigzag pattern (Figure 11-25). The zigzag pattern is a result of the total reflection. If the fiber-optic cable is bent, the light waves are reflected by total reflection at the borderline of the core coating and are guided through the bend (Figure 11-26).

Wireless Bus Networks

Wireless networks can connect modules together to transmit information without the use of physical connections by wires. For example, tire pressure information can be transmitted from a sensor in the tire to a module on the vehicle without wires. Although there are different technologies used for wireless communications, a popular one is called **Bluetooth.** Bluetooth technology allows several modules from different manufacturers to be connected using a standardized radio transmission. Laptop computers, notepads, and hands-free cell phones are all examples of devices that may be connected in the vehicle through Bluetooth.

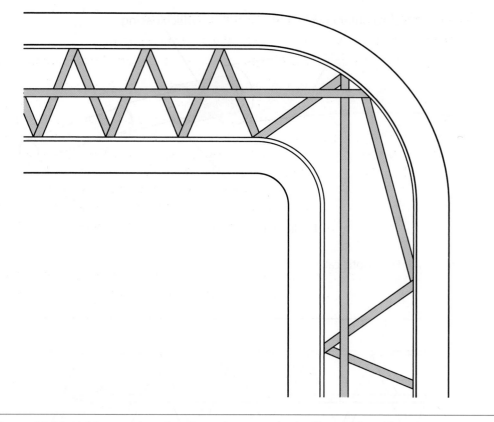

Figure 11-26 Light waves traveling through a curve in the fiber-optic cable.

The Bluetooth Special Interest Group consists of more than 2,000 companies. The name *Bluetooth* comes from the Viking King Harald Blåtand who was nicknamed Bluetooth. King Harald ruled Denmark between 940 and 985 AD. During his reign King Harald united Denmark and Norway. The name *Bluetooth* was adopted for a particular wireless communications technology because it shares the same philosophy as the king, multinational unity. Bluetooth unifies multinational companies. Bluetooth was initially a code name for the project; however, it has now become the trademark name.

Radio transmission uses the 2.40 GHz to 2.48 GHz frequency range. Transmitting on this band does not require a license or a fee. The transmission rate is up to 1 Mb/s, and the normal operating range for transmissions is about 33 feet (10 meters). The short transmitting range makes it possible to integrate the antenna, control, encryption, and transmission/receiver technology into a single module.

Since the frequency range Bluetooth uses is the same as that of other wireless devices such as garage door openers, microwaves, and many types of medical equipment, the technology uses special measures to protect against interference. These measures include:

❏ Dividing the data into short message packages using a duration of about 625 ms.

❏ Using a check sum of 16 bits to confirm the messages were not corrupted.

- Automatically repeating the transmission of faulty data.
- Using language coding that is converted into digital signals.
- Changing the transmitting/receiving frequencies at random, 1,600 times per second.

For security, the data is encrypted using a key that is 128 bits long. The receiver is checked for authenticity with the key. In addition, a secret password is used so devices can connect to each other. The key uses rolling code and is different for each connection.

To connect devices, each device is first adapted through the use of a personal identification number (PIN). When the PIN is entered, **piconets** are formed. These are small transmission cells that assist in the organization of the data. Each piconet will allow for up to eight devices to be operated at the same time. One device in each piconet will be assigned as the master. The master is responsible for establishing the connection and synchronizing the other devices. Once the PIN is entered, two Bluetooth-compatible devices will automatically establish a connection.

 AUTHOR'S NOTE: Each device has an address that is 48 bits long and is unique worldwide. This means over 281 trillion devices can be identified worldwide.

Summary

- Multiplexing (MUX) is a system in which electrical signals are transmitted by a peripheral serial bus instead of by conventional wires. This allows several devices to share signals on a common conductor.
- A MUX wiring system uses bus data links that connect each module and allow for the transporting of data from one module to another.
- Each computer connected to the data bus is called a node.
- The Society of Automotive Engineers (SAE) has defined different classes of protocols according to their baud rate (speed of communication).
- ISO 9141-2 is not a network protocol since it can be used only for diagnostic purposes. It is a Class B system that has a baud rate of 10.4 Kb/s.
- An adoption of the ISO 9141-2 protocol is the ISO-K bus that allows for bidirectional communication on a single wire.
- The ISO 14230-4 protocol uses a single-wire, bidirectional data line to communicate between the scan tool and the nodes.
- J1850 protocol is the class B standard for OBD II. The J1850 standard allows for two different versions based on baud rate.
- The controller area network (CAN) system can support baud rates up to 1 Mb/s and is designed for real-time control of specific systems.
- One of the earliest multiplexing systems was developed by Chrysler in 1988 and used through the 2003 model year. This system is called Chrysler Collision Detection (CCD).
- The CCD system uses a twisted pair of wires to transmit the data in digital form. One of the wires is called the bus (+) and the other is bus (−).
- The CCD bias voltage on the bus (+) and bus (−) circuits is approximately 2.50 volts when the system is idle (no data transmission occurring). This is accomplished through a regulated 5-volt circuit and a series of resistors.

❑ The Programmable Communication Interface (PCI) system is a single-wire, bidirectional communication bus. Each module on the bus system supplies its own bias voltage and has its own termination resistors.

❑ The modules of the PCI system send messages by a variable pulse width modulation (VPWM) voltage; between 0 and 7.75 volts is used to represent the 1 and 0 bits.

❑ Most vehicles that are following the J2284 protocol integrate either two or three controller area network (CAN) bus networks that operate at different baud rates.

❑ The lower-speed bus is typically used for vehicle body functions such as seat, window, radio, and instrumentation control. A high-speed bus is used for real-time functions as required for engine management and antilock brake operation.

❑ The circuitry of the CAN bus usually consists of a pair of twisted wires. For digital data to transfer, voltage is simultaneously pulled high on one circuit and pulled low on the other.

❑ A central gateway (CGW) module is used where all three CAN bus networks (CAN B, CAN C, and diagnostic CAN C) connect together. Similar to a router in a computer network, this module allows data exchange between the different busses. The CGW is also the gateway to the CAN network for the scan tool.

❑ The local interconnect network (LIN) was developed to supplement the CAN bus system. The term *local interconnect* refers to all of the modules in the LIN network being located within a limited area.

❑ The LIN bus is a UART single master, multiple slave, low-speed network.

❑ The Media Oriented System Transport (MOST) cooperation data bus system is based on standards established by a cooperative effort between automobile manufacturers, suppliers, and software programmers that resulted in a data system specifically designed for the data transmission of media-oriented data. MOST uses fiber optics to transmit data at an extremely fast rate, up to 25 megabits per second.

❑ Modules on the MOST data bus use an LED, a photodiode, and a MOST transceiver to communicate with light signals.

❑ Wireless networks can connect modules together to transmit information without the use of physical connection by wires.

❑ Bluetooth technology allows several modules from different manufacturers to be connected using a standardized radio transmission.

Review Questions

Short-Answer Essay

1. Explain the principle of multiplexing.

2. Briefly describe the different multiplexing communication protocols.

3. Explain the different classes of communications.

4. Briefly explain the principle of operation of the CCD bus system as an example of Class A multiplexing.

5. Briefly explain the principle of operation of the PCI bus system as an example of Class B multiplexing.

6. Briefly explain the principle of operation of the Controller Area Network (CAN) bus system.

7. Describe the purpose and operation of different supplemental data bus networks.

8. Explain the operation of the local interconnect network data bus system.

9. Describe the operation of the Media Oriented System Transport data bus using fiber optics.

10. Explain the operation of wireless networks using Bluetooth technology.

Fill in the Blanks

1. Multiplexing provides the ability to use a _____ circuit to distribute and share data between several control modules.

2. Each computer connected to the data bus is called a _____.

3. The _____ protocol is the class B standard for OBD II.

4. The CCD system uses a twisted pair of wires to transmit the data in _____ form.

5. _____ _____ are used to control induced voltages.

6. The PCI system uses a _____ pulse width modulation voltage between 0 and 7.75 volts to represent the 1 and 0 bits.

7. A _____ _____ module is used where all three CAN bus networks (CAN B, CAN C, and diagnostic CAN C) connect together.

8. In the MOST data bus system, the _____ changes light signals into voltage that is then transmitted to the MOST transceiver.

9. The modules of the MOST data bus system are connected in a _____ fashion by fiber optic cable.

10. _____ technology allows several modules from different manufacturers to be connected using a standardized radio transmission.

Multiple Choice

1. In the CAN bus system:
   A. CAN B is high speed and not fault tolerant.
   B. CAN C is fault tolerant.
   C. Can C is low speed and used for many body control functions.
   D. None of the above.

2. All of the following statements concerning multiplexing are true EXCEPT:
   A. Hard wiring and vehicle weight are reduced.
   B. A single computer controls all of the vehicle functions.
   C. Enhanced diagnostics are possible.
   D. Reduces driver requirements in the computer.

3. The purpose of the central gateway module in a CAN bus system is:
   A. To provide a means for the modules on the different CAN bus networks to communicate with each other.
   B. To provide a method for the scan tool to communicate with the modules.
   C. Both A and B.
   D. Neither A nor B.

4. *Technician A* says that some multiplexing systems use a data bus that consists of a twisted pair of wires. *Technician B* says that some multiplexing systems use a data bus that consists of a single wire. Who is correct?
   A. A only.   C. Both A and B.
   B. B only.   D. Neither A nor B.

5. On a data bus that uses two wires, why are the wires twisted?
- **A.** To increase the physical strength of the wires.
- **B.** For identification of the circuit.
- **C.** To reduce the effect of the high current flow through the bus wires on other circuits.
- **D.** To minimize the effects of an induced voltage on the data bus.

6. In the local interconnect network (LIN) bus:
- **A.** The master controller is connected to the CAN bus and controls the data transfer speed.
- **B.** The master controller translates data between the slave modules and the CAN bus.
- **C.** Supporting up to 15 slave modules is possible.
- **D.** All of the above.

7. Protocol is defined as:
- **A.** A common communication method.
- **B.** A method of reducing EMI.
- **C.** Data transmission through a single circuit.
- **D.** All of the above.

8. *Technician A* says that multiplexed circuits are used to communicate multiple messages over a single circuit.
Technician B says that multiplexed circuits communicate by transmitting serial data.
Who is correct?
- **A.** A only.
- **B.** B only.
- **C.** Both A and B.
- **D.** Neither A nor B.

9. In a single-wire bus network, EMI is controlled by:
- **A.** Using a shielded wire.
- **B.** Slowly ramping up and down the voltage levels.
- **C.** Locating the wire outside of the normal wiring harness.
- **D.** None of the above.

10. A computer that can communicate on a data bus is known as:
- **A.** Node.
- **B.** Byte.
- **C.** Protocol.
- **D.** Transceiver.

Advanced Lighting Circuits

Upon completion and review of this chapter, you should be able to:

❑ Explain the operation of common computer-controlled concealed headlight systems.

❑ Describe the function of the computer-controlled headlight system.

❑ Explain the function of automatic headlight on/off and time delay features.

❑ Explain the operation of most common types of automatic headlight dimming systems.

❑ Explain the operation of the SmartBeam™ headlight system as an example of today's sophisticated headlight systems.

❑ Describe the function of automatic headlight leveling systems.

❑ Describe the purpose and function of daytime running lamps.

❑ Describe the operation of common illuminated entry systems.

❑ Explain the operation of common instrument panel dimming systems.

❑ Explain the use and function of fiber optics.

❑ Describe the purpose and operation of lamp outage indicators.

Introduction

With the addition of solid-state circuitry in the automobile, manufacturers have been able to incorporate several different lighting circuits or modify the existing ones. Some of the refinements that were made to the lighting system include automatic headlight washers, automatic headlight dimming, automatic on/off with timed delay headlights, and illuminated entry systems. Some of these systems use sophisticated body computer–controlled circuitry and fiber optics.

Some manufacturers have included such basic circuits as turn signals into their body computer to provide for pulse width dimming in place of a flasher unit. The body computer can also be used to control instrument panel lighting based on inputs, including if the side marker lights are on or off. By using the body computer to control many of the lighting circuits, the amount of wiring has been reduced. In addition, the use of computer control has provided a means of self-diagnosis in some applications.

Computer-Controlled Concealed Headlights

Shop Manual
Chapter 12, page 451

The body control module (BCM) has been utilized by some manufacturers to operate the concealed headlight system. The BCM will receive inputs from the headlight and flash-to-pass switches (Figure 12-1). When the headlight switch is turned on, the BCM receives a signal that the headlights are being activated. To open the headlight doors, the BCM energizes the open door relay. The contacts of the open relay are closed and battery voltage is applied to the door motor. In this example, battery voltage to operate the door motor is supplied from the 30-ampere circuit breaker. The computer energizes the door open relay, which moves the normally open relay contact arm. Ground is provided through the door close relay contact.

When the headlight switch is turned off, the computer energizes the close door relay. With the door close relay energized, the contacts provide battery voltage to the door motor. The ground is supplied through the door open relay. Reversing the polarity through the door motor closes the door.

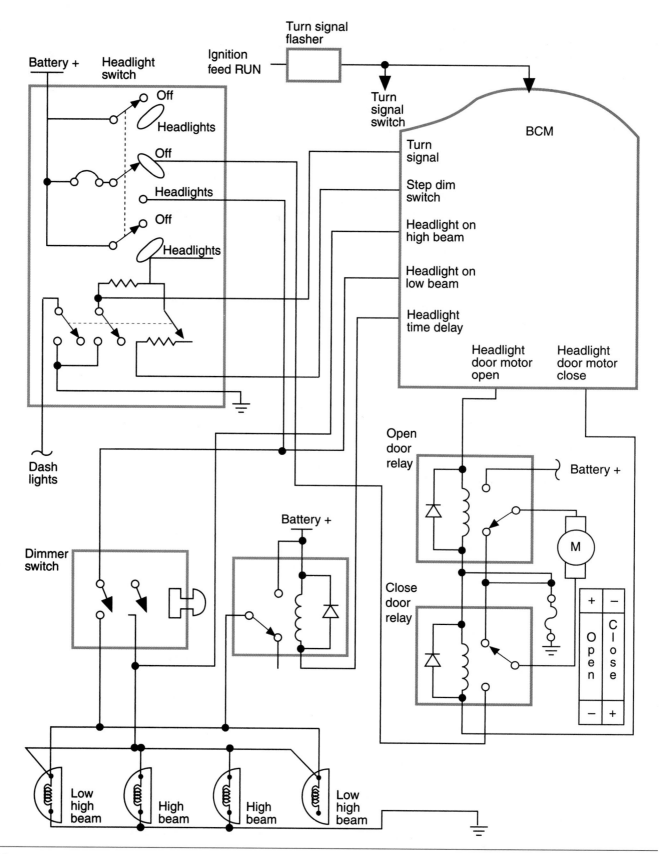

Figure 12-1 Computer-controlled concealed headlight door circuit.

If the flash-to-pass option is activated, the body computer receives a high (on) signal and energizes the door open relay. When the switch is released, the computer receives a low (off) signal and activates the door close relay. The computer delays the activation of the door close relay for 3 seconds.

 AUTHOR'S NOTE: Another module besides the BCM can be used to operate the concealed headlight system; however, the function of the system is the same as described.

Computer-Controlled Headlight Systems

Shop Manual
Chapter 12, page 456

The computer can be used to control the operation of the headlights. Simpler circuits use the headlight switch as an input to the body computer. The headlight switch is usually a **resistive multiplex switch** that provides multiple inputs over a single circuit (Figure 12-2). The BCM sends a fixed voltage to the switch through a fixed resistor. Each switch position has a different resistor value in series with the fixed resistor. As different switch positions are selected, voltage drop over the fixed resistor changes.

The BCM will control the relays associated with the exterior lighting. If input from the switch indicates that the headlights are requested, the BCM will energize the low-beam headlight relay (Figure 12-3). With this relay energized, current flows to the low-beam lamps. Voltage is also applied to the multifunction switch (dimmer switch function). If the multifunction switch is placed in the high-beam position, it will complete the path for the flow of current to the coil of the high-beam headlight relay. Since this relay is connected to ground, the coil is energized and the contacts

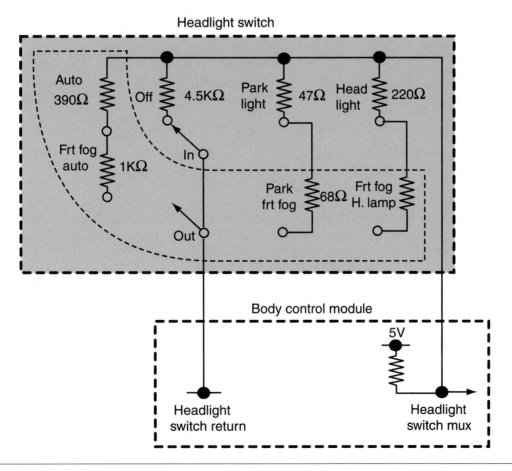

Figure 12-2 Resistive multiplex switch used as input for the headlight system.

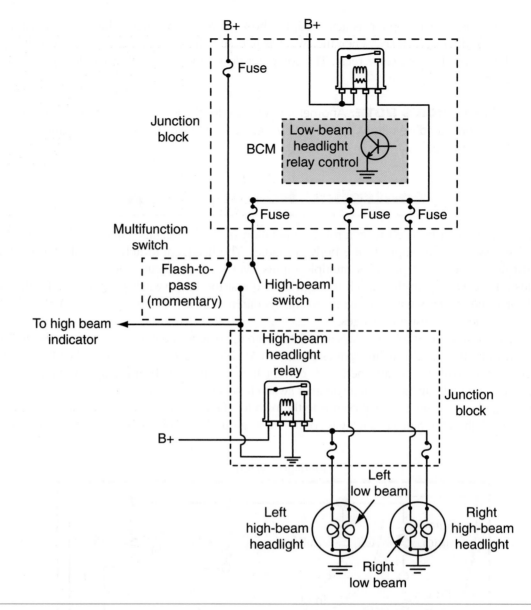

Figure 12-3 The body computer controls the relay of the headlight system.

move to supply current to the high-beam lamps. The low-beam relay must be energized to turn on the high beams unless the flash-to-pass function is being performed. The flash-to-pass function of the multifunction switch bypasses the BCM control of the circuit.

Park and fog light relays are controlled in the same fashion. Since the BCM controls the operation of the relays, it can turn off the exterior lights if the driver forgets to do so. Once the ignition switch is turned off, a timer is started. After a programmed length of time, the BCM will turn off all the relays associated with the exterior lighting system, even if the switch is still in the headlight position.

A more sophisticated system uses high side drivers (HSD) (Figure 12-4). In the example shown, the lamp and fog lamp relays are controlled, as previously described, using low side drivers (relay drivers). The difference is that the headlight switch input is sent to the BCM, then the BCM sends the request to the integrated power module (IPM). The IPM then turns on and off the relays as needed. The operation of the headlights is unique. Based on the input from the headlight switch, the BCM sends the requested state to the IPM. The IPM then turns on the headlights by supplying power from the HSDs. This system does not require relays or fuses because the

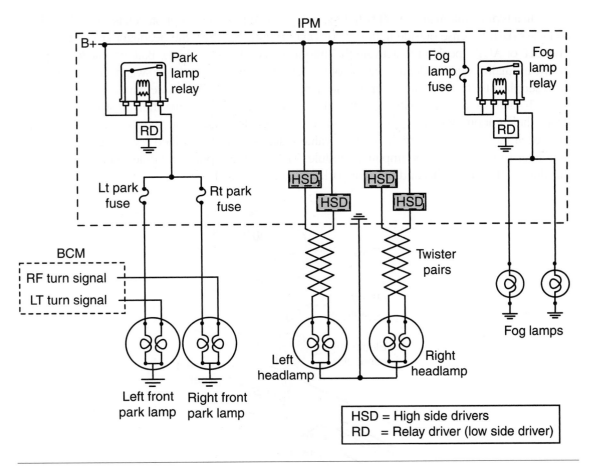

Figure 12-4 Headlight circuit that uses high side drivers.

HSDs perform these functions. If a high current condition occurs, the HSDs sense this and turn off the circuit until the cause of the high current is no longer present. The output of the HSDs is pulse width modulated at a frequency of 90 Hz in order to maintain a constant 13.5 volts to the headlight bulbs, relative to battery voltage. This is done to increase the life of the headlight bulbs. In addition, the HSDs can perform diagnostics of the system, set DTCs, and turn on indicator lights to notify the driver of a malfunction.

Automatic On/Off with Time Delay

AUTHOR'S NOTE: This system is given several different names by the various manufacturers. Some of the more common names include: Twilight Sentinel, Auto-lamp/Delayed Exit, and Safeguard Sentinel.

Shop Manual
Chapter 12, page 458

The **automatic on/off with time delay** has two functions: to turn on the headlights automatically when ambient light decreases to a predetermined level and to allow the headlights to remain on for a certain amount of time after the vehicle has been turned off. The common components of the automatic on/off with time delay include:

1. Photocell and amplifier.
2. Power relay.
3. Timer control.

In a typical automatic on/off with time delay headlight system, a photocell is located inside the vehicle's dash to sense outside light (Figure 12-5). In most systems, the headlight switch must be in the OFF or AUTO position to activate the automatic mode (Figure 12-6). Battery voltage is applied to the normally open headlight contacts of the relay through the headlight switch. Battery voltage is also supplied to the normally open exterior light contacts through the fuse panel (Figure 12-7).

To activate the automatic on/off feature, the photocell and amplifier must receive voltage from the ignition switch. As the ambient light level decreases, the internal resistance of the photocell increases. When the resistance value reaches a predetermined level, the photocell and amplifier trigger the sensor-amplifier module. The sensor-amplifier module energizes the relay, turning on the headlights and exterior parking lights (Figure 12-8).

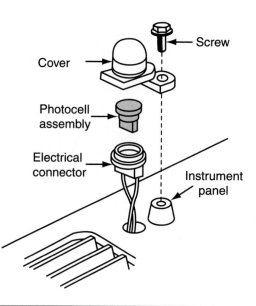

Figure 12-5 Most automatic on/off headlight systems have the photocell located in the dash to sense incoming light levels.

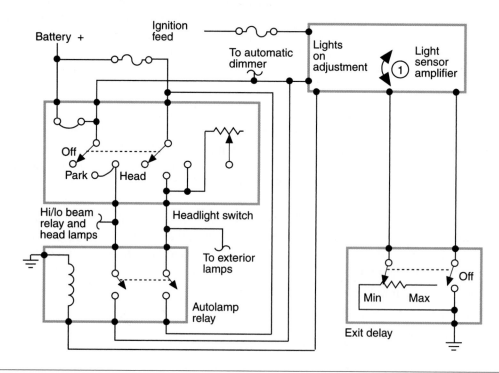

Figure 12-6 Schematic of automatic headlight on/off with time delay system.

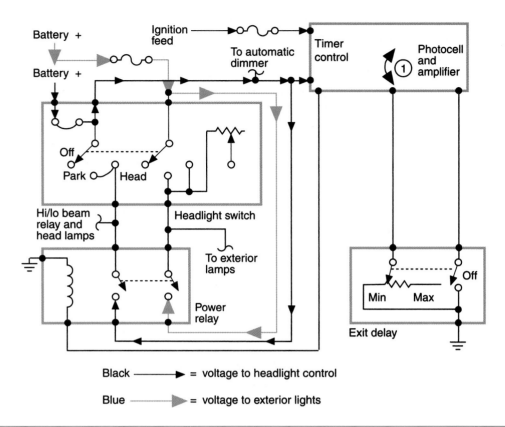

Black ──────▶ = voltage to headlight control

Blue ──────▶ = voltage to exterior lights

Figure 12-7 Current flow in automatic mode with the headlights off.

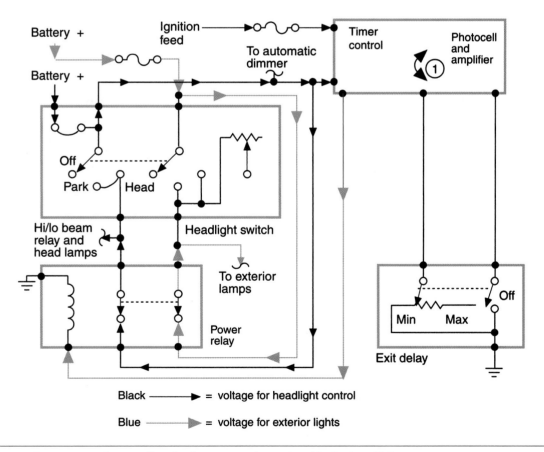

Black ──────▶ = voltage for headlight control

Blue ──────▶ = voltage for exterior lights

Figure 12-8 Current flow in the automatic mode with the headlights on.

Early systems use a **timer control** that uses a potentiometer that is part of the headlight switch. The timer control unit controls the automatic operation of the system and the length of time the headlights stay on after the ignition switch is turned off. The timer control signals the sensor-amplifier module to energize the relay for the requested length of time.

If the headlights are on when the ignition switch is turned off, the photocell and amplifier's ignition switch voltage is opened. This activates a timer circuit in the amplifier. The amplifier still receives battery voltage from the headlight switch and uses this voltage to keep the relay energized for the requested time interval. When the preset length of time has passed, the amplifier module removes power to the relay and the headlights (and exterior lights) turn off.

The driver can override the automatic on/off feature by placing the headlight switch in the ON position. This bypasses the relay and sends battery voltage directly to the headlight circuit.

 AUTHOR'S NOTE: The Twilight Sentinel System is overriden when the delay control switch is in the OFF position. The delay control switch is located on the left switch panel.

Depending on model application, General Motors' Twilight Sentinel System can use the body computer to control system operation (Figure 12-9). The body control module (BCM) senses the voltage drop across the photocell and the delay control switch. The delay control switch resistance is wired in series with the photocell. If the ambient light level drops below a specific value, the BCM grounds the headlamp and parklamp relay coils. The BCM also keeps the headlights on for a specific length of time after the ignition switch is turned off.

Most newer systems continue to use the same principle of using the photocell as the early systems for automatic headlight operation. The difference is that the time control is a function of the module (usually the BCM). An interface module such as the instrument panel or overhead console provides a means for the driver to select the desired amount of time delay after the ignition switch is turned off. The BCM keeps the headlight relay energized for the length of time programmed. When the ignition switch is turned off, all other relays controlling exterior lighting are de-energized except the headlight relay.

Automatic Headlight Dimming

Modern **automatic headlight dimming** systems use solid-state circuitry and electromagnetic relays to control the beam switching. Automatic headlight dimming automatically switches the headlights from high beams to low beams under two different conditions: when light from oncoming vehicles strikes the photocell-amplifier or light from the taillights of a vehicle being passed strikes the photocell-amplifier. Most systems consist of the following major components:

The **photocell** is like a variable resistor that uses light to change resistance.

1. Light-sensitive **photocell** and amplifier unit.
2. High-low beam relay.
3. Sensitivity control.
4. Dimmer switch.
5. Flash-to-pass relay.
6. Wiring harness.

The photocell-amplifier is usually mounted behind the front grill, but ahead of the radiator. The **sensitivity control** sets the intensity level at which the photocell-amplifier will energize. This control is set by the driver and is located next to, or is a part of, the headlight switch assembly

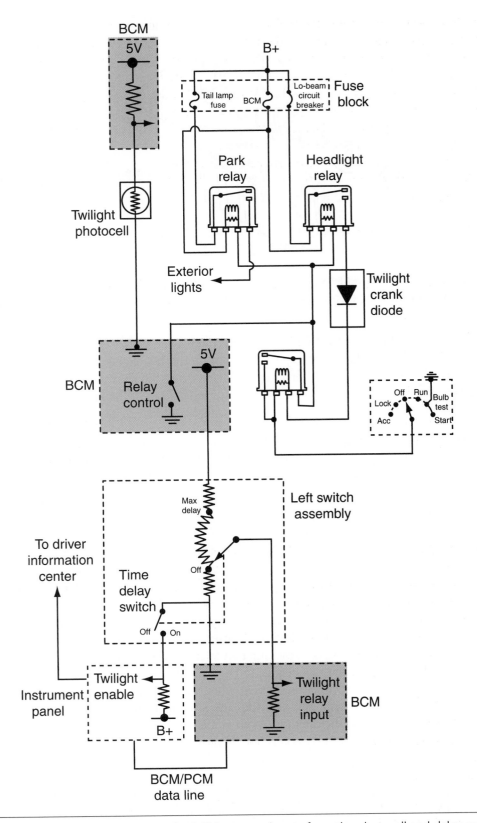

Figure 12-9 Some systems use the BCM to sense inputs from the photocell and delay control switch.

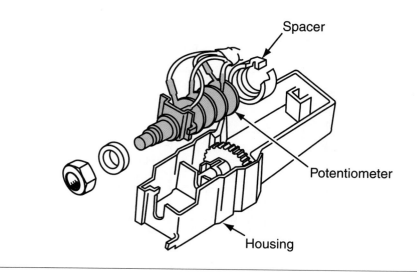

Spacer

Potentiometer

Housing

Figure 12-10 The driver sets the sensitivity of the automatic headlight dimmer system by rotating the potentiometer to change resistance values.

(Figure 12-10). The driver is able to adjust the sensitivity level of the system by rotating the control knob. An increase in the sensitivity level will make the headlights switch to the low beams sooner (approaching vehicle is farther away). A decrease in the sensitivity level will switch the headlights to low beams when the approaching vehicle is closer. If the knob is rotated to the full counterclockwise position, the system goes into manual override.

 AUTHOR'S NOTE: Many vehicle manufacturers install the sensor-amplifier in the rearview mirror support.

The high-low relay is a single-pole, double-throw unit that provides the switching of the headlight beams (Figure 12-11). The relay also contains a clamping diode for electrical transient damping to protect the photocell and amplifier assembly.

The dimmer switch is usually a flash-to-pass design. If the turn signal lever is pulled partway up, the flash-to-pass relay is energized. The high beams will stay on as long as the lever is held in this position, even if the headlights are off. In addition, the driver can select either low beams or automatic operation through the dimmer switch.

Although the components are similar in most systems, there are differences in system operations. Systems differ in how the manufacturer uses the relay to do the switching from high beams to low beams. The system can use either an energized relay to activate the high beams or an energized relay to activate the low beams. If the system uses an energized relay to activate the high beams, the relay control circuit is opened when the dimmer switch is placed in the low-beam position or the driver manually overrides the system (Figure 12-12). With the headlight switch in the ON position and the dimmer switch in the low-beam position, battery voltage is not applied to the relay coil and the relay coil is not energized. Since the dimmer switch is in the low-beam position, it opens the battery feed circuit to the relay coil and the automatic feature is bypassed.

With the dimmer switch in the automatic position, battery feed is provided for the relay coil (Figure 12-13). Ground for the relay coil is through the sensor-amplifier. The energized coil closes the relay contacts to the high beams and battery voltage is applied to the headlamps. When the photocell sensor receives enough light to overcome the sensitivity setting, the amplifier opens the relay's circuit to ground. This de-energizes the relay coil and switches battery voltage from the high-beam to the low-beam position.

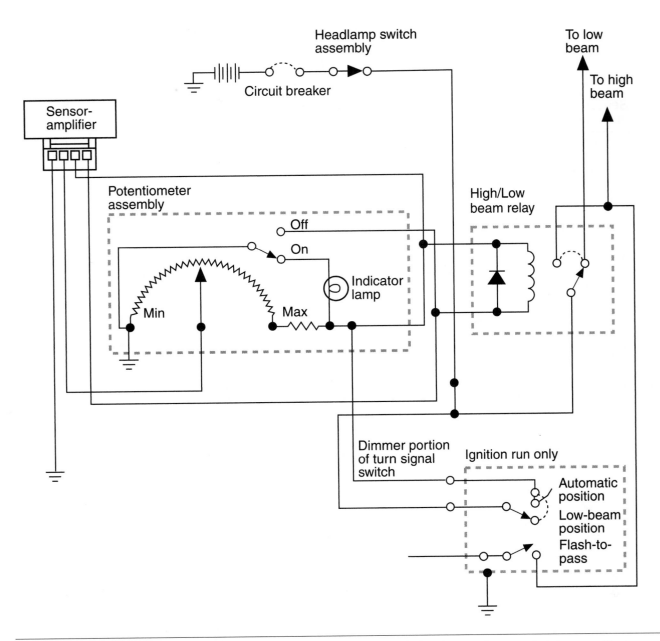

Figure 12-11 Automatic headlight dimming circuit that uses a high-low beam relay to switch beam settings.

If the system uses an energized relay to switch to low beams, placing the dimmer switch in the low-beam position will energize the relay. With the headlights turned on and the dimmer switch in the automatic position, battery voltage is applied to the photocell-amplifier, one terminal of the high-low control, and through the relay contacts to the high beams. The voltage drop through the high-low control is an input to the photocell-amplifier. When enough light strikes the photocell-amplifier to overcome the sensitivity setting, the amplifier allows battery current to flow through the high-low relay, closing the contact points to the low beams. Once the light has passed, the photocell-amplifier opens battery voltage to the relay coil and the contacts close to the high beams.

When flash-to-pass is activated, the switch closes to ground. This bypasses the sensitivity control and de-energizes the relay to switch from low beams to high beams.

Today the headlight system can be very sophisticated. The following is an example of how the SmartBeam™ system used by some manufacturers performs auto headlamp and auto high-beam operation. This system provides lighting levels based on conditions and will operate the high beams by sensing light levels.

Shop Manual
Chapter 12, page 463

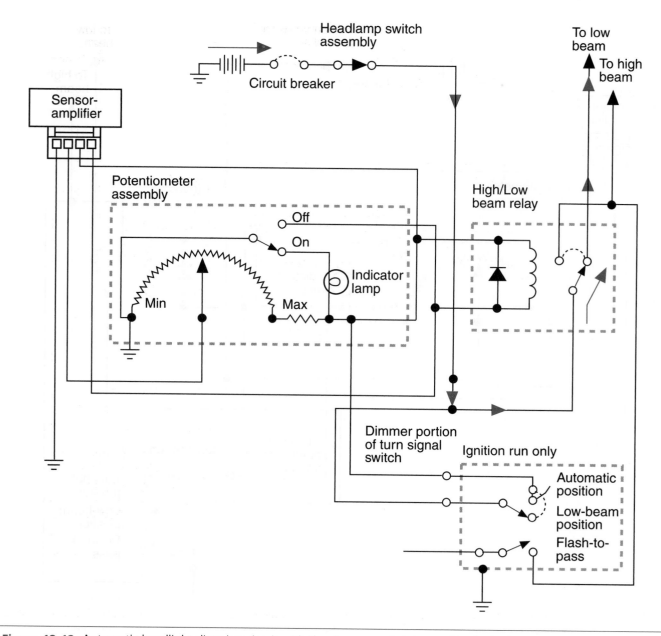

Figure 12-12 Automatic headlight dimming circuit with the dimmer switch located in the LOW-BEAM position.

The SmartBeam™ system uses a forward-facing, 5,000-pixel, digital imager camera that is attached to the rearview mirror mount (Figure 12-14). The camera's field of vision is in front of the vehicle within 2 degrees of the vehicle's centerline and 10 degrees of horizontal.

The operation of SmartBeam™ requires interaction with several vehicle modules. Figure 12-15 shows how one system interacts between modules. Ambient light levels for automatic headlight operation are provided by the light rain sensor module (LRSM), if equipped. If the LRSM is not used on the vehicle, then the system will use a photocell located on top of the dash. The headlamp switch position is signaled by the lighting multifunction switch and is an input to the steering column module (SCM). The SCM sends switch position status over the data bus. The front control module (FCM) uses HSDs to provide power to the both low- and high-beam bulbs. The cabin compartment node (CCN) controls the operation of the high-beam indicator.

The decisions made for headlight intensity are based on the sensed intensity of light, the light's location, and the light's movement. The system is capable of distinguishing between light types, such as mercury vapor used for street lighting. In addition, it can distinguish colors, so it is

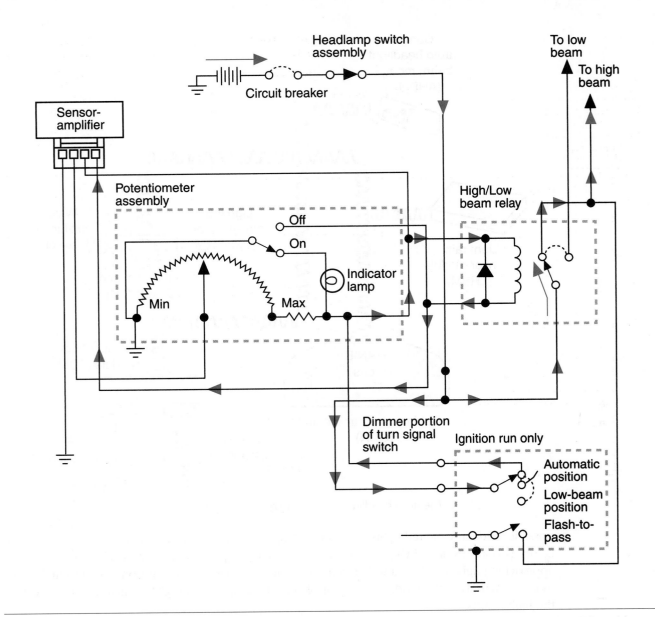

Figure 12-13 Automatic headlight dimming circuit with the dimmer switch located in the AUTOMATIC position and no oncoming light sensed.

Figure 12-14 The SmartBeam™ auto headlight system uses a digital camera to determine oncoming light intensity.

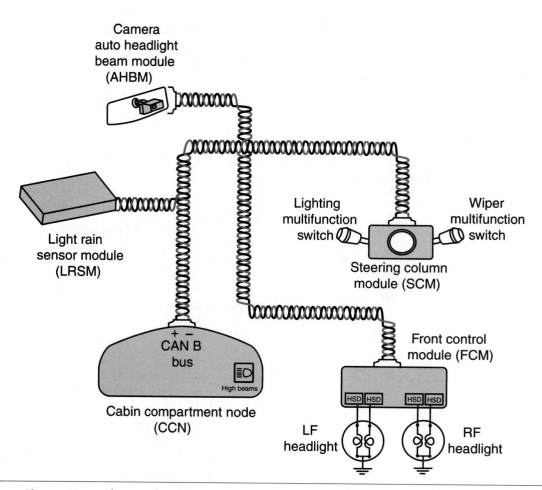

Figure 12-15 The auto headlight system integrates several modules.

possible to identify the red lights used on taillamps and sign colors. Distinguishing light types and colors is done by identifying the wave length of the light source. If it is determined that the approaching light source is another vehicle, a data bus message is sent from the auto high-beam module (AHBM) to the module that controls headlight operation (FCM in this case) to deactivate the high beams.

For this system to be operational, the headlight switch must be placed in the auto headlamp position and the "LOW/HIGH BEAM" option must be selected from the configurable display (Figure 12-16). The engine must also be running for the auto headlight function to operate. For the SmartBeam™ function to operate, the vehicle speed must be over 20 mph.

The auto headlight function will use either a photocell in the dash or a part of the mirror assembly to sense ambient light intensity. When the engine is running and the ambient light levels are less than 1000 **LUX,** the auto headlamp low beam becomes operational. LUX is the International System unit of measurement of the intensity of light. It is equal to the illumination of a surface one meter away from a single candle (one lumen per square meter).

Once vehicle speed exceeds 20 mph, if the ambient light level sensed at the SmartBeam™ camera is 5 LUX or less, a PWM voltage is applied to the high-beam circuit by the controlling module (FCM). Within 2½ to 5 seconds, the high beams will be at full intensity. By using PWM, the high beams are ramped up and down; this eliminates the usual "flash" that occurs as the high beams are turned on and off. Drivers in oncoming vehicles do not see any indication of the beam change since it is gradual and based on distance.

When another vehicle approaches, the camera determines the light intensity from its headlights. Once the light intensity reaches a predetermined level, a bus message is sent to the control module (FCM) to deactivate the high beams. The voltage to the high beams is ramped down using PWM until they are turned off.

Figure 12-16 For the auto headlight system to operate, the driver must activate it.

AUTHOR'S NOTE: Federal law specifies that the high-beam indicator is turned on at the initial start of the ramping up to high-beam operation and that it remains on during all high-beam operation. High beam is considered in operation throughout the ramp down phase; thus, the indicator light stays on until the high beams are totally turned off.

If the driver uses the high-beam switch to manually turn on the high beams, SmartBeam™ operation is defeated. Also, the system is momentarily defeated if the driver uses the flash-to-pass function. If the headlamp switch is placed in any other position other than AUTO, both the auto headlamp and SmartBeam™ functions are defeated.

Initial camera calibration and verification are performed at the factory as the vehicle is near completion. During this time the camera is precisely aligned. Once the camera is properly aligned, logic used by the AHBM will make adjustments to fine-tune the alignment based on sensed lighting inputs while the vehicle is driven. These adjustments occur as the processing logic looks for a light source that represents oncoming headlights. This would include light sources that start as low intensity and then gradually increase in brightness and have movement, indicating a gradual rate of approach. In addition, the light source must be coming from just to the left of center. These conditions indicate a vehicle is approaching from a distance. The computer logic of the AHBM will apply a weighting factor to its calibration to correct the aim. If the system is out of calibration, the LED in the mirror will flash.

Headlight Leveling

Some European and Asian markets require a **headlight leveling system (HLS).** These systems can also be found on import vehicles from these market areas. The HLS uses front lighting assemblies with a leveling actuator motor. Some systems use a leveling switch that the driver controls. The switch will allow the headlights to be adjusted into different vertical positions (usually four). This allows the driver to compensate for headlight position that can occur when the vehicle is loaded.

Electrical motors use a pushrod assembly to change the position of the headlight reflector. When different voltage levels are inputted to leveling motors from the multiplexed switch, the motors move the reflector to the selected position.

Daytime Running Lamps

Shop Manual
Chapter 12, page 470

All late-model Canadian vehicles and many domestic vehicles are equipped with **daytime running lamps (DRL).** The basic idea behind DRLs is dimly lit headlamps during the day. This allows other drivers and pedestrians to see the vehicle from a distance. Manufacturers have taken many different approaches to achieve this lighting. Most have a control module or relay (Figure 12-17) that turns the lights on when the engine is running and allows normal headlamp operation when the driver turns on the headlights. Daytime running lamps generally use the high-beam or low-beam headlight system at a reduced intensity.

The dimmer headlights can result from headlight current passing through a resistor during daylight hours (Figure 12-18). The resistor reduces the voltage available and the current flowing through the circuit to the headlights. The resistor is bypassed during normal headlamp operation.

Other systems use a control module (Figure 12-19), which uses a duty cycled output to either the high-beam or low-beam headlights (depending on manufacturer). The duty cycle reduces the output of the headlights by 50 to 75%. Most systems that use the high-beam headlights have a method of turning off the high-beam indicator lamp in the instrument cluster if DRLs are activated. However, on some systems, it is normal for the high-beam indicator to be dimly lit during DRL operation.

GM's DRL system includes a solid-state control module assembly, a relay, and an ambient light sensor assembly. The system lights the low-beam headlights at a reduced intensity when the ignition switch is in the RUN position during daylight. The DRL system is designed to light the low-beam headlamps at full intensity when low-light conditions exist.

As the intensity of the light reaching the ambient light sensor increases, the electrical resistance of the sensor assembly decreases. When the DRL control module assembly senses the low resistance, the module allows voltage to be applied to the DRL diode assembly and then to the low-beam headlamps. Because of the voltage drop across the diode assembly, the low-beam headlamps are on with a low intensity.

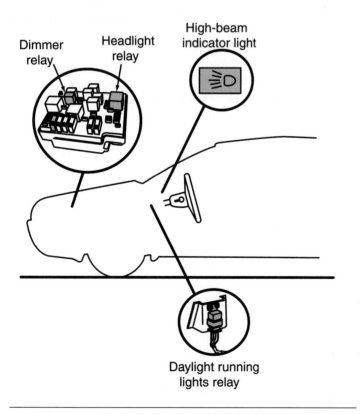

Figure 12-17 A daytime running light relay.

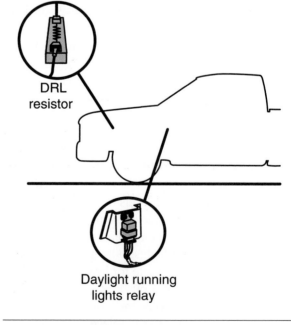

Figure 12-18 A resistor in line with the headlights reduces the current going to the headlights.

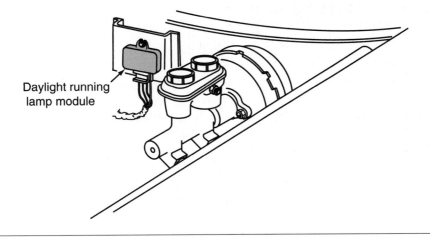

Daylight running
lamp module

Figure 12-19 A DRL system that uses a control module to pulse current to the high-beam headlights. This results in reduced illumination.

As the intensity of the light reaching the ambient light sensor decreases, the electrical resistance of the sensors increases. When the DRL module assembly senses high resistance in the sensor, the module closes an internal relay, which allows the low-beam headlamps to illuminate with full intensity.

Most DRL systems also use the parking brake switch as an input. If the parking brake is applied while the engine is running, the daytime running light feature is turned off.

Illuminated Entry Systems

Illuminated entry systems turn on the courtesy lights before the doors are opened. Most modern illuminated entry systems incorporate solid-state circuitry that includes an illuminated entry actuator and side door switches in the door handles. Illumination of the door lock tumblers can be provided by the use of fiber optics or light-emitting diodes.

When either of the front door handles is lifted, a switch in the handle will close the ground path from the actuator. This signals the logic module to energize the relay (Figure 12-20). With the

Shop Manual
Chapter 12, page 472

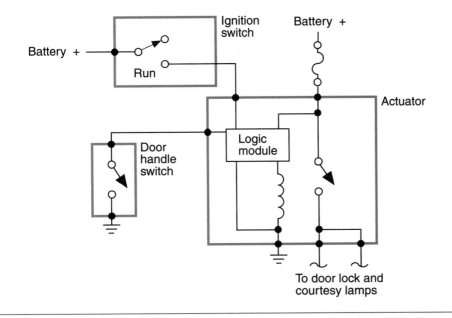

Figure 12-20 Illuminated entry actuator circuit.

relay energized, the contacts close and the interior and door lock lights come on. A timer circuit is incorporated that will turn off the lights after 25 to 30 seconds. If the ignition switch is placed in the RUN position before the timer circuit turns off the interior lights, the timer sequence is interrupted and the interior lights turn off.

Some manufacturers have incorporated the illuminated entry actuator into their BCM (Figure 12-22). Activation of the system is identical as discussed. The signal from the door handle switch can also be used as a **"wake-up" signal** to the BCM. A wake-up signal is used to notify the BCM that an engine start and operation of accessories is going to be initiated soon. This signal is used to warm up the circuits that will be processing information (Figure 12-21).

The signal from the door handle switch informs the body computer to activate the courtesy light driver. Some systems use a pair of door jamb switches that signal the body computer to keep the courtesy lights on when the door is open. When the door is closed and the ignition switch is in the RUN position, the lights are turned off.

Referring to Figure 12-22, this type of system is capable of turning off the interior lights if the driver forgets to turn them off or leaves a door open. After a programmed period of time has elapsed with the ignition in the OFF position and no other inputs received, the BCM will turn off the courtesy lamp driver and turn off the lamps. The battery saver driver is turned on whenever the BCM receives its wake-up signal and provides ground for the lamps through the switches. This allows the vehicle occupants to turn on and off individual reading lights. Once the ignition key is turned off and a programmed length of time has elapsed, the BCM will turn off the battery saver driver. This will assure that all lights are turned off while the vehicle is not being used even though a switch was left in the ON position.

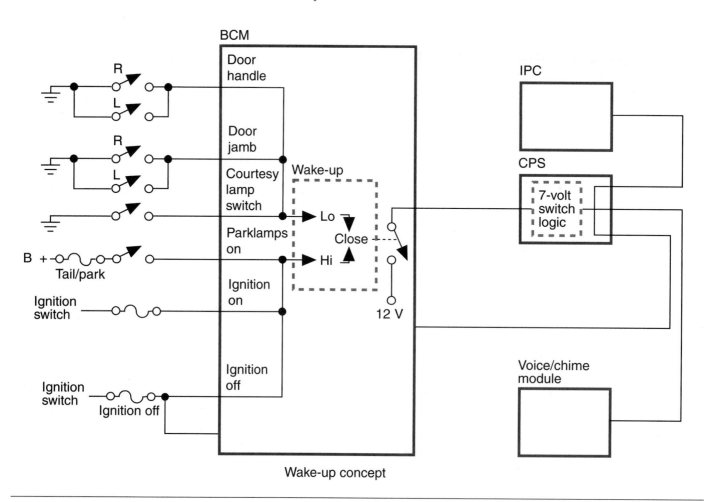

Figure 12-21 The illuminated entry system can also send a wake-up signal to the body computer.

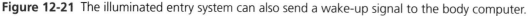

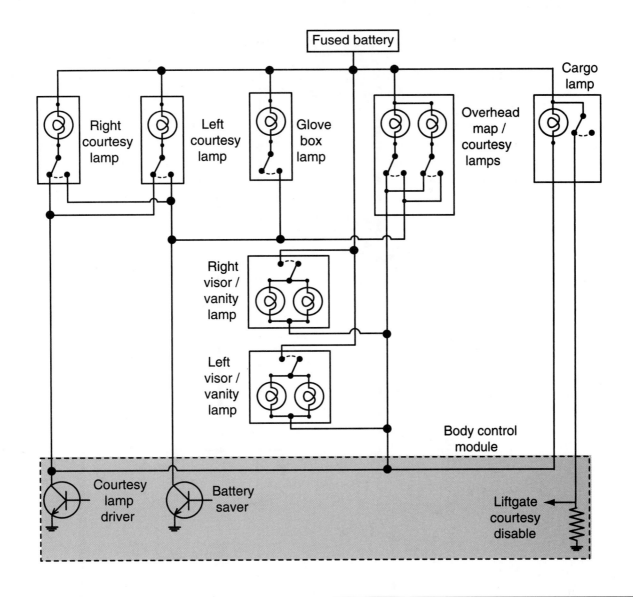

Figure 12-22 Body computer control of the illuminated entry system.

Another feature of this system is fade-to-off. Whenever the BCM determines it will turn off the courtesy lamps (either by the ignition switch being placed in the RUN position or the timer expiring), the driver circuit will gradually change the duty cycle, resulting in the lamps dimming as they go off.

Some manufacturers use the twilight photocell to inform the body computer of ambient light conditions. If the ambient light is bright, the photocell signals the BCM that courtesy lights are not required.

Instrument Panel Dimming

The BCM can also be used to control the **instrument panel dimming** feature. The body computer uses inputs from the panel dimming control and photocell to determine the illumination level of the instrument panel lights (Figure 12-23). With the ignition switch in the RUN position, a 5-volt signal is supplied to the panel dimming control potentiometer. The wiper of the potentiometer returns the signal to the BCM.

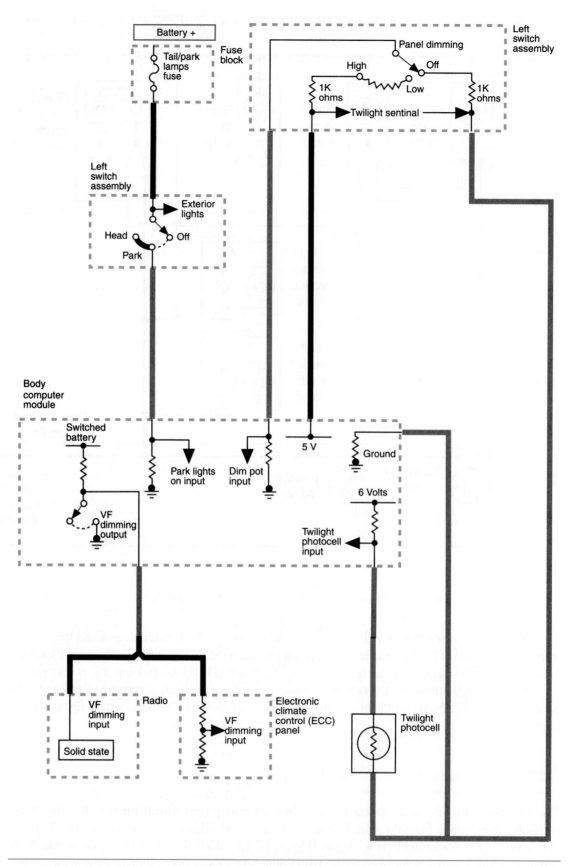

Figure 12-23 The dimming control and photocell are inputs to the BCM to control instrument panel dimming.

When the dimmer control is moved toward the dimmer positions, the increased resistance results in a decreased voltage signal to the BCM. By measuring the voltage that is returned, the BCM is able to determine the resistance value of the potentiometer. The BCM controls the intensity level of the illumination lamps by pulse-width modulation.

Some digital instrument panel modules use an ambient light sensor in addition to the rheostat. The ambient sensor will control the display brightness over a 35-to-1 range and the rheostat will control over a 30-to-1 range. When the headlights are on, the module compares the values from both inputs and determines the illumination level. When the headlights are off, the module uses only the ambient light sensor for its input.

A variation of this operation is that the BCM will receive the light intensity level request from the headlight rheostat as it did before, but the requested level is then sent to the instrument cluster by the multiplexing circuit. The instrument cluster uses this input information to control lamp intensity through its own microprocessor.

Fiber Optics

Shop Manual
Chapter 12, page 477

Fiber optics is the transmission of light through polymethyl methacrylate plastic that keeps the light rays parallel even if extreme bends are in the plastic. The invention of fiber optics has made it possible to provide illumination of several objects by a single light source (Figure 12-24). Plastic fiber-optic strands are used to transmit light from the source to the object to be illuminated. The strands of plastic are sheathed by a polymer that insulates the light rays as they travel within the strands. The light rays travel through the strands by means of internal reflections.

Fiber optics are commonly used as indicator lights to show the driver that certain lights are functioning. Many vehicles with fender-mounted turn signal indicators use fiber optics from the

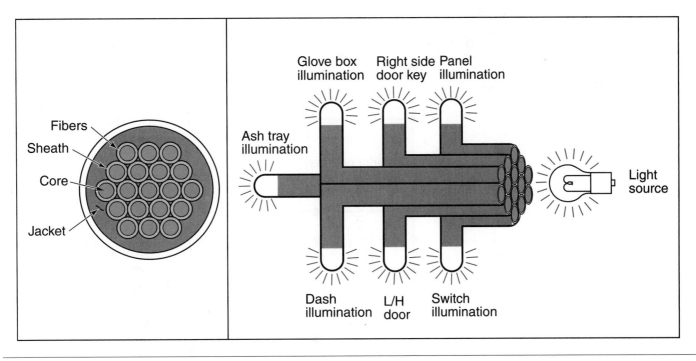

Figure 12-24 One light source can illuminate several areas by using fiber optics.

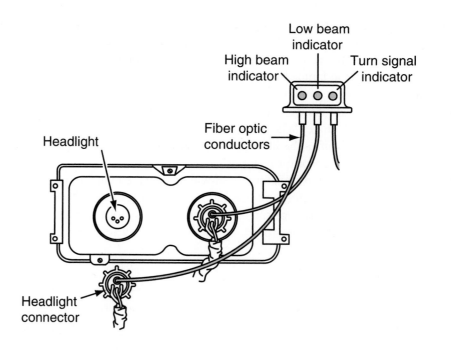

Figure 12-25 Fiber optics can be used to indicate the operation of exterior lights to the driver.

turn signal light to the indicator (Figure 12-25). The indicator will only show light if the turn signal light is on and working properly.

Some manufacturers use fiber optics to provide illumination of the lock cylinder "halo" during illuminated entry operation (Figure 12-26). When the illuminated entry system is activated, the light collector provides the source light to the fiber optics and the halo lens receives the light from the fiber optic cable.

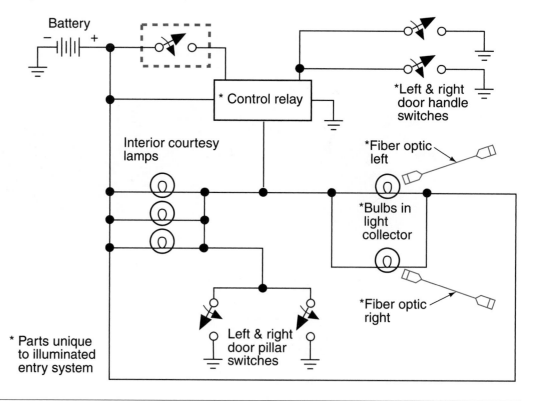

Figure 12-26 Fiber optics used in an illuminated entry system.

The advantage of fiber optics is it can be used to provide light in areas where bulbs would be inaccessible for service. Other uses of fiber optics include:

❑ Lighting ash trays.

❑ Illuminating instrument panels.

❑ Dash lighting over switches.

Optical communication systems date back to the "optical telegraph" that French engineer Claude Chappe invented in the 1790s. Alexander Graham Bell patented an optical telephone system, which he called the Photophone, in 1880. During the 1920s, John Logie Baird of England and Clarence W. Hansell of the United States patented the idea of using arrays of hollow pipes or transparent rods to transmit images for television or facsimile systems. However, the first person known to have demonstrated image transmission through a bundle of optical fibers was Heinrich Lamm, a medical student in Munich. His goal was to look inside inaccessible parts of the body and, in a 1930 paper, he reported transmitting the image of a light bulb filament through a short bundle.

Lamp Outage Indicators

A common lamp outage indicator uses a translucent drawing of the vehicle (Figure 12-27). If one of the monitored systems fails or is in need of driver attention, the graphic display illuminates a light to indicate the location of the problem.

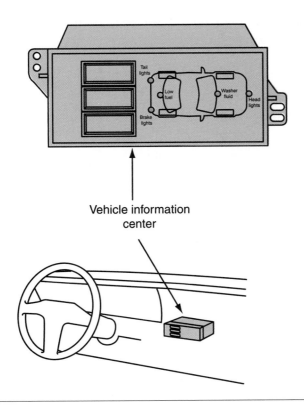

Vehicle information center

Figure 12-27 Many vehicle information systems use a graphic display to indicate warning areas to the driver.

The basic lamp outage indicator system is used to monitor the stop light circuit. This system consists of a reed switch and opposing electromagnetic coils (Figure 12-28). When the ignition switch is turned to the RUN position, battery voltage is applied to the normally open reed switch. When the brake light switch is closed, current flows through the coils on the way to the stop light bulbs. If both bulbs are operating properly, the coils create opposing magnetic fields that keep the reed switch in the open position. If one of the stop light bulbs burns out, current will only flow through one of the coils, which attracts the reed switch contacts and closes them. This completes the stop light warning circuit and illuminates the warning light on the dash. The warning light will remain on as long as the stop light switch is closed.

 AUTHOR'S NOTE: Opposing magnetic fields are created because the coils are wound in opposite directions.

Some manufacturers will use a **lamp outage module** either as a stand-alone module or in conjunction with the BCM. A lamp outage module is a current-measuring sensor that contains a set of resistors, wired in series with the power supply to the headlights, taillights, and stop lights. If the module is a "stand-alone" unit, it will operate the warning light directly. The module monitors the voltage drop of the resistors. If the circuits are operating properly, there is a 500 mv input signal to the module. If one of the monitored bulbs burns out, the voltage input signal drops to about 250 mv. The module completes the ground circuit to the warning light to alert the driver that a bulb has burned out. The module is capable of monitoring several different light circuits.

Many vehicles today use a computer-driven information center to keep the driver informed of the condition of monitored circuits (Figure 12-29). The vehicle information center usually receives its signals from the BCM (Figure 12-30). In this system, the lamp outage module is used to send signals to the BCM. The BCM will either illuminate a warning light, give a digital message, or activate a voice warning device to alert the driver that a light bulb is burned out.

A burned-out light bulb means there is a loss of current flow in one of the resistors of the lamp outage module. A monitoring chip in the module compares the voltage drop across the resistor (Figure 12-31). If there is no voltage drop across the resistor, there is an open in the circuit (burned-out light bulb). When the chip measures no voltage drop across the resistor, it signals the BCM, which then gives the necessary message to the vehicle information center (Figure 12-32).

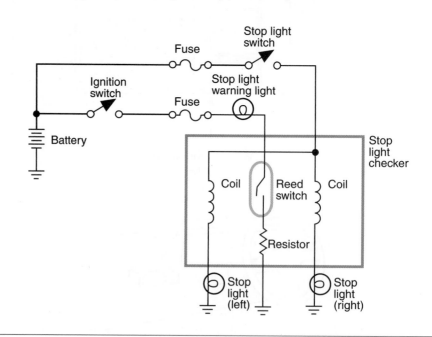

Figure 12-28 Stop light lamp outage indicator circuit.

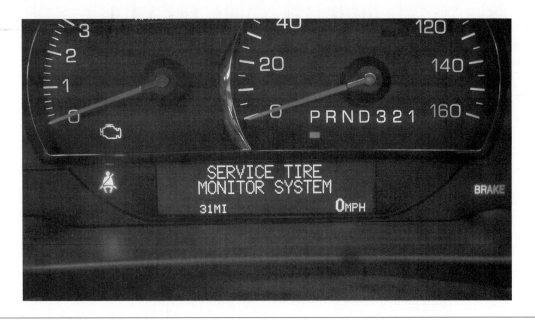

Figure 12-29 The computer-driven vehicle information center keeps the driver aware of the condition of monitored systems.

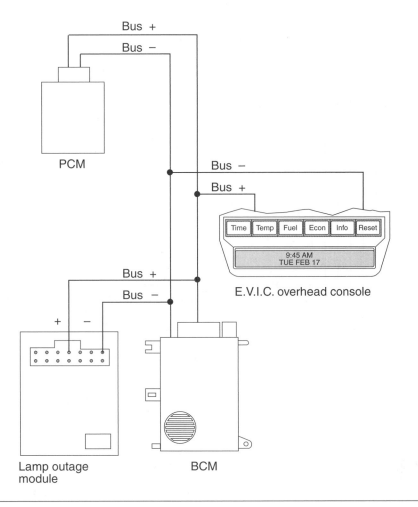

Figure 12-30 The body computer can be used to receive signals from varies inputs and to give signals to control the information center.

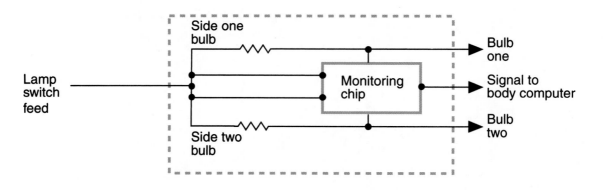

Figure 12-31 One section of the lamp outage module circuit. The monitoring chip compares voltage on both sides of the resistor to measure voltage drop changes.

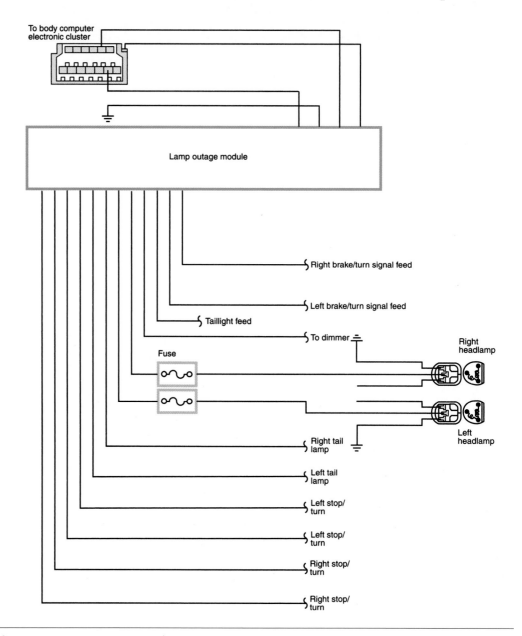

Figure 12-32 Lamp outage indicator circuit.

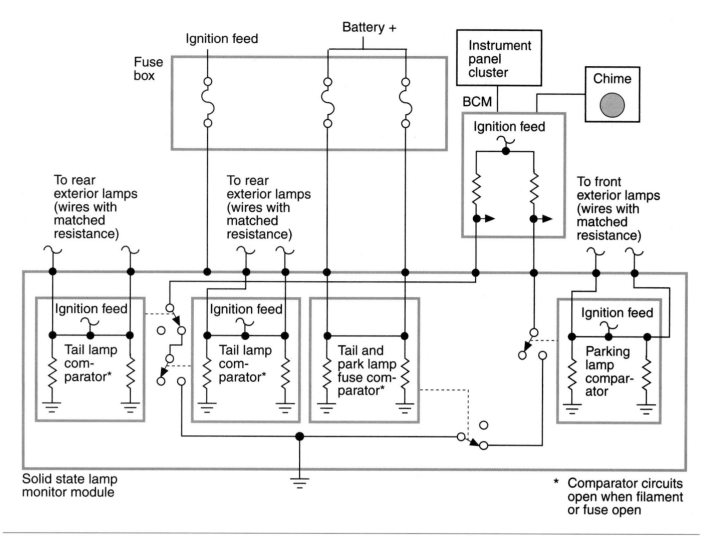

Figure 12-33 Lamp monitor module circuit.

AUTHOR'S NOTE: The bulbs are monitored only when current is supplied to them.

General Motors uses the lamp monitor module to connect the light circuits to ground (Figure 12-33). When the circuits are operating properly, the ground connection in the module causes a low circuit voltage. Input from the lamp circuits are through two equal resistance wires. The module output to the bulbs is from the same module terminals as the inputs.

If a bulb burns out, the voltage at the lamp monitor module terminal will increase. The module will open the appropriate circuit from the BCM, signaling the BCM to send a communication to the instrument panel cluster (IPC) computer, which displays the message in the information center.

Vehicles that use HSDs to control lamp illumination are also capable of detecting lamp outage. Often these systems can determine if a lamp circuit is open without the lamp system being activated. This is done by sending a small diagnostic current through the circuit. If the circuit is in tack, an expected voltage drop will occur. If the circuit is open, the voltage reading will remain high and the module will request the lamp outage indicator to come on.

Summary

❑ Some manufacturers can use the body control module (BCM) to operate the concealed headlight system.

❑ Usually the concealed headlight system will use a pair of relays that the BCM controls.

❑ Many computer-controlled headlight systems use resistive multiples headlight switches as an input.

❑ Computer-controlled headlight systems can use relays operated by the control module or high side drivers (HSDs) to illuminate the lamps.

❑ The automatic on/off with time delay has two functions: to turn on the headlights automatically when ambient light decreases to a predetermined level and to allow the headlights to remain on for a certain amount of time after the vehicle has been turned off.

❑ Most automatic headlight dimming systems consist of a light-sensitive photocell and amplifier unit, High-Low beam relay, sensitivity control, dimmer switch, flash-to-pass relay, and a wiring harness.

❑ The SmartBeam™ system uses a forward-facing, 5,000-pixel, digital imager camera.

❑ The operation of SmartBeam™ requires interaction with several vehicle modules, including the light rain sensor module (LRSM), the steering column module (SCM), the front control module (FCM), and the cabin compartment node (CCN).

❑ Decisions about headlight intensity are based on the sensed intensity of light, the light's location, and the light's movement.

❑ Once vehicle speed exceeds 20 mph, if the ambient light level sensed at the SmartBeam™ camera is 5 LUX or less, a PWM voltage is applied to the high-beam circuit by the controlling module (FCM). Within $2\frac{1}{2}$ to 5 seconds, the high beams will be at full intensity.

❑ A headlight leveling system (HLS) uses front lighting assemblies with a leveling actuator motor.

❑ Daytime running lamps can use a relay or module to illuminate the low- or high-beam lamps at a reduced output.

❑ The illuminated entry system turns on the interior lights prior to the door being opened. The system may also be capable of turning off the lights if the driver fails to shut a door when exiting.

❑ Instrument panel dimming is usually done by the BCM providing a pulse-width modulation to the illumination lamps or LEDs.

❑ Fiber optics is the transmission of light through polymethyl methacrylate plastic that keeps the light rays parallel even if there are extreme bends in the plastic.

❑ The lamp outage indicator alerts the driver, through an information center on the dash or console, that a light bulb has burned out.

Review Questions

Short-Answer Essays

1. Describe the operation of computer-controlled concealed headlight systems.

2. List the common components of the automatic headlight dimming system.

3. Explain the operation of body computer–controlled instrument panel illumination dimming.

4. What is the function of the sensitivity control in the automatic dimmer system?

5. What is the basic operation of the illuminated entry system?

6. Explain the operation of the SmartBeam™ headlight system.

7. Describe the function of automatic headlight leveling systems.

8. Describe the purpose and function of daytime running lamps.

9. Explain the use and function of fiber optics.

10. What is meant by pulse width dimming?

Fill in the Blanks

1. With body computer–controlled concealed headlights, the computer receives inputs from the _____ and _____ switches.

2. The sensitivity control used with automatic dimming sets the sensitivity at which the photocell and amplifier are _____.

3. The photocell will have _____ resistance as the ambient light level increases.

4. In some illuminated entry systems the _____ signals the body computer that the courtesy lights are not required.

5. The body computer uses inputs from the _____ _____ _____ and _____ to determine the illumination level of the instrument panel lights.

6. The body computer dims the illumination lamps by using a _____ _____ _____ signal to the panel lights.

7. Most computer-controlled headlight systems use a _____ _____ switch for an input.

8. Fiber optics are commonly used as _____ lights.

9. Lamp outage modules detect _____ _____ in a normally operating circuit.

10. HSDs supply _____ to the lamps.

Multiple Choice

1. Which of the following statements is most correct?
   A. Decreasing the sensitivity control of the automatic headlight dimming system means the headlights will switch to low beams when the approaching vehicle is farther away.
   B. Increasing the sensitivity control of the automatic headlight dimming system means the headlights will switch to low beams when the approaching vehicle is closer.
   C. All of the above.
   D. None of the above.

2. All of the following statements about the SmartBeam™ system are true, EXCEPT:
   A. The system uses a digital camera to determine light intensity.
   B. The system is capable of detecting movement of oncoming light.
   C. The system is capable of distinguishing colors.
   D. The AHBM turns off the high-beam relay when oncoming light intensity is 10 LUX or more.

3. Computer-controlled instrument panel dimming is being discussed.

   Technician A says the body computer dims the illumination lamps by varying resistance through a rheostat that is wired in series to the lights.

   Technician B says the body computer can use inputs from the panel dimming control and photocell to determine the illumination level of the instrument panel lights on certain systems.

   Who is correct?
   - **A.** A only.
   - **B.** B only.
   - **C.** Both A and B.
   - **D.** Neither A nor B.

4. Which statement about fiber optics is correct?
   - **A.** Fiber optics is the transmission of light through several plastic strands that are sheathed by a polymer.
   - **B.** Fiber optics is used only for exterior lighting.
   - **C.** Fiber optics can only be used in applications where the conduit can be laid straight.
   - **D.** All of the above.

5. The purpose of the headlight leveling system is to:
   - **A.** Reduce the need to align the light beams.
   - **B.** To allow the driver to raise or lower the light beams as vehicle loads change.
   - **C.** To allow the driver to raise or lower the light beams as the vehicle ascends and descends hills.
   - **D.** All of the above.

6. *Technician A* says computer-controlled headlight systems can use relays that the BCM operates.

   Technician B says compute-controlled headlight systems can use high side drivers to operate the lamps.

   Who is correct?
   - **A.** A only.
   - **B.** B only.
   - **C.** Both A and B.
   - **D.** Neither A nor B.

7. In the SmartBeam™ system, the headlight intensity is based on which of the following:
   - **A.** Movement of the light.
   - **B.** Intensity of the light.
   - **C.** Location of the light.
   - **D.** All of the above.
   - **E.** None of the above.

8. *Technician A* says daytime running lamps illuminate the taillights at 25% duty cycle.

   Technician B says daytime running lamps can use a resistor to reduce current to the low-beam headlight.

   Who is correct?
   - **A.** A only.
   - **B.** B only.
   - **C.** Both A and B.
   - **D.** Neither A nor B.

9. Which statement is the most correct?
   - **A.** Lamp outage modules can use voltage drop to determine circuit operation.
   - **B.** HSDs can only be used to detect opens in activated circuits.
   - **C.** Low side driver controlled lamps cannot determine lamp outage conditions.
   - **D.** All of the above.

10. *Technician A* says the computer-controlled concealed headlight system cannot support the flash-to-pass feature.

    Technician B says the computer-controlled concealed headlight system may use a pair of relays to operate the doors.

    Who is correct?
    - **A.** A only.
    - **B.** B only.
    - **C.** Both A and B.
    - **D.** Neither A nor B.

Instrumentation and Warning Lamps

Upon completion and review of this chapter, you should be able to:

❏ Describe the operation of electromagnetic gauges, including d'Arsonval, three-coil, two-coil, and air-core.

❏ Describe the operation of electronic fuel, temperature, oil, and voltmeter gauges.

❏ Describe the operation of quartz analog instrumentation.

❏ Explain the function and operation of the various gauge sending units, including thermistors, piezoresistive, and mechanical variable resistors.

❏ Describe the purpose of speedometers and odometers.

❏ Describe the purpose of the tachometer.

❏ Describe the operating principles of the digital speedometer.

❏ Explain the operation of IC chip and stepper motor odometers.

❏ Explain the use and operation of light-emitting diodes, liquid crystal, vacuum fluorescent, and CRT displays in electronic instrument clusters.

❏ Explain the operation of various warning lamp circuits.

❏ Explain the operation of various audible warning systems.

❏ Explain the operation of body computer–controlled instrument panel illumination light dimming.

Introduction

Instrument gauges and indicator lights monitor the various vehicle operating systems. They provide information to the driver of their correct operation (Figure 13-1). Warning devices also provide information to the driver; however, they are usually associated with an audible signal. Some vehicles use a voice module to alert the driver to certain conditions.

Early instrument cluster gauges were analog or swing needle type. Although many modern vehicles still use the analog gauge, they are now computer-driven. Computer-driven instruments are becoming increasingly popular on today's vehicle. These instruments provide far more accurate

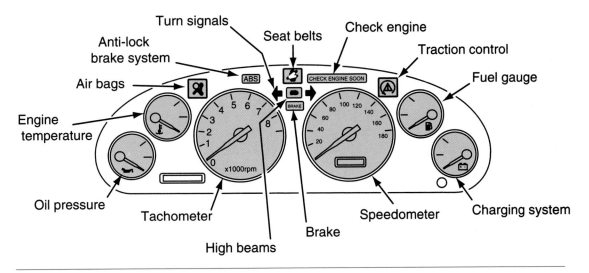

Figure 13-1 Instrument panel.

readings than their conventional analog counterparts. This chapter introduces you to the most commonly used computer-driven instrumentation systems. These systems include the speedometer, odometer, fuel, oil, tachometer, and temperature gauges.

The computer-driven instrument panel uses a microprocessor to process information from various sensors and to control the gauge display. Depending on the manufacturer, the microprocessor can be a separate computer that receives direct information from the sensors and makes the calculations, or can use the BCM to perform all functions.

In addition, there are many types of information systems used today. These systems keep the driver informed of a variety of monitored conditions, including vehicle maintenance, trip information, and navigation.

Shop Manual
Chapter 13, page 506

Electromechanical Gauges

A gauge is a device that displays the measurement of a monitored system by the use of a needle or pointer that moves along a calibrated scale. The **electromechanical** gauge acts as an ammeter since the gauge reading changes with variations in resistance. The gauge is called an electromechanical device because it is operated electrically, but its movement is mechanical. There are two basis types of electromechanical gauges: the bimetallic gauge and the electromagnetic gauge. Conventional analog instrument clusters that used these types of gauges had a direct connection to the sending unit. The resistance of the sending unit determined the location of the needle on the gauge face. A short study of the different types of gauges is provided to give a foundation to the study of computer-driven gauges.

Bimetallic gauges are not used in today's automobiles. These gauges (or thermoelectric gauges) were simple dial and needle indicators that transformed the heating effect of electricity into mechanical movement. The construction of the bimetallic gauge featured an indicating needle linked to the free arm of a U-shaped bimetallic strip (Figure 13-2). The free arm had a heater coil connected to the gauge terminal posts. When current flowed through the heater coil, it heated the bimetallic arm and caused the arm to bend and move the needle across the gauge dial. The amount the bimetallic strip bent is proportional to the heat produced in the heater coil; the greater the heat created, the more the needle would move. Current flow was controlled by the changing resistance of a sending unit.

Electromagnetic gauges produce needle movement by magnetic forces instead of heat. There are four types of electromagnetic gauges: the d'Arsonval, the three-coil, the two-coil, and the air-core.

The **d'Arsonval gauge** uses the interaction of a permanent magnet and an electromagnet, and the total field effect to cause needle movement. The d'Arsonval gauge consists of a permanent horseshoe-type magnet that surrounds a moveable electromagnet (armature) that is attached

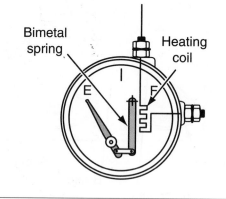

Figure 13-2 Bimetallic gauge construction.

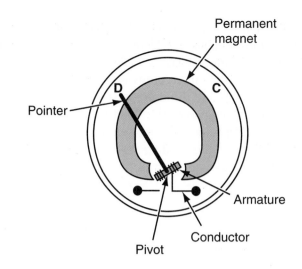

Figure 13-3 D'Arsonval gauge needle movement.

to a needle (Figure 13-3). When current flows through the armature, it becomes an electromagnet and is repelled by the permanent magnet. When current flow through the armature is low, the strength of the electromagnet is weak and needle movement is small. When the current flow is increased, the magnetic field created in the armature is increased and needle movement is greater. The armature has a small spring attached to it to return the needle to zero when current is not applied to the armature.

The **three-coil gauge** uses the interaction of three electromagnets and the total field effect upon a permanent magnet to cause needle movement. The three-coil gauge consists of a permanent magnet with a needle attached to it. The permanent magnet is surrounded by three electromagnets. There may also be a quantity of silicone dampening fluid to restrict needle movement due to vehicle movement. The current flow controlled by the resistance from the variable resistor–type sending unit determines the magnetic strength of the coils.

The three coils of fine wire are wound on a square plastic frame. The needle shaft is supported by a bearing sleeve extending from the frame. The needle shaft connects the pointer.

The three coils are the **bucking coil,** the **low-reading coil,** and the **high-reading coil** (Figure 13-4). The bucking coil produces a magnetic field that bucks or opposes the low-reading coil. The low-reading coil and the bucking coil are wound together, but in opposite directions.

The **three-coil gauge** is also known as a magnetic bobbin gauge.

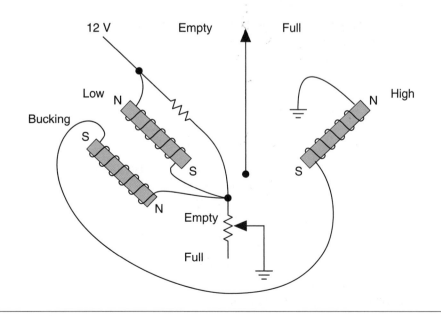

Figure 13-4 Three-coil gauge circuit.

The high-reading coil is positioned at a 90° angle to the low-reading and bucking coils. To compensate for production tolerances in the coils, a selective shunt resistor is attached to the back of the gauge housing. This selective resistor bypasses a certain amount of current past the coils.

When voltage is applied to the gauge, there are two paths in which the current can flow. One path is through the low-reading coil and through the sending unit to ground (Figure 13-5). The second path is through the low-reading coil to the bucking coil and the high-reading coil to ground (Figure 13-6). The amount of resistance that the sending unit has determines the path. If there is less resistance to ground through the sending unit than through the bucking and high-reading coils, most of the current will take the path through the sending unit. If the resistance through the sending unit is greater than the resistance through the coils, very little current will flow through the sending unit.

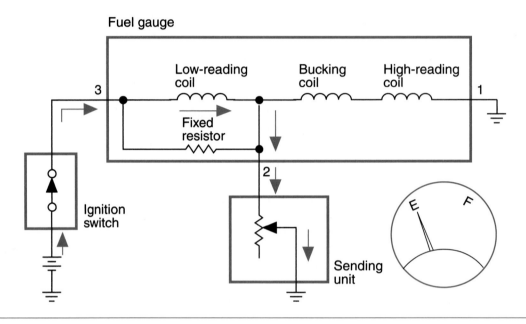

Figure 13-5 With sending unit resistance low, the needle is attracted to the low-reading coil.

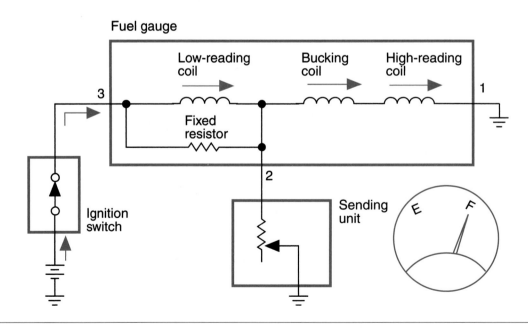

Figure 13-6 With sending unit resistance high, the needle is attracted to the high-reading coil.

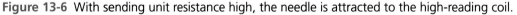

When sending unit resistance is low, more current flows through the low-reading coil than through the bucking and high-reading coils. This causes the needle to be attracted to the left, and the gauge reads toward zero. When sending unit resistance is high, very little current will flow through the sending unit. The current now flows through the three coils. The magnetic field created by the bucking coil cancels the magnetic field of the low-reading coil. The high-reading coil's magnetic field then attracts the needle and it swings toward maximum.

At an intermediate sending unit resistance value, the current can flow through both paths. If resistance was equal in the two paths, the needle would point at the midrange. As the magnetic field(s) changes, as the result of resistance change in the sending unit, the needle will swing toward the more powerful magnetic field.

The **two-coil gauge** uses the interaction of two electromagnets and the total field effect on an armature to cause needle movement. There are different designs of the two-coil gauge, depending upon the gauge application. For example, in a coolant temperature gauge (Figure 13-7), both coils receive battery voltage. One of the coils is grounded directly, while the other is grounded through the sender unit. As the resistance in the sender unit varies, as a result of temperature changes, the current flow through that coil changes. The two magnetic fields have different strengths, depending upon the amount of current flow through the sender unit. A two-coil gauge that is constructed to be used as a fuel gauge will have the E coil receiving battery voltage (Figure 13-8). At the end of the coil, the voltage is divided. One path is through the F coil to ground, while the other is through the sender unit to ground. The stronger the current flow through a coil, the more the needle will move toward that coil.

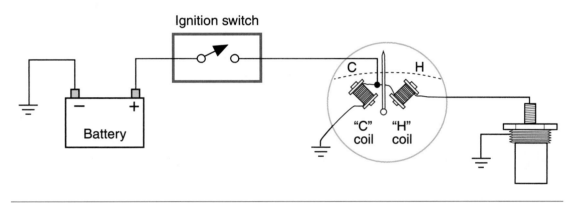

Figure 13-7 A two-coil temperature gauge.

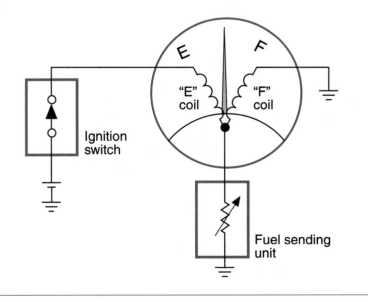

Figure 13-8 A two-coil fuel gauge.

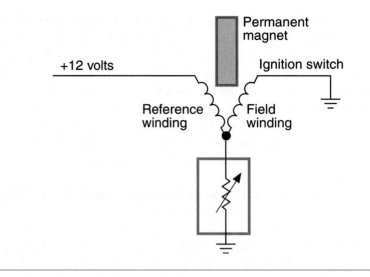

Figure 13-9 Air-core fuel gauge circuit.

The most common style of gauge used today is the **air-core gauge.** The air-core gauge works on the same principle as the two-coil by using the interaction of two electromagnets and the total field effect upon a permanent magnet to cause needle movement. This gauge has the pointer connected to a permanent magnet (Figure 13-9). Two windings are placed at different angles, one wound around the other. There is no core inside of the windings. Instead, the permanent magnet is placed inside the windings. The magnet aligns itself to a resultant field, in the field windings, according to the resistance of the sender unit. The sender unit resistance varies the strength of the field winding, which opposes the strength of the reference winding. The strength of the electromagnetic field depends upon the resistance in the sending unit.

Quartz Analog Instrumentation

Shop Manual
Chapter 13,
pages 501, 509

Computer-driven quartz swing needle displays are similar in design to the air-core electromagnetic gauges used in conventional analog instrument panels (Figure 13-10). Any, or all, of the gauges in the instrument cluster may be this type. We will look at the **speedometer** as an example of operation. The speedometer is used to indicate the speed of the vehicle.

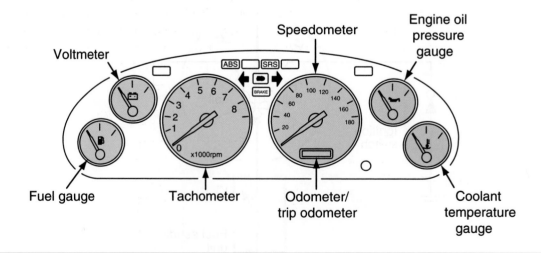

Figure 13-10 Electronic-controlled swing needle instrumentation.

Conventional speedometers used a cable that was connected to the output shaft of the transmission (or transfer case if four-wheel drive). The rotation of the output shaft caused the speedometer cable to rotate within its housing and then transferred to the speedometer assembly. This system relied on a rotating permanent magnet that produced a rotating magnetic field around a drum. The rotating magnetic field generated circulating eddy currents in the drum that produced a small magnetic field that interacted with the field of the rotating magnet. This interaction of the two magnetic fields pulled the drum and needle around with the rotating magnet. It is not hard to see that this system would not be extremely accurate. Today's vehicles use sensors and computer logic to display vehicle speed.

Shop Manual
Chapter 13, page 499

A BIT OF HISTORY

One of the early styles of speedometers used a regulated amount of air pressure to turn a speed dial. The air pressure was generated in a chamber containing two intermeshing gears. The gears were driven by a flexible shaft that was connected to a front wheel or the driveshaft. The air was applied against a vane inside the speed dial. The amount of air applied was proportional to the speed of the vehicle.

In many quartz analog speedometer gauge systems, a permanent magnet generator sensor is installed in the transaxle, transmission, or differential. As the PM generator is rotated, it causes a small AC voltage to be induced in its coil. This AC voltage signal is sent to a **buffer circuit** that changes the AC voltage from the PM generator into a digitalized signal (Figure 13-11). The signal is then sent to the processing unit (Figure 13-12). The signal is passed to a quartz clock circuit, a gain selector circuit, and a driver circuit. The driver circuit sends voltage pulses to the coils of the gauge; the coils operate like conventional air-core gauges to move the needle.

Shop Manual
Chapter 13, page 502

Often the sensor used to determine vehicle speed has multiple purposes. In this case, the signal being generated may not be an accurate representation of vehicle speed because it comes before the final drive unit. The control module may need to do additional calculations to make the speedometer accurate. For example, DaimlerChrysler vehicles equipped wit the 41TE or 42LE electronic shift transaxles use an output speed sensor that generates an AC signal from a 24-tooth tone wheel on the rear planetary unit. This signal is sent to the transmission control module (TCM). The TCM will apply **pinion factor** to the hertz count of the speed sensor signal to calculate vehicle speed. Pinion factor is a calculation using the final drive ratio and the tire circumference to obtain accurate vehicle speed signals. The TCM then transmits this information to the PCM at a set rate of 8,000 pulses per mile by pulsing the dedicated circuit. The PCM will then send the vehicle speed signal over the data bus circuit to all modules that require it.

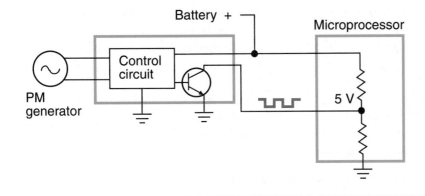

Figure 13-11 A buffer circuit.

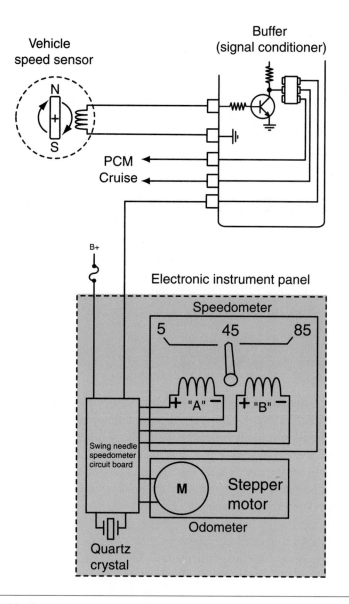

Figure 13-12 Quartz swing needle speedometer schematic. The A coil is connected to system voltage and the B coil receives a voltage that is proportional to input frequency. The magnetic armature reacts to the changing magnetic fields.

In some applications, the mechanical instrument cluster (MIC) receives the vehicle speed message from the BCM instead of receiving it directly from the PCM. The MIC then sets the needle to read the vehicle speed. Even though the MIC is on the bus system, it does not respond to the vehicle speed signal the PCM sends. It is programmed to accept messages only from the BCM.

Pinion factor information is set into the TCM at the factory. If the TCM is replaced in the field, the scan tool must be used to program the tire size used on the vehicle. In some systems, the gear ratio also has to be programmed. If the pinion factor is not programmed into the TCM, the speedometer and cruise control systems will not function.

Some manufacturers will use the wheel speed sensors from the antilock brake system (ABS) to determine vehicle speed. Usually the two front or the two rear sensor inputs are averaged. The ABS module then determines vehicle speed and broadcasts the information on the data bus.

The other air-core gauges (temperature, fuel level, and so on) work as described earlier, with conventional instrument clusters. The difference is the sending unit input goes to a module. The current flow through the gauge coils is controlled by the module, based on the sending unit resistance.

Gauge Sending Units

The **sending unit** is the sensor for the gauge. It is a variable resistor that changes resistance values with changes in the monitored conditions. There are three types of sending units that are associated with the gauges just described: (1) a **thermistor,** (2) a **piezoresistive sensor,** and (3) a mechanical variable resistor. These same types of sending units can also be used for computer-driven instrument clusters.

In the conventional coolant temperature sensing circuit, current is sent from the gauge unit into the top terminal of the sending unit, through the variable resistor (thermistor), and to the engine block (ground). The resistance value of the thermistor changes in proportion to coolant temperature (Figure 13-13). As the temperature rises, the resistance decreases and the current flow through the gauge increases. As the coolant temperature lowers, the resistance value increases and the current flow decreases.

In a computer-driven gauge or digital display, the PCM will send a 5-volt reference voltage through a pull-up resistor and then to the temperature sensor. This type of circuit was discussed in Chapter 10. As the resistance changes, the voltage dropped over the pull-up resistor changes and the voltmeter reading will indicate the engine temperature. The PCM will send the temperature information over the data bus to the instrument cluster (or BCM). The module will then send a current to the gauge coils to move the pointer to the correct temperature reading.

The piezoresistive sensor sending unit is threaded into the oil delivery passage of the engine and the pressure that is exerted by the oil causes the flexible diaphragm to move (Figure 13-14). The diaphragm movement is transferred to a contact arm that slides along the resistor. The position of the sliding contacts on the arm in relation to the resistance coil determines the resistance value, and the amount of current flow through the gauge to ground.

Another style is a transducer that operates much like a Wheatstone bridge MAP sensor, as discussed in Chapter 10. The function of the gauge is the same as that just discussed for the computer-driven temperature gauge, based on data bus messages from the PCM.

Some computer-driven instrument clusters have an oil gauge but do not use a sensor. These systems use an oil pressure switch. When oil pressure is greater than 6 psi (41 kPa), the switch opens the sense circuit. This will pull the sense voltage high. The gauge will display an oil pressure that is based on a calculated value determined by engine run time, engine temperature, engine load value, and ambient temperature. If the oil pressure drops below 6 psi (41 kPa), the switch closes the circuit to ground, pulling the sense voltage low. The instrument cluster gauge will now read 0. As long as the switch is open, the gauge will indicate normal oil pressure.

Shop Manual
Chapter 13, page 510

The **thermistor** is a resistor whose resistance changes in relation to changes in temperature; it is often used as a coolant temperature sensor.

A **piezoresistive** sensor is sensitive to pressure changes. The most common use of this type of sensor is to measure the engine oil pressure.

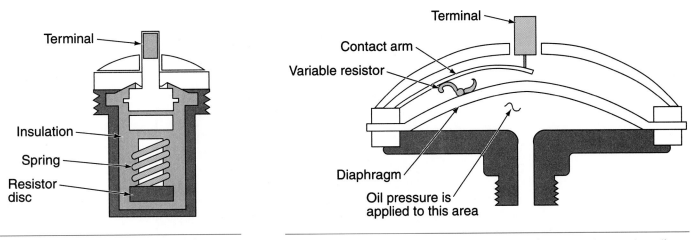

Figure 13-13 A thermistor used to sense engine temperature.

Figure 13-14 Piezoresistive sensor used for measuring engine oil pressure.

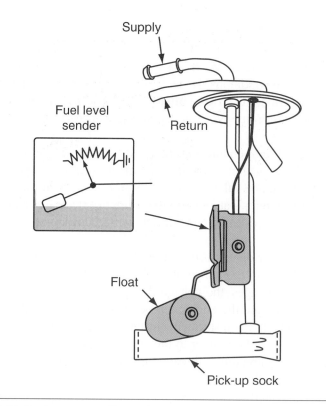

Figure 13-15 Fuel gauge sending unit.

A fuel level sending unit is an example of a mechanical variable resistor (Figure 13-15). The sending unit is located in the fuel tank and has a float that is connected to the wiper of a variable resistor. The floating arm rises and falls with the difference in fluid level. This movement of the float is transferred to the sliding contacts. The position of the sliding contacts on the resistor determines the resistor value.

Digital Instrumentation

Shop Manual
Chapter 13, page 509

Digital instrumentation is far more precise than conventional analog gauges. Analog gauges display an average of the readings received from the sensor; a digital display will present exact readings. In some systems, the information to the gauge is updated as often as 16 times per second.

Digital instrument clusters use digital and linear displays to notify the driver of monitored system conditions (Figure 13-16). Most digital instrument clusters provide for display in English or metric values. Also, many gauges are a part of a multigauge system. Drivers select which gauges they

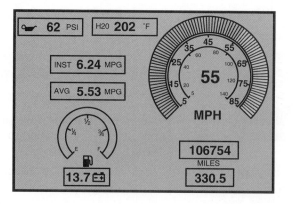

Figure 13-16 Digital instrument cluster.

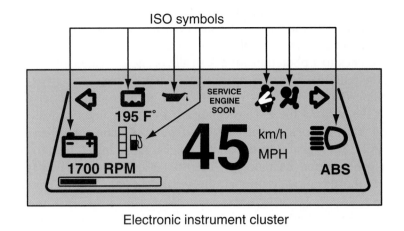

Electronic instrument cluster

Figure 13-17 A few of the ISO symbols used to identify the gauge.

wish to have displayed. Most of these systems will automatically display the gauge to indicate a potentially dangerous situation. For example, if the driver has chosen the oil pressure gauge to be displayed and the engine temperature increases above set limits, the temperature gauge will automatically be displayed to warn the driver. A warning light and/or a chime will also activate to get the driver's attention.

Most electronic instrument panels have self-diagnostic capabilities. The tests are initiated through a scan tool or by pushing selected buttons on the instrument panel. The instrument panel cluster also initiates a self-test every time the ignition switch is turned to ACC or RUN. Usually the entire dash is illuminated and every segment of the display is lighted. **International Standards Organization (ISO)** symbols are used to represent the gauge function (Figure 13-17). These symbols will usually flash during this test. At the completion of the test, all gauges will display current readings. A code is displayed to alert the driver if a fault is found.

Speedometers

Ford, GM, and Toyota have used optical vehicle speed sensors. The Ford and Toyota optical sensors are operated from the conventional speedometer cable. The cable rotates a slotted wheel between an LED and a phototransistor (Figure 13-18). As the slots in the wheel break the light, the transistor conducts an electronic pulse signal to the speedometer. An integrated circuit rectifies the analog input signal from the optical sensor and counts the pulses per second. The value is calculated into mph and displayed in the digital readout. The display is updated every 1/2 second. If the driver selected the readout to be in kilometers per hour, the computer makes an additional calculation to convert the readout. These systems may use a conventional gear-driven odometer.

Shop Manual
Chapter 13,
pages 501, 503

The electronic speedometer receives voltage signals from the vehicle speed sensor (VSS). This sensor can be a PM generator, Hall-effect switch, or an optical sensor.

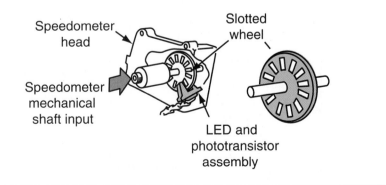

Speedometer head

Slotted wheel

Speedometer mechanical shaft input

LED and phototransistor assembly

Figure 13-18 Optical speed sensor.

The early style of GM speed sensor also operated from the conventional speedometer cable. The LED directs its light onto the back of the speedometer cup. The cup is painted black and the drive magnet has a reflective surface applied to it. As the drive magnet rotates in front of the LED, its light is reflected back to a phototransistor. A small voltage is created every time the phototransistor is hit with the reflective light.

The illustration (Figure 13-19) is a schematic of an instrument panel cluster that uses a PM generator for the VSS. As the PM generator is rotated, it causes a small AC voltage to be induced in its coil. This AC voltage signal is sent to the powertrain control module (PCM) and is shared with the BCM. The signal is rectified into a digital signal that is used to control the output to the instrument panel cluster (IPC) module. The BCM calculates the vehicle speed and provides this information to the IPC module through the serial data link. The IPC module turns on the proper display.

The microprocessor will initiate a self-check of the electronic instrument cluster any time the ignition switch is placed in the ACC or RUN position. The self-check usually runs for about 3 seconds. The most common sequence for the self-check is as follows:

1. All display segments are illuminated.

2. All displays go blank.

3. 0 mph or 0 km/h is displayed.

In addition to the methods of sensing speed mentioned earlier, Hall-effect switches are also used. The sensor is attached to a gear-driven wheel that rotates a trigger wheel. The gear is determined by tire size and the final drive gear ratio of the vehicle. As the trigger wheel rotates, it will

Shop Manual
Chapter 13, page 503

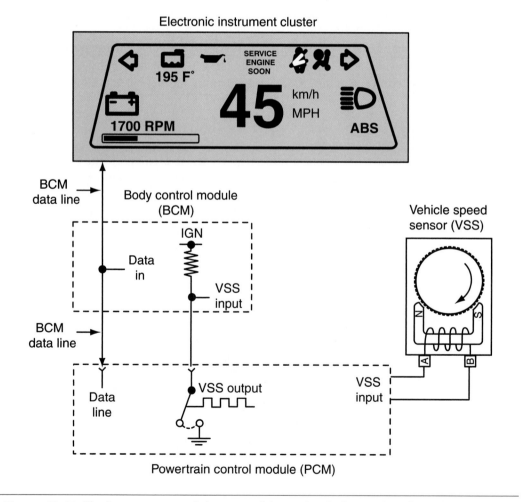

Figure 13-19 The instrument panel cluster module receives its instructions from the BCM, which shares the signals from the VSS with the ECM.

cause the Hall-effect switch to change voltages to high and low at a set amount each revolution. The amount of switches per revolution remains constant regardless of vehicle speed. Once the control module receives a programmed number of switches (8,000 for example), it knows it has traveled one mile.

Odometers

The **odometer** is a counter that uses the speedometer inputs to indicate the total miles accumulated on the vehicle. Many vehicles also have a second odometer that can be reset to zero; this is referred to as a trip odometer.

Shop Manual
Chapter 13, page 501

Early odometers were driven by the speedometer cable through a worm gear. If the speedometer uses an optical sensor, the odometer may be of conventional design. Two other types of odometer are used with electronic displays: the electromechanical type with a stepper motor and the electronic design using an IC chip.

Stepper Motor. The electromechanical odometer uses a DC stepper motor that receives control signals from the speedometer circuit (Figure 13-20). The digital signal impulses from the speedometer are processed through a circuit that will halve the signal. The stepper motor receives one-half of the VSS signals sent to the instrument panel cluster. As the stepper motor is activated, the rollers are rotated to accurately display accumulated mileage.

General Motors controls the stepper motor through the same impulses that are sent to the speedometer. The stepper motor uses these signals to turn the odometer drive IC on and off. An **H-gate** arrangement of four transistors is used to drive the stepper motor by alternately activating a pair of its coils (Figure 13-21). The H-gate is constantly reversing system polarity, causing the permanent magnet poles to rotate in the same direction.

Figure 13-20 A stepper motor is used to rotate the odometer dial.

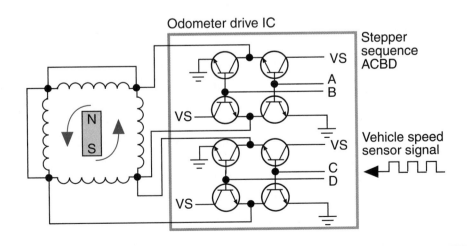

Figure 13-21 The H-gate energizes two coils at a time and constantly reverses system polarity.

In most systems, distance is updated to the RAM every 10 miles and whenever the ignition switch is turned off.

IC Chip. The IC chip–type odometer uses a nonvolatile RAM that receives distance information from the speedometer circuit or from the engine controller. The controller can update the odometer display every 1/2 second.

Many instrument panel clusters cannot display both trip mileage and odometer readings at the same time. Drivers must select which function they wish to have displayed (Figure 13-22). By depressing the trip reset button, a ground is applied as an input to the microprocessor. The microprocessor clears the trip odometer readings from memory and returns the display to zero. The trip odometer will continue to store trip mileage even if this function is not selected for display.

If the IC chip fails, some manufacturers provide for replacement of the chip. Depending on the manufacturer, the new chip may be programmed to display the last odometer reading. Most replacement chips will display an X, S, or * to indicate the odometer has been changed. If the odometer IC chip cannot be programmed to display correct accumulated mileage, a door sticker must be installed to indicate the odometer has been replaced.

AUTHOR'S NOTE: Federal Motor Vehicle Safety Standards require the odometer be capable of storing up to 500,000 miles in nonvolatile memory. Most odometer readouts are up to 199,999.9 miles.

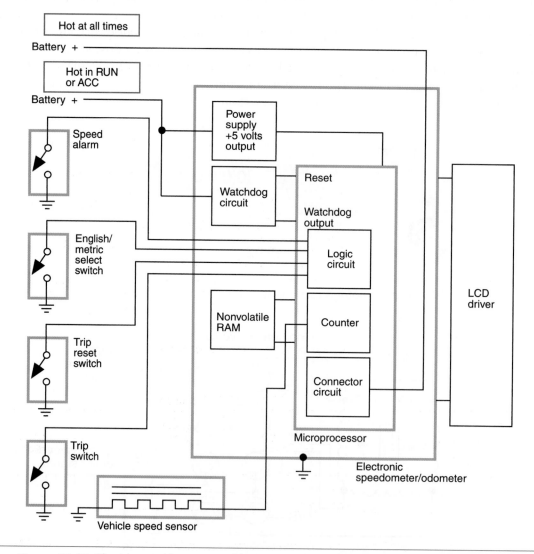

Figure 13-22 The trip reset button provides a ground signal to the logic circuit, which is programmed to erase the trip odometer memory while retaining total accumulated mileage in the odometer.

If an error occurs in the odometer circuit, the display will change to notify the driver. The form of error message differs among manufacturers. In some systems, the word "ERROR" is displayed, while others may use dashed lines.

Federal and state laws prohibit tampering with the correct mileage as indicated on the odometer. If the odometer must be replaced, it must be set to the reading of the original odometer, or a door sticker must be installed indicating the reading of the odometer when it was replaced.

Tachometers

A **tachometer** is an instrument that measures the speed of the engine in revolutions per minute (rpm). The electric tachometer receives voltage pulses from the ignition system, usually the ignition coil (Figure 13-23). The tachometer signal is picked up from the negative (−) side of the coil as the switching unit opens the primary circuit. Each of the voltage pulses represents the generation of one spark at the spark plug. The rate of spark plug firing is in direct relationship to the speed of the engine. A circuit within the tachometer converts the ignition pulse signal into a varying voltage. The voltage is applied to a voltmeter that serves as the engine speed indicator.

The digital tachometer can be a separate function that is displayed at all times, or a part of a multigauge display. The digital tachometer receives its voltage signals from the ignition module or PCM via the bus network and displays the readout in a bar graph (Figure 13-24). The multigauge system has a built-in power supply that provides a 5-volt reference signal to the other monitored systems for the gauge. Also, the gauge has a **watchdog circuit** incorporated in it. The power on/off watchdog circuit supplies a reset voltage to the microprocessor in the event that pulsating output signals from the microprocessor are interrupted.

Shop Manual
Chapter 13, page 504

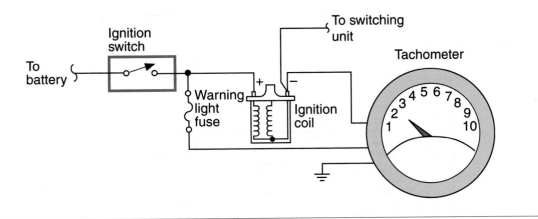

Figure 13-23 Electrical tachometer wired into the ignition system.

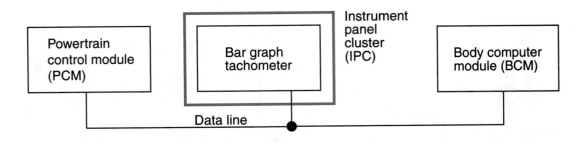

Figure 13-24 The IPC "listens in" on the communications between the PCM and the BCM to gather information on engine speed.

Electronic Fuel Gauges

Most digital fuel gauges use a fuel level sender that decreases resistance value as the fuel level decreases. This resistance value is converted to voltage values by the microprocessor. A voltage-controlled **oscillator** changes the signal into a frequency signal. The microprocessor counts the cycles and sends the appropriate signal to operate the digital display (Figure 13-25).

An F is displayed when the tank is full and an E is displayed when less than 1 gallon is remaining in the tank. Other warning signals include incandescent lamps, a symbol on the dash, or flashing of the fuel ISO symbol. If the warning is displayed by a bulb, usually a switch is located in the sending unit that closes the circuit. The microprocessor usually controls flashing digital displays.

The bar graph–style gauge uses segments that represent the amount of fuel remaining in the tank (Figure 13-26). The segments divide the tank into equal levels. The display will also include the F, 1/2, and E symbols along with the ISO fuel symbol. A warning to the driver is displayed when only one bar is lit. The gauge will also alert the driver to problems in the circuit. A common method of indicating an open or short is to flash the F, 1/2, and E symbols while the gauge reads empty.

Other Digital Gauges

Most of the gauges used to display temperature, oil pressure, and charging voltage are of bar graph design. Another popular method is to use a floating pointer (Figure 13-27).

The temperature gauge will usually receive its input from an NTC thermistor. When the engine is cold, the resistance value of the thermistor is high, resulting in a high-voltage input to the microprocessor. This input signal is translated into a low-temperature reading on the gauge. As the engine coolant warms, the resistance value drops. At a predetermined resistance level, the microprocessor will activate an alert function to warn the driver of excessive engine temperature.

The voltmeter calculates charging voltage by comparing the voltage supplied to the instrument panel module to a reference voltage signal. The oil pressure gauge uses a piezoresistive sensor that operates like those used for conventional analog gauges.

Digital gauges perform self-tests. If a fault is found, a warning signal will be displayed to the driver. A "CO" indicates the circuit is open, a "CS" indicates the circuit is shorted. The gauge will continue to display these messages until the problem is corrected.

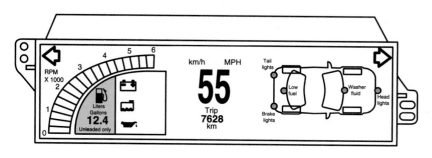

Figure 13-25 The digital fuel gauge displays remaining fuel in gallons or liters.

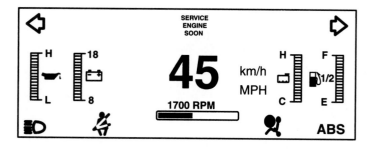

Figure 13-26 Bar graph style of electronic instrumentation. Each segment represents a different value.

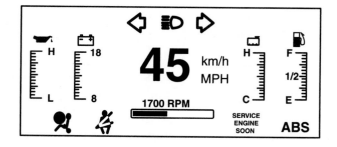

Figure 13-27 Floating pointer indicates the value received from the sensor.

Digital Displays

There are four common methods of display: light-emitting diodes (LEDs), liquid crystal display (LCDs), vacuum fluorescent display (VFD), and cathode ray tube (CRT).

LED Digital Displays

The light-emitting diode (LED) is an electroluminescent lamp that converts the energy developed during normal diode operation into light. When first used in the instrument panel, the LED would indicate an on/off status. Chrysler used the LED in conjunction with its gauges to alert the driver of conditions requiring immediate attention. Some manufacturers are using LEDs in bar graph displays (Figure 13-28). Also, the LEDs can be combined to display alphanumeric characters. There are two common methods of using the LED for digital display: (1) seven-segment display (Figure 13-29), and (2) dot **matrix** display (Figure 13-30).

A **matrix** is a group of elements arranged in columns and rows.

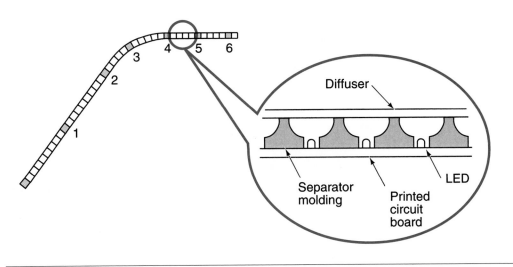

Figure 13-28 LEDs arranged to create a bar graph display.

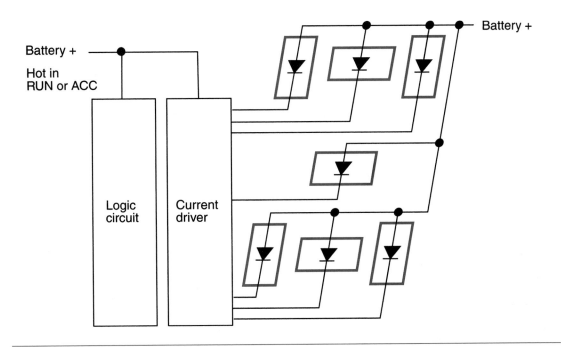

Figure 13-29 Seven-segment display panel.

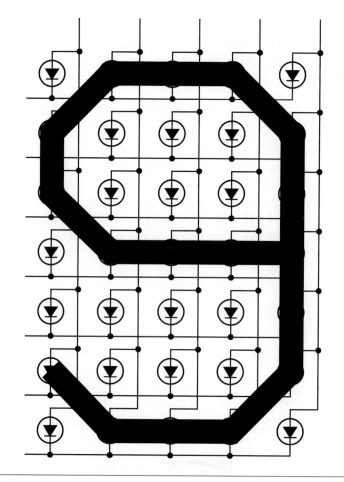

Figure 13-30 Dot matrix display panel.

By activating selected segments or dots, any number or letter can be displayed. LEDs generally are designed to produce a red, green, or yellow light. They work very well in low-light conditions. However, they are difficult to see in bright light. Although some manufacturers use LED readout instrument panels, their use is limited due to their comparatively high power requirements.

Liquid Crystal Displays

Liquid crystal displays (LCDs) require an external light source because they do not generate their own (Figure 13-31). The external light source can be supplied either by daylight or by an artificial light. In daylight, the segments are activated from the front; the artificial light activates the LCD from the back. The artificial light can be controlled by the headlight switch or turned on whenever the ignition switch is placed in the RUN or ACC position.

The LCD construction consists of a twisted **nematic** fluid that is sandwiched between two polarized glass sheets. The nematic fluid is a liquid crystal that has a threadlike form. It has light slots that can be rearranged by applying small amounts of voltage. The front **polarizer** is a vertical polarizer and the rear polarizer is a horizontal polarizer. The polarizer makes light waves vibrate in only one direction. Light is composed of waves that vibrate in several different directions. The polarizer converts the light into polarized light (Figure 13-32). The display is viewed through the vertical polarizer. The nematic fluid's molecules are arranged in such a manner that they rotate the light from the vertical polarizer 90 degrees (Figure 13-33). The light leaves the fluid in a horizontal waveform. The light continues to pass through the horizontal polarizer to the reflector. The light is then reflected back through the horizontal polarizer to the fluid. The fluid once again rotates the light wave into a vertical position and out the vertical polarizer. When light passes through the LCD in this manner, the display appears light and no pattern is seen.

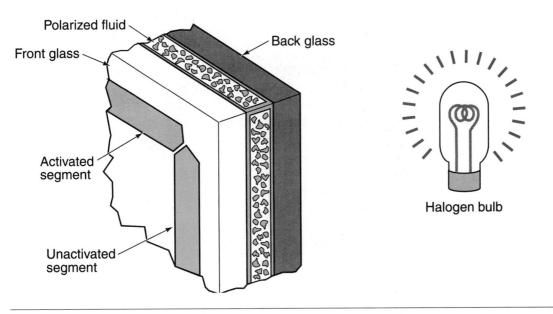

Figure 13-31 LCDs use external light source through a polarized fluid.

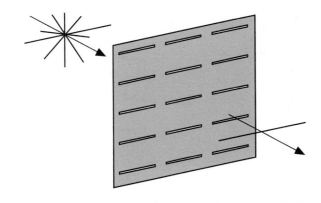

Figure 13-32 The polarizer makes the light waves vibrate in only one direction.

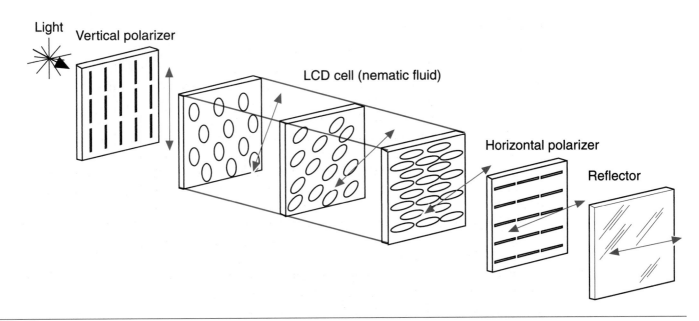

Figure 13-33 The nematic fluid rotates the polarized light wave 90 degrees.

When a small square wave voltage is applied to the fluid, its light slots are rearranged. The fluid will no longer rotate the light waves. The light waves leave the fluid in a vertical plain and cannot pass through the horizontal polarizer to get to the reflector (Figure 13-34). Because the light cannot be reflected back, the display appears dark. Characters are displayed by controlling which segments are dark and which segments remain light (Figure 13-35).

In most instrument panels, the LCD cluster is constantly backlit with halogen lights. However, a slightly different principle is used. When the segment is not activated, the light is unable to transmit through the opaque fluid and the segment appears dark (Figure 13-36). When voltage is applied to the fluid, its light slots align and allow the light to pass to the segment. The intensity of the halogen lights is controlled through pulse width modulation to provide the correct illumination levels for the LCD under different ambient light conditions. A photocell is used to sense the amount of light intensity inside the vehicle. The microprocessor processes this information to determine the correct light intensity for the display. Additional control of light intensity is provided by the driver through the headlight switch rheostat.

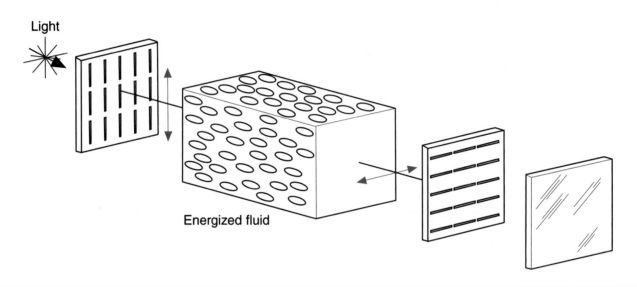

Figure 13-34 When the fluid is energized, the light wave is not rotated.

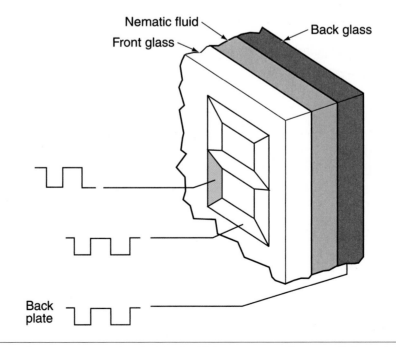

Figure 13-35 The square wave signals direct which segments will appear light and dark.

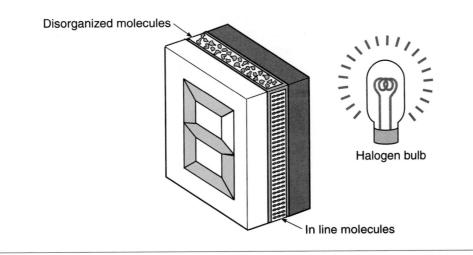

Figure 13-36 Using artificial light to display the LCD.

The voltage to the fluid is provided through the polarizers that have contacts with each segment in the display. The front glass has metallized shapes where the characters will be displayed. The back glass also is metallized. The color of the display is determined by filters placed in front of the display.

Vacuum Fluorescent Displays

The vacuum fluorescent display (VFD) is constructed of a hot cathode of tungsten filaments, a grid, and a phosphorescent screen that is the anode (Figure 13-37). The components are sealed in a flat glass envelope that has been evacuated of oxygen and filled with argon or neon gas.

A constant voltage is applied to the hot cathode, which results in tungsten electrons being released from the filament wires. The grid is at a higher positive voltage than the cathode. The freed tungsten electrons are accelerated by the positive grid wires and pass through the grid to the anode. The grid ensures that the tungsten electrons will strike the anode uniformly.

The anode is at a higher positive voltage than the grid. The phosphorescent-coated anode (screen) will be luminescent when the tungsten electrons strike it. The display is controlled by which segments of the screen are activated by a digital circuit. If the segment is activated, the screen will illuminate. If the segment is not activated, the electrons striking the screen have no effect on the phosphors and the screen remains off.

The segments of a VFD can be arranged in several different patterns. The most common are seven- or fourteen-segment patterns (Figure 13-38). The computer selects the sets of segments that are to emit light for the message.

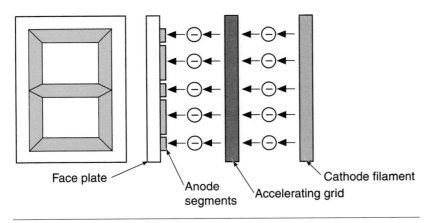

Figure 13-37 VFD construction.

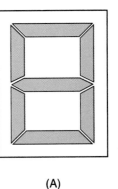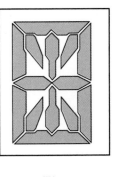

(A)   (B)

Figure 13-38 (A) Seven-segment display pattern. (B) Fourteen-segment display pattern.

The VFD display is very bright. Most manufacturers will dim the intensity of the VFD to 75% brightness whenever the headlights are turned on. To provide sufficient brightness in the daylight, with the headlights on, the headlight switch rheostat may have an additional detent to allow bright illumination of the VFD.

CRT Displays

A **cathode ray tube (CRT)** is similar to a television tube. It contains a cathode that emits electrons and an anode that attracts them. The screen will glow at the points that are hit by the electrons. Control plates dictate the direction of the electrons. A CRT display was first offered as standard equipment on the 1986 Buick Riviera (Figure 13-39). The screen of the CRT is touch sensitive. By touching the button on the screen, the menu can be changed to display different information. The menu-driven instrumentation brings up a screen with a menu of items. The driver can select a particular area of vehicle operation. The menu of items includes the radio, climate control, trip computer, and dash instrument information. The technician can access diagnostics through the CRT.

The CRT receives information from the BCM and ECM. It also provides inputs to the BCM and ECM in the form of driver commands to control the various functions (Figure 13-40).

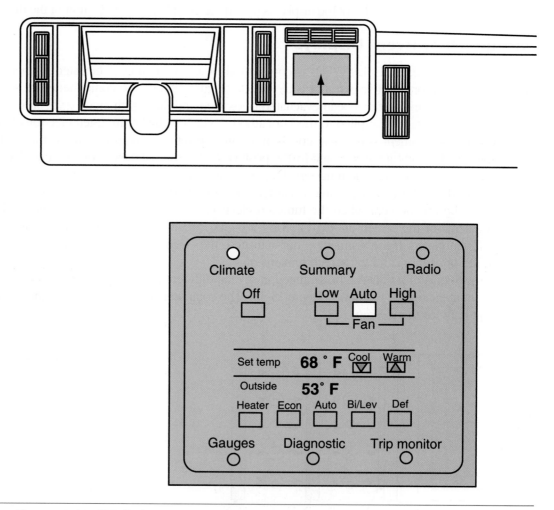

Figure 13-39 CRT display.

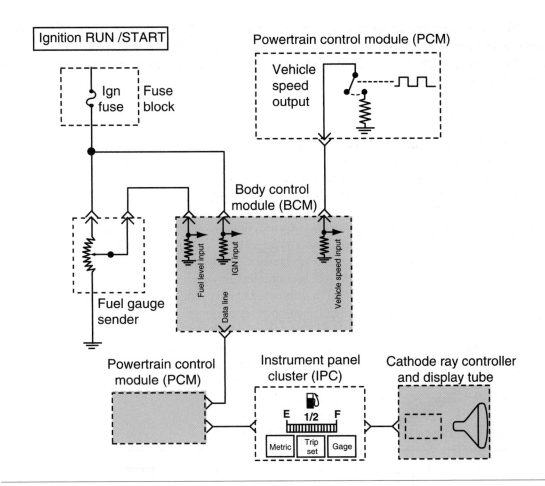

Figure 13-40 The CRT receives input information from the ECM and BCM.

Head-Up Display

Some manufacturers have equipped selected models with a **head-up display (HUD)** feature. This system displays visual images onto the inside of the windshield in the driver's field of vision (Figure 13-41). With the display located in this area, drivers do not need to remove their eyes from the road to check the instrument panel. The images are projected onto the windshield from a vacuum fluorescent light source, much like a movie projector.

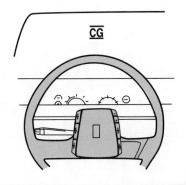

Figure 13-41 The HUD displays various information inside the windshield.

The head-up control module is mounted in the top of the instrument panel. This module contains a computer and an optical system that projects images to a holographic combiner integrated into the windshield above the module. The holographic combiner projects the images in the driver's view just above the front end of the hood. The HUD contains the following displays and warnings:

1. Speedometer reading with USC/metric indicator.

2. Turn signal indicators.

3. High-beam indicator.

4. Low-fuel indicator.

5. Check gauges indicator.

The head-up control switch contains a head-up display on/off switch, USC/metric switch, and a head-up dimming switch. The head-up dimming switch is a rheostat that sends an input signal to the head-up module. The vertical position of the head-up display may be moved with one of the switches in the control switch assembly that is connected through a mechanical cable-drive system to the head-up module. Moving the vertical position switch moves the position of the head-up module.

The vehicle speed sensor signal information is sent from the PCM to the head-up module for the speedometer display. The check gauges and low-fuel signals are sent from the instrument cluster to the head-up module. A high-beam indicator input signal is sent from the dimmer switch to the head-up module. This module also receives inputs from the signal light switch to operate the signal light indicators.

Travel Information Systems

Shop Manual
Chapter 13, page 514

The travel information system can be a simple calculator that computes fuel economy, distance to empty, and remaining fuel (Figure 13-42). Other systems provide a much larger range of functions.

Fuel data centers display the amount of fuel remaining in the tank and provide additional information for the driver (Figure 13-43). By depressing the RANGE button, the BCM calculates the distance until the tank is empty by using the amount of fuel remaining and the average fuel economy. When the INST button is depressed, the fuel data center displays instantaneous fuel economy. The display is updated every 1/2 second and is computed by the BCM.

Depressing the AVG button displays average fuel economy for the total distance traveled since the reset button was last pushed. FUEL USED displays the amount of fuel that has been used since the last time this function was reset. The RESET button resets the average fuel economy and fuel-used calculations. The function to be reset must be displayed on the fuel data center.

Deluxe systems may incorporate additional features such as outside temperature, compass, elapsed time, estimated time of arrival, distance to destination, day of the week, time, and average speed. The illustration (Figure 13-44) shows the inputs that are used to determine many of

Figure 13-42 Fuel data display panel.

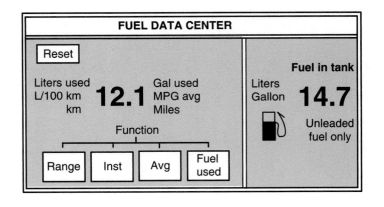

Figure 13-43 Fuel data center.

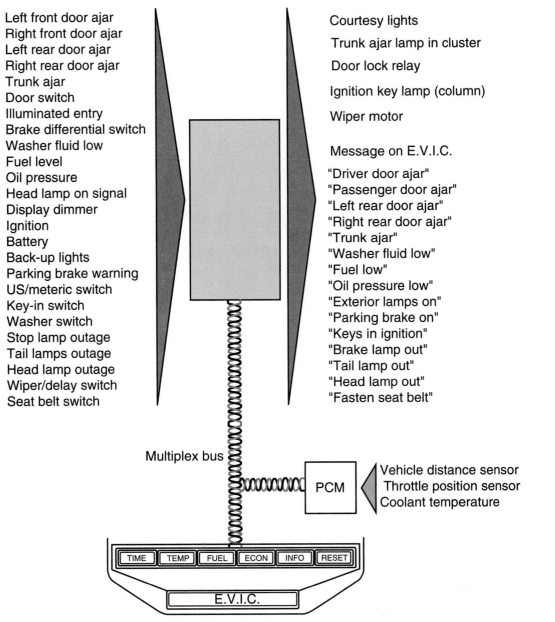

INPUTS

Left front door ajar
Right front door ajar
Left rear door ajar
Right rear door ajar
Trunk ajar
Door switch
Illuminated entry
Brake differential switch
Washer fluid low
Fuel level
Oil pressure
Head lamp on signal
Display dimmer
Ignition
Battery
Back-up lights
Parking brake warning
US/meteric switch
Key-in switch
Washer switch
Stop lamp outage
Tail lamps outage
Head lamp outage
Wiper/delay switch
Seat belt switch

OUTPUTS

Courtesy lights

Trunk ajar lamp in cluster

Door lock relay

Ignition key lamp (column)

Wiper motor

Message on E.V.I.C.

"Driver door ajar"
"Passenger door ajar"
"Left rear door ajar"
"Right rear door ajar"
"Trunk ajar"
"Washer fluid low"
"Fuel low"
"Oil pressure low"
"Exterior lamps on"
"Parking brake on"
"Keys in ignition"
"Brake lamp out"
"Tail lamp out"
"Head lamp out"
"Fasten seat belt"

Multiplex bus

PCM

Vehicle distance sensor
Throttle position sensor
Coolant temperature

TIME TEMP FUEL ECON INFO RESET

E.V.I.C.

Figure 13-44 Inputs used for the electronic vehicle information center.

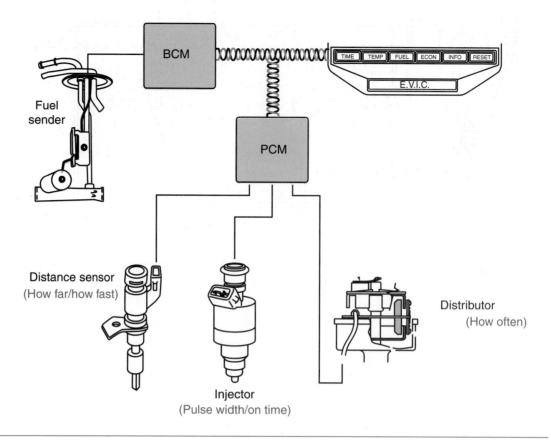

Figure 13-45 Fuel data system inputs. The injector on time is used to calculate the rate of fuel flow.

these functions. The sensors shown in Figure 13-45 determine fuel system calculations. Injector on time and vehicle speed pulses determine the amount of fuel flow. Some manufacturers use a fuel flow sensor that provides pulse information to the microprocessor concerning fuel consumption (Figure 13-46).

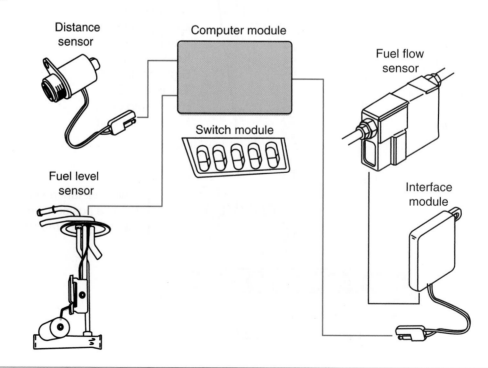

Figure 13-46 Some information centers use a fuel flow sensor.

Warning Lamps

Shop Manual
Chapter 13, page 512

A **warning lamp** is a lamp that is illuminated to warn the driver of a possible problem or hazardous condition. A warning lamp may be used to warn of low oil pressure, high coolant temperature, defective charging system, or a brake failure. A warning lamp can be operated by two methods: a sending unit circuit hardwired to the instrument cluster, or computer-controlled lamp drivers.

Sending Unit-Controlled Lamps

Unlike gauge sending units, the sending unit for a warning lamp is nothing more than a simple switch. The style of switch can be either normally open or normally closed, depending on the monitored system.

Most oil pressure warning circuits use a normally closed switch (Figure 13-47). A diaphragm in the sending unit is exposed to the oil pressure. The switch contacts are controlled by the movement of the diaphragm. When the ignition switch is turned to the RUN position with the engine not running, the oil warning lamp turns on. Because there is no pressure to the diaphragm, the contacts remain closed and the circuit is complete to ground. When the engine is started, oil pressure builds and the diaphragm moves the contacts apart. This opens the circuit and the warning lamp goes off. The amount of oil pressure required to move the diaphragm is about 3 psi. If the oil warning lamp comes on while the engine is running, it indicates that the oil pressure has dropped below the 3 psi limit.

Most coolant temperature warning lamp circuits use a normally open switch (Figure 13-48). The temperature sending unit consists of a fixed contact and a contact on a bimetallic

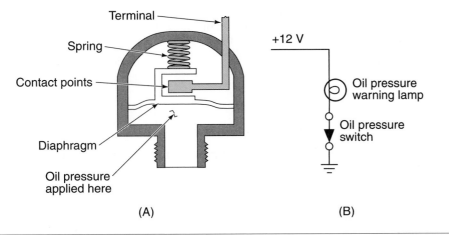

Figure 13-47 (A) Oil pressure light sending unit. (B) Oil pressure warning lamp circuit.

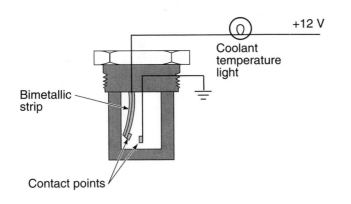

Figure 13-48 Temperature indicator light circuit.

The prove-out function is also known as "Bulb Test" or "Bulb Check" position.

strip. As the coolant temperature increases, the bimetallic strip bends. As the strip bends, the contacts move closer to each other. Once a predetermined temperature level has been exceeded, the contacts are closed and the circuit to ground is closed. When this happens, the warning lamp is turned on.

With normally open–type switches, the contacts are not closed when the ignition switch is turned to ON. In order to perform a bulb check on normally open switches, a **prove-out circuit** is included (Figure 13-49). A prove-out circuit completes the warning light circuit to ground through the ignition switch when it is in the START position. The warning light will be on during engine cranking to indicate to the driver that the bulb is working properly.

It is possible to have more than one sending unit connected to a single bulb. The illustration (Figure 13-50) shows a wiring circuit of a dual-purpose warning lamp. The lamp will come on whenever oil pressure is low or coolant temperature is too high.

Another system that is monitored with a warning lamp is the braking system. The illustration (Figure 13-51) shows a brake system combination valve. The center portion of the valve senses differences in the hydraulic pressures on both sides of the valve. With the differential valve centered, the plunger on the warning lamp switch is in the recessed area of the valve. If the pressure drops in either side of the brake system, the differential valve will be forced to move by hydraulic pressure. When the differential valve moves, the switch plunger is pushed up and the switch contacts close.

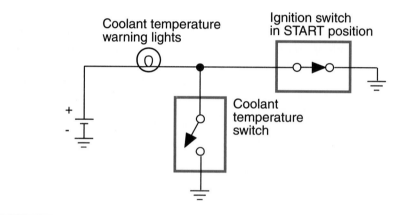

Figure 13-49 A prove-out circuit included in a normally open (NO) coolant temperature light system.

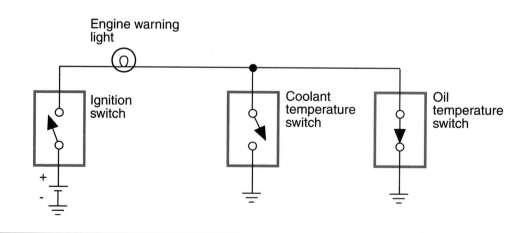

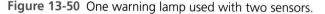

Figure 13-50 One warning lamp used with two sensors.

404

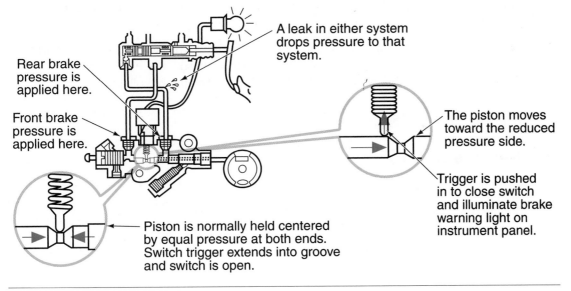

A leak in either system drops pressure to that system.

Rear brake pressure is applied here.

Front brake pressure is applied here.

The piston moves toward the reduced pressure side.

Trigger is pushed in to close switch and illuminate brake warning light on instrument panel.

Piston is normally held centered by equal pressure at both ends. Switch trigger extends into groove and switch is open.

Figure 13-51 Brake warning light switch as part of the combination valve.

Computer-Driven Warning Lamp Systems

The computer-driven warning lamp system uses either high side or low side drivers to illuminate the warning lamp. Usually the driver module (instrument cluster or BCM) will receive a data bus message from the module that monitors the effected system that the warning lamp needs to be turned on. The driver module will then command the lamp on. For example, the PCM monitors the engine temperature. If the engine temperature reaches the upper threshold, the PCM will send a data bus message to the instrument cluster to turn on the warning lamp. The instrument cluster will activate its driver to illuminate the lamp. Usually the instrument cluster must receive a bus message from the monitoring module at a set time interval. If the message is not received, the warning lamp is illuminated. Some systems will illuminate a CHECK GAUGES lamp if the cluster uses a gauge and the gauge indicated a condition that the driver needs to be notified of.

Shop Manual
Chapter 13, page 513

Summary

❑ Through the use of gauges and indicator lights, the driver is capable of monitoring several engine and vehicle operating systems.

❑ The gauges include speedometer, odometer, tachometer, oil pressure, charging indicator, fuel level, and coolant temperature.

❑ The most common types of electromechanical gauges are the d'Arsonval, three-coil, two-coil, and air-core.

❑ Computer-driven quartz swing needle displays are similar in design to the air-core electromagnetic gauges used in conventional analog instrument panels.

❑ All gauges require the use of a variable resistance sending unit. Styles of sending units include thermistors, piezoresistive sensors, and mechanical variable resistors.

❑ Digital instrument clusters use digital and linear displays to notify the driver of monitored system conditions.

❑ The most common types of displays used on electronic instrument panels are: light-emitting diodes (LED), liquid crystal displays (LCDs), vacuum fluorescent displays (VFD), and a phosphorescent screen that is the anode.

Terms to Know
Air-core gauge
Bucking coil
Buffer circuit
Cathode ray tube (CRT)
d'Arsonval gauge
Digital instrument clusters
Electromagnetic gauges
Electromechanical
Head-up display (HUD)
H-gate

❏ A head-up display system displays visual images onto the inside of the windshield in the driver's field of vision.

❏ In the absence of gauges, important engine and vehicle functions are monitored by warning lamps. These circuits generally use an on/off switch–type sensor. The exception would be the use of voltage-controlled warning lights that use the principle of voltage drop.

Review Questions

Short-Answer Essays

1. What are the most common types of electromagnetic gauges?

2. Describe the operation of the piezoresistive sensor.

3. What is a thermistor used for?

4. What is meant by *electromechanical*?

5. Describe the operation of the air-core gauge.

6. What is the basic difference between conventional analog and computer-driven analog instrument clusters?

7. Describe the operating principles of the digital speedometer.

8. Explain the operation of IC chip–type odometers.

9. Describe the operation of the electronic fuel gauge.

10. Describe the operation of quartz analog speedometers.

Fill in the Blanks

1. The purpose of the tachometer is to indicate _____ _____.

2. A piezoresistive sensor is used to monitor _____ changes.

3. The most common style of fuel level sending unit is _____ variable resistor.

4. The brake warning light is activated by _____ pressure in the brake hydraulic system.

5. In a three-coil gauge, the _____ _____ produces a magnetic field that bucks or opposes the low-reading coil. The _____ _____ coil and the bucking coil are wound together, but in opposite directions. The _____ _____ coil is positioned at a 90° angle to the low-reading and bucking coils.

6. A _____ _____ circuit completes the warning light circuit to ground through the ignition switch when it is in the START position.

7. Digital instrument clusters use _____ and _____ displays to notify the driver of monitored system conditions.

8. _____ _____ is a calculation using the final drive ratio and the tire circumference to obtain accurate vehicle speed signals.

9. Most digital fuel gauges use a fuel level sender that _____ resistance value as the fuel level decreases.

10. Computer-driven quartz swing needle displays are similar in design to the _____ _____ electromagnetic gauges used in conventional analog instrument panels.

Multiple Choice

1. Odometer replacement is being discussed.
   Technician A says that it is permissible to turn back the reading on a odometer as long as the customer is notified.
   Technician B says that if an odometer is replaced, it must be set to the same reading as the original odometer.
   Who is correct?
   A. A only
   B. B only
   C. Both A and B
   D. Neither A nor B

2. Electromagnetic gauges are being discussed.
   Technician A says that the d'Arsonval gauge uses the interaction of a permanent magnet and a electromagnet, and the total field effect to cause needle movement.
   Technician B says that the three-coil gauge uses the interaction of three electromagnets and the total field effect upon a permanent magnet to cause needle movement.
   Who is correct?
   A. A only
   B. B only
   C. Both A and B
   D. Neither A nor B

3. *Technician A* says that the three-coil gauge uses the principle that electricity seeks the path of least resistance.
   Technician B says that the three coils used are the low-reading coil, a bucking coil, and a high-reading coil.
   Who is correct?
   A. A only
   B. B only
   C. Both A and B
   D. Neither A nor B

4. Warning light circuits are being discussed.
   Technician A says that most oil pressure warning circuits use a normally closed switch.
   Technician B says most conventional coolant temperature warning light circuits use a normally open switch.
   Who is correct?
   A. A only
   B. B only
   C. Both A and B
   D. Neither A nor B

5. The brake failure warning system is being discussed.
   Technician A says if the pressure drops in either side of the brake system, the switch plunger is pushed up and the switch contacts close.
   Technician B says if the pressure is equal on both sides of the brake system, the warning light comes on.
   Who is correct?
   A. A only
   B. B only
   C. Both A and B
   D. Neither A nor B

6. The IC chip odometer is being discussed.
   Technician A says if the chip fails, some manufacturers provide for replacement of the chip.
   Technician B says depending on the manufacturer, the new chip may be programmed to display the last odometer reading.
   Who is correct?
   A. A only
   B. B only
   C. Both A and B
   D. Neither A nor B

7. Computer-driven quartz swing needle displays are being discussed.
   Technician A says the A coil is connected to system voltage and the B coil receives a voltage that is proportional to input frequency.
   Technician B says the quartz swing needle display is similar to air-core electromagnetic gauges.
   Who is correct?
   A. A only
   B. B only
   C. Both A and B
   D. Neither A nor B

8. *Technician A* says digital instrumentation displays an average of the readings received from the sensor.
   Technician B says conventional analog instrumentation gives more accurate readings but is not as decorative.
   Who is correct?
   A. A only
   B. B only
   C. Both A and B
   D. Neither A nor B

9. The microprocessor-initiated self-check of the electrical instrument cluster is being discussed.
   Technician A says during the first portion of the self-test all segments of the speedometer display are lit.
   Technician B says the display should not go blank during any part of the self-test.
   Who is correct?
   A. A only
   B. B only
   C. Both A and B
   D. Neither A nor B

10. *Technician A* says bar graph–style gauges do not provide for self-tests.
    Technician B says the digital instrument panel will display CO to indicate the circuit is shorted.
    Who is correct?
    A. A only
    B. B only
    C. Both A and B
    D. Neither A nor B

CHAPTER 14

Electronic Chassis Control and Accessory Systems

Upon completion and review of this chapter, you should be able to:

❑ Explain the purpose of the automatic temperature control (ATC).

❑ Explain the purpose and operation of the control module in ATC systems.

❑ List and describe the types of sensors used in ATC systems.

❑ Define the purpose of the cruise control system.

❑ Explain the operating principles of the electronic cruise control system.

❑ Explain the operating principles of laser cruise control systems.

❑ Explain the operating principles of the memory seat feature.

❑ Describe the control concepts of electronically controlled sunroofs.

❑ Detail the operation of common antitheft systems.

❑ Explain the function of immobilizer security systems.

❑ Explain the purpose and operation of automatic door lock systems.

❑ Detail the operation of the keyless entry system.

❑ Explain the operation of keyless start.

❑ Explain the operating principles of Ford and GM's heated windshield systems.

❑ Explain the purpose of electronic shift transmissions.

❑ Explain the purpose of variable assist steering.

❑ Describe the purpose of electronic suspension and stabilizing systems.

❑ State the purpose of antilock braking systems.

❑ Describe the function of brake assist systems.

❑ Describe the function of electronic stability systems.

❑ Describe the basic purpose of automatic traction control systems.

❑ Describe the operation of the tire pressure monitoring system.

❑ Describe the purpose and configuration of vehicle audio and entertainment systems.

Introduction

In today's automobile there is no system that computers cannot control. The vehicle may be equipped with computer-controlled wipers, transmissions, locking differentials, brakes, suspensions, all-wheel drive systems, and so on. It would be impossible to cover the various operations of all these systems. It is important for today's technician to have an understanding of how electronics work and a basic knowledge of the control system. Whether you are working on a domestic or foreign-built automobile, electricity and electronics work the same. Always refer to the proper service information to get an understanding of a system that may be new to you.

This chapter discusses several of the electronic systems found in today's automobiles. Some of these systems are covered in greater detail in other *Today's Technician* series books. Refer to these textbooks for more information.

In this chapter, you will learn the operation of the automatic temperature control systems, cruise control systems, and the many electrical accessory systems that have electronic controls added to them for added features and enhancement. These accessories include memory seats, electronic sunroofs, antitheft systems, automatic door locks, keyless entry, and electronically heated windshields.

The comfort and safety of the driver and/or passengers depend on the technician properly diagnosing and repairing these systems. As with all electrical systems, the technician must have a basic understanding of the operation of these systems before attempting to perform any service.

Introduction to Automatic Temperature Control

Shop Manual
Chapter 14, page 528

Electronic automatic air conditioning systems operate with the same basic components as the conventional systems. The major difference is the automatic temperature control (ATC) system is capable of maintaining a preset level of comfort as selected by the driver. Sensors are used to determine the present temperatures and the system can adjust the level of heating or cooling as required.

The system uses actuators that open and close air-blend doors to achieve the desired in-vehicle temperature. In addition, they control fan motor speeds to keep the temperature very close to that requested by the driver.

Though the systems differ in methods of operation, they are all designed to provide in-car temperatures and humidity conditions at a preset level. The in-car humidity and temperature levels are maintained, regardless of the climate conditions outside the vehicle. The in-car humidity level is maintained at 45 to 55%.

A BIT OF HISTORY

In 1939, Packard introduced a car with air conditioning. It used refrigeration coils in an air duct behind the rear seat.

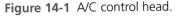

Common Components

Although not all systems will have every component that is described here, most will have a combination of several of them.

Shop Manual
Chapter 14, page 531

Control Panel Assembly. The **control panel assembly** is located in the instrument panel and provides the means for driver input to the ATC system. The vehicle driver or passengers can input the desired cabin temperatures and select operating modes (A/C, heat, defogger, and vent) along with fan motor speeds through push button selection on the control assembly (Figure 14-1). If the AUTO function is selected, then mode and fan speed are determined by the control module.

Sensors. There can be several different sensors used on the ATC system. The most common are the in-vehicle and ambient temperature sensors, sunload, and evaporator temperature sensors.

Figure 14-1 A/C control head.

The **in-vehicle sensor** contains a temperature sensing NTC thermistor to measure the average temperature inside the vehicle (Figure 14-2). The in-vehicle sensor is located in the **aspirator** unit, which is usually mounted in the dashboard (Figure 14-3). Most aspirators are tubular devices that use a venturi effect to draw air from the passenger compartment over the in-vehicle sensor (Figure 14-4). An aspirator tube is connected from the in-vehicle sensor to the heater and

Figure 14-2 The in-car sensor is an NTC thermistor located in the aspirator unit.

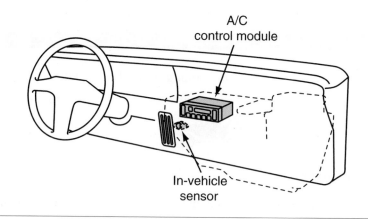

Figure 14-3 In-vehicle sensors are usually in the dash behind a small grill.

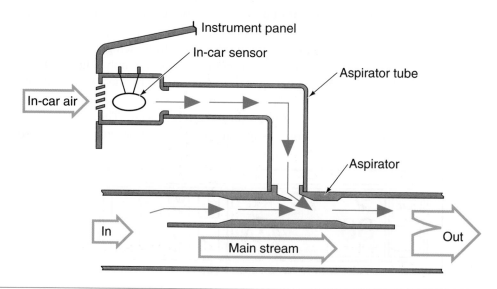

Figure 14-4 A typical aspirator. The main airstream creates a low pressure at the inlet of the aspirator, drawing in-car air over the sensor.

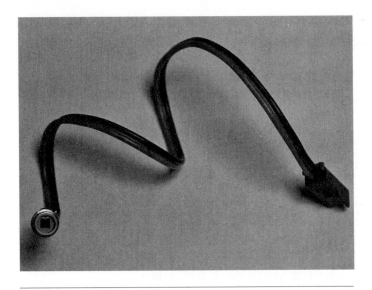

Figure 14-5 Ambient temperature sensor.

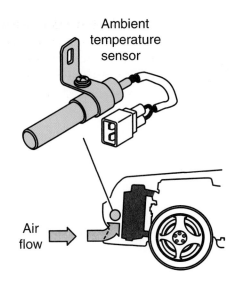

Figure 14-6 Ambient temperature sensor located behind the grill.

air conditioning duct. The rush of air past the aspirator tube in the heater duct creates a slight vacuum, which pulls a small amount of air through the in-vehicle sensor and aspirator tube. Some A/C systems have a small electric fan motor to move air through the in-vehicle sensor.

The in-vehicle sensor resistance changes in relation to temperature. When the sensor is cold, it has a higher resistance, whereas an increase in temperature decreases the sensor resistance. Two terminals on the in-vehicle sensor are connected to the control module. As the resistance of the in-vehicle sensor changes, the programmer senses the change in voltage drop across the sensor.

The **ambient temperature sensor** is an NTC thermistor used to measure the temperature outside the vehicle (Figure 14-5). The ambient temperature sensor is usually located behind the front grill (Figure 14-6). Due to its location, and possible influence by engine temperatures, the sensor circuit has several memory features that prevent false input.

Some systems use a sunload sensor (Figure 14-7). The sunload sensor is a **photovoltaic diode** that sends signals to the programmer concerning the extra generation of heat as the sun beats through the windshield. Photovoltaic diodes are capable of producing a voltage when

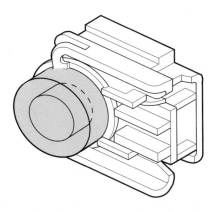

Figure 14-7 The sunload sensor produces a signal proportional to the heat intensity of the sun's heat through the windshield.

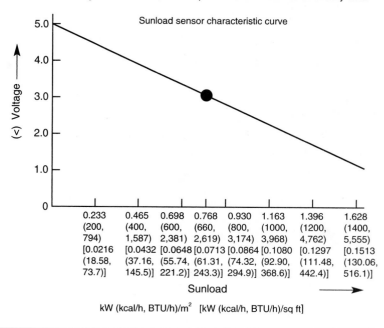

● When checking sunload sensor, select a place where the sun shines directly on it.

Figure 14-8 The conversion of sunlight intensity to a current value used by the programmer.

exposed to radiant energy. The sensor converts the light signal into a voltage value that is sent to the control module (Figure 14-8). This sensor is usually located on the dash next to a speaker grill (Figure 14-9).

The evaporator temperature sensor (or FIN sensor) is mounted in the cooling unit. Many evaporator sensors have a pickup that is mounted in the evaporator fins, enabling it to measure evaporator temperature accurately (Figure 14-10). Some evaporator sensors are mounted so they measure air temperature as it leaves the evaporator. These sensors are usually NTC-type thermistors.

Some manufacturers use an **infrared temperature sensor** to determine the surface temperature of the occupants. This information is used to determine the best climate control for the desired temperature set by the driver. Infrared temperature sensors are used to measure the temperature of an object without contact. Infrared temperature sensors use the principle that all objects emit energy and as the temperature of an object rises, so does the amount of energy it emits.

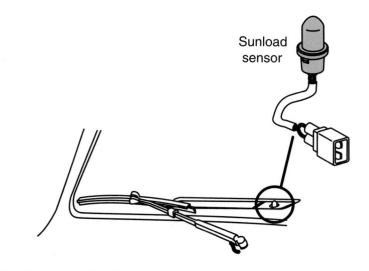

Sunload sensor

Figure 14-9 Sunload sensor mounted in one of the defroster ducts.

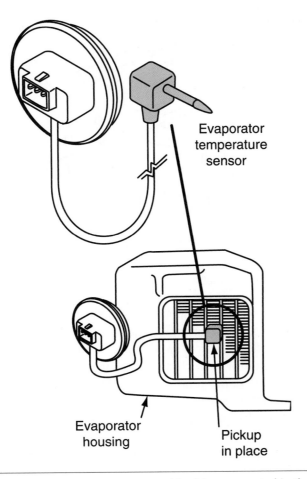

Evaporator temperature sensor

Evaporator housing

Pickup in place

Figure 14-10 Evaporator temperature sensor with pickup mounted in the evaporator fins.

An infrared temperature sensor determines temperature by measuring the intensity of the energy given off by an object. The sensor collects the energy from a target through a lens system. The energy is then focused onto a detector. The detector generates a voltage signal that is sent to the control module. Since the control module has been programmed to know the relationship between the voltage signal and corresponding temperature, the control module knows the surface temperature of the occupant.

Some ATC systems will also use engine speed and throttle position sensors as inputs. If these input sensors indicate a wide-open throttle condition or high engine rpm, the control module does not energize the compressor clutch.

On some A/C systems, the driver's and passenger's door ajar switches send input signals to the control module. When either door is initially opened, after the car has been sitting, the door ajar switch signals to the control module, causing it to turn on the aspirator motor and flush hot air out of the in-vehicle sensor. This action only occurs above a specific in-vehicle temperature.

On many systems, some input signals are sent to the PCM and then relayed to the ATC control module. These inputs include engine coolant temperature, vehicle speed, and A/C system pressure. On some vehicles, these input signals are transmitted on data links between the PCM and ATC control module. A **cold engine lock-out switch** may be used to signal the ATC control module to prevent blower motor operation until the air entering the passenger compartment reaches a specified temperature.

Outputs. Common outputs of the system include the blend air door actuator, heater core flow control valve and solenoid, mode door actuator, recirc/air inlet door actuator, and the compressor clutch control.

Shop Manual
Chapter 14, page 530

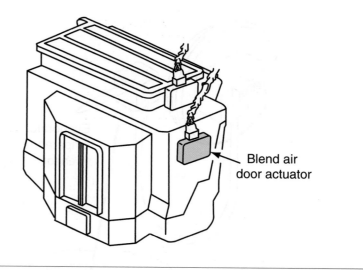

Figure 14-11 Blend air door actuator.

The system may have several actuators to perform the commands that are input by the driver. Actuators can be either vacuum motors, electrically controlled vacuum solenoids, or electrical motors. Electrically controlled actuators are discussed here.

The **blend air door actuator** is an electric motor that controls the position of the blend air door to supply the temperature in the vehicle selected by the driver (Figure 14-11). If the driver selects 72°F (22°C) on the control panel and the in-vehicle temperature is 32°F (0°C), the control module operates the blend air door motor to move the door so it blocks airflow through the evaporator and allows airflow through the heater core into the vehicle interior. This action brings the in-vehicle temperature up to the selected temperature as quickly as possible.

When the selected temperature is 72°F (22°C) and the in-vehicle temperature is 98°F (37°C), the control module requests the A/C clutch be energized to operate the A/C compressor. The module then operates the blend air door motor to move the door so it blocks airflow through the heater core and moves air through the evaporator into the vehicle interior. This action quickly cools down the vehicle interior to supply the selected temperature.

Under most operating conditions, the actuator motor positions the blend air door to allow a blend of warm air through the heater core and cold air through the evaporator to supply the in-vehicle temperature selected by the driver.

Some systems use a **heater core flow valve** and solenoid to shut off the coolant flow through the heater core (Figure 14-12). When the A/C system is in the max air mode, the control module grounds the coolant control solenoid winding. This action moves the solenoid plunger and supplies vacuum through the solenoid to the vacuum diaphragm connected to the valve in the heater hose. Under this condition, the valve closes to stop the coolant flow through the heater core. In other operating modes, the control module opens the coolant control solenoid ground circuit. Under this condition, the solenoid shuts off vacuum to the diaphragm, and the coolant flow valve in the heater hose is open.

The **mode door actuator** is an electric motor that is linked to the mode door. The control module operates the mode door to supply airflow to the floor ducts, A/C panel ducts, or defrost ducts. In the bi-level mode, the mode door is positioned to supply air to the floor and A/C panel ducts. The mix mode supplies airflow from the defrost and floor ducts. The control module positions the mode door to supply the airflow selected by the driver or by its programming based on driver inputs in AUTO mode. For example, if the driver presses the defrost button on the A/C controls, the control module commands the mode door actuator to supply airflow from the defrost ducts.

In some systems, if the AUTO button is pressed, the mode door is positioned automatically. For example, if A/C is selected, the programmer operates the mode door so airflow is directed from the panel ducts in the dashboard. In these systems, the driver can press the A/C control buttons and override the auto function. The AUTO button may be pressed to return the system to

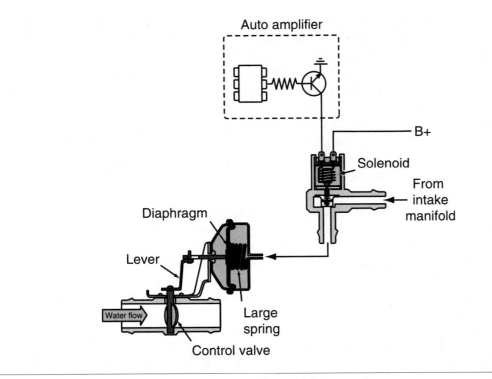

Figure 14-12 Solenoid and coolant control valve in the heater hose.

automatic operation. When the system is in the auto mode, auto fan mode is displayed in the A/C control display. If the temperature control is placed in the full HOT or full COLD manual override position, the system will not enter the auto mode. The control panel or the control module will illuminate an LED in each A/C control button to indicate the operating mode to the driver.

The **recirc/air inlet door actuator** is also an electric motor that is linked to the recirculation door. The control module operates the recirc/air inlet door actuator to position the door to move outside air or in-vehicle air into the A/C heater case. If the A/C system is in the max air mode, the control module operates the recirc/air door actuator to position the door so in-vehicle air is moved into the A/C heater case. This action provides faster cooling of the passenger compartment. In most other modes, the programmer positions the recirc/air door so outside air is moved into the A/C heater case.

When the driver selects the A/C mode, the control module grounds the compressor clutch relay winding. On some A/C systems, the PCM grounds this winding. When the A/C compressor clutch relay winding is grounded, voltage is supplied through the closed relay contacts to the compressor clutch winding. This action energizes the compressor clutch, and the drive belt begins turning the compressor shaft.

ATC Operation

There are two categories of ATC systems: BCM-controlled systems and stand-alone systems that use their own computers.

If the vehicle uses a BCM-controlled ATC system, climate control is one of the primary functions of the BCM. The BCM monitors all system sensors and switches, compares the data with programmed instructions, and commands the actuators to provide accurate control of the system.

The climate control panel (CCP) contains a circuit board that translates driver inputs into electrical signals. The CCP and BCM communicate with each other over a data bus circuit.

The driver uses the panel buttons or knobs to set the desired temperature control by one-degree increments between a range of 65°F (18°C) and 85°F (29°C). If the buttons are held down, the set temperature will change until the button is released or is at the end of the scale.

Shop Manual
Chapter 14,
pages 531, 541

The driver can select temperature settings of 60°F (16°C) or 90°F (32°C) that will override the automatic temperature control. The other buttons enable operating mode selection. The display will indicate which mode has been selected.

Special BCM programming controls the ambient temperature sensor input. During idle and low-speed conditions, engine heat may influence the sensor. At vehicle speeds less than 20 mph (1.6 km/h), the BCM will use the last known good value of the ambient temperature sensor. Between 20 and 45 mph (1.6 and 72 km/h), the update occurs only after a 2-minute delay. At speeds above 45 mph (72 km/h), the update is continuous. The BCM monitors the air conditioning system through several sensors. High side temperature is monitored through a sensor in the pressure line. By monitoring the high side temperature, the BCM is capable of making calculations that translate into pressure. The calculations are based on the pressure-temperature relationship of R-12 or R-134a. The BCM will also monitor low side pressure in the same manner through a low side temperature sensor. The BCM also receives a signal from the low-pressure switch. The BCM will shut down the compressor clutch if system operation is not within set parameters.

The BCM uses bidirectional motors to adjust the blend, mode, and recirculation doors. The BCM uses different methods to determine the actual position of these doors. The most common method is to use a potentiometer feedback signal (Figure 14-13). The resistance of the

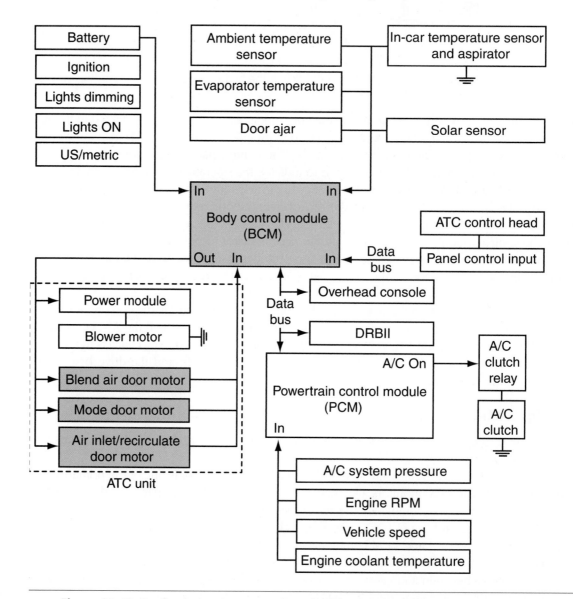

Figure 14-13 Feedback signals are sent from the blend air door, mode door, and recirc door motors to the BCM.

potentiometer in each door changes in relation to the door opening. The BCM senses the voltage drop across the potentiometer as its resistance changes. The second method of tracking door positions is counting the commutator pulses from the door motors. As the armature rotates in the coil windings of the motor, the fields change directions and an electrical pulse is generated. The pulse tells the BCM how far the door has moved.

A power module (Figure 14-14) controls the blower motor based on drive signals from the BCM. The body computer sends a pulse-width-modulated (PWM) signal to the power module and the power module amplifies the signals to provide the variable fan blower speed (Figure 14-15). If the auto button is pressed on the A/C control head, the blower speed is automatically controlled by the body computer, depending on the temperature setting, ambient temperature, in-vehicle temperature, amount of sunload, evaporator temperature, and blend air door position. The blower speed may be manually controlled by pressing the blower speed switch in the A/C control head.

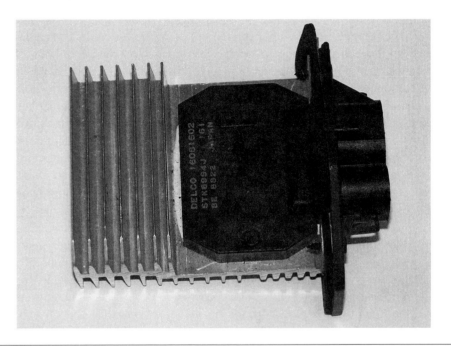

Figure 14-14 Power module amplifies BCM signals to control the blower motor speed.

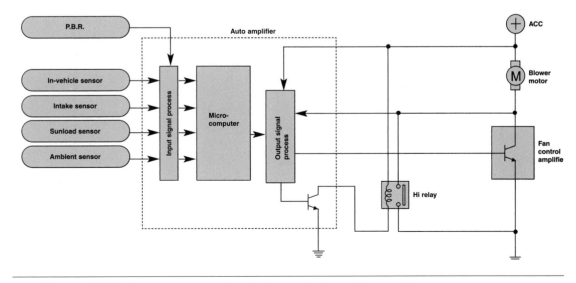

Figure 14-15 Blower motor circuit with fan control amplifier and high relay.

The BCM calculates a **program number,** which represents the amount of heating or cooling required to obtain the temperature set by the driver. A zero represents maximum cooling and 100 represents maximum heating. The program number is based on inputs from the control assembly (driver input), ambient temperature, and in-car temperature. Based on this number, air delivery mode, fan blower speed, and blend door positioning are determined.

To provide the proper mix of inside air temperature entering the passenger compartment, the BCM monitors the ambient and in-car temperatures, the average low side temperature, and the coolant temperature. These inputs are combined with the program number. The BCM commands the programmer to position the blend door for the correct temperature of incoming air. The programmer feedback potentiometer is also monitored by the BCM.

Blower speed is determined by a combination of the program number and driver input temperature. The CCP will signal the BCM over a data line (Figure 14-16). The signal is sent from the

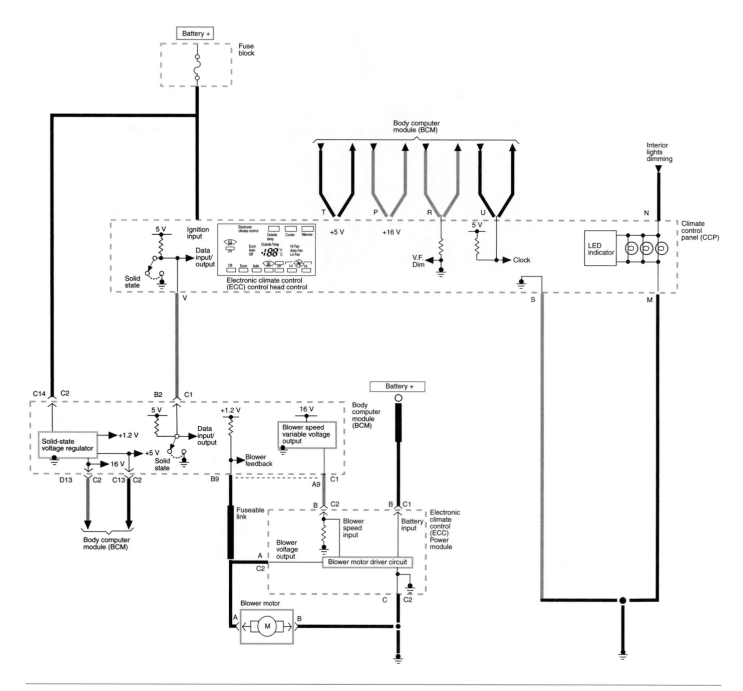

Figure 14-16 Blower control schematic of BCM-controlled ATC.

418

BCM to the power module. This signal is a constantly variable voltage proportional to blower speed. The power module will amplify the signal then apply it to the blower motor. The BCM monitors the blower speed through a feedback voltage from the motor.

The engine control module (ECM) performs air conditioner compressor clutch control through inputs from the BCM (Figure 14-17). The ECM cycles the compressor on and off based on input signals from the high side and low side temperature sensors to the BCM. In addition, the ECM will anticipate clutch cycling and will adjust engine idle speed accordingly.

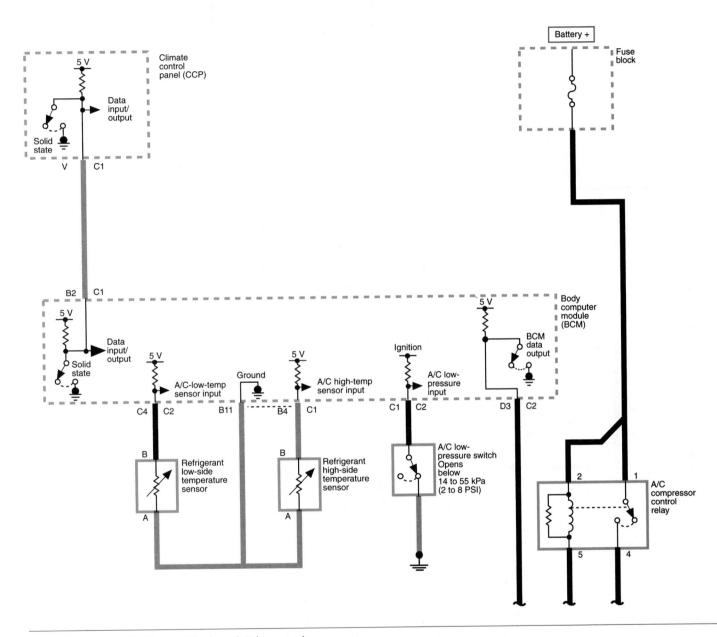

Figure 14-17 Air conditioning clutch controls.

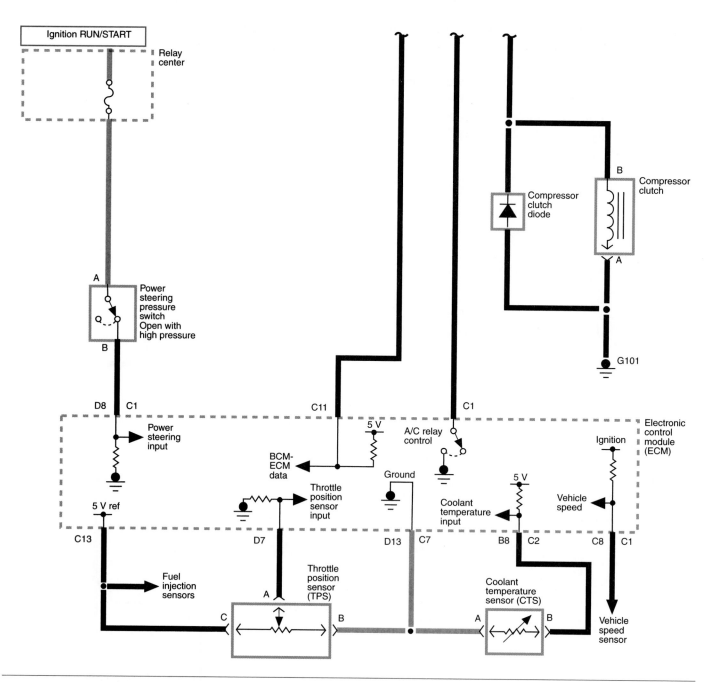

Figure 14-17 *(continued)*

The BCM monitors system inputs and feedback signals. If these voltage signals fall outside programmed parameters, the BCM will store a DTC.

Several ATC systems use a separate computer for the sole purpose of climate control. Usually this computer is integrated with the control panel (Figure 14-18). The control head receives inputs from ambient temperature, in-car temperature, and setting selections to determine position of the blend door, mode door, and fan blower speed.

Depending on the model of vehicle the system is installed on, the mode door and water flow actuators can be controlled by either vacuum or electric servomotors. All systems use an electrically operated blend door.

The control head signals a separate power module to provide appropriate blower speeds (Figure 14-19). The power module uses PWM to provide infinite fan speed between full off and full on.

Many manufacturers offer **dual climate control.** Dual climate control provides separate temperature settings for the driver and the front-seat passenger. This system is similar to previous systems except two blend doors are used to control separate temperature settings. Often this system uses infrared sensors to control comfort temperatures. This system is also referred to as automatic zone control (AZC).

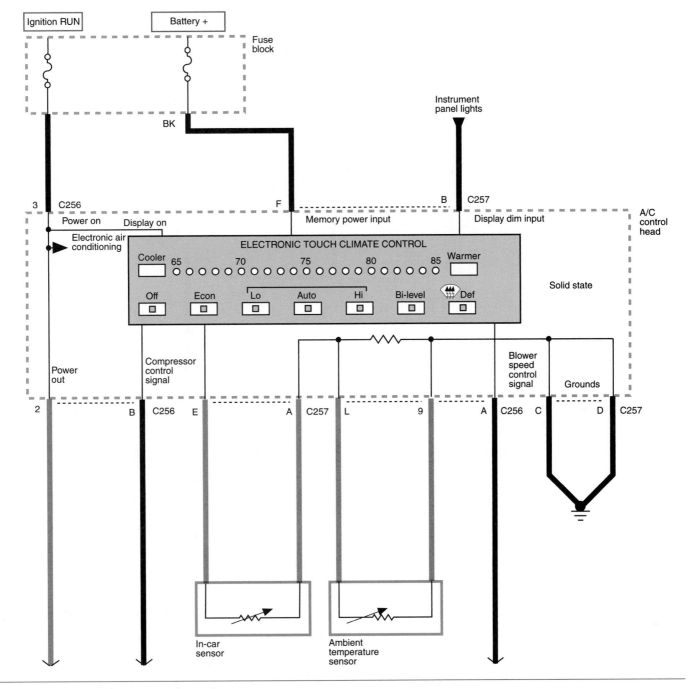

Figure 14-18 ATC wiring schematic.

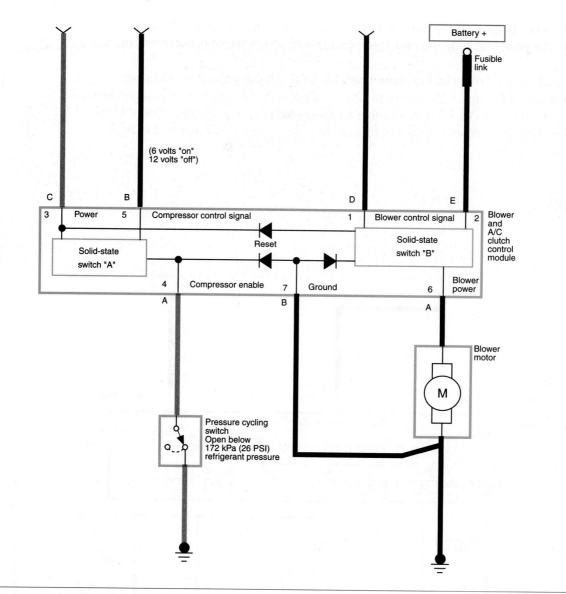

Figure 14-18 *(continued)*

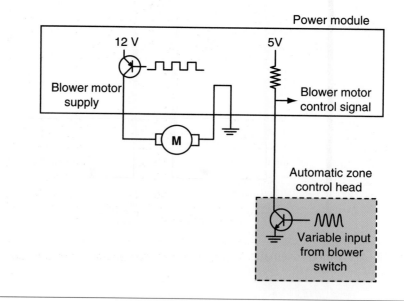

Figure 14-19 The control sends fan speed requests to the power module.

Figure 14-20 AZC control head.

The AZC control head (Figure 14-20) contains the input buttons and control unit. The front-seat occupants set their desired temperatures into the control head. Also, the control head contains two infrared sensors (located between the two temperature control dials). Unlike a system that use an in-vehicle temperature sensor to measure air temperature, the infrared sensors measure the surface temperature of the front-seat occupants. This means the system does not maintain a set air temperature but a set comfort temperature. The infrared sensors are used to maintain this set comfort level under changing conditions.

Based on the inputs from the driver and front-seat passenger, the control head will move the doors to the required position by sending a voltage signal to the actuator motor (Figure 14-21). The control head runs the actuators for a number of commutator pulses that correspond to the desired door position. The actuators will periodically recalibrate themselves at known zero and full travel conditions.

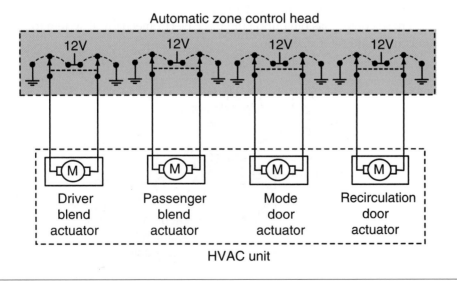

Figure 14-21 The control head uses a series of drivers to operate the bidirectional actuator motors.

Electronic Cruise Control Systems

Ralph R. Teetor was born August 17, 1890. At the age of 10, Ralph built a dynometer. At the age of 12, he designed and built his own gasoline-powered automobile. Also at age 12, he built a generator to supply electricity not only to his home but for every house on the block. His most famous automotive invention was the Speedostat, which is now known as cruise control. He also designed and patented one of the first automatic gear shifts. However, the story of Ralph R. Teetor becomes more amazing once it is learned that he was totally blind from the age of 5. In 1902, a newspaper reporter wrote a story on Ralph but never noticed his blindness.

Cruise control is a system that allows the vehicle to maintain a preset speed with the driver's foot off the accelerator. Cruise control was first introduced in the 1960s for the purpose of reducing driver fatigue. When engaged, the cruise control system sets the throttle position to maintain the desired vehicle speed.

Most cruise control systems are a combination of electrical and mechanical components. The components used depend on manufacturer and system design. However, the operating principles are similar.

Early cruise control systems were generally electromechanical systems (Figure 14-22). These systems used a transducer that received vehicle speed signals through the speedometer cable. Electrical signals from the control switch, brake switch, or clutch switch were sent to the transducer. In addition, the transducer received the engine manifold vacuum. It regulated the vacuum to the servo through the electrical signals received.

The servo controls the throttle plate position. It is connected to the throttle plate by a rod, bead chain, or Bowden cable. The servo maintains the set speed by receiving a controlled amount of vacuum from the transducer. When a vacuum is applied to the servo, the spring is compressed and the throttle plate is moved to increase speed (Figure 14-23). When the vacuum is released, the spring returns the throttle plate to reduce engine speed.

Electronic cruise control has built upon the basic electromechanical system to provide more accurate speed control and safer operation.

Electromechanical cruise control receives its name from two subsystems: the electrical and mechanical portion.

Some manufacturers combine the transducer and servo into one unit. They usually refer to this unit as a servomotor.

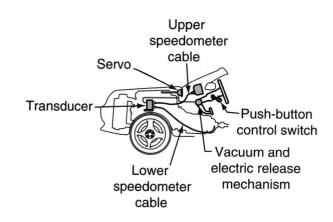

Figure 14-22 Components of typical electromechanical cruise control system.

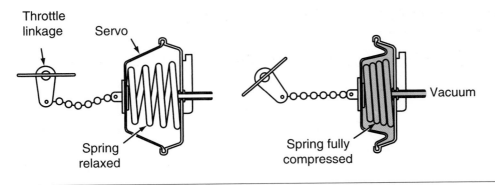

Figure 14-23 Cutaway view of the servo. Vacuum is used to compress the spring and open the throttle.

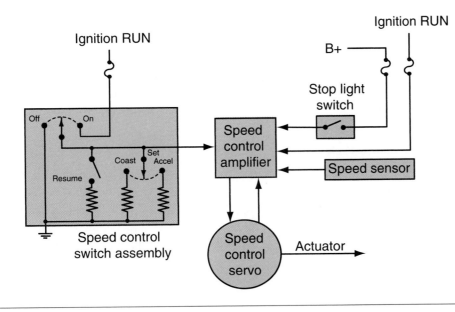

Figure 14-24 The control switch can be mounted on the turn signal stock or into the steering wheel. The switch is used to provide driver inputs for the system.

The electronic cruise control system uses an electronic module to operate the actuators that control throttle position (Figure 14-24). Other benefits include:

❑ More frequent throttle adjustments per second.

❑ More consistent speed increase/decrease when using the tap-up/tap-down feature.

❑ Greater correction of speed variation under loads.

❑ Rapid deceleration cutoff when deceleration rate exceeds programmed rates.

❑ Wheelspin cutoff when acceleration rate exceeds programmed parameters.

❑ System malfunction cutoff when the module determines there is a fault in the system.

Common Components

Common components of the electronic cruise control system include:

1. The control module: The module can be a separate cruise control module, the PCM, or the BCM. The operation of the systems are similar regardless of the module used.

Shop Manual
Chapter 14, page 555

425

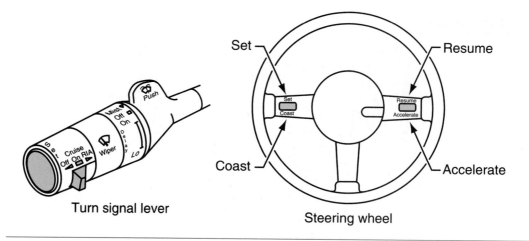

Set — | — Resume
Turn signal lever
Coast — | — Accelerate
Steering wheel

Figure 14-25 Block diagram of electronic cruise control system.

2. The control switch (Figure 14-25): Depending on system design, the control switch contacts apply the ground circuit through resistors. Because each resistor has a different value, a different voltage is applied to the control module. In some systems, the control switch will send a 12-volt signal to different terminals of the control module.

3. The brake or clutch switch.

4. Vacuum release switch.

5. Servo: The servo operates on a vacuum that is controlled by supply and vent valves. These operate from controller signals to solenoids.

Depending on system design, the sensors used as inputs to the control module include the vehicle speed sensor, servo position sensor, and throttle position sensor. Other inputs are provided by the brake switch, instrument panel switch, control switch, and the park-neutral switch.

The control module receives signals from the speed sensor and the control switch. When the vehicle speed is fast enough to allow cruise control operation and the driver pushes the SET button on the control switch, an electrical signal is sent to the controller. The voltage level received by the controller is set in memory. This signal is used to create two additional signals. The two signal values are set at 1/4 mph above and below the set speed. The module uses the comparator values to change vacuum levels at the servo to maintain set vehicle speed.

Three safety modes are operated by the control module:

1. Rapid deceleration cutoff: If the module determines that deceleration rate is greater than programmed values, it will disengage the cruise control system and return operation back over to the driver.

2. Wheelspin cutoff: If the control module determines that the acceleration rate is greater than programmed values, it will disengage the system.

3. System malfunction cutoff: The module checks the operation of the switches and circuits. If it determines there is a fault, it will disable the system.

The vacuum-modulated servo is the primary actuator. Vacuum to the servo is controlled by two solenoid valves: supply and vent. The vent valve is a normally open valve and the supply valve is normally closed (Figure 14-26). The servo receives signals from the controller to operate the solenoid valves to maintain a preset throttle position.

Principles of Operation

Shop Manual
Chapter 14, page 556

When the driver sends a SET signal to the controller, it sets the voltage signals received from the vehicle speed sensor (VSS) into memory. It then determines the high and low comparators.

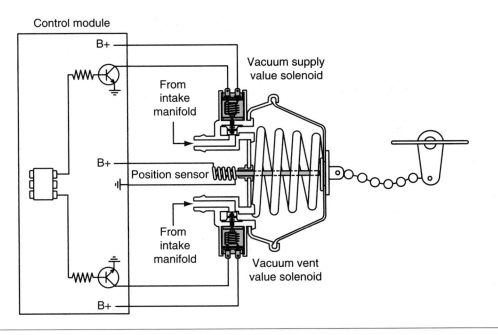

Control module

B+

Vacuum supply
value solenoid

From
intake
manifold

B+

Position sensor

From
intake
manifold

Vacuum vent
value solenoid

B+

Figure 14-26 Servo valve operation in electronic control system. The servo position sensor informs the controller of servo operation and position.

The controller energizes the supply and vent valves to allow manifold vacuum or atmospheric pressure to enter the servo. The servo uses the vacuum and pressure to move the throttle and maintain the set speed. The vehicle speed is maintained by balancing the vacuum in the servo. The vacuum used to move the servo may be an engine vacuum or may be supplied by a vacuum pump.

If the voltage signal from the VSS drops below the low comparator value, the control module energizes the supply valve solenoid to allow more vacuum into the servo and increases the throttle opening. When the VSS signal returns to a value within the comparator levels, the supply valve solenoid is de-energized.

If the VSS signal is greater than the high comparator value, the control module de-energizes the vent solenoid valve to release vacuum in the servo. The vehicle speed is reduced until the VSS signals are between the comparator values, at which time the control module will energize the vent valve solenoid again. This constant modulation of the supply and vent valves maintains vehicle speed.

During steady cruise conditions, both valves are closed and a constant vacuum is maintained in the servo.

Many manufacturers now use electronic throttle control (ETC) systems instead of cables to operate the throttle body. The ETC throttle body (Figure 14-27) uses a motor to actuate the throttle

Figure 14-27 An electronic throttle control throttle body.

plate. The driver's movement of the accelerator pedal is an input to the PCM. The PCM then directly controls the placement of the ETC throttle body plates. When ETC is used, cruise control servos are not necessary. Inputs and operation are the same as just described, except the output is to the ETC throttle body motor instead of the servo solenoids.

Laser Cruise Control Systems

One of the last improvements to the cruise control system is the use of lasers. The **laser radar sensor** determines the vehicle-to-vehicle distances and relational speeds. Like other cruise control systems, the laser system maintains a fixed speed that has been set by the driver. However, the system also allows the driver to set a distance between his vehicle and any vehicle in front of him in the same lane, along with the set speed function. If a vehicle ahead is traveling slower and the distance is closing between the vehicles, then the ETC throttle plate is closed to decelerate. If additional deceleration is necessary, then the transmission is shifted into a lower gear. If further deceleration is still necessary, the system controls the brake actuator to apply the brakes.

The system will continue to maintain the set distance between vehicles until the vehicle in front is no longer there (due to lane change). At this time the vehicle is accelerated slowly until the set vehicle speed is reached, and then the system will maintain operation of driving at the fixed speed.

A switch that is located on the steering wheel provides the driver with a selection of distances between his vehicle and the vehicle in front of him. Three distances are provided. The first button push sets the system to long, which is approximately 245 feet (75 m) between vehicles. The second button push selects a middle range of approximately 165 feet (50 m). A third button push sets the distance range to short, which is approximately 100 feet (30 m).

The laser radar sensor is mounted on the right side of the vehicle in the front grill (Figure 14-28). The sensor has three parts (Figure 14-29): the laser emitter that radiates laser rays forward, the laser receiving portion that receives the laser beams as they are reflected back by the vehicle that is ahead, and the processing unit that determines the length of time it takes for the reflected beams to return to the sensor; the unit calculates the distance to the vehicle ahead and the relative speed. This data is then transmitted to the distance control ECU (Figure 14-30).

The distance and relative speed information is sent to a control module (either a separate distance ECU or the PCM) that determines which vehicle to follow based on the information provided by the laser radar sensor. The control module will also calculate the target acceleration signals for following the vehicle. To maintain the set distance, the control module will send acceleration or deceleration requests. If necessary, it will also request a transmission downshift and brake application.

If brake control is necessary, the distance control module will determine a target deceleration rate. The maximum rate will be set at 0.20 G. The target deceleration rate is determined by current vehicle speed, the distance to the vehicle in front, and the relative speed. The brake apply request signal is sent to the antilock brake system (ABS) control module, which controls the brake actuator to apply the brakes. If the vehicle is not decelerating at a fast enough rate to avoid a collision, a warning buzzer is sounded to prompt the driver to apply the brakes.

If the vehicle is slowed down due to a slower vehicle in front, then the system will enter follow mode and match the speed of the vehicle that is ahead. With the speeds matched, the set distance is maintained. While in this mode, if either vehicle changes lanes, the system will request an acceleration mode. The throttle plates are opened to provide a gradual increase in speed until the set speed requested by the driver is reached. At this time the system enters fixed speed mode and maintains the set speed.

Figure 14-28 Laser radar sensor located on front of vehicle.

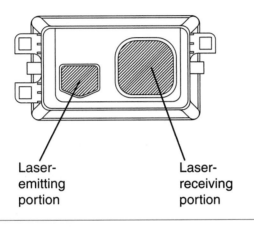

Laser-emitting portion

Laser-receiving portion

Figure 14-29 The laser radar sensor.

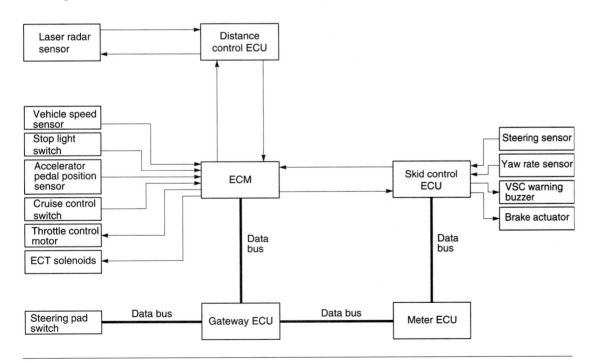

Figure 14-30 The laser radar system.

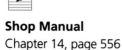

Memory Seats

The memory seat feature is an addition to the basic power seat system that allows the driver to program different seat positions that can be recalled at the push of a button. Most memory seat systems share the same basic operating principles. The difference is in programming methods and the number of positions that can be programmed.

The power seat system may operate in any gear position. However, the memory seat function will operate only when the transmission is in the PARK position. The purpose of the memory disable feature is to prevent accidental seat movement while the vehicle is being driven. In the PARK position, the seat memory module will receive a 12-volt signal that will enable memory operation. In any other gear selection, the 12-volt signal is removed and the memory function is disabled. This signal can come from the gear selector switch or the neutral safety switch (Figure 14-31).

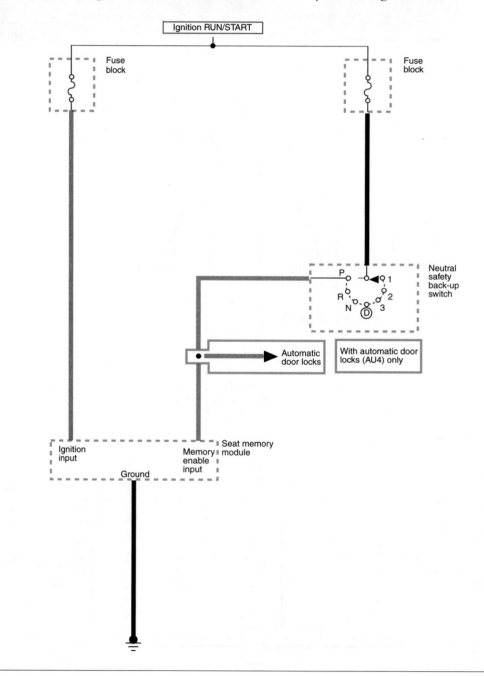

Figure 14-31 The neutral safety back-up switch signals the seat memory module when the transmission is in park.

Most systems provide for two seat positions to be stored in memory. Some systems allow for three positions by pushing both position 1 and 2 buttons together. With the seat in the desired position, depressing the SET button and moving the memory select switch to either the memory 1 or 2 position will store the seat position into the module's memory (Figure 14-32).

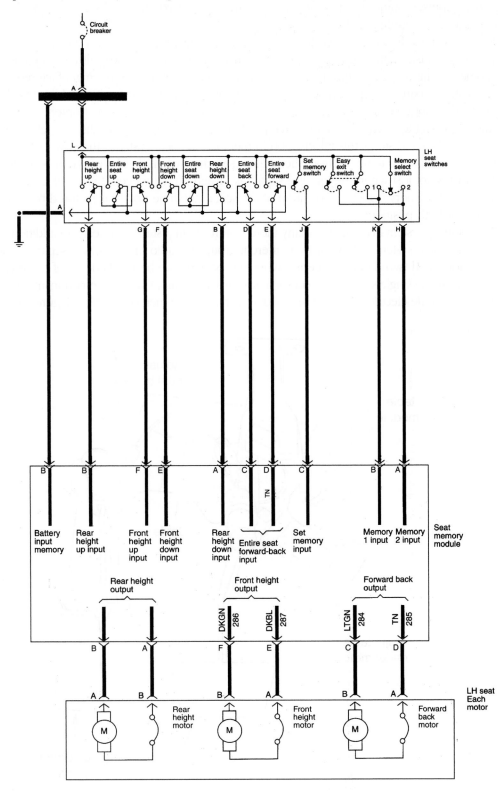

Figure 14-32 Memory seat circuit.

When the seat is moved from its memory position, the seat memory module transmits the voltage applied from the switch to the motors. The module counts the pulses produced by motor operation and then stores the number of pulses and direction of movement in memory. When the memory switch is closed, the module will operate the seat motors until it counts down to the preset number of pulses. Some systems use a potentiometer or a two-wire Hall sensor to monitor seat position instead of counting pulses.

Some systems offer an easy exit feature that is an additional function of the memory seat; the feature provides easier entrance and exit of the vehicle by moving the seat all the way back and down. some systems also move the steering wheel up and to full retract. Some manufacturers use an easy exit switch as part of the power seat switch assembly. When the easy exit switch is closed, voltage is applied to both memory 1 and 2 inputs of the module. This signal is interrupted by the module to move the seat to its full down and back position. As the seat moves to the easy exit position, it counts the pulses and stores this information in memory. In some systems, the easy exit feature is activated when the key is removed from the ignition switch or when the driver's door is opened.

Memory is not lost when the ignition switch is turned off. However, it is lost if the battery is disconnected. If memory is lost, the position of the seat at the time power is restored becomes set in memory for both positions.

The memory system may include many more features than just moving seats. In addition to the seat positions, some systems allow for two different groups of radio presents, two separate outside mirror positions, and two different steering wheel tilt positions to be recalled. All of these can be recalled by using the single switch and, on some systems, the remote keyless entry transmitter. Some systems even include electrically adjustable pedals as part of the memory system (Figure 14-33).

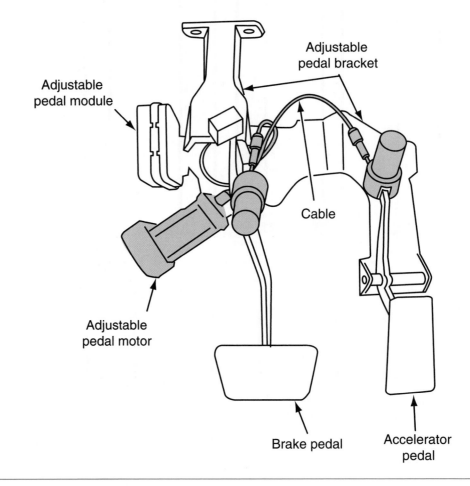

Figure 14-33 Electrically adjustable pedal assembly.

Climate-Controlled Seats

Many manufacturers provide the option of **climate-controlled seats.** Climate-controlled seats provide additional comfort by heating and/or cooling the seat cushion and seat back.

Figure 14-34 is a schematic of a heated seat system that uses heating element grids. This heated seat system uses four heated seat elements. Two elements are integral to each seat, one in the seat back and the other in the seat cushion. The heated seat module (HSM) contains the control logic and software for the heated seat system. Two heated seat switches are used, one for each heated seat. The cabin compartment node (CCN) is part of the instrument cluster on this vehicle and is the link between the heated seat switches and the HSM.

The heated seat system operates on battery current received through a fused ignition switch output. Since ignition feed is used, the heated seat system will operate only when the ignition switch is in the RUN position. When either of the heated seat switches is depressed, the CCN receives a multiplexed resistance signal. This requested heating level is then sent to the HSM over the data bus. The HSM controls the 12-volt output to the heating elements through the use of high side drivers (HSDs) based on the heat level requested. The HSD uses PWM to control the current flow through the seat elements. The carbon fiber heated seat elements consist of multiple heating circuits wired in parallel. The heated seat elements are located between the leather trim cover and the seat cushion. As electrical current passes through the heated seat element, the resistance of the wire used in the element converts the electrical energy into heat energy.

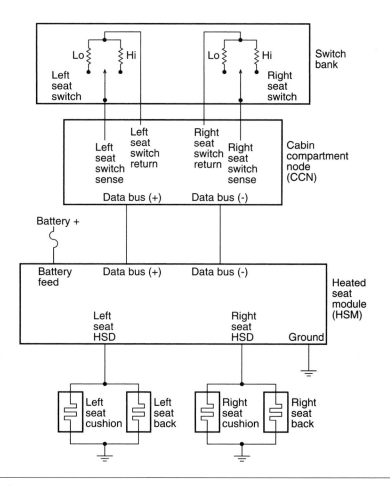

Figure 14-34 Heat seat system using seat grids.

The heat is then radiated through the seat cushion and seat back trim covers. Since the element grid is wired in parallel, if an open occurs in one or more of the individual carbon fiber circuits, the others will continue to operate.

When the switch selection indicates a high-temperature heating request, the HSM will provide a boosted heat level during the first four minutes of operation. During this time, the HSM will use a 95% duty cycle. The heat output then drops to the normal high-temperature level of a 30% duty cycle. This will maintain a temperature of about 107.6°F (42°C). If high-level heating is activated, the HSM will automatically switch to the low level after two hours of continuous operation. The low heat level is maintained using a 15% duty cycle to hold the seat temperature to about 100.4°F (38°C). The HSM will turn off operation of the low heat level after 2 hours.

The HSDs in the HSM monitor the operation of the heater element circuits. The HSM will turn off the heating elements if it detects an open or short in the heating element circuit.

Some systems will use a thermistor to measure the seat temperature. This system wil energize the seat elements at 100% duty cycle until the desired heat level is reached. Once it is reached, the HSM will stop current flow. If the temperature of the seat drops a programmed amount, the HSM will re-energize the seat elements. This process is repeated throughout the heated seat operation.

Some manufacturers use air flow to warm or cool the seat surface. A **Peltier element** operates similar to the way a bimetal switch operates. The element consists of two different types of metals. The two metals are joined together. This joint area will generate or absorb heat when an electric current is applied to the element at a specified temperature. The Peltier element is integral to the climate controller (Figure 14-35). The climate controller cools or warms the airflow from the climate control fan motor based on the climate control ECU activation. The ECU output is based on the setting selection of the control switch. The control switches provide for cool air in three stages, airflow, and warm air in three stages. A temperature sensor is used to monitor the surface temperature of the seat back and cushion.

The climate control fan motor provides airflow to the seat cushion and seat back. The air flow passes through grooves in the seat pad, is passed to the nonwoven cloth layer, and is dissipated through the seat cover (Figure 14-36). The Peltier element will either heat or cool the airflow as it passes through the controller.

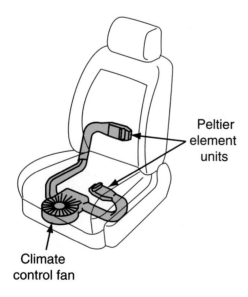

Peltier element units

Climate control fan

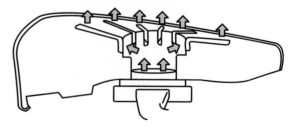

The airflow passes through the grooves on the seat pad surface

Figure 14-35 Climate-controlled seat using air. The Peltier element heats or cools the air.

Figure 14-36 The air flow is sent through the seat cushion.

Electronic Sunroof Concepts

Shop Manual
Chapter 14, page 561

Many manufacturers have introduced electronic control of their electric sunroofs. These systems incorporate a pair of relay circuits and a timer function into the control module. Although there are variations between manufacturers, the systems discussed here provide a study of the two basic types of systems.

Electronically Controlled Toyota Sunroof

Refer to schematic (Figure 14-37) of a sunroof control circuit used by Toyota. The movement of the sunroof is controlled by the motor that operates a drive gear. The drive gear either pushes or pulls the connecting cable to move the sunroof.

Motor rotation is controlled by relays that are activated according to signals received from the slide, tilt, and limit switches. The limit switches are operated by a cam on the motor (Figure 14-38).

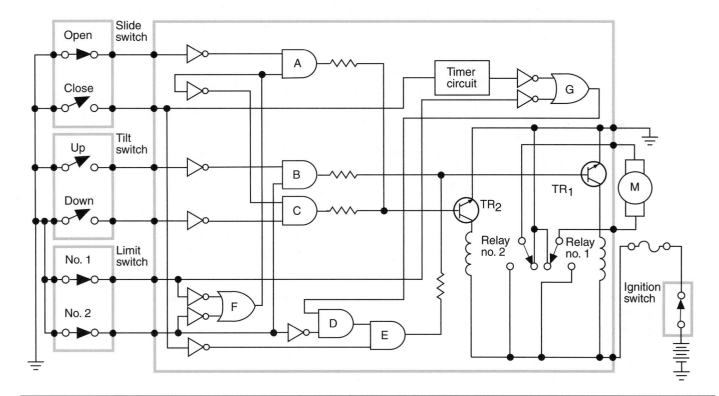

Figure 14-37 Toyota sunroof circuit using electronic controls.

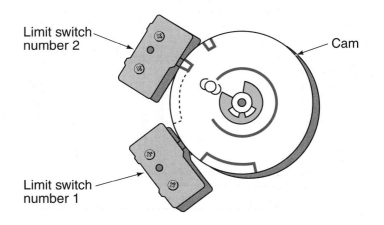

Figure 14-38 The limit switches operate off a cam on the motor.

AUTHOR'S NOTE: The schematics used to explain the operation of the Toyota sunroof use logic gates. If needed, refer to Chapter 12 of this manual to review the operation of the gates.

The logic gates of this system operate on the principle of **negative logic,** which defines the most negative voltage as a logical 1 in the binary code. When the slide switch is moved to the OPEN position, either limit switch 1 or both limit switches are closed (Figure 14-39). Limit switches 1 and 2 provide a negative side signal to the OR gate labeled F. The output from gate F is sent to gate A. Gate A is an AND gate, requiring input from gate F and the open slide switch. The output signal from gate A is used to turn on TR_2. This provides a ground path for the coil in relay 2. Battery voltage is applied to the motor through relay 2; the ground path is provided through the de-energized relay 1. Current is sent to the motor as long as the OPEN switch is depressed. If the OPEN switch is held in this position too long, a clutch in the motor disengages the motor from the drive gear.

Operation of the system during closing depends on how far the sunroof is open. If the sunroof is open more than 7.5 inches and the slide contact is moved to the CLOSE position, an input signal is sent to gate E (Figure 14-40). The other input signal required at gate E is received from the limit switches. The limit switch 1 signal passes through the OR gate G to the AND gate D. Limit switch 2 provides the second signal required by gate D. The output signal from D is the second input signal required by gate E. The output signal from E turns on TR_1. This energizes relay 1 and reverses the current flow through the motor. The motor will operate until the slide switch is opened or limit switch 2 opens.

If the sunroof is open less than 7.5 inches and the slide switch is placed in the CLOSE position, the timer circuit is activated (Figure 14-41). The CLOSE switch signals the timer and provides an input signal to gate E. Limit switch 1 is open when the sunroof is opened less than 7.5 inches. The second input signal required by gate D is provided by the timer. The timer is activated for 0.5 second. This turns on TR_1 and operates the motor for 0.5 second, or long enough for rotation of the motor to close limit switch 1. When limit switch 1 is closed, the operation is the same as described when the sunroof is closed after it is more than 7.5 inches open.

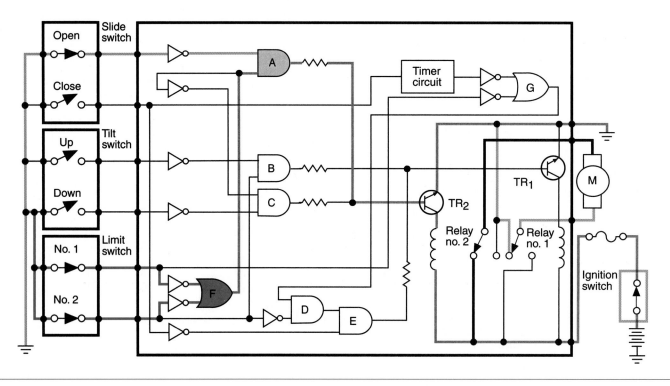

Figure 14-39 Circuit operation when the switch is in the OPEN position.

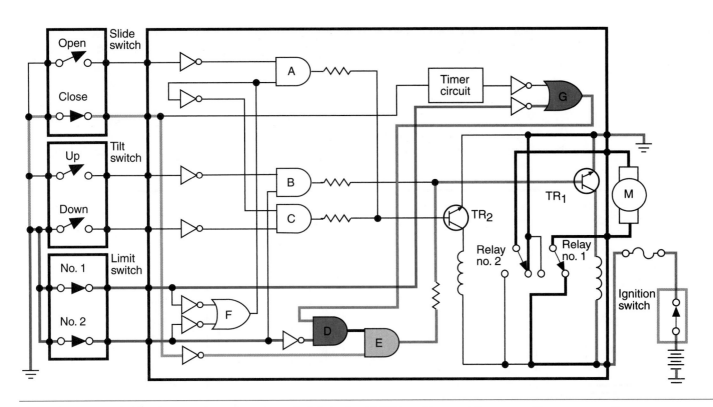

Figure 14-40 Circuit operation when the switch is in the CLOSE position and the sunroof is open more than 7.5 in. (19 cm).

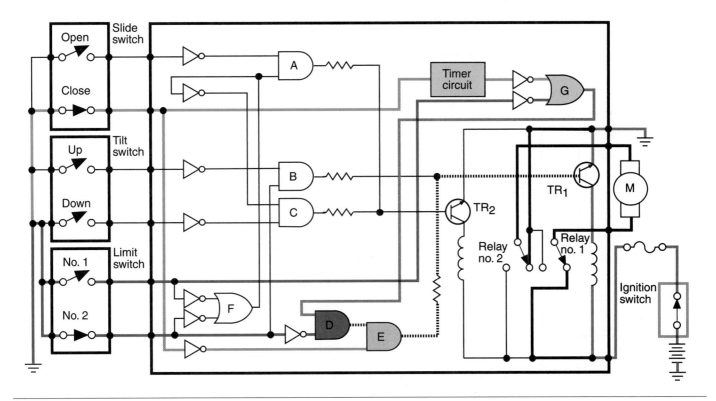

Figure 14-41 Circuit operation when the switch is in the CLOSE position and the sunroof is open less than 7.5 in. (19 cm).

When the tilt switch is located in the UP position, a signal is imposed on gate B (Figure 14-42). This signal is inverted by the NOT gate and is equal to the value received from the opened number 2 limit switch. The output signal from gate B turns on TR_1, which energizes relay 1 to turn on the motor. The motor clutch will disengage if the switch is held in the closed position longer than needed.

When the tilt switch is placed in the DOWN position, a signal is imposed on gate C (Figure 14-43). The second signal to gate C is received from the limit switches (both are open) through gate F. The signal from gate F is inverted by the NOT gate and is equal to that from the

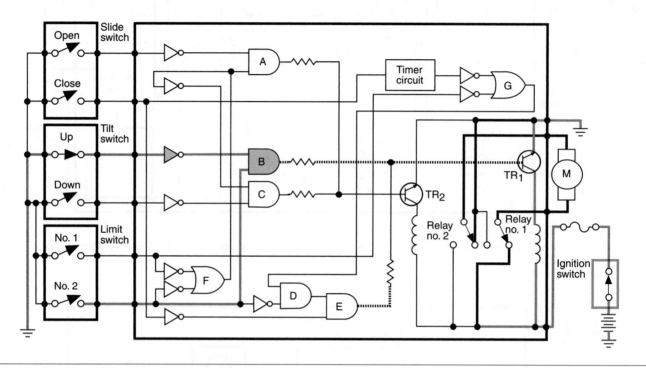

Figure 14-42 Circuit operation in the TILT UP position.

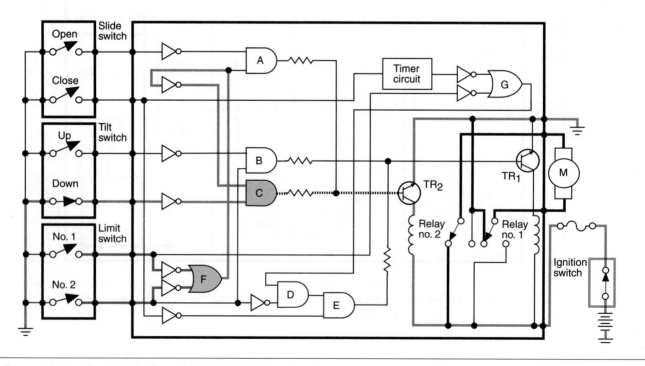

Figure 14-43 Circuit operation during TILT DOWN.

DOWN switch. The output signal from gate C turns on TR$_2$ and energizes relay 2 to lower the sunroof. If the DOWN switch is held longer than necessary, limit switch 1 closes. When this switch is closed, the signals received by gate F are not opposite. This results in a mixed input to gate C and turns off the transistor.

Electronically Controlled General Motors' Sunroof

See the schematic (Figure 14-44) of the sunroof system used on some GM model vehicles. The timing module uses inputs from the control switch and the limit switches to direct current flow to the motor. Depending on the inputs, the relays will be energized to rotate the motor in the proper direction. When the switch is located in the OPEN position, the open relay is energized, sending current to the motor (Figure 14-45). The sunroof will continue to retract as long as the switch is held in the OPEN position. When the sunroof reaches its full open position, the limit switch will open and break the circuit to the open relay.

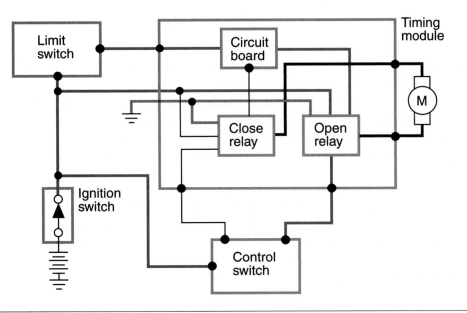

Figure 14-44 Block diagram of the GM sunroof.

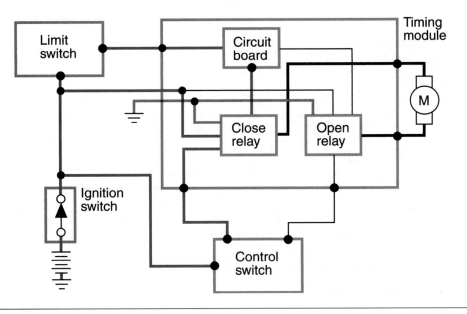

Figure 14-45 Sunroof circuit operation when the control switch is placed in the OPEN position.

Placing the switch in the CLOSE position will energize the close relay. The current sent to the motor is in the opposite direction to close the sunroof. If the close switch is held until the sunroof reaches the full closed position, the limit switch will open.

Shop Manual
Chapter 14, page 562

Antitheft Systems

A vehicle is stolen in the United States every 26 seconds. In response to this problem, vehicle manufacturers are offering antitheft systems as optional or standard equipment. These systems are deterrents designed to scare off would-be thieves by sounding alarms and/or disabling the ignition system. The illustration (Figure 14-46) shows many of the common components that are used in an antitheft system. These components include:

1. An electronic control module.
2. Door switches at all doors.
3. Trunk key cylinder switch.
4. Hood switch.
5. Starter inhibitor relay.
6. Horn relay.
7. Alarm.

In addition, many systems incorporate the exterior lights into the system. The lights are flashed if the system is activated.

For the system to operate, it must first be **armed.** This is done when the ignition switch is turned off and the doors are locked. When the driver's door is shut, a security light will illuminate for approximately 30 seconds to indicate that the system is armed and ready to function. If any other door is open, the system will not arm until it is closed. Once armed the system is ready to detect an illegal entry.

The control module monitors the switches. If the doors or trunk are opened or the key cylinders are rotated, the module will activate the system. The control module will sound the alarm and flash the lights until the timer circuit has counted down. At the end of the timer function, the system will automatically rearm itself.

Some systems use ultrasonic sensors that will signal the control module if someone attempts to enter the vehicle through the door or window. The sensors can be placed to sense the parameter of the vehicle and sound the alarm if someone enters within the protected parameter distance.

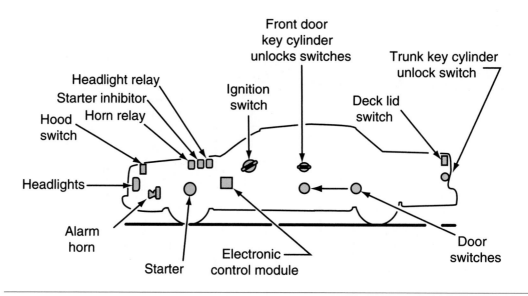

Figure 14-46 Typical components of an antitheft system.

The system can also use current sensitive sensors that will activate the alarm if there is a change in the vehicle's electrical system. The change can occur if the courtesy lights come on or if an attempt is made to start the engine.

The following systems are provided to give you a sample of the types of antitheft systems used. Figure 14-47 illustrates an antitheft system that uses a separate control module and an inverter relay. If the system is triggered, it will sound the horn, flash the low-beam headlights, the taillights, and parking lamps, and disable the ignition system.

The arming process is started when the ignition switch is turned off. Voltage provided to the module at terminal K is removed. When the door is opened, a voltage is applied to the courtesy lamp circuit through the closed switch to terminal 2 of the inverter relay. This voltage energizes the inverter relay and provides a ground for module terminal J. This signal is used by the control module to provide an alternating ground at terminal D, causing the indicator lamp to blink. The flashing indicator light alerts the driver that the system is not armed. When the door lock switch is placed in the LOCK position, battery voltage is applied to terminal G of the module. The module uses this signal to apply a steady ground at terminal D, causing the indicator light to stay on continuously. When the door is closed, the door switch is opened. The opened door switch de-energizes the inverter relay coil. Terminal J is no longer grounded, and the indicator light goes out after a couple of seconds.

To disarm the system, one of the front doors must be opened with a key or by pressing the correct code into the keyless entry keypad. Unlocking the door closes the lock cylinder switch and grounds terminal H of the module. This signal disarms the system.

Once the system is armed, if terminal J and C receive a ground signal, the control module will trigger the alarm. Terminal C is grounded if the trunk tamper switch contacts close. Terminal J is grounded when the inverter relay contacts are closed. The inverter relay is controlled by the doorjamb switches. If one of the doors is opened, the switch closes and energizes the relay coil. The contacts close and ground is provided to terminal J.

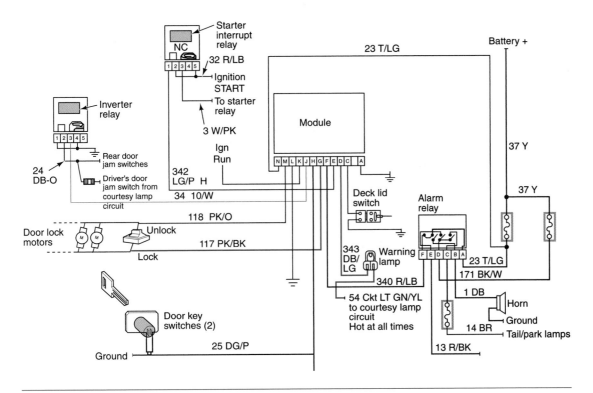

Figure 14-47 Circuit schematic of Ford's antitheft system.

When the alarm is activated, a pulsating ground is provided at module terminal F. This pulsating ground energizes and de-energizes the alarm relay. As the relay contacts open and close, a pulsating voltage is sent to the horns and exterior lights.

At the same time, the start interrupt circuit is activated. The start interrupt relay receives battery voltage from the ignition switch when it is in the START position. When the alarm is activated, the module provides a ground through terminal E, causing the relay coil to be energized. The energized relay opens the circuit to the starter system, preventing starter operation.

Often the BCM will control the functions of the antitheft system. In this case the inputs and outputs are wired to the BCM (Figure 14-48). The BCM will monitor the arming process and will then trigger the alarm if an unauthorized entry is attempted. At this time the BCM will control the exterior lamps to cause them to flash; it also will cycle the horn on and off. The BCM will also send a data bus message to the PCM not to start the engine. The PCM is programmed that it must receive a data bus message that it is alright to start. If the data bus circuit should fail, the engine may not start, since this message cannot be received.

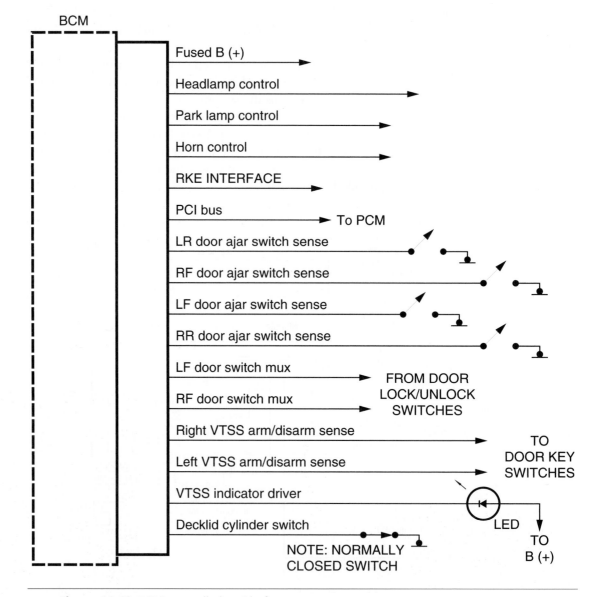

Figure 14-48 BCM-controlled antitheft system.

Systems that incorporate an intrusion monitor will detect movement of a person or object inside the passenger compartment. The intrusion monitor uses a single sensor that transmits 40-kHz ultrasonic sound waves. The sensor will also receive the ultrasonic sound waves. If an object moves within the coverage area of the sensor, the received ultrasonic signals are distorted. This distortion is detected and processed by the module, which then signals the antitheft system to trigger.

Immobilizer Systems

Shop Manual
Chapter 14, page 565

The **immobilizer system** is designed to provide protection against unauthorized vehicle use by disabling the engine if an invalid key is used to start the vehicle or an attempt to hot-wire the ignition system is made. Manufacturers have several different names for this system. The following are examples of system operation.

The primary components of the system include the immobilizer module, ignition key with transponder chip, the PCM, and an indicator lamp. The immobilizer module is either mounted to the steering column or in the instrument panel. It may include an integral antenna that surrounds the ignition switch lock cylinder (Figure 14-49). If the module is remote mounted in the instrument panel, then it is connected to an antenna by a cable.

The system includes special keys that have a transponder chip under the covering (Figure 14-50). Most systems are capable of recognizing up to eight different keys. Any additional keys that are to be used with the vehicle require programming to the immobilizer module. Most systems provide a procedure that allows the customer to program additional keys if they have two valid immobilizer keys that are programmed to the vehicle already.

System Operation

The immobilizer module contains a radio frequency transceiver and a microprocessor. When the ignition switch is placed in the RUN position, the immobilizer module begins to transmit a radio frequency signal to the transponder in the key. The transponder in the key then sends its coded message to the immobilizer module. If the message properly identifies the key as being valid, the immobilizer module sends a message over the data bus to the PCM indicating that the engine may be started. If the response received from the key transponder is missing or identifies the key as invalid, the immobilizer module sends an "invalid key" message to the PCM over the data bus.

AUTHOR'S NOTE: The default condition in the PCM is "invalid key." If the PCM does not receive any messages from the immobilizer module, the engine is prevented from starting.

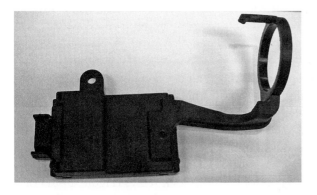

Figure 14-49 The immobilizer module with halo antenna.

Figure 14-50 The immobilizer key has an internal transponder chip. The key cover was cut off for this illustration.

AUTHOR'S NOTE: When the system is preventing the engine from running, it will allow the engine to start and run for about 2 seconds and then will shut the engine off. Some systems will prevent the starter from engaging after a set number of attempts with an invalid key are made. In this case, the vehicle will not recover until a valid key is used.

The immobilizer module is programmed with a unique secret key code and also retains in memory the unique ID number of all keys that are programmed to the system. The secret key code is transmitted to the keys during the programming function and is stored in the transponder. In addition, the secret key code is programmed into the PCM. Another identification code called a PIN is used to gain secured access to the immobilizer module for service. The immobilizer module also is programmed with the vehicle identification number (VIN). All messages transmitted by the immobilizer module are scrambled to reduce the possibility of unauthorized immobilizer module access or disabling.

The immobilizer module also sends indicator lamp status messages. The indicator lamp will normally illuminate for 3 seconds for a bulb check when the ignition switch is first placed in the RUN position. After the bulb check is complete, the lamp should go out. If the lamp remains on after the bulb check, this indicates that the immobilizer module has detected a system malfunction or that the system has become inoperative. If the lamp flashes after the bulb check is completed, this indicates that an invalid key is detected or that a key-related fault is present.

Each key has a unique transponder identification code that is permanently programmed into it. When a key is programmed into the immobilizer module, the transponder identification code is then stored in the immobilizer's memory. In addition, the key learns the secret key code from the immobilizer module and permanently programs it into its transponder memory. For the engine to start and run, all of the following must be in place:

❏ Each key's transponder ID must be programmed into the immobilizer module.

❏ The immobilizer's secret key code must be programmed into each key.

❏ The VIN number programmed in the immobilizer module must match the VIN in the PCM.

❏ The data bus network must be intact to allow messages to be sent and received between the immobilizer module and the PCM.

❏ The immobilizer module and the PCM must have properly functioning power and ground circuits.

GM Pass-Key Antitheft System

In 1996, GM introduced an immobilizer system that does not use a chip in the key. This system has a Hall-effect sensor in the key cylinder that measures the magnetic properties of the key as it is inserted into the cylinder. The cut pattern of every key has its own magnetic identity. If the wrong key is inserted into the lock cylinder, the car will not start, even if the lock cylinder turns.

Automatic Door Locks

Shop Manual
Chapter 14, page 567

Automatic door locks (ADL) is a passive system used to lock all doors when the required conditions are met. The ADL system is an additional safety and convenience system that uses the existing power door function. Most systems lock the doors when the gear selector is placed in drive, the ignition switch is in RUN, and all doors are shut. Some systems will lock the doors when the gear shift selector is passed through the reverse position; others do not lock the doors unless the vehicle is moving 15 mph or faster.

The system may use the body computer to control the door lock relays (Figure 14-51) or a separate controller (Figure 14-52). The controller (or body computer) takes the place of the door lock switches for automatic operation.

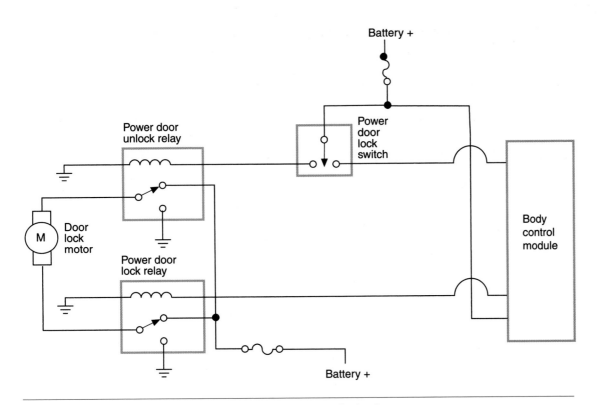

Figure 14-51 Automatic door lock system utilizing the body computer.

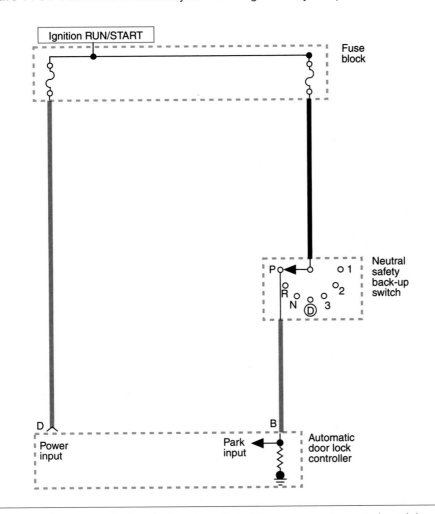

Figure 14-52 Automatic door lock system that utilizes a separate control module.

When all of the doorjamb switches are open (doors closed), the ground is removed from the doorjamb input circuit to the controller (Figure 14-53). This signals the controller to enable the lock circuit. When the gear selection is moved from the PARK position, the neutral safety switch removes the power signal from the controller. The controller sends voltage through the LH seat switch to the lock relay coil. Current is sent through the motors to lock all doors.

When the gear selector is returned to the PARK position, voltage is applied through the neutral safety switch to the controller. The controller then sends power to the unlock relay coil to reverse current flow through the motors (Figure 14-54).

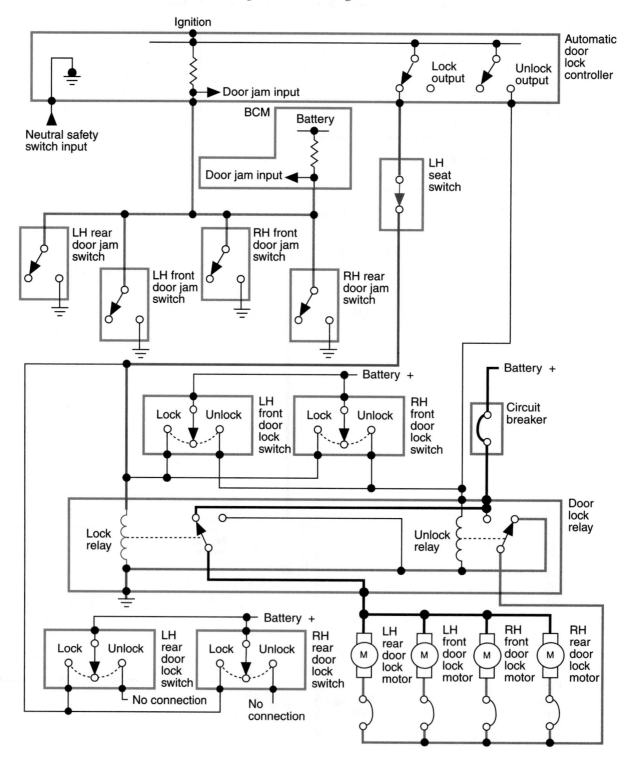

Figure 14-53 Automatic door lock system circuit schematic indicating operation during the lock procedure.

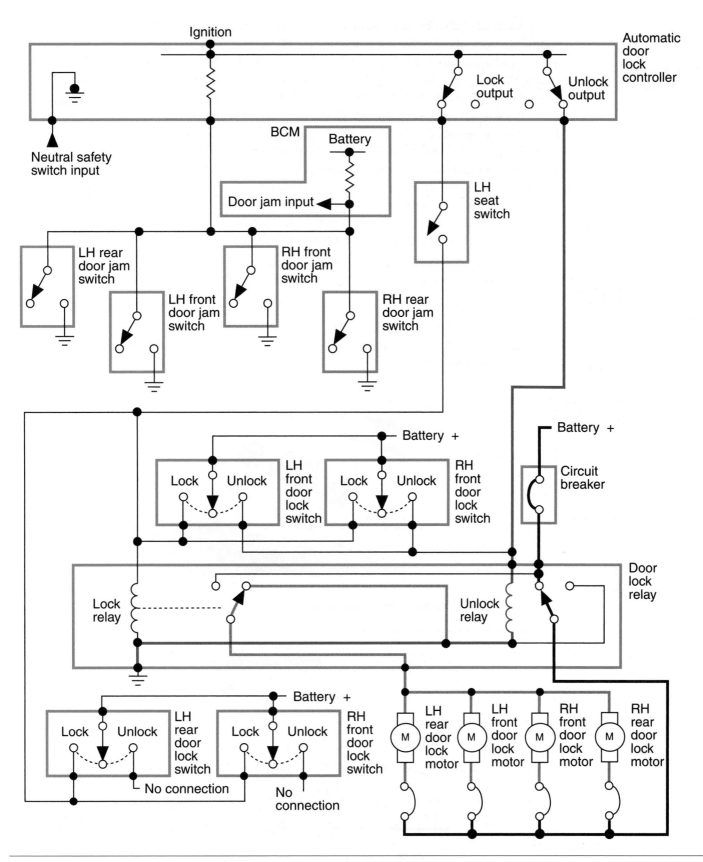

Figure 14-54 Circuit operation when the system is unlocking the doors.

Keyless Entry

The keyless entry system allows the driver to unlock the doors or the deck lid (trunk) from outside the vehicle without the use of a key. The main components of the keyless entry system are the control module, a coded-button keypad located on the driver's door, and the door lock motors.

The keypad consists of five normally open, single-pole, single-throw switches. Each switch represents two numbers: 1-2, 3-4, 5-6, 7-8, and 9-0 (Figure 14-55).

The keypad is wired into the circuit to provide input to the control module (Figure 14-56). The control module is programmed to lock the doors when the 7-8 and 9-0 switches are closed at the same time. The driver's door can be unlocked by entering a five-digit code through the keypad. The unlock code is programmed into the controller at the factory. However, the driver may enter a second code. Either code will operate the system.

In addition to the aforementioned functions, the keyless entry system also:

1. Unlocks all doors when the 3-4 button is pressed within 5 seconds after the five-digit code has been entered.

2. Releases the deck lid lock if the 5-6 button is pressed within 5 seconds of code entry.

3. Activates the illuminated entry system if one of the buttons is pressed.

4. Operates in conjunction with the automatic door lock system and may share the same control module.

Figure 14-55 Keyless entry system keypad.

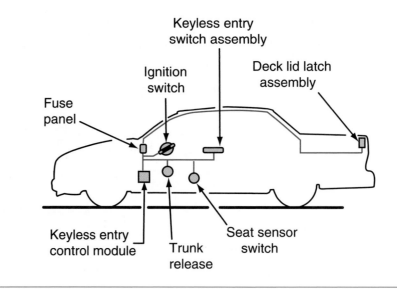

Figure 14-56 Components of a typical keyless entry system.

See the schematic (Figure 14-57) of the keyless entry system used by Ford. When the 7-8 and 9-0 buttons on the keypad are pressed, the controller applies battery voltage to all motors through the lock switch (Figure 14-58).

When the five-digit code is entered, the controller closes the driver's switch to apply voltage in the opposite direction to the driver's door motor (Figure 14-59). If the driver presses the 3-4 button, the controller will apply reverse voltage to all motors to unlock the rest of the doors.

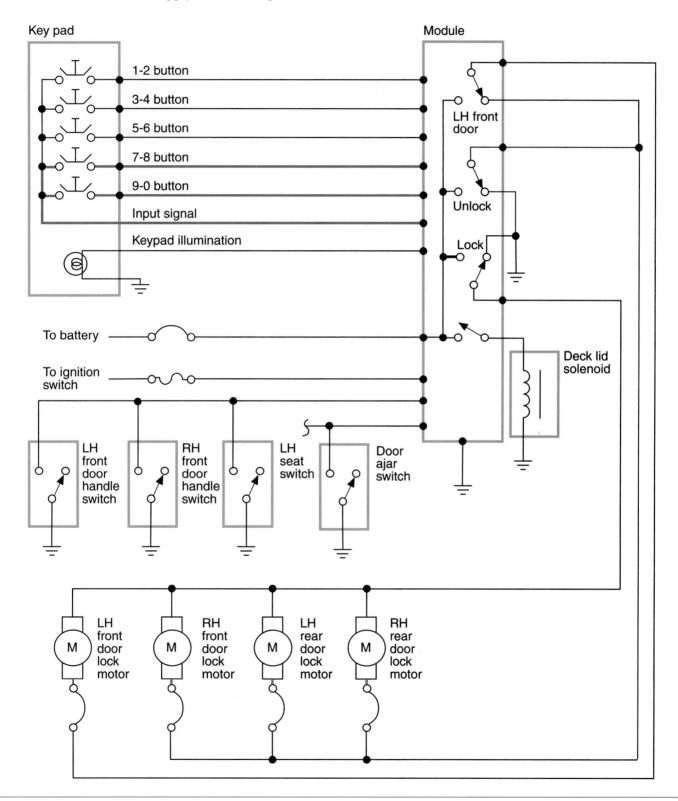

Figure 14-57 Simplified keyless entry system schematic.

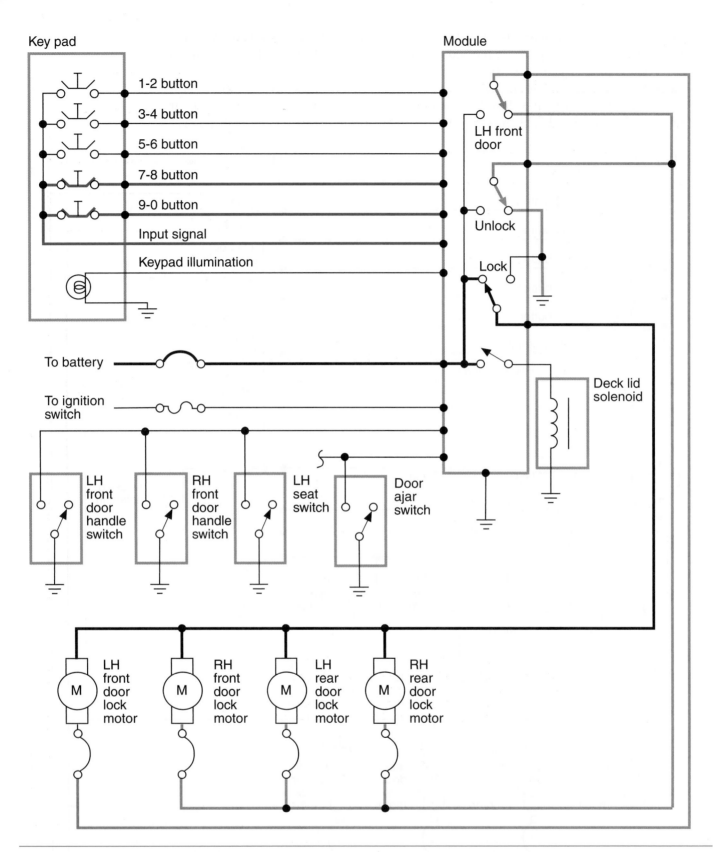

Figure 14-58 Circuit operation when the 7-8 and 9-0 buttons are pressed to lock all doors.

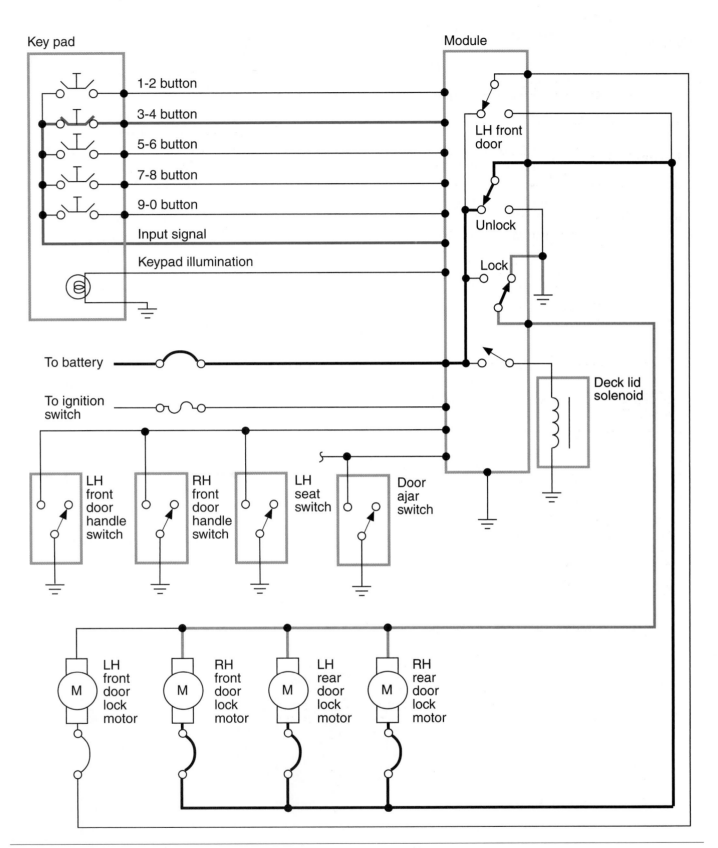

Figure 14-59 Circuit operation when the correct five-digit code is entered to unlock the driver's door.

Remote Keyless Entry

Many vehicles are equipped with a remote keyless entry system that is used to lock and unlock the doors, turn on the interior lights, and release the trunk latch. A small receiver is installed in the vehicle. The transmitter assembly is a hand-held item attached to the key ring (Figure 14-60). Pressing a button on a hand-held transmitter will allow operation of the system from a distance of 25 to 50 feet (7.6 to 15.2 m). When the UNLOCK button is pressed, the driver's door unlocks and the interior lights are illuminated. If a theft deterrent system is installed on the vehicle, it is also disarmed when the unlock button is pressed. A driver exiting the vehicle can activate the door locks and arm the security system by pressing the lock button. Many transmitters also have a third button for opening the deck lid.

The system operates at a fixed radio frequency. If the unit does not work from a normal distance, check for two conditions: weak batteries in the remote transmitter or a stronger radio transmitter close by (radio station, airport transmitter, etc.).

Keyless Start

An enhancement to the remote keyless entry system is the **keyless start system.** This system allows the vehicle to be started without the use of an ignition key. Early versions of these systems use the key fob to unlock the vehicle doors and disarm the antitheft system. At the same time, a message is sent that the engine can be started. Instead of the ignition key, the pointed end of the fob is inserted into the ignition switch. This activates an infrared data exchange that unlocks the steering column and starts the engine.

Newer versions have improved on this. They no longer require the use of a key fob to unlock the doors or to start the engine. The following is the Keyless Go system used on Mercedes-Benz vehicles.

This system will use a transmitter card (Figure 14-61), signal acquisition and actuation modules (SAMs) that are installed throughout the vehicle, an electronic ignition switch (EIS) (Figure 14-62), and a total of seven antennas. There are two antennas on the left side, two on the right side, and three in the trunk area.

The Keyless Go module will communicate with the transmitter card at a frequency of 125 kHz. In order to receive messages, the transponder card must be within 5 feet (1.5 m) of the vehicle. The electromagnetic field of the antennas causes the transmitter card to send its code by RF signal at a frequency of 433 MHz to the rear SAM. The antennas will determine if the transmitter card is located within the vehicle or outside of the vehicle.

Figure 14-60 Remote keyless entry system transmitter.

Figure 14-61 Transponder card used for entry and vehicle starting.

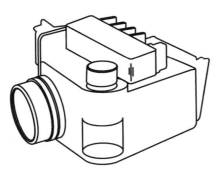

Figure 14-62 Electronic ignition switch.

When the driver approaches the vehicle with the transponder in his possession and attempts to open the door with the door handle, a capacitive sensor or microswitch will signal the Keyless Go module to active the left-front door antenna. Messages are exchanged between the module and the card to determine authorized entry. If the entry is authorized, the doors are unlocked and the door will open.

When the vehicle doors are to be locked, the antennas will determine if the card's signal is coming from inside or outside of the vehicle. This is determined by a decrease in antenna fields in the interior due to the sheet metal of the body. The radio range is limited to a defined range. Also, the ranges overlap between antennas that further refine the location of the card. If it is determined that the card is in the vehicle or trunk, the doors cannot be locked. A warning message will be displayed in the instrument cluster notifying the driver that the card is still in the vehicle.

The engine starting function can be performed only if the antenna determines the card is within the vehicle. Also, the transmission must be in park and the driver's foot on the brake pedal. Once these conditions are met, the driver simply presses the START/STOP on the gear shift handle to start the engine (Figure 14-63). To shut off the engine, the driver simply exits the vehicle with the card in possession. The module looks for the card every 10 seconds; if the card is no longer in the vehicle, the engine shuts off.

Figure 14-63 This start/stop switch is located on the gear shift handle.

Electronic Heated Windshield

Shop Manual
Chapter 14, page 570

The heated windshield system is designed to melt ice and frost from the windshield three to five times faster than conventional defroster systems (Figure 14-64). The windshield undergoes a special process during manufacturing to allow for current flow through the glass without interfering with the driver's vision.

Figure 14-64 The heated windshield removes ice and frost from the windshield in just a few minutes.

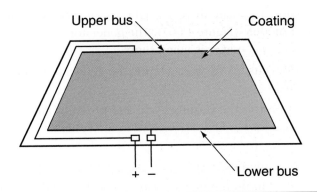

Figure 14-65 The power and ground circuits are connected to the silver and zinc coating through busbars.

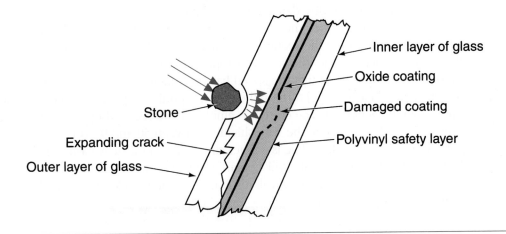

Figure 14-66 An open in the circuit can be caused by a chip or crack in the windshield. A sensor is used to prevent operation if the windshield is damaged.

There are two basic methods used to make the heated windshield:

1. Use a layer of plastic laminate that is between two layers of glass. The back of the outer layer is fused with a silver and zinc oxide coating. The coating carries the electrical current. Busbars are attached to the coating at the top and bottom of the windshield (Figure 14-65). A sensor is used to check the condition of the windshield coating. If the windshield has a crack or chip that will affect heating (Figure 14-66), the voltage drop across the resistor will indicate this condition to the control module. If the windshield is damaged, the controller will not allow heated windshield operation.

2. Use a layer of resistive coating sprayed between the inner and outer windshield layers. The coating is transparent and does not provide any tint. A sensor is used to indicate if the coating has been damaged. If a chip or crack is not deep enough to penetrate the coating, it will not affect the system operation.

The two systems discussed here are representative of the methods used to heat the windshield.

General Motors' Heated Windshield

General Motors' heated windshield consists of the following components:

1. The heated windshield: Contains a transparent internal resistive coating that heats when current is applied to it.

The silver and zinc coating gives the windshield a gold tint.

The film used on the heated windshield will block some radio or microwave signals. This may reduce the effective range of garage door openers and radar detectors.

Shop Manual
Chapter 14, page 570

2. Special CS 144 generator: There are three special phase terminals to provide AC power to the system's power module (Figure 14-67). The generator can continue to supply its normal DC voltage while AC power is being supplied.

3. The power module: Converts the AC voltage from the generator to a higher DC voltage for use by the windshield.

4. The control module: Controls the heating cycle and provides automatic shutdown at the end of the time cycle, or if a fault is detected in the system.

5. The control switch.

See the schematic (Figure 14-68) of the GM heated windshield. When the driver activates the system, the control module starts its turn-on sequence. First, it checks that there is more than 11.2 volts present at terminal B6. This assures there will be sufficient voltage to operate other circuits.

The second step for the control module is to check the vehicle's inside temperature. For the system to operate, the temperature must be below 65°F (18°C). Next, the controller checks the windshield sensor to see if there is any damage to the film coating.

If all of these conditions are met, the control module sends a signal to the BCM to increase the engine speed. The BCM passes the request on to the PCM. If the gear selector is in PARK or NEUTRAL, the ECM will increase the idle speed to approximately 1,400 rpm. The PCM will send a signal back to the BCM to indicate that the speed has been increased. When this feedback signal is received, the control module will turn on the power module relays. The power module will draw AC current from the generator. The current is amplified and rectified by the power module, then sent to the windshield. Voltage at the windshield is between 50 and 90 volts.

Incorporated into the control module is a timer circuit. When the activation switch is turned on for the first time, the control module will operate the system for 3 minutes. If the switch is pressed again, at the end of the first cycle, it will result in a 1-minute cycle. If the switch is pressed while the cycle is still in operation, the system is turned off.

The BCM and PCM are used to provide certain functions when the system is activated.

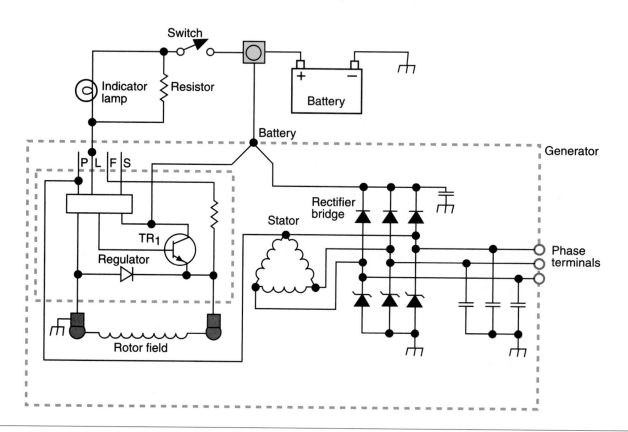

Figure 14-67 Schematic of CS 144 alternator used on vehicles equipped with heated windshields. The three terminals provide AC current to the power module.

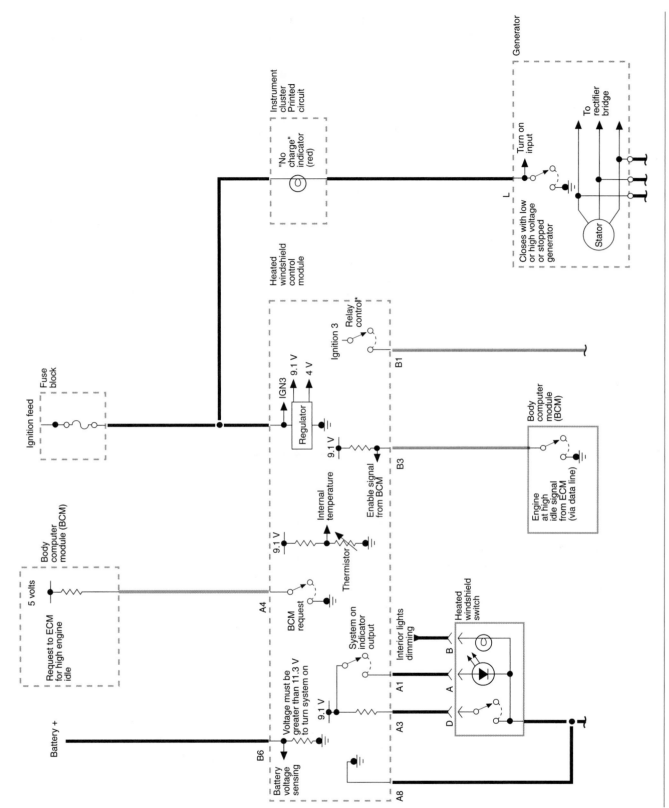

Figure 14-68 General Motors' heated windshield schematic.

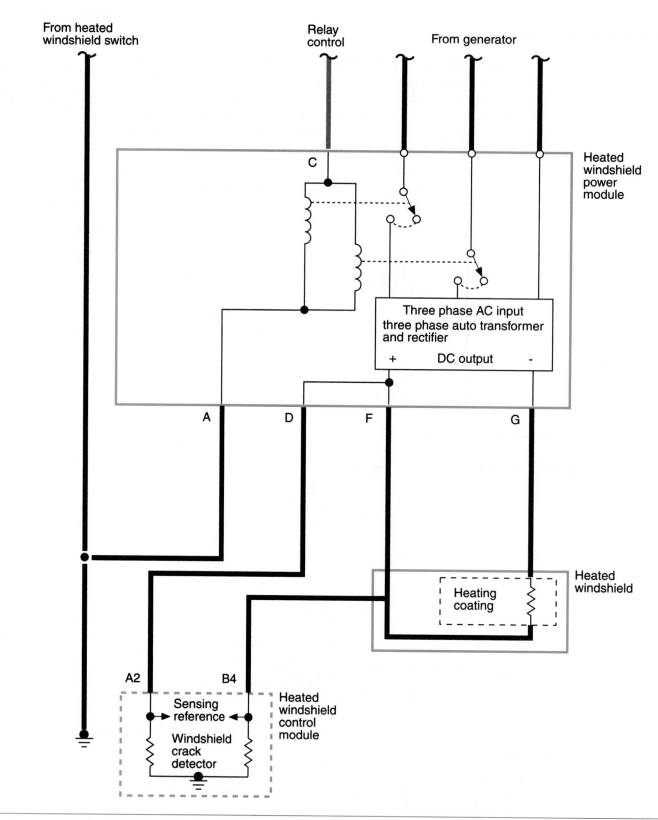

Figure 14-68 (continued)

458

Ford's Heated Windshield

The illustration (Figure 14-69) shows the major components of the Ford heated windshield system. For the system to be activated, the engine must be running and inside temperature must be 40°F or less. When the driver activates the system, the control module shuts off the voltage regulator and energizes the generator output control relay. This switches the generator output from the electrical system to the windshield circuit (Figure 14-70). After the switch has been completed, the control module turns on the voltage regulator to restore generator output.

For the system to be activated, the engine must be running.

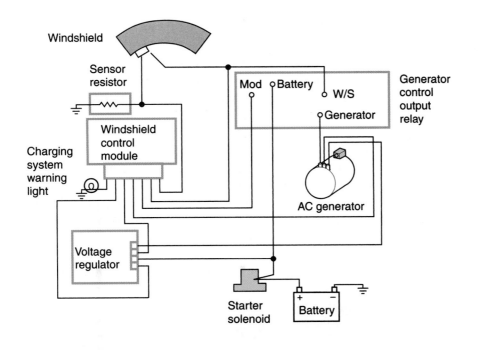

Figure 14-69 Components of Ford's heated windshield system.

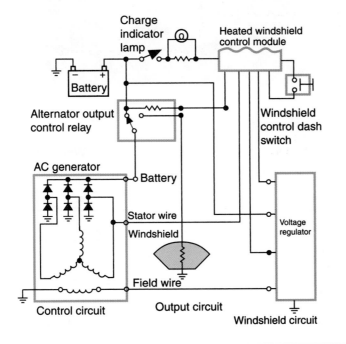

Figure 14-70 Simplified circuit schematic of Ford's heated windshield system.

With the generator output disconnected from the battery, battery voltage drops below 12 volts. The voltage regulator attempts to charge the battery by full fielding the generator. Because the battery does not receive the generator output, full field voltage reaches 30 to 70 volts. All of the full field power is sent to the windshield.

The control module will monitor the battery voltage and generator output. It will prevent the output from increasing over 70 volts to protect the system. To prevent damage to the battery, if its voltage drops below 11 volts, the control module reconnects the electrical system to the generator.

When the system is activated, the control module sends a signal to the EEC controller to increase the idle speed to about 1,400 rpm. If the transmission is placed into a gear selection other than PARK or NEUTRAL, the EEC will return the idle speed to the normal setting.

Intelligent Windshield Wipers

To avoid making the driver select the correct speed of the windshield wipers according to the amount of rain, manufacturers have developed intelligent wiper systems. Two intelligent wiper systems will be discussed here: one senses the amount of rainfall and the other adjusts wiper speed according to vehicle speed.

The automatic wiper system selects the wiper speed needed to keep the windshield clear by sensing the presence and amount of rain on the windshield. The system relies on a series of LEDs that shine at an angle onto the inside of the windshield glass and an equal number of light collectors (Figure 14-71). The outer surface of a dry windshield will reflect the lights from the LEDs back into a series of collectors. The presence of water on the windshield will refract some of the light away from the collectors (Figure 14-72). When this happens, the wipers are turned on. If the water is not cleared by one complete travel of the wipers, the wipers operate again. The frequency and speed of wiper operation is determined by the amount of water sensed on the windshield.

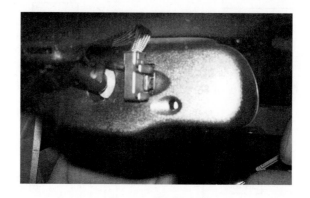

Figure 14-71 The rain sensor is mounted to the rearview mirror.

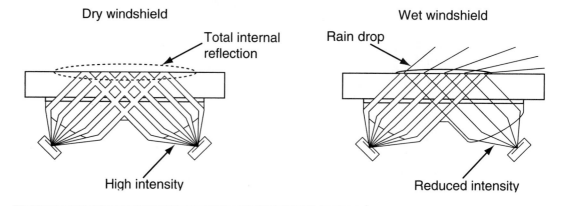

Figure 14-72 The light beams are deflected when water is on the windshield.

Speed-sensitive wipers do not require additional components to operate since most use the BCM. Speed-sensitive wipers compensate for extra moisture that normally accumulates on the windshield at higher speeds in the rain. At higher speeds, the delay between wipers shortens when the wipers are operating in the interval mode. This delay is automatically adjusted at speeds between 10 and 65 miles per hour. Basically, this system functions according to the input the computer receives about vehicle speed.

Electronic Shift Transmissions

The use of solenoids and relays in controlling the operation of the engine has been expanded to include the drivetrain. Many of today's vehicles are equipped with electronic shift automatic transmissions. The control module uses several inputs to determine torque converter clutch operation, hydraulic pressure levels, and shift points. The use of electronics within the transmission has improved shift quality and fuel economy. The Ford **AXODE** transaxle is discussed as an example of the principles used in electronic shift transmissions. AXODE is the model designation given to a version of Ford's electronically shifted automatic translaxle with overdrive.

The AXODE transaxle is a fully automatic electronically controlled unit with a lock-up torque converter (Figure 14-73). All major transaxle operations are controlled through the EEC-IV electronic control assembly (ECA). These functions include transaxle shifting, torque converter clutch operation, and line pressure regulation.

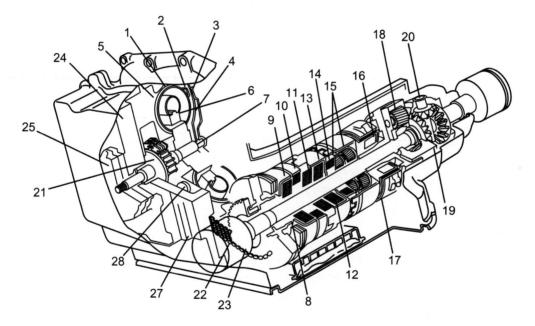

1. Torque converter	14. Reverse clutch
2. Plate clutch and damper	15. Planetary gears
3. Converter cover	16. Parking gear
4. Turbine	17. Low/intermediate band
5. Impeller	18. Final drive sun gear
6. Reactor	19. Final drive planet
7. Oil pump drive shaft	20. Differential assembly
8. Forward clutch	21. Drive sprocket
9. Low one-way clutch	22. Drive link assembly
10. Overdrive band	23. Driven sprocket
11. Direct clutch	24. Valve body
12. Direct one-way clutch	25. Oil pump
13. Intermediate clutch	

Figure 14-73 AXODE transaxle main components.

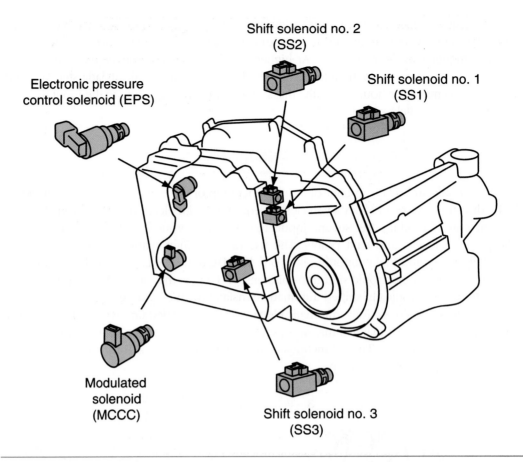

Shift solenoid no. 2
(SS2)

Electronic pressure
control solenoid (EPS)

Shift solenoid no. 1
(SS1)

Modulated
solenoid
(MCCC)

Shift solenoid no. 3
(SS3)

Figure 14-74 Output actuators and their locations.

The ECA receives inputs concerning throttle position, engine speed, torque converter turbine speed, and other drivetrain operations. The ECA processes the information then controls transaxle operation through activation of five solenoids located within the valve body (Figure 14-74). Refer to *Today's Technician Automatic Transmissions and Transaxles* for a detailed explanation of electronic shift transmissions.

Shop Manual
Chapter 14, page 574

Speed-Sensitive Steering

Most vehicles are available with some form of power steering, either as an option or as standard equipment. Conventional power steering systems provide a certain degree of assist to the driver when turning the steering wheel. The disadvantage of conventional power steering is the reduced road feel that the system offers during medium and high speeds. At these speeds, it is desirable for a feeling of increased control and performance. Through the use of electronic controls, the advantages of high power assist and excellent road feel can be achieved.

Manufacturers have chosen different methods of accomplishing this task. The most common is to use a means of varying the output of the power steering pump through an electronic variable orifice. Other methods include the use of electric motor–driven power steering pumps and electric rack and pinion steering (Figures 14-75 and 14-76). Refer to *Today's Technician Automotive: Suspension and Steering Systems* for a detailed explanation of this system.

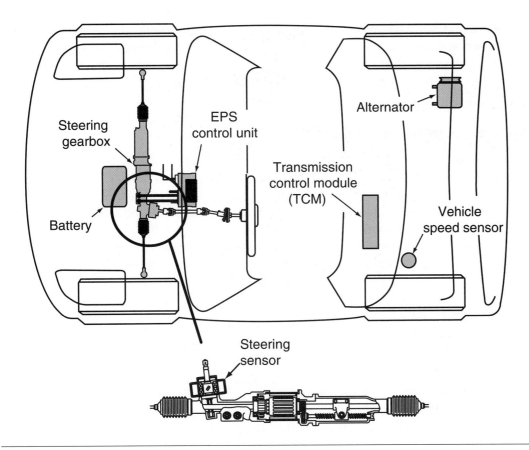

Figure 14-75 The Honda/Acura NSX electric rack and pinion system.

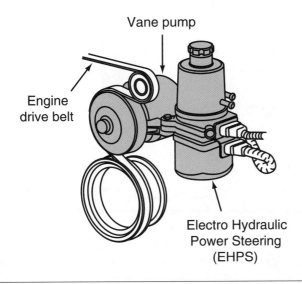

Figure 14-76 Toyota's electrohydraulic steering system has a self-contained motor in the pump.

Electronic Suspension Systems

Passive suspension systems have fixed spring rates and shock valving. Adaptive suspension systems are able to change ride characteristics by altering shock damping and ride height continuously. Active suspension systems are controlled by double-acting hydraulic cylinders or solenoids (actuators) mounted at each wheel. The actuators, instead of conventional springs or air springs, support the vehicle's weight.

Adaptive and active suspension systems provide additional benefits over conventional passive suspension systems. They are able to change ride height, shock damping, and spring rates in response to changing road and driving conditions. Electronic suspension system types can vary from basic shock-damping variations to a complex system of height and ride control that utilizes extensive computer programming.

Adaptive Suspension Systems

Sensors monitor vehicle speed, steering angle, ride height, brake pressure, and vehicle acceleration. They provide this information to the suspension control module. The control module then signals actuators to change shock rates in response to the changing conditions. A small actuator on top of each shock allows a variable range from firm to soft, and solenoids on each air spring enable the computer to maintain constant ride height (Figure 14-77).

Depending on system design, other sensors used by the computer include G-sensors, throttle position sensors, speed sensors, roll sensors, lateral acceleration sensors, and brake switches (Figure 14-78).

Active Suspension Systems

There are several differences between adaptive and active suspension systems. The active system is capable of eliminating body roll and it is faster in its reaction time. The adaptive system is able to reduce body roll but it cannot eliminate it.

Truly active suspension systems use high-pressure hydraulic actuators to carry the vehicle's weight. The actuators combine the function of the shock absorber and the spring into one unit. Through control of the fluid pressure inside the actuators, ride height, body roll, damping, and

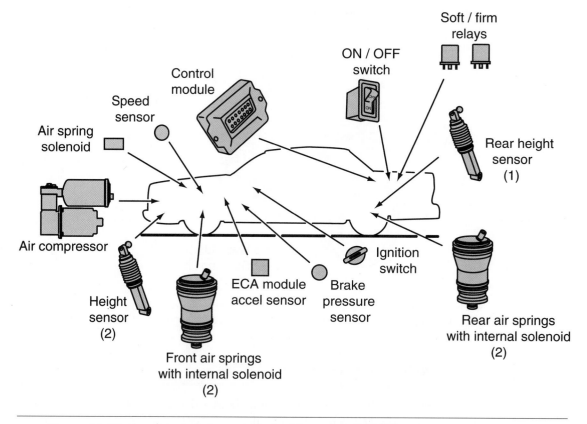

Figure 14-77 Lincoln's air spring automatic ride control (ASARC) components.

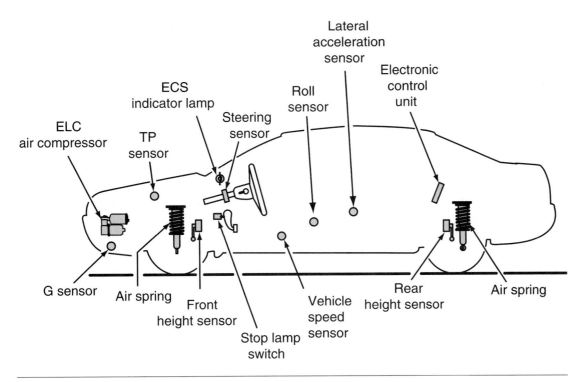

Figure 14-78 Several sensors are used to determine the level of the vehicle at any given time.

spring rate can be controlled by the computer. These systems can be programmed to respond almost perfectly to all driving conditions (Figure 14-79).

The system is also able to eliminate body roll on turns by stiffening one side of the vehicle (Figure 14-80). All of these attitude control functions improve vehicle stability and increase tire traction and driver control. Refer to *Today's Technician Automotive Suspension and Steering Systems* for a detailed explanation of electronic suspension systems.

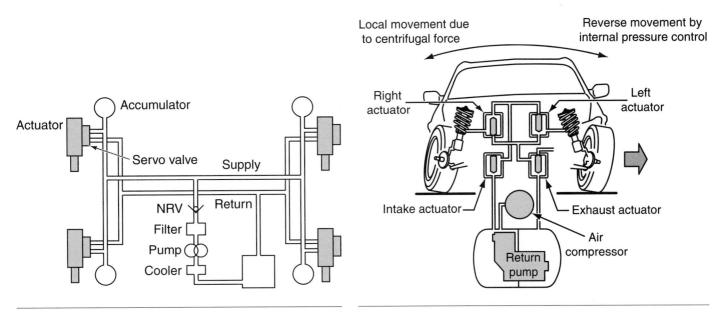

Figure 14-79 Hydraulic schematic of an active suspension system.

Figure 14-80 System correction to eliminate body roll during a turn.

Introduction to Antilock Brake Systems

Shop Manual
Chapter 14, page 575

The modern brake system is more than adequate to stop the vehicle under normal conditions. However, in approximately 1% of its use, it will fail to stop the vehicle safely. This failure is generally the fault of drivers who allow their vehicle to enter an uncontrollable skid. Wheel lockup during braking will increase the stopping distance. A good driver knows that "pumping" the brakes during emergency stops keeps the vehicle from entering into an uncontrollable skid. A tire that is on the verge of slipping produces more friction with respect to the road than one that is locked. The antilock braking system (ABS) is designed to act in a similar manner as a driver pumping the brakes but with much more control and at a much faster rate.

The ABS system is capable of pumping each brake up to 15 times per second. The control module can pulse the two front brakes separately and the rear brakes separately or as a pair, depending on system configuration. ABS automatically stops the vehicle in the shortest possible distance without locking a wheel. In addition, ABS maintains directional control on almost any type of road surface or condition. Even though ABS improves vehicle braking, it cannot compensate for worn brake components, worn tires, excessive speed, or driver error.

The exact components of a system depend on manufacturer and system design. Nonintegrated ABS systems use conventional brake master cylinders with a separate hydraulic unit (Figure 14-81). Integrated ABS systems combine the brake master cylinder, hydraulic brake booster, and the ABS components in a single hydraulic assembly (Figure 14-82). Refer to *Today's Technician: Automotive Brake Systems* for more information.

Brake Assist

Some ABS systems have an additional function that assists in the application of the brakes during a panic-stop situation. The following is an example of this function that is incorporated within Teves Mark 25 ABS system. The **brake assist system (BAS)** uses a brake pedal travel sensor, a brake pedal release switch, a special vacuum booster, a pedal force pressure sensor in the brake line from the master cylinder, and a BAS solenoid (Figure 14-83).

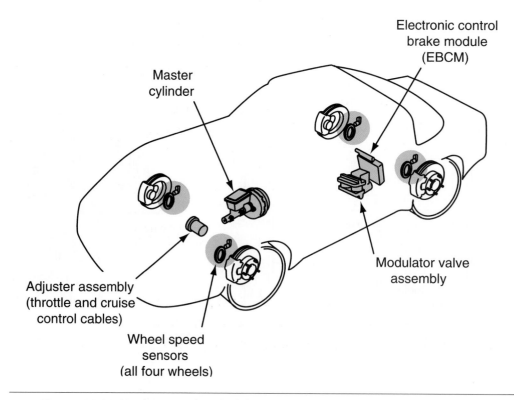

Figure 14-81 Nonintegrated antilock brake system.

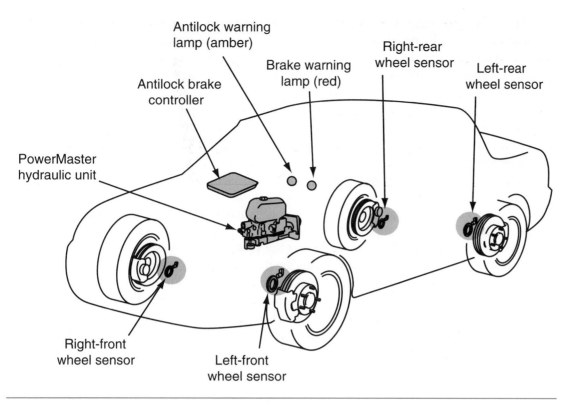

Figure 14-82 Integrated antilock brake system.

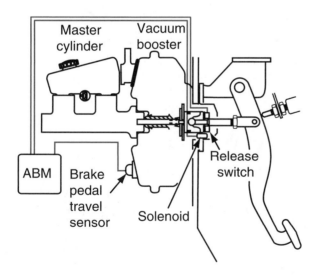

Figure 14-83 The brake assist system components.

For an emergency stop to cause activation of the BAS, all of the following conditions must be present at the same time:

❑ No faults are detected in the system.

❑ Vehicle speed must be greater than 8 mph (13 kph).

❑ The release switch indicates the brakes are being applied.

❑ The threshold of the pedal force indicated by the pressure sensor has been exceeded.

When an emergency stop is initiated, the brake pedal travel sensor monitors the rate of application of the brake pedal. If the acceleration of the travel sensor and the input from the pressure sensor indicate that the driver is in an emergency-stop situation, the BAS programming within the antilock brake module (ABM) energizes the solenoid. The actuation of the solenoid opens a larger orifice to atmospheric pressure. This orifice provides for faster "fill" of atmospheric pressure to the diaphragm of the booster assembly than what is normally provided. As a result, brake application occurs sooner and the stopping distance is reduced. When the driver releases the brake pedal, the brake pedal release switch input informs the BAS to de-energize the solenoid and return the system to normal brake assist.

Additional programming in the ABM is used to determine when a vehicle is braking in a turn. This function is active for any braking event during turns; it does not need an ABS braking event. This system will use the wheel speed sensors to make the determination that a turn is being negotiated. If the brakes are applied during this time, the system will attempt to balance the side-to-side brake forces at the wheels to counteract the **yaw** effect. Yaw is the tendency for the vehicle to rotate around its center of gravity. Balancing the brake forces improves the vehicle's stability. The ABM calculates the outside wheel speed and compares it to the inside wheel speed. If the speed differences indicate activation of the system, the ABM reduces the front inside wheel braking pressure in relation to the front outside wheel braking pressure. This will result in less **oversteer** (Figure 14-84). Oversteer is the tendency of the back of the vehicle to turn on the vehicle's center of gravity and come around the front of the vehicle.

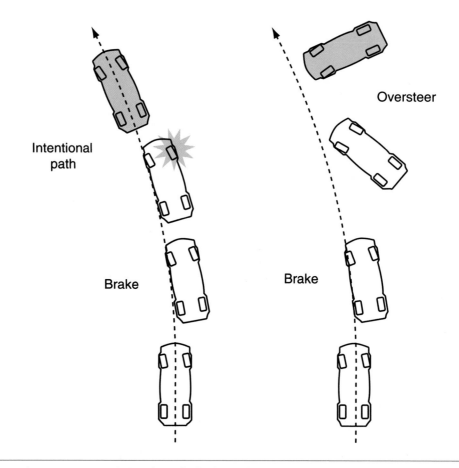

Figure 14-84 Assisting the vehicle through a turn to prevent oversteer.

Reducing braking pressure is accomplished by energizing the inlet valve for the front inside wheel's brake caliper and PWM the outlet valve. This vents the front inside caliper brake pressure. The ABM then attempts to maintain the front braking balance side to side by energizing the inlet valve and de-energizing the outlet valve. The ABM continues to control the brake balance through the inlet and outlet valve positions to maintain front wheel braking balance.

Automatic Traction Control

The same technology used for ABS systems is also applied to automatic traction control. An electronic control monitor monitors the wheel speed sensors. If it determines that one of the drive wheels is spinning faster than the other, it will automatically apply the brakes to the spinning wheel. With the brake applied, it requires a greater amount of torque to spin the wheel. Because a differential delivers equal torque to each drive wheel, the greater torque requirement is also transferred to the stationary or slower moving wheel. This allows the wheel that has the greatest amount of traction ability to move the vehicle.

Some manufacturers integrate some engine control functions into the automatic traction control system instead of, or in addition to, applying the brakes. If the control module senses a loss of traction, it signals the PCM to retard the engine timing and to decrease the throttle plate position. This action will reduce the engine output in an attempt to reduce the amount of power to the drive wheels. If this action fails to reduce tire slippage, the module will cut fuel delivery to one or more engine cylinders. The reduction of engine power is to prevent overspeeding the engine.

Electronic Stability Control

Electronic stability control is an additional function of the ABM. A popular system that is used by several manufacturers is the electronic stability program (ESP) designed by Continental Teves AG & Co. The system is designed to maintain proper vehicle tracking through the driver's intended path. In addition to standard components used with ABS, a steering angle sensor, a brake pressure sensor, a yaw sensor, and a lateral acceleration sensor are used to determine the intended path and the vehicle's actual path.

The driver's intended course is determined by the steering angle and wheel speed sensors. These inputs are then compared to lateral acceleration and yaw sensor inputs to determine the actual path. If there is a deviation, the ESP system will activate the brake calipers to individual front or rear wheels. In addition, the ABM may send a data request to the PCM to reduce engine torque. In an extreme situation, all engine torque can be taken away in an effort to bring the vehicle under control.

When deviation is detected, the ABM activates the brake assist solenoid that opens a port to atmospheric pressure. This moves the diaphragm in the booster and moves the master cylinder pushrod to create braking pressure in the master cylinder.

 AUTHOR'S NOTE: The brake pedal does not need to be depressed to activate ESP. Once activated, the brake pedal will drop by itself.

To counteract the yaw tendency of the vehicle, brake pressures are applied to the wheel calipers that oppose the yawing motion. For example, if the vehicle travel indicates a counterclockwise yaw motion, the brake calipers of one or both wheels of the right side of the vehicle are applied. The braking pressure at each wheel is modulated by the inlet and outlet valves.

Figure 14-85 One style of steering wheel angle sensor using LEDs and photo cells.

Steering wheel angle sensors are usually an optical-type sensor. An example would by one that has a series of LEDs situated across from the same number of photocells (Figure 14-85). A shutter wheel that rotates with the steering wheel will travel between the LEDs and the photocells, breaking the light beams. Based on this input, the ABM can determine the direction the steering wheel is being turned and how fast it is being turned.

Tire Pressure Monitoring Systems

The **tire pressure monitoring system** is a safety system that notifies the driver if a tire is under-inflated or overinflated. Most systems use a pressure sensor transmitter in each wheel. Wheel-mounted tire pressure sensors (TPSs) are attached to the rim by a sleeve nut on the valve stems (Figure 14-86). The TPS's internal transmitter uses a 315-MHz signal to broadcast tire pressure and temperature information. In addition, it transmits its transmitter ID. Each TPS has an internal battery that is designed to last approximately 10 years. The battery is not serviceable.

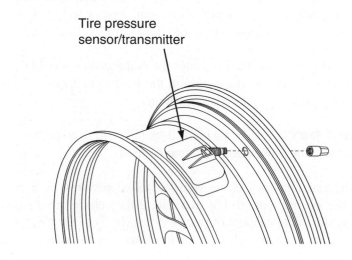

Tire pressure
sensor/transmitter

Figure 14-86 The TPS is attached to the rim as part of the valve stem.

Some systems will only transmit tire inflation information when the vehicle is in motion, while others continue to monitor tire pressure even with the vehicle stationary. The TPSs transmit tire pressure data by RF once per minute at speeds over 20 mph (32 kph) and up to once every hour when the vehicle is parked. A receiver module receives the tire inflation information and then sends the message over the data bus. If a tire pressure is out of the programmed parameters, a warning chime is sounded and a warning lamp is illuminated (Figure 14-87).

Although the receiver module stores each TPS's ID, some systems do not associate the ID with a particular location on the vehicle. If a tire was to exceed the acceptable pressure limits, the warning would indicate a problem but would not identify which tire had the problem. The actual tire causing the condition would need to be determined using a tire pressure gauge.

The TPS will send pressure and temperature data to the receiver module when any of the following occur:

❑ If the vehicle is parked for more than 15 minutes, the TPS monitors the pressure every minute but sends tire pressure data only once every 13 hours, unless there is a change in pressure.

❑ While the vehicle is in motion, the TPS sends data once every 15 seconds for the first 30 data blocks and then once per minute after that.

❑ If there is a change of pressure greater than 1 psi (6.9 kPa) when the vehicle is stationary, the TPS will send data once per minute.

❑ If there is a change of pressure greater than 1 psi (6.9 kPa) when the vehicle is in motion, the TPS will send data once every 5 seconds.

Logic in the receiving module prevents false warning displays to the driver. For example, since the signal transmission frequency is dependent on wheel speed, if a signal from one of the TPSs is not received as often as a signal from the other sensors, the receiving module will assume that the tire has been relocated and the spare is in use. Also, rapid tire pressure changes can occur due to changes in temperature. If a vehicle is moved from a heated garage and is driven on a cold day, a low-tire warning may be indicated. To prevent this, the system uses temperature information along with pressure data to filter the data and compensate for rapid pressure changes that occur as a result of temperature changes.

Figure 14-87 Low tire pressure warning symbol.

The receiver module can set a DTC when one of the transmitters fails to produce a signal. If a replacement transmitter is installed on the vehicle, the system learns the new transmitter when the vehicle is driven. When an unrecognized signal is received at the same transmission intervals as the recognized transmitters, the receiver module stores the ID and begins monitoring that transmitter. The receive module will relearn the TPS IDs every time the vehicle is driven after being stopped for at least 15 minutes.

 AUTHOR'S NOTE: It may be necessary to drive the vehicle up to 10 miles before the replacement transmitter ID is learned.

Premium systems use the same basic TPSs as just described, but they use additional components or programming to identify which tire is out of limits. The premium system will record TPS ID locations, so this system is capable of displaying the actual tire pressures to the driver (Figure 14-88). Some systems require that the ID location be trained into the receiving module. This is usually done by entering training mode and placing a magnet around the tire's valve stem in a specified order. As each location is learned, the ID is locked to that position. This procedure needs to be performed each time the tire is rotated. When the magnet is placed around the valve stem, it pulls a reed switch closed and causes the TPS to send it data.

Some systems do not require training. These use a transponder (Figure 14-89) located behind the wheel splash shields at three locations: left front, right front, and right rear. The left rear is inferred when a fourth TPS ID is received while the vehicle is in motion. Tire position is relearned each time the vehicle is in motion.

The transponders use two ground terminals. One of the grounds terminals is common for all transponders. The other three ground terminals are used to identify the transponder location. This arrangement allows the receiving module to determine the location of the transponder.

When the vehicle is in motion, all four TPSs will send periodic signals to the receiving module. The receiving module will then send a signal to one of the transponders. This causes the transponder to emit a 125-kHz signal in the area surrounding the wheel. This signal will excite the TPS at that location. Under this condition, the TPS will send a constant signal to the receiving module. The receiving module will then identify the TPS and its location. This procedure is repeated for the other two transponder locations. Once three wheel locations have been identified, the fourth location can be inferred and identified. This learning process occurs any time the vehicle is driven after being stationary for at least 15 minutes.

 AUTHOR'S NOTE: Because of this procedure to learn TPS locations, training magnets are not required.

Figure 14-88 Premium systems will display the tire pressures of each tire.

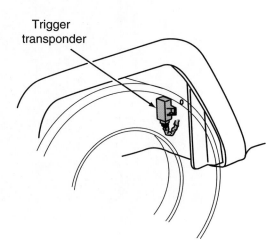

Figure 14-89 Transponder used to identify sensor location on the vehicle.

Vehicle Audio Entertainment Systems

Shop Manual
Chapter 14, page 583

When radios were first introduced to the automotive market, they produced little more than tinny noise and static. Today's audio sound systems produce music and sound that rivals the best that home sound systems can produce. And with nearly the same or even greater amounts of volume or sound power!

The most common sound system configuration is the all-in-one unit called a receiver. Housed in this unit is the radio tuner, amplifier, tone controls, and unit controls for all functions. These units may also include internal capabilities such as cassette players, compact disc players, digital audiotape players, and/or graphic equalizers. Most will be electronically tuned with a display that shows all functions being accessed/controlled and digital clock functions (Figure 14-90).

Recent developments by the manufacturers have been made to take individual functions (tape, disc, equalizer, control head, tuner, amplifier, etc.) and put them in individual boxes and call them components. This would allow owners greater flexibility in selecting options to suit their needs and tastes (Figure 14-91). Componentizing has allowed greater dash design flexibility. Some components, such as multiple-CD changers, can be remotely mounted in a trunk area for greater security.

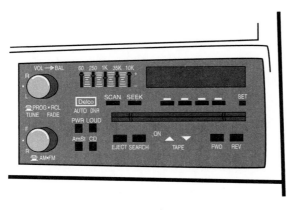

Figure 14-90 Radios are actually receivers. They contain the basic elements of a tuner, an amplifier, and a control assembly in one housing and can also contain a tape or CD player.

Figure 14-91 Components, like those shown here, allow for more options to the sound system.

Wiring diagrams for component systems will be more complex (Figure 14-92). In addition to power and audio signal wires, note that some systems will have a serial data wire for microprocessor communication between components for the controlling of unit functions. Some functions are shared and integrated with factory-installed cellular phones, such as radio mute. Some systems allow remote control of functions through a control assembly mounted in the steering wheel or alternate passenger compartment location (Figure 14-93).

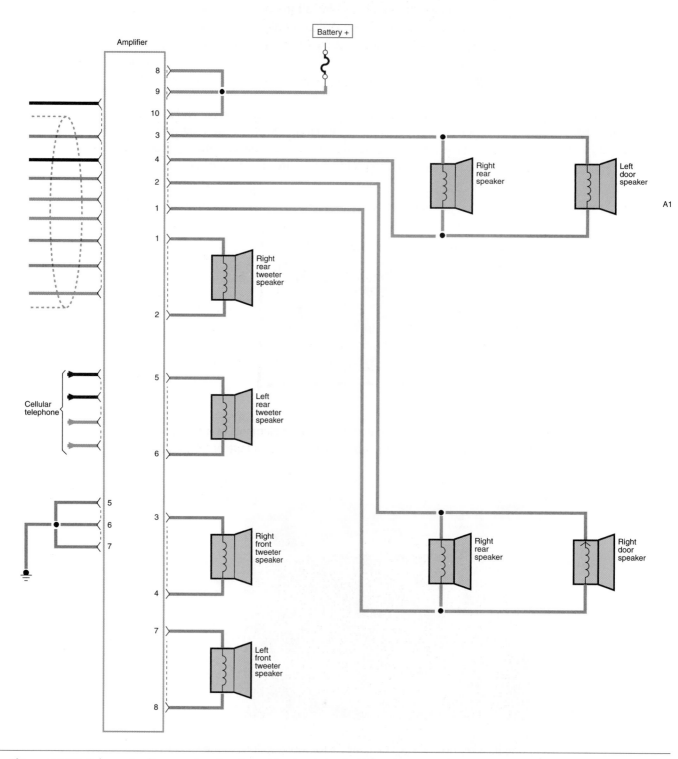

Figure 14-92 Schematic diagram showing the wiring hookups for remote-mounted components.

Figure 14-93 Audio system with steering wheel–mounted remote control.

The remote controls can be either resistive multiplexed switches or use a supplementary bus system such as LIN. These inputs go to the controlling module. The switches send different voltage signals to the module, corresponding to which switch is pressed. The module responds by sending a request message via the data bus to the radio.

An antenna is needed to collect amplitude modulation (AM) and frequency modulation (FM) radio signal waves. The radio station's broadcast tower transmits electromagnetic energy through the air that induces an AC voltage that averages 50 microvolts in the antenna. The radio receiver processes this AC voltage signal and converts it to an audio output.

Some vehicles have the antenna incorporated into the rear window defogger grid. A rear window defogger/antenna module (Figure 14-94) separates the RF signal used by the radio from the electrical current used by the grid. When the radio is turned on, a 12-volt signal is sent to the defogger/antenna module (Figure 14-95). A coaxial cable from the module provides the AM and FM tuners the RF signal input. Usually the top grid lines are unheated. These are used to receive the AM signals. The heated grid lines are used to receive the FM signals.

Sound system amplifiers can be either remotely mounted or integrated with the speakers. Most remotely mounted amplifiers are connected to the data bus. This allows for configuration to match it to the vehicle. Amplifier configuration includes amplification output power, the number of output channels, and the equalizer curve.

Functions that are supported by data bus inputs to the amplifier include speed-proportional volume increase, designating speakers for shared audio functions, and fixed audio output if an amplified speaker system is used.

Integrated amplifiers are mounted in the speakers. Power to the amplifiers is supplied through the amplifier relay. When the radio is turned on, 12 volts are supplied to energize the relay.

Figure 14-94 Defogger/antenna module.

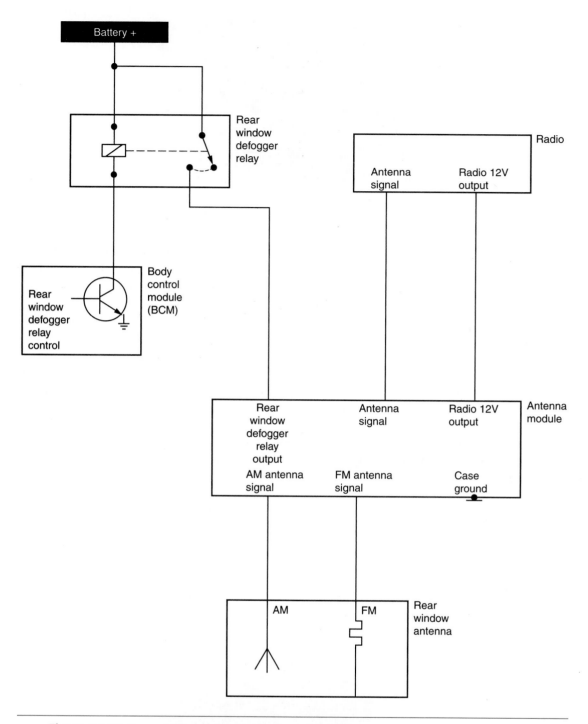

Figure 14-95 Rear window defogger/antenna circuit.

Reception to the radio receiver from the antenna can be interfered with by noise resulting from RFI. This is especially true of the AM band. The noise is picked up by the radio receiver and amplified through the audio circuits. The FM band is susceptible to EMI, but usually is not as noticeable compared to AM. Control of RFI and EMI noise is done by proper radio antenna base, proper radio receiver, proper engine-to-body ground, and proper heater core grounds. In addition, resistor-type spark plugs and radio suppression secondary ignition wiring are used. The radio itself will also have internal suppression devices, such as capacitors that shunt AC noise to ground and slow sudden changes of voltage in a circuit. Some systems will use a **radio choke.** The choke is a winding of wire. In a DC circuit the choke acts like a short, but in an AC circuit it represents high resistance. The choke blocks the noisy AC current but allows the DC current to pass normally.

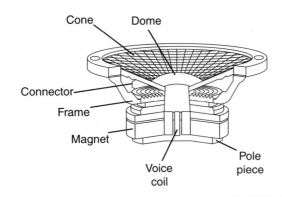

Figure 14-96 Speaker construction.

Speakers turn the electrical energy from the radio receiver amplifier into acoustical energy. The acoustical energy moves air to produce sound. A speaker moves the air using a permanent magnet and electromagnet (Figure 14-96). The electromagnet is energized when the amplifier delivers current to the voice coil at the speaker. The coil forms magnetic poles that cause the voice coil and speaker cone to move in relation to the permanent magnet. The current to the speaker is rapidly changing AC current that results in the speaker cone moving rapidly in and out and producing sound.

Since one speaker cannot reproduce the entire hearing frequency range of approximately 20 Hz to 20 KHz, speakers are designed to reproduce only parts of the desired frequency. Large speakers, called **woofers,** produce the low frequencies of midrange and bass better than small speakers. Smaller speakers, called **tweeters,** produce the high frequencies of treble better than large speakers. Coaxial speakers have two separate speakers combined in one speaker frame and cover a broader frequency range than a single-cone speaker. Subwoofer speakers can be coupled with the coaxial speaker (or a separate tweeter and midrange speaker) to cover the hearing range and maximize sound quality.

Satellite Radio

Satellite radios are the latest technology in radios (Figure 14-97). Satellite radio provides several commercial-free music and talk show channels using orbiting satellites to provide a digital signal. An in-vehicle receiver receives the digital signal, which travels to the conventional vehicle radio as an auxiliary audio input and plays through the normal speaker system. In this case the satellite radio function becomes an additional mode of the radio. Satellite radio operation is available only when the owner purchases the subscription service.

 AUTHOR'S NOTE: If all satellite signal is lost, the radio becomes unlike the loss of an AM or FM signal where some static or hiss may be heard.

Figure 14-97 Satellite radio system.

The satellite digital audio receiver (SDAR) is an additional audio receiver that is installed separate from the vehicle's radio. The satellite's signal is processed by an SDAR, which provides an input signal to the conventional radio over dedicated circuits for right and left channels. The usual radio controls for mode selection and tuning are bused to the SDAR. This allows the radio controls to operate the SDAR. Any text information such as channel numbers, track, and artist that the stations transmit is sent to the radio over the data bus and displayed on the radio screen.

The SDAR signals the satellite antenna located on the centerline of the roof (Figure 14-98), and the tuning of the antenna is calibrated to use the ground plane of the vehicle. Signal reception is possible only when the antenna and the satellite are in a direct line. Obstacles such as buildings, overpasses, and tunnels may temporarily disrupt the signal. A buffer is included that prevents brief interruptions when the vehicle passes under bridges.

Some radios have only one audio input. In order to add multiple audio sources such as SDAR, CD/DVD, and hands-free cell phone, a multiplexer may be used. The **multiplexer** acts as an electronic switch to switch between the different audio sources (Figure 14-99). Depending on which source requires use of the system, the multiplexer will make the switch and send the output to the radio.

Figure 14-98 Satellite radio antenna.

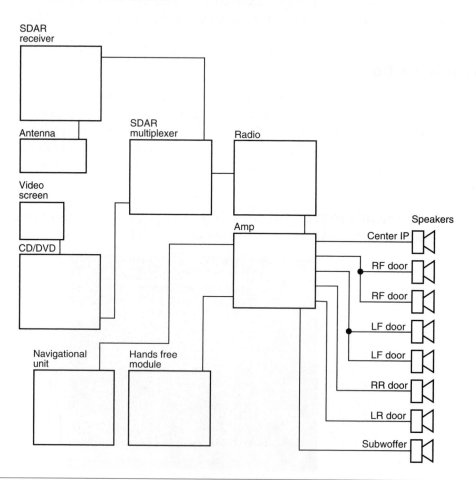

Figure 14-99 Some systems use a multiplexer to switch between audio inputs.

DVD Systems

There are several digital video disc/video entertainment systems (DVD/VES) available on today's vehicles. Most DVD systems display the video on a flip-down, roof-mounted monitor (Figure 14-100) or on a monitor attached to the back of the front seats (Figure 14-101). Most will play the audio through the vehicle's regular audio system. If a DVD is playing, the audio system automatically switches to a "surround sound" mode that is biased toward the rear of the vehicle. In addition, remote control and auxiliary input jacks that will permit the display of video cameras and video games may also be included.

When headphones are supported, the speakers may play audio from one source while the headphones can play audio from another source. If the headphones and the radio are playing in the same mode, the radio is the master of the system. Headphones use either a 900-MHz signal or infrared. The headphones enable rear-seat occupants to listen to audio while front-seat occupants listen to cassettes, CDs, or the radio on the front speakers. The headphones receive their signals from the radio or the CD changer. Two separate channels are used to minimize interference. If another audio/video system is within range, the headphones automatically switch to the strongest channel.

Figure 14-100 Flip-down, roof-mounted, DVD monitor.

Figure 14-101 DVD monitor located in the seat back.

Hands-Free Cellular Telephone

The hands-free cellular telephone system can use different wireless technologies. Bluetooth™ wireless technology is discussed here. As discussed in Chapter 12, this technology allows for communication between modules. In this case, the communication is between a compatible cellular telephone and the vehicle's on-board receiver. The system communicates with a cell phone that is anywhere within the vehicle. The system recognizes several cell phones. Each cell phone is given an identification number, name, and priority by the user during the setup process. The assigning of this information pairs the cell phone to the system. The paring process stores the cell phone's IP address in the hands-free module (HFM) and the HFM's IP address into the cell phone.

The system uses voice recognition technology to control operation. When a programmed cell phone is within range of the HFM, communication between the two components is established. When a cell phone call is initiated, the voice is broadcast through the radio speakers. If any other audio device is using the speakers at that time, the hands-free system automatically overrides it.

When the HFM broadcasts a data bus message that hands-free operation is about to be initiated, the vehicle radio stores its current volume level and mode. The radio then switches to the stored hands-free volume level. The amplifier fades all of the speakers and then transmits the hands-free audio through the speakers. When the HSM broadcasts a data bus message that the cell phone call has been terminated, the radio stores the HFM volume level and returns to the previous radio mode and volume level. At this time, the amplifier returns all speakers to their previous audio level.

A dual-element microphone module (DEMM) is located in either the dash, center console, or rearview mirror. The DEMM consists of microphone elements and electrical circuitry that includes a preamplifier network. The DEMM is capable of tuning the microphone frequency response to improve the voice recognition function.

Navigation Systems

Many manufacturers offer **navigational systems** as an option on their higher line vehicles (Figure 14-102). These systems use satellites to direct drivers to their desired destinations. Many navigational systems are integrated in the vehicle radio. Regardless of the display method, most navigational systems use a GPS antenna, to determine the vehicle's location by latitude and longitude coordinates, and a gyroscope to determine vehicle turns. Usually map data, provided on a DVD, and navigation information are displayed on a thin film transistor, liquid crystal display (TFT, LCD) color screen. Voice prompts can be sent through the audio system speakers.

Figure 14-102 Navigational system monitor and controls.

Most systems will provide at least some of the following features:

- ❑ Full-screen map display.
- ❑ Vehicle location and route guidance.
- ❑ Turn-by-turn distance in feet or meters.
- ❑ Points of interest en route.
- ❑ The storage of favorite routes and locations.
- ❑ The storage of recent routes.

During most conditions, the gyroscope, data bus information, and the map data locate the vehicle on the displayed map. If the vehicle is traveling in an unmapped area, the system uses the GPS data.

The navigational system can also be incorporated into a vehicle-tracking system. If the vehicle is stolen, its whereabouts can be tracked using the satellite. If the navigational system is tied to the hands-free cell phone system, the driver can get on-road assistance in the event of a problem. For example, if the driver locks the keys in the vehicle, he can call the assistance line and the representative can send a signal to unlock the doors. If the air bags deploy, the system can automatically send a signal of this event. A representative from the tracking subscription company will attempt to get in touch with the vehicle occupants to see if assistance is required. Emergency personnel can be dispatched because the satellite system informs the representative of the exact location of the vehicle.

A BIT OF HISTORY

Radios were introduced in cars by Daimler in 1922. Cars were equipped with Marconi wireless receivers.

Summary

- ❑ An automatic temperature control (ATC) system is capable of maintaining a preset level of comfort control as selected by the driver.

- ❑ The in-car sensor measures the average temperature inside the vehicle. The ambient sensor measures the temperature outside the vehicle.

- ❑ The sunload sensor produces a signal proportional to the heat intensity of the sun through the windshield.

- ❑ Cruise control is a system that allows the vehicle to maintain a preset speed with the driver's foot off the accelerator.

- ❑ The cruise control module energizes the supply and vent valves to allow manifold vacuum to enter the servo. The servo moves the throttle to maintain the set speed. The vehicle speed is maintained by balancing the vacuum in the servo.

- ❑ The memory seat feature allows the driver to program different seat positions that can be recalled at the push of a button.

- ❑ The easy exit feature is an additional function of the memory seat that provides for easier entrance and exit of the vehicle by moving the seat all the way back and down.

- ❑ Antitheft systems are deterrent systems designed to scare off would-be thieves by sounding alarms and/or disabling the ignition system.

Terms to Know
Ambient
   temperature
   sensor
Armed
Aspirator
AXODE
Blend air door
   actuator
Brake assist system
   (BAS)
Climate-controlled
   seats
Cold engine lock-
   out switch
Control panel
   assembly

❑ The antitheft control module monitors the switches. If the doors or trunk are opened or the key cylinders are rotated, the module will activate the system.

❑ The immobilizer system acts as an engine disable system by using an ignition key that has a transponder.

❑ Automatic door locks is a passive system used to lock all doors when the required conditions are met. Many automobile manufacturers are incorporating the system as an additional safety and convenience feature.

❑ The keyless entry system allows the driver to unlock the doors or the deck lid from outside the vehicle without the use of a key.

❑ The heated windshield system is designed to melt ice and frost from the windshield three to five times faster than conventional defroster systems.

❑ With electronically controlled transmissions, shifting, torque converter clutch operation, and line pressure regulation are controlled through computer operation.

❑ The shift solenoids provide gear selection, based on engine controller commands, by controlling the pressure to the shift valves.

❑ In speed-sensitive steering systems, steering effort is controlled based on vehicle speed and rate of steering wheel rotation.

❑ Adaptive and active suspension systems provide additional benefits over conventional passive suspension systems by varying ride height, shock damping, and spring rates in response to changing road and driving conditions.

❑ Adaptive suspension systems are able to change ride characteristics by altering shock damping and ride height continuously.

❑ Active suspension systems are controlled by double-acting hydraulic cylinders or solenoids (actuators) mounted at each wheel. Instead of conventional springs or air springs, the actuators support the vehicle's weight.

❑ Nonintegrated ABS systems use conventional brake master cylinders with a separate hydraulic unit.

❑ Integrated ABS systems combine the brake master cylinder, hydraulic brake booster, and the ABS components in a single hydraulic assembly.

❑ Automatic traction control limits the amount of tire spin on slippery road conditions by applying the brakes automatically or reducing engine power automatically.

❑ Tire pressure monitoring systems notify the driver if a tire is underinflated or overinflated.

❑ Vehicle audio entertainment systems are generally only a single component (receiver). Recently, manufacturers have been developing multiple components (tuner, amplifier, control head, etc.) to allow for greater flexibility.

❑ Some audio component systems utilize a serial data line to provide for communication and control of functions between those components.

Review Questions

Short-Answer Essays

1. Explain the operation of the immobilizer system.

2. Explain the purpose of the brake assist system.

3. Explain the basic operating principles of the electronic cruise control system.

4. List and describe the safety modes incorporated into electronic cruise control systems.

5. List the main components of common antitheft systems.

6. What functions does the electronic shift transmission control?

7. Describe the active suspension system.

8. Explain two methods that the memory seat control module uses to determine seat position.

9. What is the purpose of antilock braking systems?

10. Explain the difference between a radio receiver unit system and a component radio system.

Fill in the Blanks

1. The ATC system uses _____ that will open and close blend air doors to achieve the desired in-vehicle temperature.

2. The sunload sensor is a _____ diode that sends signals to the programmer concerning the extra generation of heat as the sun beats through the windshield.

3. By monitoring the high side temperature, the BCM is capable of making calculations that translate into _____ .

4. The _____ _____ feature is an additional function of the memory seat that provides for easier entrance and exit of the vehicle.

5. The _____ _____ sensor determines the vehicle-to-vehicle distances and relational speeds.

6. The generator in the General Motors heated windshield system provides _____ power to the system's power module.

7. In the Ford-style heated windshield system, the voltage regulator _____ _____ the generator to supply voltage to the windshield.

8. _____ suspension systems are able to change ride characteristics by altering shock damping and ride height continuously. _____ suspension systems are controlled by double-acting hydraulic cylinders or solenoids mounted at each wheel.

9. Adaptive and active suspension systems provide additional benefits over conventional passive suspension systems. They are able to change _____ , shock _____ and _____ rates in response to changing road and driving conditions.

10. Radio receiver units usually contain at least the following: an _____ , a _____ , and function _____ .

Multiple Choice

1. The brake assist system:
   A. Uses electrically controlled brake calipers to apply the brakes faster.
   B. Uses an increased rate of fill of atmospheric pressure into the brake booster to apply the brakes faster.
   C. Prevents the driver from applying the brakes if speeds are above a calibrated threshold.
   D. All of the above.

2. Laser-guided cruise control:
   A. Can maintain a set distance from the vehicle in front.
   B. Can maintain a set speed.
   C. Can resume set speed if the vehicle in front is no longer detected.
   D. All of the above.

3. Electronic cruise control systems are being discussed.
   Technician A says if the voltage signal from the VSS drops below the low comparator value, the control module energizes the vent valve solenoid.
   Technician B says if the VSS signal is greater than the high comparator value, the control module energizes the supply solenoid valve.
   Who is correct?
   A. A only
   B. B only
   C. Both A and B
   D. Neither A nor B

4. Electrionic cruise control is being discussed.
   Technician A says that the electronic cruise control system offers more precise speed control than the electromechanical system.
   Technician B says that the throttle position sensor is used to provide smooth throttle changes while the cruise control is engaged.
   Who is correct?
   A. A only
   B. B only
   C. Both A and B
   D. Neither A nor B

5. Memory seats are being discussed.
   Technician A says the power seat and memory seat functions can only be operated when the transmission is in the PARK position.
   Technician B says when the seat is moved from its memory position, the module stores the number of pulses and direction of movement in memory.
   Who is correct?
   A. A only
   B. B only
   C. Both A and B
   D. Neither A nor B

6. The operation of Toyota's electronically controlled sunroof is being discussed.
   Technician A says that it is not necessary to understand logic gate operation to understand the control of the sunroof.
   Technician B says the movement of the sunroof is controlled by a motor that operates a drive gear.
   Who is correct?
   A. A only
   B. B only
   C. Both A and B
   D. Neither A nor B

7. The keyless entry system is being discussed.
   Technician A says an additional function of the system is that the deck lid lock can be released by pressing the 5-6 button.
   Technician B says a second code can be entered into the system.
   Who is correct?
   A. A only
   B. B only
   C. Both A and B
   D. Neither A nor B

8. The heated windshield system is being discussed.
   Technician A says if the windshield has a crack or chip that will affect heating, the voltage drop across the resistor will indicate this condition to the control module.
   Technician B says if the windshield is damaged, the controller reduces the voltage to the windshield to 20 volts.
   Who is correct?
   A. A only
   B. B only
   C. Both A and B
   D. Neither A nor B

9. Electronic transmission actuators are being discussed.
   Technician A says the shift solenoids regulate the amount of line pressure in the transmission.
   Technician B says the electronic pressure control solenoid controls provide gear selection by controlling pressure to the shift valves.
   Who is correct?
   A. A only
   B. B only
   C. Both A and B
   D. Neither A nor B

10. Traction control systems are being discussed.
    Technician A says, in some systems, if the control unit senses that one drive wheel is spinning faster than the other, it will cause the brakes of that wheel to be applied.
    Technician B says the control unit of some systems will order changes to the engine control system to increase its power, if the control unit senses that one drive wheel is spinning faster than the other.
    Who is correct?
    A. A only
    B. B only
    C. Both A and B
    D. Neither A nor B

Passive Restraint Systems

Upon completion and review of this chapter, you should be able to:

❏ Explain the purpose of passive restraint systems.

❏ Describe the basic operation of passive seat belt systems.

❏ Describe the common components of an air bag system.

❏ List the components and explain the function of the air bag module.

❏ Describe the function of the clockspring.

❏ Explain the functions of the diagnostic module used in air bag systems.

❏ List and describe the operation of the different types of air bag system sensors.

❏ List the sequence of events that occur during air bag deployment.

❏ Describe normal operation of the air bag system warning light.

❏ Describe the operation of a hybrid inflator module, and explain the advantages of this type of module.

❏ Explain the function of multistage air bags.

❏ Explain the function of the side-impact air bags and describe the locations of the modules and sensors.

❏ Describe the operation and purpose of factory-installed air bag on/off switches.

❏ Explain the procedure required to install air bag deactivation kits and retrofit on/off switches.

❏ Describe the purpose and operation of seat belt pretensioners.

❏ Describe the function of occupant classification systems (OCS).

Introduction

Federal regulations have mandated the use of automatic passive restraint systems in all vehicles sold in the United States after 1990. Passive restraint systems operate automatically, with no action required on the part of the driver or occupant. Two- or three-point automatic seat belt and air bag systems are currently offered as a means of meeting this requirement.

In this chapter, you will learn the operation of the automatic passive restraint and air bag systems. The safety of the driver and/or passengers depends on the technician properly diagnosing and repairing these systems. As with all electrical systems, the technician must have a basic understanding of the operation of the restraint system before attempting to perform any service.

There are many safety cautions associated with working on air bag systems. Safe service procedures are accomplished through proper use of the service manual and by understanding the operating principles of these systems.

In a two-point system, the occupant must manually lock the lap belt.

Passive Seat Belt Systems

Shop Manual
Chapter 15, page 611

The passive seat belt system automatically puts the shoulder and/or lap belt around the driver or occupant (Figure 15-1). The automatic seat belt system operates by means of DC motors that move the belts by means of **carriers** on tracks (Figure 15-2). The carriers are attached to the shoulder anchor to move or carry the anchor from one end of the track to the other.

Figure 15-1 Passive automatic seat belt system operation.

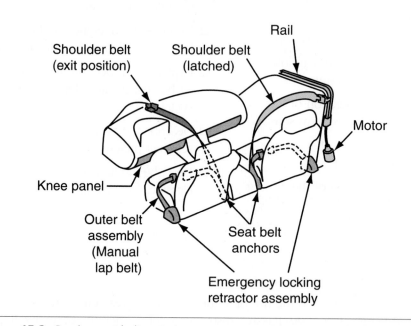

Figure 15-2 Passive seat belt restraint system uses a motor to put the shoulder harness around the occupant.

One end of the seat belt is attached to the carrier; the other end is connected to the **inertia lock retractors** (Figure 15-3). Inertia lock retractors use a pendulum mechanism to lock the belt tightly during sudden movement. When the door is opened, the outer end of the shoulder harness moves forward (to the A-pillar) to allow for easy entry or exit (Figure 15-4). When the door is closed and the ignition switch is placed in the RUN position, the motor moves the outer end of the harness to the locked position in the B-pillar (Figure 15-5).

The automatic seat belt system uses a control module to monitor operation (Figure 15-6). The monitor receives inputs from door ajar switches, limit switches, and the emergency release switch.

The door ajar switches signal the position of the door to the module. The switch is open when the door is closed. This signal is used by the control module to activate the motor and move the harness to the lock point behind the occupant's shoulders. If the module receives a signal that the door is open, regardless of ignition switch position, it will activate the motor to move the harness to the FORWARD position.

The limit switches inform the module of the position of the harness. When the harness is moved from the FORWARD position, the front limit switch (limit A) closes. When the harness is located in the LOCK position, the rear limit switch (limit B) opens and the module turns off the

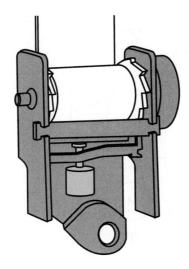

Figure 15-3   Inertia lock seat belt retractor.

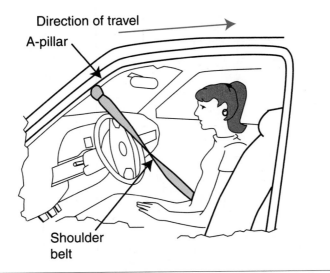

Direction of travel

A-pillar

Shoulder belt

Figure 15-4   When the door is opened, the motor pulls the harness to the A-pillar.

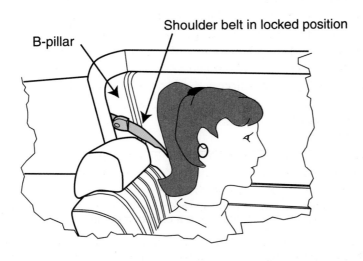

Shoulder belt in locked position

B-pillar

Figure 15-5   When the door is closed and the ignition switch is in the RUN position, the motor draws the harness to its lock position.

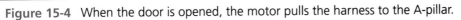

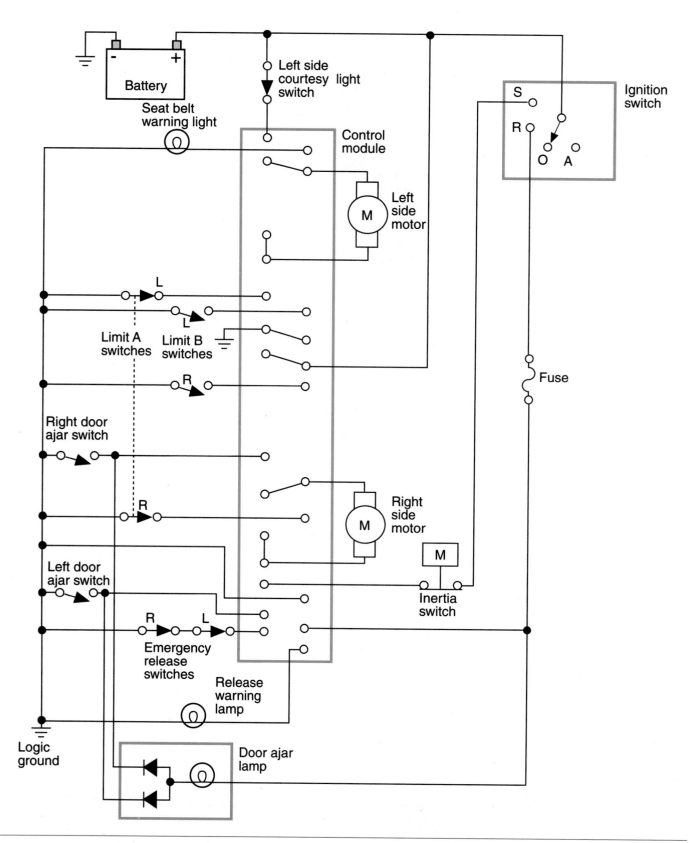

Figure 15-6 Typical circuit diagram of automatic seat belt system using a control module.

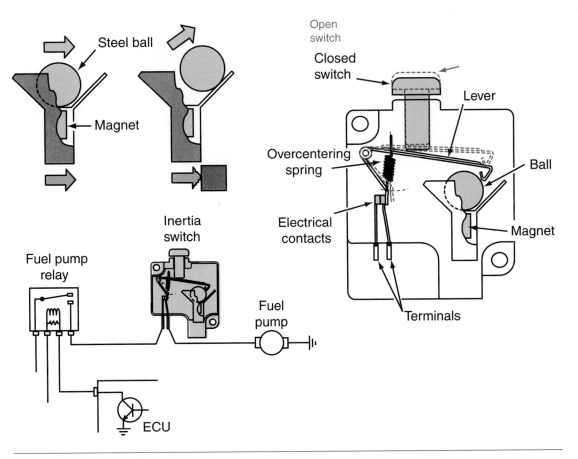

Figure 15-7 Fuel pump inertia switch.

power to the motor. When the door is opened, the module reverses the power feed to the motor until the A switch is opened.

An emergency release mechanism is provided in the event that the system fails to operate. The normally closed emergency release switch is opened whenever the release lever is pulled. The module will turn on the warning lamp in the instrument panel and sound a chime to alert the driver. The opened switch also prevents the harness retractors from locking.

Ford incorporates the **fuel pump inertia switch** into the automatic seat belt system. The fuel pump inertia switch is a normally closed switch that will open if the vehicle is involved in an impact at speeds over 5 mph or if it rolls over. When the switch opens, it turns off power to the fuel pump. This is a safety feature to prevent fuel from being pumped onto the ground or hot engine components if the engine dies. The switch has to be manually reset if it is triggered (Figure 15-7). If the seat belt module receives a signal that the switch is open, it prevents the harness from moving to the forward position if the door opens.

Air Bag Systems

The need to supplement the existing restraint system during frontal collisions has led to the development of the supplemental inflatable restraint (SIR) or air bag systems (Figure 15-8).

Today the most common name for the air bag system is supplemental restraint systems (SRS). The air bags are considered supplemental restraints because the seat belts must be worn at all times in order to provide maximum occupant protection. The air bag is a supplement and

Shop Manual
Chapter 15,
pages 616, 621

Figure 15-8 Air bag deployment sequence.

the seat belt is the primary restraint system. Seat belts must be worn in an air bag–equipped vehicle for the following reasons:

1. Seat belts hold the occupants in the proper position when the air bag inflates.
2. Seat belts reduce the risk of injury in less severe accidents in which the air bag does not deploy.
3. Seat belts reduce the risk of occupant ejection from the vehicle, thus reducing the possibility of injury.

The air bag system contains an inflatable air bag module that is designed into the steering wheel. Collapsible steering columns are used with the air bag system and tilt steering wheels are still optional. If the vehicle is involved in a frontal collision, the air bag inflates rapidly to keep the driver's body from flying ahead and hitting the steering wheel or windshield. The frontal impact must be within 30 degrees of the vehicle centerline to deploy the air bag. The air bag system helps to prevent head and chest injuries during a collision. The air bag system may be referred to as a passive restraint because it does not require active participation by the driver.

A BIT OF HISTORY

Although there were several early attempts at developing air bags, it was not until the mid-1980s that many manufacturers made air bags available as an option. In 1988, Chrysler was the first automotive manufacturer to offer the driver-side air bag as standard equipment.

Shop Manual
Chapter 15, page 623

Common Components

A typical air bag system consists of sensors, a diagnostic module, a clockspring, and an air bag module. See the typical location (Figure 15-9) of the common components of the SRS system.

> **AUTHOR'S NOTE:** Many of the components used for driver-side air bags are similar to those used in passenger-side air bags. The basic operation of the two systems is the same.

The air bag is made of neoprene-coated nylon.

Air Bag Module. The air bag module contains the **air bag** and inflator assembly packaged into a single module. This module is mounted in the center of the steering wheel (Figure 15-10).

The purpose of the air bag module is to inflate the air bag in a few milliseconds when the vehicle is involved in a frontal collision. A typical fully inflated driver's-side air bag has a volume of 2.3 cu. ft. (65 L).

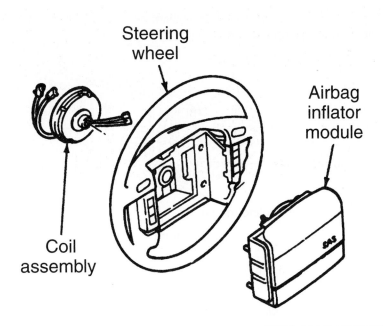

Figure 15-9 Typical location of components of the air bag system.

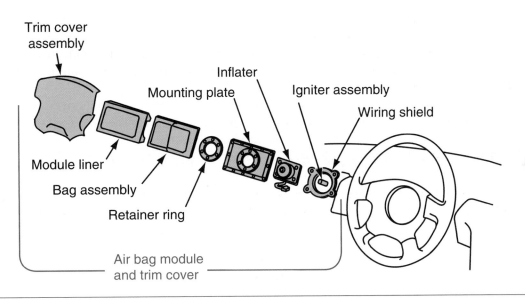

Figure 15-10 Air bag module components.

The air bag module uses pyrotechnology (explosives) to inflate the air bag. The **igniter** is an integral component of the inflator assembly (Figure 15-11) because it starts a chemical reaction to inflate the air bag. The igniter is a combustible device that converts electric energy into thermal energy to ignite the inflator propellant.

At the center of the igniter assembly is the **squib,** which contains zirconic potassium percolate (ZPP). The squib is similar to a blasting cap. Squib is a pyrotechnic term used for a fire cracker that burns but does not explode. The squib starts the process of air bag deployment. When as little as 400 ma is supplied through the squib, the air bag deploys.

Three components are required to create an explosion: fuel, oxygen, and heat. The squib and the igniter charge of barium potassium nitrate (a very fast-reacting explosive) provide the heat necessary for inflator module explosion. Fuel is supplied by the generant, which contains

The air bag module cannot be serviced. If it has been deployed, or is defective, it must be replaced.

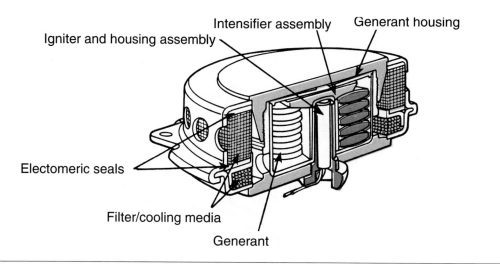

Intensifier assembly Generant housing
Igniter and housing assembly

Electomeric seals

Filter/cooling media

Generant

Figure 15-11 IIgniter assembly.

sodium azide and cupric oxide. The sodium azide provides hydrogen and the cupric oxide provides oxygen. When the chemicals in the inflator module explode, large quantities of hot, expanding nitrogen gas are produced very quickly. This expanding nitrogen gas flows through the igniter assembly diffuser, where it is filtered and cooled before inflating the air bag. Four layers of screen are positioned on each side of the ceramic in the filter. Sodium oxide dust is trapped by the filter. Sodium hydroxide is an irritating caustic. Therefore, automotive technicians are always warned to wear safety goggles and protective gloves when servicing deployed air bags. Within seconds after air bag deployment, the sodium hydroxide changes to sodium carbonate.

Tear seams in the steering wheel cover and in the instrument panel cover above the passenger's-side air bag split easily and allow the air bag to exit from the module. Large openings under the air bag, where it attaches to the module, allow the air bag to deflate in 1 second so it does not block the driver's view or cause a smothering condition.

Combustion temperature in the inflator module reaches about 2,500°F (1,371°C), but the air bag will remain slightly above room temperature. Typical by-products from inflator module combustion are:

1. Nitrogen—99.2%.

2. Water—0.6%.

3. Hydrogen—0.1%.

4. Sodium oxide—less than 1/10 of 1 part per million (ppm).

5. Sodium hydroxide—very minute quantity.

Many air bags pack corn starch into the inflator module. This, along with other combustion by-products, may appear as a white dust during and after air bag deployment.

Not all air bags systems use nitrogen gas to inflate the bag; some use compressed argon gas to inflate the air bag.

Shop Manual
Chapter 15, page 625

Clockspring. The clockspring conducts electrical signals to the module while allowing steering wheel rotation (Figure 15-12). The clockspring is a winding of special electric conductor tape housed in a plastic retainer. The clockspring maintains continuity between the air bar (and any steering wheel–mounted switches) and body wiring harness as the steering wheel is rotated. The clockspring is located between the column and the steering wheel. The clockspring electrical connector contains a long conductive ribbon. The wires from the air bag electrical system are connected through the underside of the clockspring electrical connector to one end of the

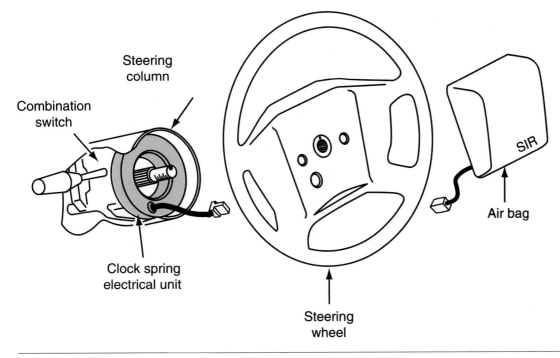

Steering
column

Combination
switch

Clock spring
electrical unit

Steering
wheel

SIR

Air bag

Figure 15-12 The clockspring provides for electrical continuity in all steering wheel positions.

conductive ribbon. The other end of the conductive ribbon is connected through wires on the top side of the clockspring electrical connector to the air bag module. When the steering wheel is rotated, the conductive ribbon winds and unwinds, allowing steering wheel rotation while completing electrical contact between the system and the air bag module.

Diagnostic Module. The air bag control module, called the occupant restraint controller (ORC), constantly monitors the readiness of the air SRS electrical system. If the ORC determines that there is a fault in the system, it will illuminate the indicator light and store a diagnostic trouble code. Depending on the fault, the SRS system may be disarmed until the fault is repaired.

The ORC also supplies backup power to the air bag module in the event the battery or cables are damaged during an accident. The stored charge can last for up to 30 minutes after the battery is disconnected.

It is important for the technician to understand that not all SRS system ORCs are capable of turning off the system in the event of a fault. A typical ORC performs the following functions:

1. Controls the instrument panel warning lamp.

2. Continuously monitors all air bag system components.

3. Controls air bag system diagnostic functions.

4. Provides an energy reserve to deploy the air bag if battery voltage is lost during a collision.

On some systems, the ORC is responsible for deploying the air bag when appropriate signals are received from the sensors. These systems may be able to turn off the air bags if a fault is detected.

Most air bag sensors contain a resistor connected in parallel with the sensor contacts. When the ignition switch is in the RUN position, the ORC supplies a small amount of current through these resistors to monitor the system. If a short, ground, or open circuit occurs in the wiring or sensors, the current flow changes. When the ORC senses this condition, it illuminates the air bag warning light in the instrument panel.

The clockspring is also known as the coil assembly, the cable reel assembly, the coil spring unit, and the contact reel.

Shop Manual
Chapter 15, page 617

The ORC is also called the air bag control module (ACM).

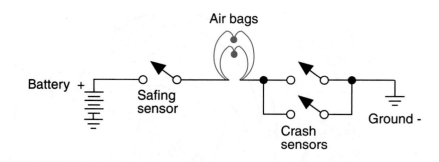

Figure 15-13 Typical sensor wiring circuit diagram.

Sensors. To prevent accidental deployment of the air bag, most systems require that at least two sensor switches be closed to deploy the air bag (Figure 15-13). The number of sensors used depends on the system design. Some systems use only a single sensor and others use up to five. The name used to identify the different sensors also varies among manufacturers.

On most three-sensor systems, the **crash sensors** are usually located in the engine compartment or below the headlights. Crash sensors are normally open electrical switches designed to close when subjected to a predetermined impact. A single **safing sensor** is usually located inside the ORC, which in turn is mounted on the centerline of the vehicle in the passenger compartment. The safing sensor determines if the collision is severe enough to inflate the air bag. When one of the crash sensors and the safing sensor closes, the electrical circuit to the igniter is complete. The igniter starts the chemical chain reaction that produces heat. The heat causes the generant to produce nitrogen gas, which fills the air bag.

There are several different types of sensors used. Common sensors include mass-type, roller-type, and accelerometers. The mass-type sensor contains a normally open set of gold-plated switch contacts and a gold-plated ball that acts as a sensing mass (Figure 15-14). The gold-plated ball is mounted in a cylinder coated with stainless steel. A magnet holds the ball about 1/8 inch (3.2 mm) away from the contacts. When the vehicle is involved in a frontal collision of sufficient force, the sensing mass (ball) moves forward in the sensor and closes the switch contacts.

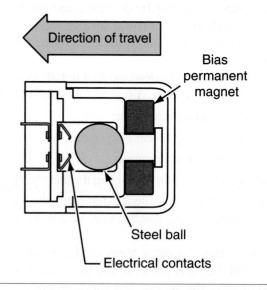

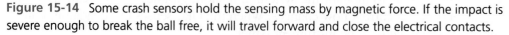

Figure 15-14 Some crash sensors hold the sensing mass by magnetic force. If the impact is severe enough to break the ball free, it will travel forward and close the electrical contacts.

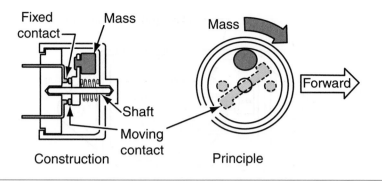

Figure 15-15 Mass-type air bag sensor with pivoted weight connected to a moving contact.

For proper operation, sensors must be mounted with the forward marking on the sensor facing toward the front of the vehicle and in the original position designed by the manufacturer. Sensor brackets must not be bent or distorted.

Some mass-type air bag sensors contain a pivoted weight connected to a moving contact. When the vehicle is involved in a frontal collision with sufficient impact to deploy the air bag, the sensor weight moves in a circular path until the moving contact touches a fixed contact (Figure 15-15).

The roller-type sensor has a roller mass mounted on a ramp (Figure 15-16). One sensor terminal is connected to the ramp. The second sensor terminal is connected to a spring contact extending through an opening in the ramp without contacting the ramp. A 10,000-Ω resistor is connected in parallel to the sensor contacts. The roller is held against a stop by small retractable springs on each side. These springs are similar to a retractable tape measure. If the vehicle is involved in a frontal collision at a high enough deceleration rate to deploy the air bag, the roller moves up the ramp and strikes the spring contact. In this position, the roller completes the circuit between the ramp and the spring contact.

In many air bag systems, **accelerometers** are used to sense deceleration forces. An accelerometer generates an analog voltage in relation to the severity of deceleration forces. The accelerometer also senses the direction of impact force. The accelerometer contains a piezoelectric element that is distorted during a collision. This element generates an analog

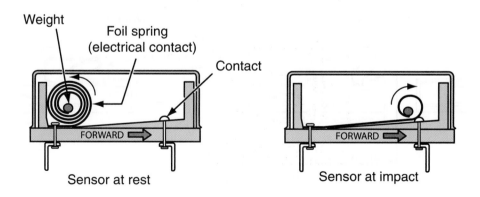

Figure 15-16 Roller-type air bag sensor.

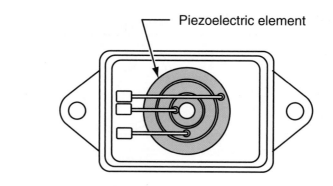

Figure 15-17   Accelerometer air bag sensor with piezoelectric element.

voltage in relation to the impact force (Figure 15-17). Usually, the accelerometer-type sensor is inside the air bag computer. The analog voltage from the piezoelectric element is sent to a collision-judging circuit in the air bag computer. The accelerometer is capable of determining the direction and severity of impact. If the collision impact is great enough, the computer deploys the air bag.

Air bag systems do not deploy the air bag based on vehicle speed information. Most systems do not even have vehicle speed input. The sensors close or provide an electrical signal based on the rate of deceleration. Most systems require about 30 g of force before the air bag will be deployed.

Shorting Bars. The SRS electrical system is a dedicated system that is not interconnected with other electrical systems on the vehicle. All wiring harness connectors in the system are the same color for easy identification. Shorting bars are located in some of the component wiring harness connectors in the air bag system, such as the inflator module connector at the steering column base. The shorting bars connect terminals together when the wiring connectors are disconnected (Figure 15-18). Since the terminals are shorted, there is no way to have electrical potential, which prevents accidental air bag deployment. If the ORC connector is disconnected, the shorting bars in the connector illuminate the air bag light.

When connector is disconnected   When connector is connected

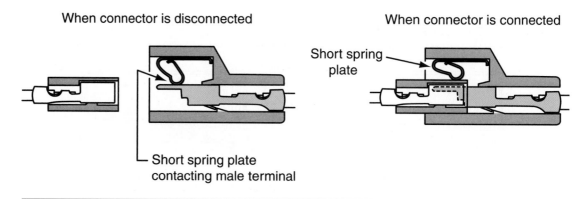

Short spring plate contacting male terminal

Short spring plate

Figure 15-18   Shorting bars on air bag system wiring connectors.

Air Bag Deployment

The sequence of events occurring during an impact of a vehicle traveling at 30 mph (48 kmh) is as follows:

1. When an accident occurs, the arming sensor is the first to close. It will close due to sudden deceleration caused by braking or immediately upon impact. One of the crash sensors will then close. The amount of time required to close the switches is within 15 milliseconds.

2. Within 40 milliseconds, the igniter module burns the propellant and generates the gas to completely fill the air bag.

3. Within 100 milliseconds, the driver's body has stopped forward movement and the air bag starts to deflate. The air bag deflates by venting the nitrogen gas through holes in the back of the bag.

4. Within 2 seconds, the air bag is completely deflated.

Shop Manual
Chapter 15, page 623

Air Bag Warning Lamp

The air bag system warning lamp indicates the system condition to the driver. The warning lamp is operated by the ORC. Ignition on and crank signals are received by the ORC. When the ignition switch is placed in the RUN position, the air bag warning lamp should illuminate for a bulb check. In some systems, the lamp will flash seven to nine times and then remain steadily illuminated while the engine is cranking. Once the engine starts, the air bag warning lamp should be extinguished. An air bag system failure may be indicated by any of the following warning lamp conditions:

1. If the lamp remains on but does not flash when the ignition is turned on.

2. If the lamp flashes seven to nine times and then remains on when the ignition is turned on.

3. If the lamp comes on when the engine is running.

4. If the lamp does not come on at any time.

5. If the lamp does not come on steadily while the engine is cranking.

If any of these lamp conditions are present, the driver should have the air bag system checked.

Passenger-Side Air Bags

Federal law expanded to require that all passenger vehicles produced after 1995 be equipped with front passenger-side air bags (Figure 15-19). Since there is a greater distance between the passenger and the instrument panel compared to the distance between the driver and the steering wheel, the passenger-side air bag is much larger. A typical passenger-side air bag has a fully inflated volume of 7 cubic feet (198 L). In most systems, the passenger-side air bag deploys with the driver-side air bag.

 AUTHOR'S NOTE: Reference here to passenger-side air bags means the front-seat passenger air bag that is deployed from the instrument panel. This system is not to be confused with side-impact air bags, which will be discussed later.

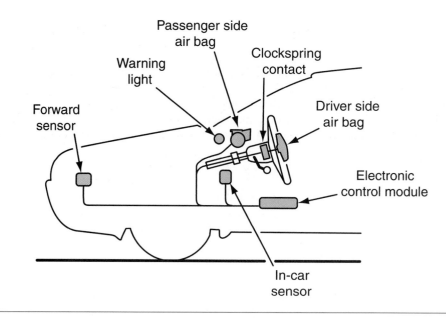

Figure 15-19 Components of typical driver and passenger air bag system.

Hybrid Air Bag Types

Up to this point, the discussion of air bags has centered around the conventional sodium azide type. However, there are **hybrid air bag** systems that use compressed gas to fill the air bag. There are three common types of hybrid air bag modules.

Solid Fuel with Argon Gas

The first use of solid-fuel hybrid air bags was on the passenger-side air bags. They are now used for driver- and passenger-side air bags. The hybrid inflator module contains an initiator similar to the squib in other inflator modules. However, the hybrid inflator module also has a container of pressurized argon gas (Figure 15-20). The same method is used to energize the

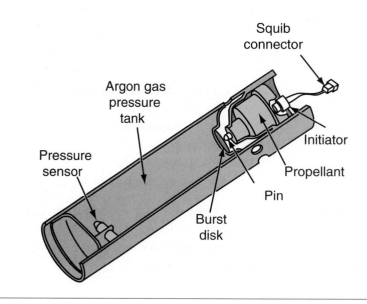

Figure 15-20 Hybrid inflator module with argon gas pressure chamber.

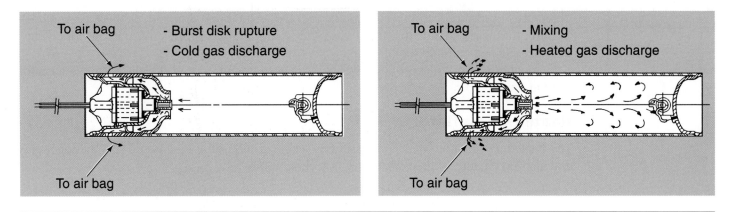

Figure 15-21 When the initiator is energized, the propellant explodes and punctures the container and allows pressurized argon gas to fill the air bag.

initiator in the hybrid inflator as in conventional systems. When the initiator is energized, the propellant surrounding the initiator explodes and pushes out the burst disc. As the pressurized argon escapes through the exhaust holes and fills the air bag, the burning propellant heats the argon gas (Figure 15-21). Heating of the gas makes it expand quickly to fill the air bag.

The early version of the hybrid system used a pressure sensor mounted in the end of the argon gas chamber opposite from the initiator and propellant. This sensor sends a signal to the ORC in relation to the argon gas pressure. If the gas pressure decreases below a preset value, the module illuminates the air bag warning light.

Liquid Fuel with Argon Gas

The liquid-fueled air bag module uses a small quantity of ethanol alcohol to blow out the burst disc (Figure 15-22). Although similar to the hybrid system previously discussed, there are some

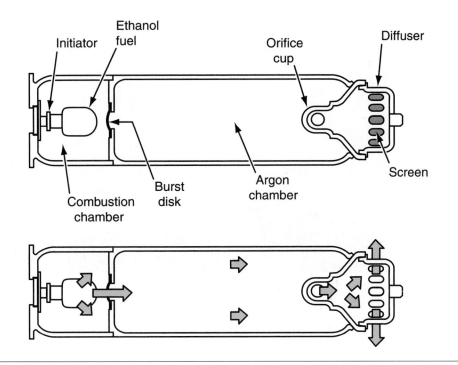

Figure 15-22 Liquid-fueled hybrid inflator operation.

differences in methods. In this system, the fuel blows the burst disc and, as the fuel continues to burn, the heat expands the argon gas. The expanding gas pushes through an orifice cup, over the diffuser, and into the air bag.

Alcohol is used because it ignites at a lower temperature and does not leave a harmful residue.

Heated Gas Inflator

The heated gas inflator (HGI) is pressurized with a combustible mixture of 12% hydrogen gas and air. The mixture is ignited with a pyrotechnic squib.

Multistage Air Bag Deployment

One recent development to the air bag system is the development of **multistage air bags.** Multistage air bags are hybrid air bags that use two squibs to control the rate of inflation. Multistage air bags are used for both driver- and passenger-side air bags. These modules apply the principle of using heat to expand the argon gas to fill the air bag. The bag fills faster as the heat increases.

The air bag module of the multistage system uses two squibs (Figure 15-23). When one of the squibs is fired, the air bag will begin to deploy. The second squib is then fired to generate more heat so the air bag fills faster (Figure 15-24). The length of time between the firing of the two squibs determines the rate of air bag deployment. In a minor accident that requires air bag deployment at a slow rate, only the first squib is ignited. The second squib may be ignited 160 ms later to use up the other igniter charge but this is too late to fill the air bag. As the severity of the deceleration forces indicate faster air bag deployment is needed, the firing of the squibs will get closer together.

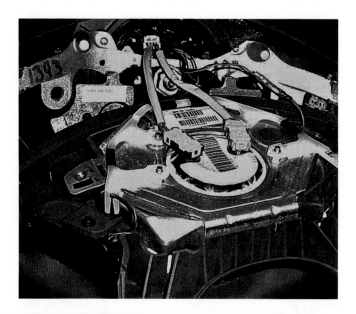

Figure 15-23 The multistage air bag module uses two squibs.

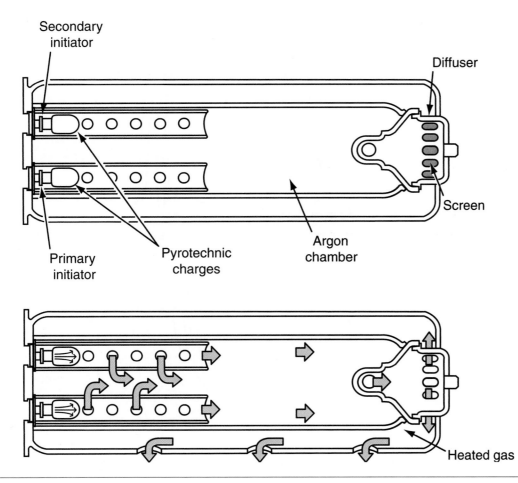

Figure 15-24 Sequence of events within the multistage inflator.

Side-Impact Air Bags

Many manufacturers are now offering side-impact air bag systems. Most of these are a single-stage hybrid design. The location of the air bag varies depending on the vehicle. Some are designed to come out of the door panel (Figure 15-25), from the seat back (Figure 15-26), between the A-pillar and the headliner (Figure 15-27), or from a roof-mounted curtain in the headliner that protects both the front- and rear-seat occupants (Figure 15-28).

The side-impact air bags deploy separately from the front air bags. The system may have a separate control module mounted in the B-pillars of the vehicle or sensors that relay impact information to the ORC. The ORC then deploys the side air bags.

Air Bag On/Off Switches

For the air bags to perform safely, there should be at least 10 inches (25 cm) between the air bag module and the occupant. Also, children should not sit in the front seat with an air bag, and a rearward-facing infant seat should *NEVER* be used in the front seat with an air bag. Some vehicles with limited rear seating may be factory equipped to turn off the passenger air bag if it would not

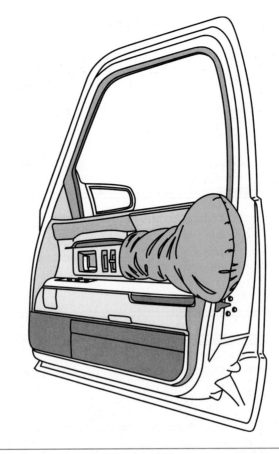

Figure 15-25 Side-impact air bag located in the door panel.

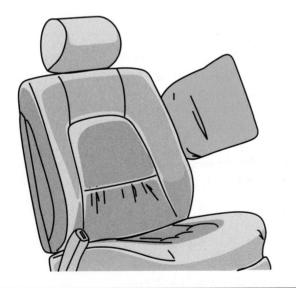

Figure 15-26 Side-impact air bag located in the seat back cushion.

Figure 15-27 Side-impact air bag designed to protect the occupant's head from injury.

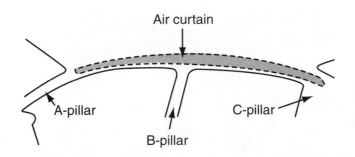

Air curtain

A-pillar

B-pillar

C-pillar

Figure 15-28 Side-impact air curtains protect both front- and rear-seat occupants.

Figure 15-29 The BabySmart system used by Mercedes automatically deactivates the passenger-side air bag when a child safety seat is placed in the front seat.

be safe to have it deploy. One such type of automatic system is used by Mercedes (Figure 15-29). This system uses a resonator built into the child seat to recognize the seat is in place. A light on the dash will confirm that the passenger-side air bag is turned off.

Most systems rely on input from the driver or passenger to turn a switch (Figure 15-30). Early systems placed a resistor in the passenger-side air bag circuit when the switch was turned to the OFF position (Figure 15-31). Turning the switch opens the circuit between the ORC and the passenger-side air bag so it will not deploy. The resistor is used to trick the ORC into believing the circuit is still intact so it will not set false fault codes.

Newer systems use a MUX signal from an on/off module (Figure 15-32). The occupant restraint controller (ORC) provides a pulsed signal to the on/off module at a frequency of 10 Hz with a 3% duty cycle. The ORC monitors the voltage drop across the switch. If the switch is in the ON position, 4 volts will be monitored. In the OFF position, about 10 volts will be monitored. A reading of 20 volts will be considered an open, while a reading of zero volts would set a short-circuit fault. If the switch is in the OFF position, the ORC will deactivate the passenger air bag internally. This system does not interrupt the circuit as earlier systems did.

Figure 15-30 Passenger-side air bag on/off switch.

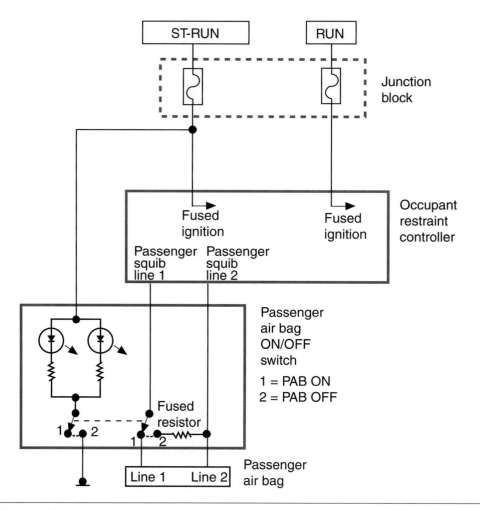

Figure 15-31 Wiring diagram of a hardwired passenger-side air bag on/off switch.

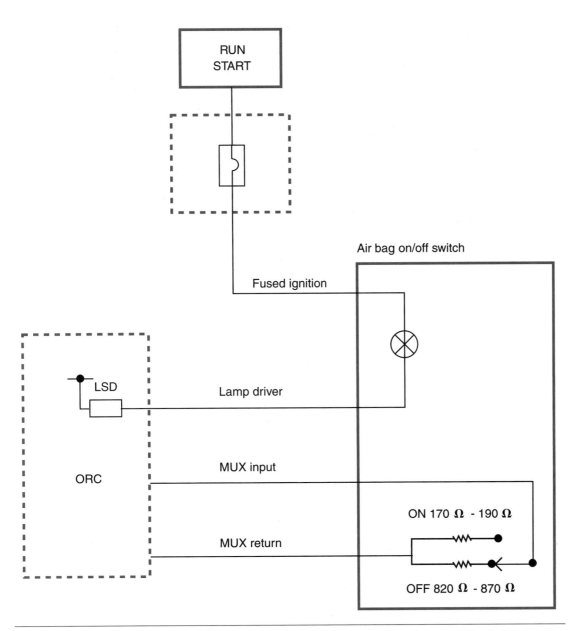

Figure 15-32 Wiring diagram of MUX circuit for passenger-side air bag on/off switch.

Deactivation and Retrofit On/Off Switches

Some owners may desire to have their air bag systems deactivated or to have a switch installed that will allow them to deactivate the system as needed. At this time, in order to perform this service, the vehicle owner must first obtain permission from the National Highway Traffic Safety Administration (NHTSA). Permission may be obtained to turn off one or both of the front air bags. The vehicle owner must supply a letter of approval before the deactivation or switch kit is installed. Deactivation may be approved due to medical reasons, size of the driver (not able to sit at least 10 inches [25 cm] from the air bag), or because a child must sit in the front seat.

The customer must also sign a waiver form written on the shop's letterhead. This form releases the shop owner and technician from any liability that may occur as a result of the air bag systems being turned off. The waiver also gives the technician permission to put the required warning labels on the vehicle.

Seat Belt Pretensioners

To assure that the driver and/or passenger stay in position during an accident, some vehicles are equipped with seat belt **pretensioners**. There are two ways of mounting the pretensioner.

The first is to mount the pretensioner on the buckle side of the seat belt (Figure 15-33). At the same time the front air bags are deployed, the control module will also deploy the pretensioner. A small piston is attached to a cable connected to the buckle. There is a charge below the piston. When the pretensioner is fired, the piston will travel up the cylinder, pulling the buckle tight by the cable.

The pretensioner can also be mounted on the retractor side of the seat belt (Figure 15-34). The system type shown has a retractor assembly with a fan wheel–type unit attached to one end (Figure 15-35). When the pretensioner is fired, a series of balls shoot out of the channel and hit the fan wheel. As the balls hit the fan wheel, the retractor rotates and pulls the seat belt tight. The last ball is a little bigger than the others and will lodge into the fan wheel, causing the seat belt to lock.

Figure 15-33 Buckle-mounted pretensioner.

Figure 15-34 Retractor-mounted pretensioner.

Figure 15-35 The balls are shot at the fan wheel, which causes the retractor to wind up the seat belt tight against the occupant.

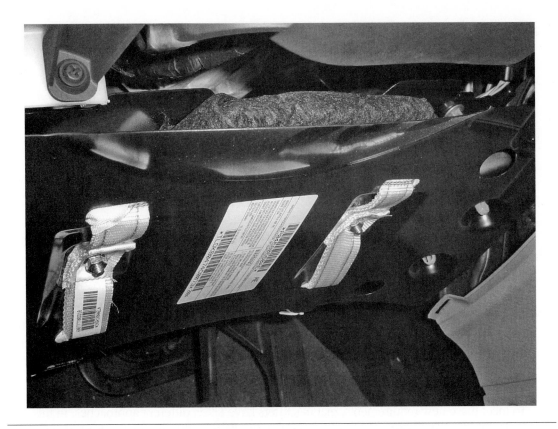

Figure 15-36 Inflatable knee blocker.

Inflatable Knee Blockers

Manufacturers are now incorporating a driver's-side **inflatable knee blocker (IKB)** into their air bag systems (Figure 15-36). The IKB is located on the driver's side of the vehicle beneath the instrument panel cover and is attached to the instrument panel reinforcement. The IKB deploys simultaneously with the driver's-side air bag to increase driver impact protection. The IKB provides upper-leg protection and positioning of the driver. When the IKB is deployed, it pushes a tethered plate against the driver's knees. This keeps the driver in the correct upright position during a collision, so the air bag is more efficient.

Occupant Classification Systems

As the result of an amendment to the Federal Motor Vehicle Safety Standard 208, manufacturers are presently designing and installing air bag systems that reduce the risk of injuries resulting from air bag deployment. The goal of this amendment is to reduce injuries suffered by children and small adults that are in the **fifth percentile female** weight classification. The fifth percentile female is classified to be those who weigh less than 100 pounds (45 kg). The amendment mandates that the passenger-side air bag be suppressed when an infant in a rear-facing infant seat (RFIS) occupies the front passenger seat. The amendment also mandated that a passenger air bag disable lamp (PADL) be illuminated whenever the passenger seat is occupied and the passenger-side air bag has been suppressed (Figure 15-37). If the front passenger seat is not occupied, the lamp is not illuminated.

Shop Manual
Chapter 15, page 626

Figure 15-37 The PADL illuminates if the air bag is suppressed.

AUTHOR'S NOTE: The fifth percentile female is determined by averaging all potential occupants by size and then plotting the results on a graph. The middle of the bell curve on the graph would indicate the majority of occupants. At the far right of the bell curve would be those occupants who are very large, while on the left side of the curve would be occupants that are very small. At the fifth percentile range on the left of the curve, the majority of occupants would be female.

To meet these new requirements, manufacturers have taken different approaches. In this section, the Delphi and TRW systems are presented to provide two different methods of meeting this regulation. The Delphi system uses a bladder to determine weight, and the TRW system uses strain gauges. Both systems use an occupant classification module (OCM) that determines the weight classification of the front passenger and sends this information to the ORC. Also, both system use multistage passenger-side air bags.

Delphi Bladder System

The bladder system uses a silicone-filled bladder that is positioned between the seat foam and the seat support (Figure 15-38). A pressure sensor is connected by a hose to the bladder (Figure 15-39).

Figure 15-38 Bladder used to determine occupant classification.

Figure 15-39 The pressure sensor changes voltage signals as weight is added to the bladder.

The three-wire pressure sensor operates similar to the way an MAP sensor operates. When the seat is occupied, pressure that is applied to the bladder disperses the silicone and the pressure sensor reads the increase in pressure. The pressure reading is inputted to the ORC (Figure 15-40).

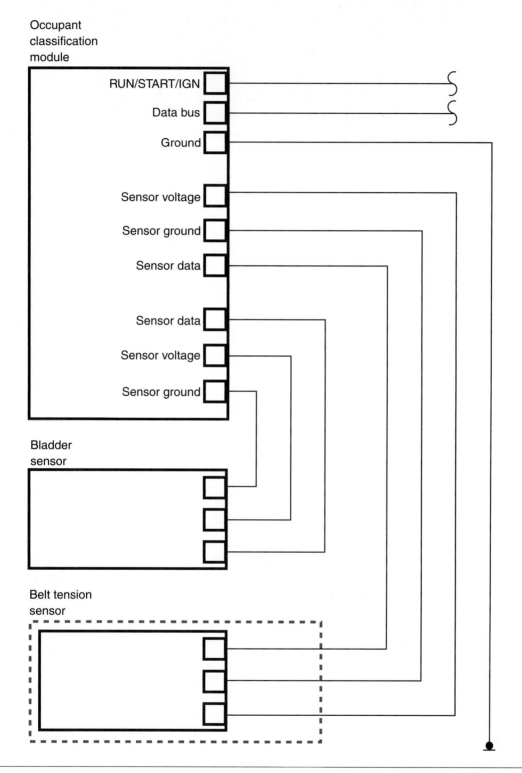

Figure 15-40 Schematic of bladder occupant classification system.

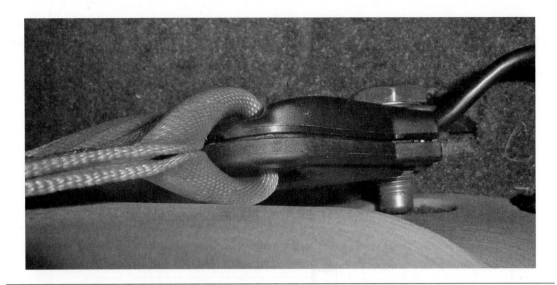

Figure 15-41 Belt tension sensor.

Shop Manual
Chapter 15, page 628

Since pressure is used to determine seat occupation and ultimately occupant classification, the system will correct for changes in atmospheric pressures. In addition, natural aging of the seat foam is also learned by monitoring gradual changes. The OCM stores seat aging and calibration information in the ORC. If a new OCM is installed, it will retrieve this information from the ORC so the system will continue to function properly.

When the seat is not occupied, the OCM compares the sensor voltage with the value stored in memory. The voltage values measures by the sensor will change as weight is added to the seat. Based on the change of voltage from the sensor, the OCM can determine the weight of the occupant.

Based on the weight information that the OCM sends to the ORC, the following is determined:

❏ An empty seat. The PADL is off and the air bag is suppressed.

❏ Weight equivalent to or less than that of a six-year-old child. The PADL is illuminated and the air bag is suppressed.

❏ Weight equivalent to or greater than that of a fifth percentile female. The PADL light is off and the front passenger air bag is enabled. Deployment rate is based on the severity of the impact.

Since an infant seat that is securely strapped into the seat will cause an increase of downward pressures on the bladder, the system uses a **belt tension sensor (BTS).** The BTS is a strain gauge–type sensor located on the seat belt anchor (Figure 15-41). The increase in pressures can be great enough to indicate a weight greater than that of a fifth percentile female is in the seat. In this case the air bag will not be suppressed. However, the BTS will indicate that the belt is tight around an object (about 24–26 lbs. [11–12 kg] of force). The reading will indicate that the belt is tighter than what it normally would be for a belt around a person.

As seat belt tension is increased, the sensor voltage changes. Based on the change of voltage from the BTS, the OCM can estimate how much of the sensed load results from the cinched seat belt. If the BTS indicates a cinched seat belt load over a certain threshold, the OCM determines a rear-facing infant classification.

TRW Strain Gauge System

The TRW system uses four strain gauges to determine weight classification (Figure 15-42). One strain gauge is located at each corner of the seat frame where the frame attaches to the seat riser. The strain gauges support the weight of the seat. Data from each sensor is sent to the OCM (Figure 15-43).

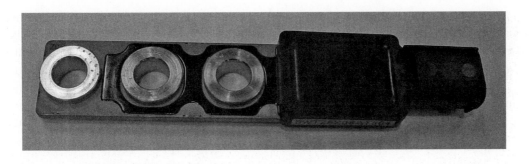

Figure 15-42 Strain gauge.

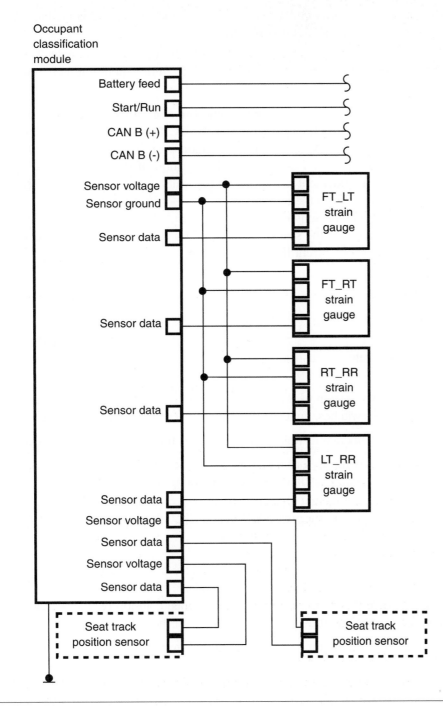

Figure 15-43 Schematic of strain gauge occupant classification system.

The OCM compares current voltage readings with the values stored in memory. The electrical resistance of the strain gauge changes based on the amount of strain against it. A circuit board is bonded to the frame of the gauge. The circuit board has a grid made of metallic foil that changes in resistance when strain is applied. As weight is added to the seat, the voltage values of the sensors change. Based on the change of voltage from each sensor, the OCM can determine the occupant's weight. The OCM will attempt to calibrate every key off if the seat is not occupied.

Based on the weight information that the OCM sends to the ORC, the following is determined:

❑ An empty seat. The PADL is off and the air bag is suppressed.

❑ Weight equivalent to or less than that of a RFIS. The PADL is illuminated and the air bag is suppressed.

❑ Weight equivalent to a child. The PADL is off and the air bag is enabled. Deployment is at the low-risk deployment level.

❑ Weight equivalent to or greater than that of a fifth percentile female. The PADL is off and the front passenger air bag is enabled. Deployment rate is based on the severity of the impact.

The TRW system also utilizes a **seat track position sensor (STPS)** on both the driver and passenger seats. The STPS provides information to the OCM concerning the position of the seat in relation to the air bag. The OCM sends this information over the data bus to the ORC. The ORC modifies the deployment strategy based on this information. If the occupant's seat position is closer to the air bag, the deployment rate of the air bag will be slower than for an occupant's seat that is position farther away from the air bag.

The STPS uses a Hall-type sensor. The STPS is mounted onto a seat rail while a steel plate is mounted to the seat track. As the seat is moved, the steel plate covers or uncovers the sensor's magnetic field. This alters the current flow in the sensor.

Summary

Terms to Know

Accelerometers

Air bag

Belt tension sensor
(BTS)

Carriers

Crash sensors

Fifth percentile
female

Fuel pump inertia
switch

Hybrid air bag

Igniter

Inertia lock retractors

Inflatable knee
blocker (IKB)

❑ Passive restraints operate automatically with no action required on the part of the driver or occupant.

❑ The automatic seat belt system uses a control module to monitor operation by receiving inputs from door ajar switches, limit switches, and the emergency release switch.

❑ The air bag is a supplemental restraint. The seat belt is the primary restraint system.

❑ The air bag module is composed of the air bag and inflator assembly. It is packaged in a single module and mounted in the center of the steering wheel.

❑ The diagnostic module constantly monitors the readiness of the SRS electrical system. If the battery or cables are damaged during an accident, it supplies backup power to the air bag module.

❑ The igniter is a combustible device that converts electric energy into thermal energy to ignite the inflator propellant.

❑ Air bags will deploy if the vehicle is involved in a frontal collision of sufficient impact and the collision force is within 30 degrees on either side of the vehicle centerline.

❑ The total air bag deployment time from the instant of impact until the air bag is inflated is less than 160 ms.

❑ An accelerometer-type air bag sensor generates an analog voltage in relation to deceleration forces. The accelerometer also senses the direction of impact force.

❑ The clockspring electrical connector maintains electrical contact between the inflator module and the air bag electrical system.

❑ The air bag warning light indicates an inoperative air bag system.

❑ A hybrid inflator module contains a pressurized argon gas cylinder, which is punctured by the exploding propellant to inflate the air bag.

❑ Shorting bars connect air bag system squib terminals together when the terminal is disconnected. This will prevent accidental air bag deployment.

❑ Federal law expanded to require all passenger vehicles produced after 1995 be equipped with front passenger-side air bags.

❑ The air bag module of the multistage system uses two squibs. When one of the squibs is fired, the air bag will begin to deploy. The second squib is then fired to generate more heat so the air bag is filled faster.

❑ Side-impact air bags can be designed to come out of the door panel, from the seat back, between the A-pillar and the headliner, or from a roof-mounted curtain in the headliner that protects both the front- and rear- seat occupants.

❑ Some vehicles with limited rear seating may be factory equipped to turn off the passenger air bag if it would not be safe to have it deploy.

❑ At this time, in order to install air bag deactivation kits or on/off switches, the vehicle owner must first obtain permission from the National Highway Traffic Safety Administration (NHTSA).

❑ To assure the driver and/or passenger stay in position during an accident, some vehicles are equipped with seat belt pretensioners.

❑ There are two ways of mounting the pretensioner—on the buckle side or retractor side of the seat belt.

❑ The inflatable knee blocker (IKB) deploys simultaneously with the driver's-side air bag to provide upper-leg protection and positioning of the driver.

❑ Occupant classification systems are a mandated requirement designed to reduce the risk of injuries resulting from air bag deployment.

❑ The Delphi system uses a bladder to determine weight, and the TRW system uses strain gauges. Both systems use an occupant classification module (OCM) that determines the weight classification of the front passenger and sends this information to the ORC. Also, both systems use multistage passenger-side air bags.

❑ The belt tension sensor (BTS) is a strain gauge–type sensor located on the seat belt anchor that is used to indicate if an infant seat is cinched into the passenger-side front seat.

❑ The seat track position sensor (STPS) provides information to the OCM concerning the position of the seat in relation to the air bag.

Terms to Know
(continued)
Multistage air bags
Pretensioners
Safing sensor
Seat track position
   sensor (STPS)
Squib

Review Questions

Short-Answer Essay

1. Define the term *passive restraint*.

2. Describe the basic operation of automatic seat belts.

3. List and describe the design and operation of three different types of air bag system sensors.

4. List and explain two of the functions of the ORC used in air bag systems.

5. List the sequence of events that occur during air bag deployment.

6. What is the purpose of the clockspring?

7. Describe the deployment of the hybrid inflator module.

8. What is the purpose of multistage air bags?

9. Where is the side-impact air bag control module or sensor usually located?

10. List the common mounting locations of the seat belt pretensioner.

Fill in the Blanks

1. The _____ conducts electrical signals to the air bag module while permitting steering wheel rotation.

2. In the automatic seat belt system, the _____ switches inform the module of the position of the harness.

3. The _____ is a combustible device that converts electric energy into thermal energy to ignite the inflator propellant.

4. The diagnostic module supplies _____ _____ to the air bag _____ in the event that the battery or cables are damaged during an accident.

5. To prevent accidental deployment of the air bag, most systems require that at least _____ sensor switches be closed to deploy the air bag.

6. The frontal collision force must be within _____ degrees of the vehicle centerline to deploy the air bag.

7. An accelerometer-type air bag sensor produces an analog voltage in relation to _____ _____.

8. The current flow through the squib required to deploy the air bag is approximately _____ amperes.

9. A hybrid inflator module contains a cylinder filled with compressed _____ gas.

10. In order to deactivate the air bag system, the vehicle owner must first obtain permission from the _____ _____ _____ _____ _____.

Multiple Choice

1. The input signals to the control module of the automatic seat belt system are being discussed.
   Technician A says the door ajar switches signal the position of the harness.
   Technician B says that the limit switches signal when the emergency release switch is opened.
   Who is correct?
   A. A only
   B. B only
   C. Both A and B
   D. Neither A nor B

2. All of the following are characteristics of the occupant classification system EXCEPT:
   A. The OCM is responsible for the deployment of the passenger-side air bag.
   B. The system is designed to determine if a rear-facing infant seat is being used in the front passenger seat.
   C. The PADL illuminates if the passenger-side air bag is suppressed.
   D. Weight classification of fifth percentile female and greater allows air bag deployment.

3. Air bag components are being discussed.
   Technician A says the igniter is a combustible device that converts electric energy into thermal energy.
   Technician B says the inflation of the air bag is done through an explosive release of compressed air.
   Who is correct?
   A. A only
   B. B only
   C. Both A and B
   D. Neither A nor B

4. The air bag system is being discussed.
   Technician A says the clockspring is located at the bottom of the steering column.
   Technician B says the clockspring conducts electrical signals to the module while permitting steering wheel rotation.
   Who is correct?
   A. A only
   B. B only
   C. Both A and B
   D. Neither A nor B

5. The air bag system components are being discussed.
   Technician A says the ORC constantly monitors the readiness of the air SRS electrical system.
   Technician B says a crash sensor may be composed of a gold-plated ball held in place by a magnet.
   Who is correct?
   A. A only
   B. B only
   C. Both A and B
   D. Neither A nor B

6. Air bag sensors are being discussed.
   Technician A says the arrow on each sensor must face toward the rear of the vehicle.
   Technician B says air bag sensor brackets must not be bent or distorted.
   Who is correct?
   A. A only
   B. B only
   C. Both A and B
   D. Neither A nor B

7. Accelerometer-type air bag sensors are being discussed.
   Technician A says an accelerometer senses collision force and direction.
   Technician B says an accelerometer produces a digital voltage.
   Who is correct?
   A. A only
   B. B only
   C. Both A and B
   D. Neither A nor B

8. Which of the following is a characteristic of the strain gauge–type occupant classification system?
   A. If the seat is empty, the PADL is illuminated.
   B. The system uses a belt tension sensor to determine the presence of an infant seat.
   C. The air bag will deploy if a child is determined to be sitting in the seat.
   D. None of the above.

9. The air bag deployment loop is being discussed.
   Technician A says if the arming sensor contacts close, this sensor completes the circuit from the inflator module to ground.
   Technician B says if the contacts close in two crash sensors, the air bag is deployed.
   Who is correct?
   A. A only
   B. B only
   C. Both A and B
   D. Neither A nor B

10. Hybrid inflator modules are being discussed.
    Technician A says a pressure sensor in the argon gas cylinder sends a signal to the ASDM in relation to gas pressure in the cylinder.
    Technician B says when the initiator is energized, the propellant explodes and pierces the propellant container, allowing the pressurized argon gas to escape into the air bag.
    Who is correct?
    A. A only
    B. B only
    C. Both A and B
    D. Neither A nor B

Vehicles with Alternative Power Sources

Upon completion and review of this chapter, you should be able to:

❏ Explain the basic operation of an electric vehicle.

❏ Describe the typical operation of a hybrid vehicle.

❏ Explain the difference between parallel and series hybrids.

❏ Explain the purpose of regenerative braking.

❏ Describe the purpose of the 42-volt system.

❏ Explain the operating principles of an integrated starter generator (ISG) system.

❏ Describe how a proton exchange membrane produces electricity in a fuel cell system.

❏ List and describe the different fuels that can be used in a fuel cell system.

❏ Describe the purpose of the reformer.

❏ Explain how different types of reformers operate.

Introduction

Due to the increase in regulations concerning emissions and the public's desire to become less dependent on foreign oil, most major automotive manufacturers have developed alternative-fuel or alternate-power vehicles. This chapter explores several alternative power sources and includes a study of common hybrid systems. Also included is a discussion of the 42-volt system and its influence on the new technology of integrated starter generator idle stop systems. In addition, the final section of this chapter covers fuel cell theories and some of the methods that manufacturers are using to approach this alternative power source. These power sources are being sold in limited numbers, or they are still in the research and development stage.

Electric Vehicles

Since the 1990s, most major automobile manufacturers have developed an electric vehicle (EV). The EV powers its motor from a battery pack. The primary advantage of an EV is a drastic reduction in noise and emission levels. In California during the late 1990s, the California Air Resource Board (CARB) established a low-emission vehicles/clean-fuel program to further reduce mobile source emissions. This program established emission standards for five vehicle types (Figure 16-1): conventional vehicle (CV), transitional low-emission vehicle (TLEV), low-emission vehicle (LEV),

	CV	TLEV	LEV	ULEV	ZEV
NMOG	0.25*	0.125	0.075	0.040	0.0
CO	3.4	3.4	3.4	1.7	0.0
NOx	0.4	0.4	0.2	0.2	0.0

(*) Emission standards of NMHC

Figure 16-1 California tailpipe emission standards in grams per mile at 50,000 miles.

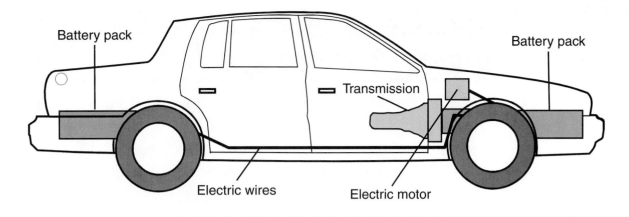

Figure 16-2 The electric vehicle is powered by an electric motor that receives its energy from battery packs.

ultra-low-emission vehicle (ULEV), and zero-emission vehicle (ZEV). The electric vehicle meets ZEV standards. Figure 16-2 shows the basic components of an electric vehicle.

General Motors introduced the EV1 electric car to the market in 1996. The original battery pack in this car contained twenty-six 12-volt batteries that delivered electrical energy to a three-phase, 102-kilowatt (kW) AC electric motor. The electric motor is used to drive the front wheels. The driving range is about 70 miles (113 km) of city driving or 90 miles (145 km) of highway driving. Temperature, vehicle load, and speed affect this range. A 1.2-kW charger in the vehicle's trunk can be used to recharge the batteries. This charger takes about 15 hours to fully recharge the batteries. An external Delco Electronics MAGNE CHARGE 6.6-kW inductive charger operating on 220 volts/30 amperes can recharge the batteries in 3 to 4 hours (Figure 16-3). The weatherproof plastic paddle is inserted into the charging port located at the front of the vehicle. Magnetic fields transfer the power.

Figure 16-3 Recharging the EV.

Figure 16-4 Positioning the EV's battery pack for installation. (Reprinted with permission)

If the batteries are fully charged, the EV1 accelerates from 0 to 60 mph (97 kmh) in 9 seconds and has a top speed of 80 mph (129 kmh). The EV1 is equipped with a Galileo electronic brake system that employs a computer and sensors at each wheel to direct power-assist braking, regenerative/friction brake blending, four-wheel antilock braking, traction assist, tire pressure monitoring, and system diagnostics. In 1998, GM installed nickel/metal/hydride batteries in the EV1 vehicles, which extended the driving range between battery charges to 160 miles (257 km).

The driving range of electric vehicles is their biggest disadvantage. Much research is being done to extend the range and to decrease the required recharging times. Currently, the use of nickel-metal-hydride or lead-acid batteries and permanent magnet motors has extended the operating range. Another disadvantage is that the battery pack adds substantial weight to the vehicle (Figure 16-4). Other features, such as **regenerative braking** and highly efficient accessories (such as a heat pump for passenger heating and cooling), are also being installed on electric vehicles. Regenerative braking means that the braking energy is turned back into electricity instead of heat. Other disadvantages are the cost of replacement batteries and the danger associated with the high voltage and high frequency of the motors.

Hybrid Vehicles

Shop Manual
Chapter 16, page 645

The first alternative vehicle was the electric vehicle, which has zero emissions and runs primarily on battery power. However, the battery has a limited energy supply and restricts the traveling distance. This limitation was a major stumbling block to most consumers; thus the use of EVs is mainly limited to fleets since they can be equipped with recharging stations. One method to improve the electric vehicle is the addition of an on-board power generator that is assisted by an internal combustion engine. The result is the hybrid electric vehicle (HEV). An HEV has two different power sources. In most hybrid vehicles, the power sources consist of a small displacement gasoline or diesel engine and an electric motor. The addition of the internal combustion engine means the vehicle cannot be classified as a ZEV. However, the engine does reduce emission levels significantly and increases fuel economy.

Figure 16-5 Hybrid power system. (Reprinted with permission)

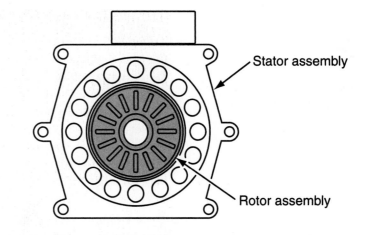

Stator assembly

Rotor assembly

Figure 16-6 Engine power is also used to rotate a generator that recharges the storage batteries and to drive the vehicle. The rotor assembly is a very powerful magnet that induces voltage into the stator windings as it is rotated.

Basically, the hybrid electric vehicle relies on power from the electric motor, the engine, or both (Figure 16-5). When the vehicle moves from a stop and has a light load, the electric motor moves the vehicle. Power for the electric motor comes from stored electricity in the battery pack. During normal driving conditions, the engine is the main power source. Engine power is also used to rotate a generator that recharges the storage batteries (Figure 16-6). The output from the generator can also be used to power the electric motor, which is run to provide additional power to the powertrain (Figure 16-7). A computer controls the operation of the electric motor, based on the power needs of the vehicle. During full-throttle or heavy-load operation, the computer sends additional electricity from the battery to the motor to increase the output of the powertrain.

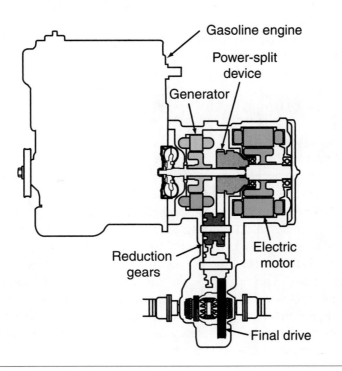

Gasoline engine

Power-split device

Generator

Reduction gears

Electric motor

Final drive

Figure 16-7 Hybrid power system with gasoline engine and electric propulsion motor.

The components of a typical hybrid vehicle include:

❑ *Batteries*. Some types of battery that are being used or experimented with now are the lead acid battery, the nickel-metal-hydride battery, and the lithium ion battery. In the development of the battery, thermal management must be taken into consideration. The temperature can vary from module to module, so the performance of the battery is dependent on the temperature. An imbalance in temperature will affect the power and capacity of the battery, the charge acceptance during regenerative braking, and vehicle operation. Also, passenger safety is a major concern. The batteries must be kept in sealed containers in order to ensure complete protection. The Toyota Prius (Figure 16-8) seals its noncaustic, nonflammable nickel-metal hydride battery in a carbon composite case positioned in the rear of the vehicle (Figure 16-9).

❑ *Electric motors*. One of the sources of power is the electric motor. The motor converts electrical energy to mechanical energy. This mechanical energy is what drives the wheels of the vehicle. This motor is designed to allow for maximum torque at low rpm. This gives the electric motor the advantage of having better acceleration than the conventional motor.

❑ *Regenerative braking*. About 30 percent of the kinetic energy lost during braking is in heat. When decreasing acceleration, regenerative braking helps to minimize energy loss by recovering the energy used to brake. It does this by converting rotational energy into electrical energy through a system of electric motors and generators. Regenerative braking assumes some of the stopping duties from the conventional friction brakes and uses the electric motor to help stop the car. To do this, the electric motor operates as a generator when the brakes are applied, recovering some of the kinetic energy and converting it into electrical energy. The motor becomes a generator by using the kinetic energy of the vehicle to store power in the battery for later use.

❑ *Ultra capacitors*. The **ultra capacitor** is a device that stores energy as electrostatic charge. It is the primary device in the power supply during hill climbing, acceleration, and the recovery of braking energy. To create a larger storage capacity for the ultra capacitors, the surface area must be increased and, in turn, the voltage is increased. However, because the voltage drops as energy is discharged, additional electronics are required to maintain a constant voltage.

Figure 16-8 The Toyota Prius was the world's first mass-produced hybrid vehicle.

Figure 16-9 Battery pack. (Reprinted with permission)

Propulsion

There are two typical ways to arrange the flow of power in an HEV. If the combustion engine is capable of turning the drive wheels as well as the generator, then the vehicle is referred to as a **parallel hybrid** (Figure 16-10). In a parallel hybrid configuration, there is a direct mechanical connection between the engine and the wheels. Both the engine and the electric motor can turn the transmission at the same time.

There is a further distinction between a **mild parallel hybrid** and a **full parallel hybrid** vehicle. A mild parallel hybrid vehicle has an electric motor that is large enough to provide regenerative braking, instant engine startup, and a boost to the combustion engine. A full parallel hybrid vehicle uses an electric motor that is powerful enough to propel the vehicle on its own.

Other configurations of the parallel hybrid vehicle include the use of an engine to power one axle and an electric motor to power the other (Figure 16-11). Another concept is to use a combination where the engine, coupled with an electric motor, powers one axle, and another electric motor powers the other axle (Figure 16-12).

Most parallel hybrid vehicles use the electric motor to accompany the engine to help drive the wheels. For example, the engine is used for long driving periods while the electric motor is used for

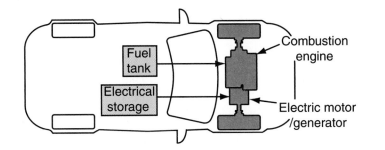

Figure 16-10 Parallel hybrid configuration.

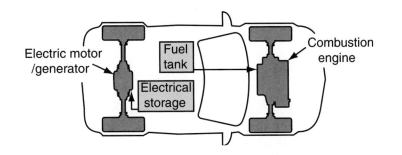

Figure 16-11 Parallel hybrid configuration using an engine to power one axle, and an electric motor to power the other.

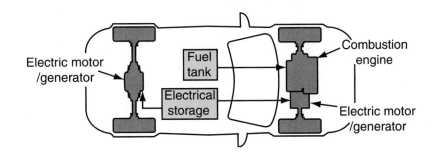

Figure 16-12 Parallel hybrid configuration using a combination where the engine coupled with an electric motor powers one axle and another electric motor powers the other axle.

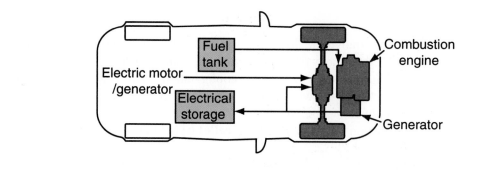

Figure 16-13 Series hybrid configuration.

short, low-intensity drives. In other words, the engine is ideal for highway driving and the electric motor is ideal for a trip around town. The electric motor also provides the vehicle with acceleration. The motor sustains this acceleration, however, only until the vehicle reaches a certain speed. Once it reaches this speed, the engine is started and takes over for the electric motor. The parallel hybrid combines the alternator, starter, and wheels to create a system that will start the engine, electronically balance it, take power from the engine and turn it into electricity, and provide extra power to the driveline when a power assist is needed for hill climbing or quick acceleration.

In the **series hybrid** vehicle, there is no mechanical connection between the engine and the wheels. The engine turns a generator, and the generator will either charge the batteries or power the electric motor, which in turn drives the transmission. Therefore, the engine never directly powers the automobile (Figure 16-13).

The power used to give the vehicle motion is transformed from chemical energy into mechanical energy, then into electrical energy, and finally back to mechanical energy to drive the wheels. This configuration is efficient in that it never idles. The automobile turns off completely at rest, such as at a stop sign or traffic light. This feature greatly reduces emissions. There is a variety of options in the configuration and mounting of all the components. Some series hybrid vehicles do not use a transmission.

HEV Examples

The Toyota Prius is considered a super ultra-low-emissions vehicle (SULEV), meaning that it is 90 percent cleaner than an ultra-low-emissions vehicle. The Prius uses a combined hybrid-electric structure to supply power. This system uses a combination of an internal combustion engine and an electric motor to turn the electrically controlled continuous variable transmission (ECCVT). However, the Prius can also accelerate using both the engine and the electric motor and can run solely on the electric motor. This is called a power split device (Figure 16-14).

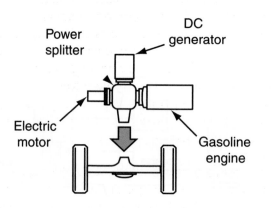

Figure 16-14 The power splitter allows for acceleration using both the engine and the electric motor for the vehicle; it also allows to run solely on the electric motor.

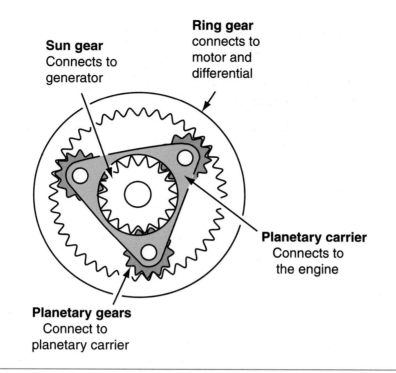

Sun gear
Connects to
generator

Ring gear
connects to
motor and
differential

Planetary carrier
Connects to
the engine

Planetary gears
Connect to
planetary carrier

Figure 16-15 Planetary gears are used to transfer power to the drive wheels.

Using a set of planetary gears (Figure 16-15), the vehicle can operate like a parallel vehicle in that either the electric motor or the gasoline engine powers the vehicle or they both do. However, the vehicle can also operate as a series hybrid where the engine can operate independently of the vehicle speed, either charging the batteries or providing power to the wheels when needed.

All the propulsion and auxiliary components of the Prius are packaged under the hood. The electric motor is rated at 33 kW and battery power is 21 kW. The nickel-metal-hydride battery pack is located between the rear seatback and the trunk.

The Prius is equipped with a variable valve timing four-cylinder engine. The generator also works as the starter of this engine. Gear shifting is not required because the planetary system acts as a continuously variable transmission to keep engine rpm in the range of best efficiency. Instead of a normal transmission, the engine drives the planet carrier of a planetary gear set. The sun gear of that set connects to a motor/generator, and the ring gear drives both the front wheels and a second motor/generator.

The Prius is also equipped with a "drive-by-wire" accelerator. The driver inputs to the motor management how much speed is requested. The management system then decides where the necessary power should come from: the engine, the battery, or both. The same thing occurs in the brake-by-wire system: the driver calls for the appropriate amount of retardation and the motor management coordinates this between the wheel brakes and the regenerative braking system. The computer also makes sure that the generator runs frequently enough to keep the battery charged.

When the battery is fully charged and the engine temperature is acceptable, the engine can be shut off (if the vehicle speed is low enough). If the driver gently presses the accelerator, the vehicle is moved by battery power. If the driver requests a quicker acceleration, the engine is started and powers the vehicle. In normal driving, the Prius maintains the battery state-of-charge within a narrow window. However, driving the vehicle under heavy loads for an extended time may deplete the battery charge.

The Honda Insight is also a parallel hybrid vehicle. In this system, the gasoline engine provides the majority of the power. The electric motor is used to help the gasoline engine provide additional power during acceleration. The Insight uses regenerative braking technology to capture energy lost during braking. The Insight also has a lightweight engine that uses **lean burn technology** to maximize its efficiency. Lean burn technology, developed in the 1960s, uses high air-fuel ratios to increase fuel efficiency.

Operating the engine independently of the vehicle speed means that even though the vehicle is traveling at highway speeds, the engine can be close to idle speed since it is only acting as a generator.

The Honda Civic Hybrid uses an integrated motor assist (IMA) system to power the vehicle. The system comprises a gasoline and electric motor combination. The electric motor is a source of additional acceleration and functions as a high-speed starter. The electric motor also acts as a generator for the charging system used during regenerative braking. This way the Civic Hybrid ensures efficiency by capturing lost energy by using regenerative braking, much like the Prius and the Insight do.

42-Volt Systems

The idea of 42-volt systems has been around for more than 20 years. The first production vehicle that used the 42-volt system was the 2002 Toyota Crown Sedan, which is only sold in the Japanese market. Although the use of 42-volt systems is very limited at the current time, the technology learned has been applied to the hybrid systems. In this section, we initially will discuss the 42-volt system as it is used for vehicle electrical systems then progress to the use of the integrated starter generator (ISG) on some mild-hybrid vehicles.

As you have probably come to realize, the increased use of electrical and electronic accessories in today's vehicles has about tapped the capabilities of the 12-volt/14-volt electrical system. Electronic content in vehicles has been rising at a rate of about 6% per year. It's been estimated that by the end of the decade the electronic content will be about 40% of the total cost of a high-line vehicle. The electrical demands on the vehicle have risen from about 500 watts in 1970 to about 4,000 watts in 2005. It is estimated that in 10 years the demand may reach 10,000 watts.

One way to meet the higher electrical demands would be to increase the amperage. This has been the approach over the past 30 years. Years ago it was common for a generator to have a rating of 35 to 50 amperes. Today, most manufacturers use a 150-ampere generator. However, just increasing the amperage output of the generator will not suffice to meet the 10,000-watt demand. Using Ohm's law, in order to obtain 10,000 watts with a 14-volt generator, the output would have to be 714 amperes. The more realistic alternative is to increase the voltage. As a result, vehicle manufacturers are developing a 36/42-volt system that incorporates a 42-volt generator charging a 36-volt battery. To meet the 10,000-watt demand in a 42-volt system will require 238 amps.

An additional benefit that may be derived from the use of a 42-volt system is it allows manufacturers to electrify most of the inefficient mechanical and hydraulic systems that are currently used. The new technology will allow electromechanical intake and exhaust valve control, active suspension, electrical heating of the catalytic converters, electrically operated coolant and oil pumps, an electric air conditioning compressor, brake-by-wire, steer-by-wire, and so on to be utilized. Studies have indicated that as these mechanical systems are replaced, fuel economy will increase by about 10 percent. In addition, emissions will decrease.

Additional fuel savings can also be realized due to the more efficient charging system used for the 42-volt system. Current 14-volt generators have an average efficiency across the engine speed range band of less than 60 percent. This translates to about 0.5 gallons (1.9 L) of fuel for 65 miles (104.6 km) of driving, to provide a continuous electrical load of 1,000 watts.. With a 42-volt generator, the fuel consumption can be reduced the equivalent of up to 15 percent. The 42-volt generator will be discussed later in this chapter.

Changing to 42-volt system provides three times as much generator power as the current system and will deliver as much as 30,000 watts. Since the increase in voltage results in a two-thirds reduction in amperage, the size of components can be reduced. Also, a significant reduction in the vehicle wiring size and weight can be realized.

There are several methods that are being used and developed for the electrical architecture of the 42-volt system. One method is a simple, single 42-volt system that is very similar to the current 12-volt/14-volt system (Figure 16-16). The challenge with this type of system is that all of the vehicle's electrical system will require redesigning to handle the 42 volts.

Due to the initial costs of implementing a single 42-volt system, a **dual-voltage system** is also being designed. With this architecture, the 42-volt system would power those electrical accessories that would require or benefit from the higher voltage. The remainder of the loads would remain on 14 V. There are several ways to implement a dual-voltage system.

The system is called 12-volt/14-volt because the battery is 12 volts but the charging system delivers 14 volts. The 42-volt system is actually a 36-volt/42-volt system.

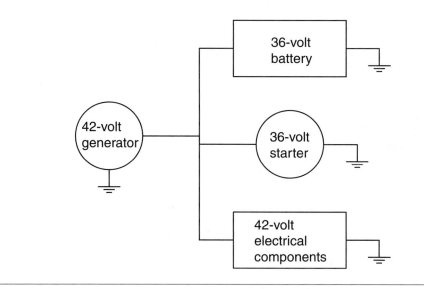

Figure 16-16 The single-voltage system is similar to the conventional 12-volt/14-volt system.

One method is to use a **dual-generator system** (Figure 16-17). One generator operates at 42 volts while the other operates at 14 volts. Another system is the **dual-stator, dual-voltage system** (Figure 16-18). In this system, dual voltage is produced from a single alternator that has two output voltages.

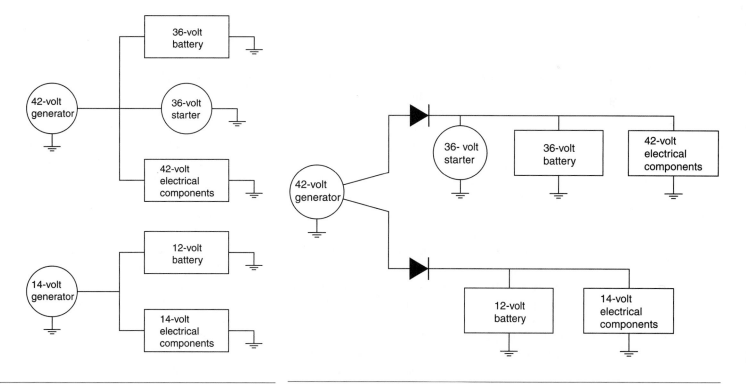

Figure 16-17 The dual-voltage system separates the electrical systems.

Figure 16-18 The dual-stator, dual-voltage system splits the electrical systems by using two outputs from a single generator.

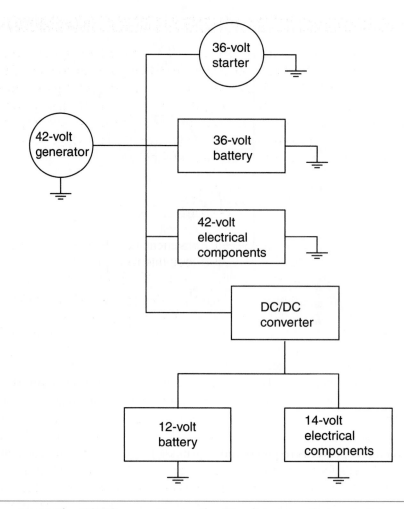

Figure 16-19 The DC/DC converter uses the 42-volt input and converts it to a 14-volt output.

Another design uses a **DC/DC converter** (Figure 16-19). The DC/DC converter is configured to provide a 14-V output from the 42-volt input. The 14-V output can be used to supply electrical energy to those components that do not require 42 volts.

As simple as it may seem to convert from a 12-volt/14-volt system to a 36-volt/42-volt system, many challenges need to be overcome. It is not as simple as adding a higher voltage output generator and expecting the existing electrical components to work. One of the biggest hurdles is the light bulb. Current 12-volt lighting filaments can't handle 42 volts. The dual-voltage systems can be used as a step toward full 42-volt implantation. However, dual-voltage systems are expensive to design. Another aspect of the 42-volt system is service technician training to address aspects of arcing, safety, and dual-voltage diagnostics.

Arcing is perhaps the greatest challenge facing the design and use of the 42-volt system. In fact, some manufacturers have abandoned further research and development of the system because of the problem with arcing. In the conventional 14-volt system, the power level is low enough that it is almost impossible to sustain an arc. Since there isn't enough electrical energy involved, the arcs collapse quickly and there is less heat buildup. Electrical energy in an arc at 42 volts is significantly greater and is sufficient to maintain a steady arc. The arc from a 42-volt system can reach a temperature of 6,000°F (3,316°C).

Voltage regulation also presents a challenge, especially in dual-voltage systems. As mentioned earlier, one method of dual-voltage control is the use of DC/DC converters. Another method is pulse-width modulation (PWM). An advantage of this method is it reduces most of the arcing problems associated with higher voltage. This is because there is not a true steady stream of current with PWM. This results in the arc collapsing quickly and in the reduction of heat.

In 1955, automobile manufactures started to move from 6-volt electrical systems to the present 12-volt system. The change was due to the demand for increased power to accommodate a greater number of electrical accessories. In 1955, the typical car wiring harness weighed 8 to 10 pounds and required approximately 250 to 300 watts. In 1990, the typical car wiring harness weighed 15 to 20 pounds and required more than 1,000 watts. In 2000, the typical car wiring increased to weigh between 22 and 28 pounds and required more than 1,800 watts. The conventional 14-volt generator is capable of producing a maximum output of 2,000 watts.

One of the newest technologies to emerge from the research and development of the 42-volt system is the **integrated starter generator (ISG).** Although this system can be used in conventional engine-powered vehicles, one of the key contributors to the hybrid's fuel efficiency is its ability to automatically stop and restart the engine under different operating conditions. A typical hybrid vehicle uses a 14 killowatt (kW) electric induction motor or ISG between the engine and the transmission (Figure 16-20). The ISG performs many functions such as fast, quiet starting; automatic engine stops/starts to conserve fuel; recharging the vehicle batteries; smoothing driveline surges, and providing regenerative braking.

Hybrid vehicles utilize the automatic stop/start feature to shut off the engine whenever the vehicle is not moving or when power from the engine is not required. Usually this feature is activated when the vehicle is stopped, no engine power is required, and the driver's foot is on the brake pedal. On vehicles with manual transmissions, this feature may be activated when the vehicle is stopped, no engine power is required, the transmission is in neutral, and the clutch pedal is released. Once the driver's foot is removed from the brake pedal (or the clutch is engaged) the starter automatically restarts the engine in less than one-tenth of a second. To further save fuel and reduce emissions, the engine is accelerated to idle speed by the starter generator prior to the start of the combustion process and the injection of fuel.

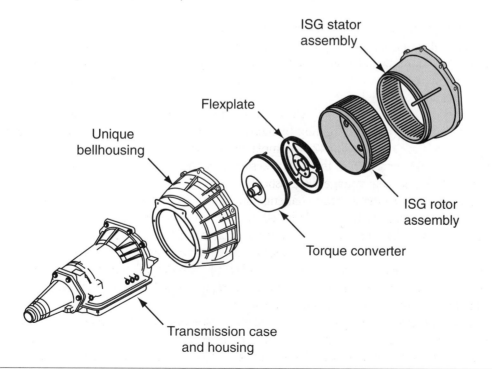

Figure 16-20 The ISG is usually located between the engine and the transmission in the bell housing.

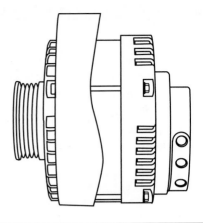

Figure 16-21 The belt-driven starter generator looks very similar to a conventional generator.

The ISG also can convert kinetic energy to DC voltage. When the vehicle is traveling downhill and there is zero load on the engine, the wheels can transfer energy through the transmission and engine to the ISG. The ISG then sends this energy to the battery for storage and use by the electrical components of the vehicle.

Most systems use a 42-volt ISG system, since the power requirements of automatic stop/start are higher than the 12-volt system can provide. Currently there are two main system designs.

The first design uses a **belt alternator starter (BAS)** that is about the same size as a conventional generator and is mounted in the same way (Figure 16-21). BASs have a maximum power output of around 5 kW. Two types of BASs are being designed: permanent magnet and induction BAS.

The second design is to mount the ISG at either end of the crankshaft. Most designs have the ISG mounted at the rear of the crankshaft between the engine and transmission. In some systems, the ISG may take the place of the engine flywheel. The ISG mounted in this method is larger than the BSG and is able to produce an output of 6 to 15 kW.

The IGS is a three-phase AC motor. At low vehicle speeds, the ISG provides power and torque to the vehicle. It also supports the engine, when the driver demands more power. During vehicle deceleration, ISG regenerates the power that is used to charge the traction batteries.

The ISG includes a rotor and stator that are located inside the transmission bell housing (Figure 16-22). The stator is attached to the engine block and coils that are formed by laser-welding

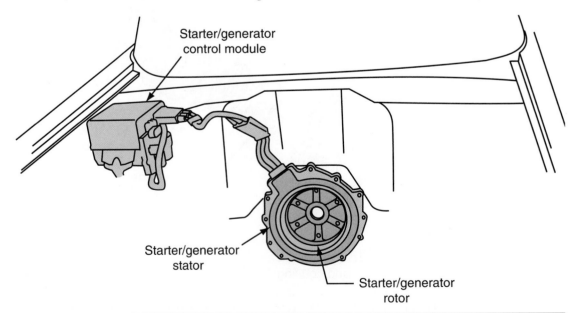

Figure 16-22 The ISF stator and rotor assembly.

copper bars. A conventional generator uses winding of copper wire for its stator. The rotor is bolted to the engine crankshaft.

Both the BAS and the ISG use the same principle to start the engine. Current flows through the stator windings and it generates magnetic fields in the rotor. This will cause the rotor to turn, thus turning the crankshaft and starting the engine. In addition, this same principle is used to assist the engine as needed when the engine is running.

A **starter generator control module (SGCM),** also called a **high-voltage ECU (HV ECU),** is used to control the flow of torque and electrical energy. Remember that torque and energy can go into or out of the ISG. The function of the SGCM is to control the engine cranking, torque, speed, and active damping functions.

Our discussion up to this point has centered on the use of 42 volts for the ISG system. It is important to note that the hybrid system could work on 42 volts or 300 volts. Also, the voltage used to start the engine and to operate the electric motors is AC current.

Fuel Cells

A fuel cell produces current from hydrogen and aerial oxygen. Fuel cell–powered vehicles have a very good chance of becoming the vehicles of the future. They combine the reach of conventional internal combustion engines with high efficiency, low fuel consumption, and minimal or no pollutant emissions. At the same time, they are extremely quiet. Because they work with regenerative fuel such as hydrogen, they reduce the dependence on crude oil and other fossil fuels.

A fuel cell–powered vehicle (Figure 16-23) is basically an electric vehicle. Like the electric vehicle, it uses an electric motor to supply torque to the drive wheels. The difference is that the fuel cell, instead of a battery, produces and supplies electric power to the electric motor. Most vehicle manufacturers and several independent laboratories are involved in fuel cell research and development programs. Manufacturers have produced a number of prototype fuel cell vehicles, with many being placed in fleets in North America and Europe.

Fuel cells electrochemically combine oxygen from the air with hydrogen to produce electricity. The oxygen and hydrogen are fed to the fuel cell as "fuel" for the electrochemical reaction.

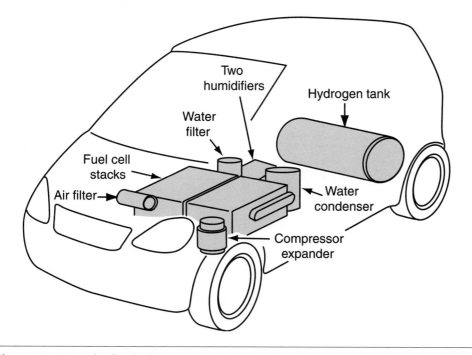

Figure 16-23 Fuel cell vehicle components. Technology has allowed engineers to design fuel cell vehicles without the loss of passenger and cargo space.

There are different types of fuel cells but the most common type is the **proton exchange membrane (PEM).** Normally, hydrogen and oxygen bond with a loud bang, but in fuel cells a special PEM impedes the oxyhydrogen gas reaction by ensuring that only protons (H+), and not elemental hydrogen molecules (H_2), react with the oxygen.

How the Fuel Cell Works

The PEM fuel cell is constructed like a sandwich (Figure 16-24). The electrolyte is situated between two electrodes of gas-permeable graphite paper. The electrolyte is a polymer membrane. Hydrogen is applied to the anode side of the PEM and ambient oxygen is applied to the cathode side (Figure 16-25). The membrane keeps the distance between the two gases and provides a controlled chemical reaction.

The anode is the negative post of the fuel cell. It conducts the electrons that are freed from the hydrogen molecules so that they can be used in an external circuit. It has channels etched into it that disperse the hydrogen gas evenly over the surface of the catalyst. The cathode is the positive post of the fuel cell. It also has channels etched into it that distribute the oxygen to the surface of the catalyst. It also conducts the electrons back from the external circuit to the catalyst, where they can recombine with the hydrogen ions and oxygen to form water.

A fine coating of platinum is applied to the foil to act as a catalyst. This is used to accelerate the decomposition of the hydrogen atoms into electrons and protons (Figure 16-26). The catalyst is rough and porous so that the maximum surface area of the platinum can be exposed to the hydrogen or oxygen. The platinum-coated side of the catalyst faces the PEM.

The PEM is the electrolyte. This specially treated material (which looks similar to ordinary kitchen plastic wrap) allows only the protons to move across from the anode to the cathode (Figure 16-27). As a result, the anode will have a surplus of electrons and the cathode has a surplus

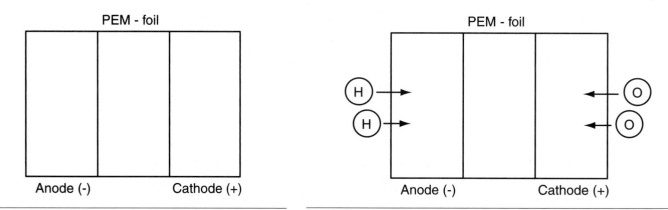

Figure 16-24 The PEM foil.

Figure 16-25 Hydrogen is applied to the anode side of the PEM, while oxygen is applied to the cathode side.

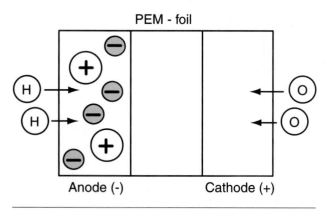

Figure 16-26 The catalyst breaks down the H_2 into protons and electrons.

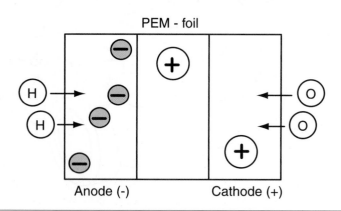

Figure 16-27 The PEM foil allows only the protons to migrate to the cathode, leaving the electrons on the anode.

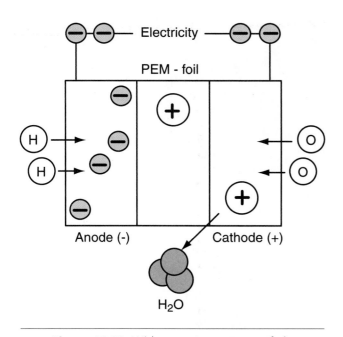

Figure 16-28 With an excess amount of electrons on the anode and an excess amount of protons on the cathode, current flows through an external conductor. The electrons then react with the protons and oxygen to produce water.

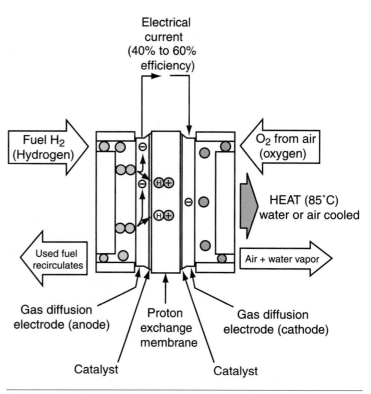

Figure 16-29 PEM fuel cell.

of protons. If the anode and cathode are connected outside of the cell, current flows through the conductor (Figure 16-28). The electrons will move through the conductor to the cathode. The electrons then recombine with the protons and the oxygen to produce water.

The entire process is illustrated in Figure 16-29. Pressurized hydrogen gas (H_2) enters the fuel cell on the anode side. This gas is forced through the catalyst by the pressure. When an H_2 molecule comes in contact with the platinum on the catalyst, it splits into two H+ ions and two electrons (e–). The electrons are conducted through the anode, where they make their way through the external circuit (doing useful work such as turning a motor) and return to the cathode side of the fuel cell.

Meanwhile, on the cathode side of the fuel cell, oxygen gas (O_2) is being forced through the catalyst, where it forms two oxygen atoms. Each of these atoms has a strong negative charge. This negative charge attracts the two H+ ions through the membrane, where they combine with an oxygen atom and two of the electrons from the external circuit to form a water molecule (H_2O).

This reaction in a single fuel cell produces only about 0.7 volt. For this voltage to become high enough to be used to move the vehicle, many separate fuel cells must be combined in series to form a fuel cell stack (Figure 16-30).

Fuels for the Fuel Cell

A fundamental problem with fuel cell technology concerns whether to store hydrogen or convert it from other fuels on board the vehicle. All four principal fuels that automotive manufacturers are considering (hydrogen, methanol, ethanol, and gasoline) pose some challenges.

Hydrogen. One solution is to store hydrogen on board the vehicle. The ability to use hydrogen directly in a fuel cell provides the highest efficiency and zero tailpipe emissions. However, hydrogen has a low energy density and boiling point; thus on-board storage requires large, heavy tanks. There are three types of hydrogen storage methods under development: compressed hydrogen, liquefied hydrogen, and binding hydrogenate to solids in metal hydrides or carbon compounds.

Compressed hydrogen offers the least expensive method for on-board storage. However, at normal CNG operating pressures of 3,500 psi (241 bar), reasonably sized, commercially available

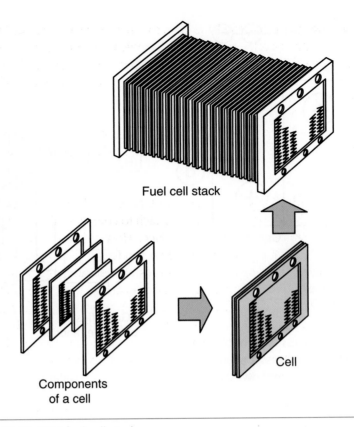

Fuel cell stack

Cell

Components
of a cell

Figure 16-30 PEM fuel cell stack.

pressure tanks will provide limited range for a fuel cell vehicle (about 120 miles or 193 km). Daim-lerChrysler and Hyundai are now using pressure tanks that are capable of 5,000 psi (345 bar). Quantum is conducting research of high-performance hydrogen storage systems, looking at pressure tanks that are capable of up to 10,000 psi (689 bar). This capability would permit a 400-mile (644 km) driving range.

Liquefied hydrogen can be stored in large cylinders containing a hydride material (something like steel wool). Liquefied hydrogen does not require the high storage capacity as that of compressed hydrogen for the same amount of driving range. However, the very low boiling point of hydrogen requires that the tanks have excellent insulation. Maintaining the extreme cold temperature of –423°F (–253°C) during refueling and storage is difficult. It is estimated that up to 25 percent of the liquid hydrogen may be boiled off during the refueling process. In addition, about 1 percent is lost per day in on-board storage. Storing liquid hydrogen on a vehicle also involves some safety concerns. As the fuel tank warms, the pressure increases and may activate the pressure relief valve. This action discharges flammable hydrogen into the atmosphere, creating a source of danger and pollution.

Methanol. Several automotive manufacturers are using methanol to power their fuel cells. It is believed that methanol fuel cells could bridge the gap over the next few decades while a hydrogen distribution infrastructure is being built. Using "methanolized" hydrogen as fuel has the advantage that it can be stored on the vehicle like gasoline is. For the reaction in the fuel cell, a **reformer** on board the vehicle produces hydrogen from the methanol fuel. A reformer is a high-temperature device that converts hydrocarbon fuels to CO and hydrogen. To produce the hydrogen, the methanol fuel is mixed with water. When it evaporates, it is decomposed into hydrogen and carbon dioxide. Prior to sending the hydrogen and carbon dioxide to the fuel cell, it is purified in additional steps.

Methanolized hydrogen contains more hydrogen atoms and has an energy density that is greater than that of liquid hydrogen. Like hydrogen, methanolized hydrogen is also independent of mineral oil. Compared with hydrogen vehicles, vehicles driven by methanol are not completely emissions-free but they produce very few pollutants and much less carbon dioxide than internal combustion engines.

A special type of PEM fuel cell, called the direct methanol air fuel cell (DMAFC), utilizes methanol combined with water directly as a fuel and ambient air for oxygen. This technology enables use of a liquid fuel without the need for an on-board reformer, while still providing a zero-emissions system. However, current research has demonstrated that the power density is lower than that for other PEM fuel cells.

Ethanol. Ethanol is considered less toxic than either gasoline or methanol. An ethanol system requires adding a reformer to the vehicle, similar to that in a methanol system. The fuel cell could use E100, E95, or E85.

Gasoline. Fuel cells can be driven with a special, more pure gasoline. Using reformers for on-board extraction of hydrogen from gasoline is one approach to commercialization of fuel cell vehicles, since the gasoline infrastructure is already in place. However, producing hydrogen from gasoline in a vehicle system is much more difficult than producing hydrogen from methanol or ethanol. Gasoline reforming requires higher temperatures and more complex systems than the methanol or ethanol reforming. The reformation reactions occur at 1,562°F to 1,823°F (850°C to 995°C), making the devices slow to start and the chemistry temperamental. Thus, the drive would work less efficiently and produce more emissions. Moreover, the capabilities for cold starts would be restricted. The size of the reformer is also an issue, making it difficult to fit under the hood of a standard-sized vehicle. Furthermore, there is concern about the sulfur levels in current gasoline and carbon monoxide in the reformer poisoning the fuel cell.

Reformer. As mentioned, some fuel cell systems may require the use of an on-board reformer to extract hydrogen from liquid fuels such as gasoline, methanol, or ethanol. On-board reformation of a hydrocarbon fuel into hydrogen allows the use of more established infrastructures but adds additional weight and cost and reduces vehicle efficiency. In addition, the reformer does create some emissions.

PEM fuel cell reformers combine fuel and water to produce additional H_2 and convert the CO to CO_2. The CO_2 is then released to the atmosphere. Reformer technologies include steam reforming (SR), partial oxidation, and high-temperature electrolytes reforming.

Steam reforming (SR) uses a catalyst to convert fuel and steam to H_2, CO, and CO_2. The CO is further reformed with steam to form more H_2 and CO_2. A purification step then removes CO, CO_2, and any impurities to achieve a high hydrogen purity level (97 to 99.9 percent). SR of methanol is the most developed and least expensive method to produce hydrogen from a hydrocarbon fuel on a vehicle, resulting in 45 to 70 percent conversion efficiency.

Partial oxidation reforming is similar to steam reforming since both technologies combine fuel and steam, but this process adds oxygen in an additional step. The process is less efficient than steam reforming, but the heat-releasing nature of the reaction makes it more responsive than steam reforming to variable load. Heavier HC can be used in POX, but it has lower carbon-to-hydrogen ratios, which limits hydrogen production.

Hydrogen can also be obtained by **electrolysis** from water. Electrolysis is the splitting of water into hydrogen and oxygen. The drawback to this process is it requires a great deal of electrical energy. Recently, the development of high-temperature electrolytes that can operate at temperatures in excess of 212°F (100°C) has shown some positive results. The benefits of high-temperature electrolytes include:

❑ *Improved CO tolerance.* This allows the manufacturer to reduce or remove the need for an oxidation reactor and for air-bleed. Since these requirements can be reduced, system efficiency is increased by 5 to 10 percent. There is also a considerable reduction in startup time. The remaining CO will be combusted in a catalytic tail-gas burner to prevent emissions of CO.

❑ *Facilitated stack cooling.* This reduces the size of the radiator and reduces the fuel cell stack cooling plate requirements.

❏ *Humidity-independent operation.* Generally, high-temperature membranes require humidifiers and water recovery, whereas this system does not.

Solid Oxide Fuel Cells. Planar solid oxide fuel cells (SOFCs) operate at high temperatures of 932°F to 1,472°F (500°C to 800°C) and can use CO and H_2 fuel. SOFCs have a good tolerance to fuel impurities and use ceramic as an electrolyte. BMW is currently developing an auxiliary power unit (APU) with Delphi and Global Thermoelectric using an SOFC. Currently, SOFCs use gasoline fuel and require a reformer.

Sodium Borohydride. DaimlerChrysler introduced a new idea in fuel technologies with a concept minivan called the Chrysler Town & Country Natrium (Latin for sodium). This research was driven by the lack of a safe, compact way to contain hydrogen. In addition, most fuel cell vehicles are not "on demand," meaning that there is a startup time required before the reformer can begin producing hydrogen.

A fairly simple chemical process mixes sodium borohydride powder with water. The sodium borohydride powder produces free hydrogen for the fuel cell (Figure 16-31). Unlike the fuel cell systems that use methanol or gasoline, the Natrium produces no pollution and no carbon dioxide.

The sodium borohydride powder holds more hydrogen than the most densely compressed hydrogen tank. The prototype minivan attains a top speed of 80 mph (129 kmh) and an operating range of almost 300 miles (487 km). The by-products of this process are water and sodium borate, which is basically laundry detergent. After use, the spent powder goes into a storage tank where it can be pumped out and reclaimed.

As the name implies, sodium borohydride ($NaBH_4$) is a salt. The salt that powers this system is not ordinary table salt (sodium chloride—NaCl). Rather, it is a white salt whose molecules contain a relatively large amount of hydrogen. Through the use of a chemical catalyst, the sodium borohydride reaction with water results in elemental hydrogen. A slurry of sodium perborate ($NaBO_2$) forms during this reaction as well. This compound is chemically related to borax, which is used as a bleaching agent in conventional detergents. The $NaBH_2$ slurry is collected in a special tank and can be effectively recycled in a chemical process. In this reverse reaction, the $NaBH_2$ is reclaimed back to $NaBH_4$, which can then be reused as an energy source for the fuel cells.

> The advantage of sodium borohydride is that it can be stored easily and transported in lightweight plastic tanks at normal temperatures and pressures when dissolved in water. Also, the borate solution is not poisonous, flammable, or explosive.

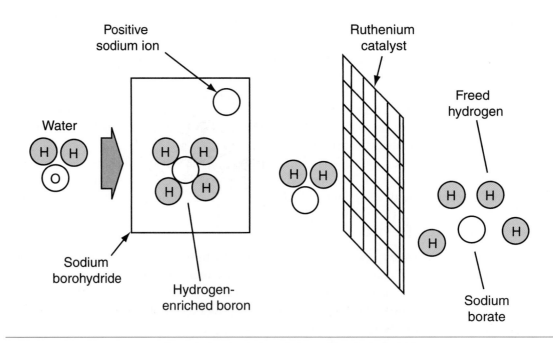

Figure 16-31 Process of extracting hydrogen from sodium borohydride.

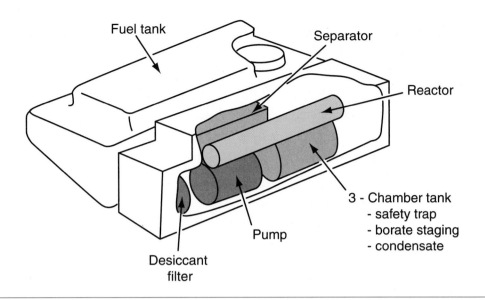

Figure 16-32 The Hydrogen on Demand™ catalyst system is the heart of the drive train. The hydrogen required for the fuel cell to generate power is extracted from sodium borohydride in a reactor.

The catalyst system is called Hydrogen on Demand ™ and was developed by Millennium Cell (Figure 16-32). When stepping on the accelerator, the minivan driver is actually "throttling" a fuel pump. The $NaBH_4$ is pumped into the catalyst, which immediately generates hydrogen. This dynamic process slows or ceases entirely when the fuel pump is throttled or completely shut off. When the driver accelerates once more, hydrogen is immediately produced again and the fuel cells immediately generate electricity.

An AC motor with an output of 35 kW powers the vehicle. A 55 kW lithium-ion battery serves as a storage unit for electrical energy. The battery is recharged by the fuel cell unit and by regenerative braking. If the fuel cell system should fail, the battery would be sufficient to drive the electric motor and move the vehicle. All fuel cell components are contained within the frame rail of the vehicle (Figure 16-33).

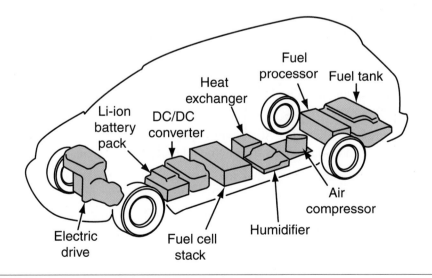

Figure 16-33 Components of the sodium borohydride fuel cell minivan are all located between the frame rails.

Summary

❏ Electric vehicles powered by an electric motor run from a battery pack.

❏ The hybrid electric vehicle relies on power from the electric motor, the engine, or both.

❏ The hybrid vehicle can be configured as a parallel, series, or combination hybrid system.

❏ Regenerative braking recovers the energy used to brake by converting rotational energy into electrical energy through a system of electric motors and generators. When the brakes are applied, the motor becomes a generator by using the kinetic energy of the vehicle to store power in the battery for later use.

❏ To meet the electrical demands of the automobile in the future, manufacturers are presently developing 42-volt systems.

❏ An additional benefit that may be derived from the use of a 42-volt system is it allows manufacturers to electrify most of the inefficient mechanical and hydraulic systems that are currently used.

❏ There are several methods that are being used and developed for the electrical architecture of the 42-volt system. One method is a simple, single 42-volt system that is very similar to the current 12-volt/14-volt system. Another method is the dual-voltage system.

❏ The dual-generator system uses two generators, one that operates at 42 volts while the other operates at 14 volts.

❏ In the dual-stator, dual-voltage system, voltage is produced from a single alternator that has two output voltages.

❏ Another design uses a DC/DC converter that is configured to provide a 14-V output from the 42-volt input.

❏ Arcing is perhaps the greatest challenge facing the design and use of the 42-volt system.

❏ One of the newest technologies to emerge from the research and development of the 42-volt system is the integrated starter generator (ISG).

❏ Hybrid vehicles utilize the automatic stop (or idle stop) feature to shut off the engine whenever the vehicle is not moving or when power from the engine is not required.

❏ The ISG can also convert kinetic energy to storable electric energy. When the vehicle is traveling downhill and there is zero load on the engine, the wheels can transfer energy through the transmission and engine to the ISG. The ISG then sends this energy to the battery for storage and use by the electrical components of the vehicle.

❏ The belt alternator starter (BAS) is about the same size as a conventional generator and is mounted in the same way.

❏ The IGS is a three-phase AC motor. At low vehicle speeds, the ISG provides power and torque to the vehicle. It also supports the engine when the driver demands more power.

❏ The ISG includes a rotor and a stator that are located inside the transmission bell housing and is formed by laser-welding copper bars. The rotor is bolted to the engine crankshaft.

❏ Both the BAS and the ISG use the same principle to start the engine. Current flows through the stator windings generate magnetic fields in the rotor. This will cause the rotor to turn, thus turning the crankshaft and starting the engine.

❏ A fuel cell–powered vehicle is basically an electric vehicle except the fuel cell, instead of a battery, produces and supplies electric power to the electric motor.

❏ Fuel cells electrochemically combine oxygen from the air with hydrogen to produce electricity.

❑ The most common type of fuel cell is the proton exchange membrane (PEM).

❑ There are several methods being explored for storage of fuel for the fuel cell. These include hydrogen, methanol, ethanol, and gasoline.

❑ Most fuel cell systems require the use of a reformer to extract hydrogen from liquid fuels such as gasoline, methanol, or ethanol.

Review Questions

Short-Answer Essays

1. Explain the meaning of regenerative braking.

2. Describe the basic operation of a typical hybrid vehicle.

3. Explain the difference between parallel and series hybrids.

4. Briefly describe how the proton exchange membrane (PEM) fuel cell produces electrical energy.

5. What is the purpose of the reformer?

6. Explain why increasing amperage output of a generator is not able to meet the demands of the electrical system on future vehicles.

7. Explain the advantages that can be derived from the 42-volt electrical system.

8. Describe the basics of the three common dual-voltage systems.

9. What are some of the challenges of implementing the 42-volt system?

10. Describe the basic function of the integrated starter generator system.

Fill in the Blanks

1. _____ _____ recovers the heat energy used to brake by converting rotational energy into _____ energy through a system of electric motors and generators.

2. Fuel cells _____ combine oxygen from the air with hydrogen to produce electricity.

3. Most fuel cell systems require the use of a _____ to extract hydrogen from liquid fuels such as gasoline, methanol, or ethanol.

4. In a _____ hybrid configuration, there is a direct mechanical connection between the engine and the wheels.

5. In the _____ hybrid vehicle, there is no mechanical connection between the engine and the wheels.

6. The _____ is the negative post of the fuel cell.

7. In the ISG system, the _____ is attached to the engine block and coils that are formed by laser-welding copper bars.

8. The ISG is a _____ AC motor. At low vehicle speeds, the ISG provides power and torque to the vehicle. It also supports the engine, when the driver demands more power. During vehicle deceleration, ISG regenerates the power that is used to charge the traction batteries.

9. A _____ is used to control the flow of torque and electrical energy.

10. In a hybrid system, the _____ _____ is a device that stores energy as electrostatic charge and is the primary device in the power supply during hill climbing, acceleration, and the recovery of braking energy.

Multiple Choice

1. *Technician A* says regenerative braking recovers the energy used to brake by converting rotational energy into electrical energy through a system of electric motors and generators.
   Technician B says when the brakes are applied, the motor becomes a generator by using the kinetic energy of the vehicle to store power in the battery for later use.
   Who is correct?
   A. A only
   B. B only
   C. Both A and B
   D. Neither A nor B

2. Electric vehicles power the motor by:
   A. A generator.
   B. A battery pack.
   C. An engine.
   D. None of the above.

3. *Technician A* says in a parallel hybrid vehicle, propulsion comes directly from the electric motor.
   Technician B says in a series hybrid vehicle, both the engine and the electric motor can turn the transmission at the same time.
   Who is correct?
   A. A only
   B. B only
   C. Both A and B
   D. Neither A nor B

4. *Technician A* says a fuel cell produces electrical energy by breaking down H_2 atoms into electrons and protons.
   Technician B says the compressed hydrogen system requires a reformer to cool the fuel cell.
   Who is correct?
   A. A only
   B. B only
   C. Both A and B
   D. Neither A nor B

5. The main advantage of the ISG system is:
   A. It can be used to provide accessory power if the battery fails.
   B. It allows for automatic stop/start functions.
   C. It increases the life of the brake linings.
   D. All of the above.

6. The splitting of water into hydrogen and oxygen is an example of:
   A. Steam reforming.
   B. Hydrocarbon reforming.
   C. Partial oxidation reforming.
   D. Electrolysis.

7. In the ISG, how does current flow to make the system perform as a starter?
   A. Through the rotor to create an electromagnetic field that excites the stator, which causes the rotor to spin.
   B. Through the rotor coils, which cause the magnetic field to collapse around the stator and rotate the crankshaft.
   C. Through the stator windings, which generate magnetic fields in the rotor, causing the rotor to turn the crankshaft.
   D. From the starter generator control module to the rotor coils that are connected to the delta-wound stator.

8. On which side of a PEM fuel cell does the pressurized hydrogen gas enter?
   A. Anode.
   B. Cathode.
   C. Drain.
   D. Gate.

9. All of the following are benefits of the 42-volt electrical system EXCEPT:
   A. Electrifying most of the mechanical and hydraulic systems.
   B. Light bulbs last longer.
   C. Reduction in size of electrical components and wires.
   D. Increased efficiency of the charging system.

10. The dual-voltage system can use:
    A. A dual-generator system.
    B. A dual-stator, dual-voltage system.
    C. A DC/DC converter.
    D. All of the above.

GLOSSARY

Accelerometer Generates an analog voltage in relation to the severity of deceleration forces. The accelerometer also senses the direction of impact force.

Acelerómetro Genera un voltaje análogo en relación a la severidad de las fuerzas de deceleración. El acelerómetro tambien detecta la dirección de la fuerza de un impacto.

A circuit A generator regulator circuit that uses an external grounded field circuit. In the A circuit, the regulator is on the ground side of the field coil.

Circuito A Circuito regulador del generador que utiliza un circuito inductor externo puesto a tierra. En el circuito A, el regulador se encuentra en el lado a tierra de la bobina inductora.

Active suspension systems Suspension systems that are controlled by double-acting hydraulic cylinders or solenoids (actuators) mounted at each wheel. The actuators support the vehicle's weight, instead of conventional springs or air springs.

Sistemas activos de suspensión Sistemas de suspensión controlados por cilindros hidráulicos de doble acción o por solenoides (accionadores) montados en cada rueda. Los accionadores apoyan el peso del vehículo, en vez de muelles convencionales o muelles de aire.

Actuators Devices that perform the actual work commanded by the computer. They can be in the form of a motor, relay, switch, or solenoid.

Accionadores Dispositivos que realizan el trabajo efectivo que ordena la computadora. Dichos dispositivos pueden ser un motor, un relé, un conmutador o un solenoide.

Adaptive suspension systems Suspension systems that are able to change ride characteristics by continuously altering shock damping and ride height.

Sistemas adaptadores de suspensión Sistemas de suspensión que pueden cambiar las características del viaje al alterar continuamente el amortiguamiento y la altura del viaje.

After top dead center (ATDC) Piston travel once top dead center has been achieved.

Después de punto muerto superior (ATDC) El viaje del pistón después de que se ha cumplido el punto muerto superior.

Air bag Inflates in a few milliseconds when the vehicle is involved in a frontal collision. A typical fully inflated air bag has a volume of 2.3 cu. ft.

Bolsa de aire Infla en unos milisegundos cuando un vehículo se ha involucrado en una colisión delantera. Una bolsa de aire típica tiene un volúmen de 2.3 pies cúbicos al estar completamente inflada.

Air bag module Composed of the air bag and inflator assembly, which is packaged into a single module.

Unidad del Airbag Formada por el conjunto del Airbag y el inflador. Este conjunto se empaqueta en una sola unidad.

Air bag system Designed as a supplemental restraint that, in the case of an accident, will deploy a bag out of the steering wheel or passenger-side dash panel to provide additional protection against head and face injuries.

Sistema de bolsa de aire Diseñada como una restricción suplemental que, en el caso de un accidente, desplegará una bolsa del volante de dirección o del tablero lateral del pasajero para proveer la protección adicional contra los daños a la cabeza y a la cara.

Air core gauge Gauge design that uses the interaction of two electromagnets and the total field effect upon a permanent magnet to cause needle movement.

Calibrador de núcleo de aire Calibrador diseñado para utilizar la interacción de dos electroimanes y el efecto inductor total sobre un imán permanente para generar el movimiento de la aguja.

Alternating current Electrical current that changes direction between positive and negative.

Corriente alterna Corriente eléctrica que recorre un circuito ya sea en dirección positiva o negativa.

Ambient temperature The temperature of the outside air.

Temperatura ambiente Temperatura del aire ambiente.

Ambient temperature sensor A thermistor used to determine the ambient temperature.

Sensor de la temperatura ambiente Termostato que se usa para determinar la temperatura ambiente.

American wire gauge (AWG) System used to determine wire sizes based on the cross-sectional area of the conductor.

Calibrador americano de alambres Sistema utilizado para determinar el tamaño de los alambres, basado en el área transversal del conductor.

Ammeter A test meter used to measure current draw.

Amperímetro Instrumento de prueba utilizado para medir la intensidad de una corriente.

Ampere-hour rating Indicates the amount of steady current a battery can supply for 20 hours.

Límite de amperio-hora Indica la cantidad de corriente fijo que un puede proveer una batería durante 20 horas.

Amperes *See* current.

Amperios *Véase* corriente.

Analog A voltage signal that is infinitely variable or that can be changed within a given range.

Señal analógica Señal continua y variable que debe traducirse a valores numéricos discontinuos para poder ser tratada por una computadora.

Anode The positive charge electrode in a voltage cell.

Ánodo Electrodo de carga positiva de un generador de electricidad.

Antilock brakes (ABS) A brake system that automatically pulsates the brakes to prevent wheel lockup under panic stop and poor traction conditions.

Frenos antibloqueo Sistema de frenos que pulsa los frenos automáticamente para impedir el bloqueo de las ruedas en casos de emergencia y de tracción pobre.

Antitheft device A device or system that prevents illegal entry or driving of a vehicle. Most are designed to deter entry.

Dispositivo a prueba de hurto Un dispositivo o sistema que previene la entrada o conducción ilícita de un vehículo. La mayoría se diseñan para detener la entrada.

Armature The movable component of an electric motor, which consists of a conductor wound around a laminated iron core and is used to create a magnetic field.

Armadura Pieza móvil de un motor eléctrico, compuesta de un conductor devanado sobre un núcleo de hierro laminado y que se utiliza para producir un campo magnético.

Arming sensor A device that places an alarm system into "ready" to detect an illegal entry.

Sensor de armado Un dispositivo que pone "listo" un sistema de alarma para detectar una entrada ilícita.

Aspirator Tubular device that uses a venturi effect to draw air from the passenger compartment over the in-car sensor. Some manufacturers use a suction motor to draw the air over the sensor.

Aspirador Dispositivo tubular que utiliza un efecto Venturi para extraer aire del compartimiento del pasajero sobre el sensor dentro del vehículo. Algunos fabricantes utilizan un motor de succión para extraer el aire sobre el sensor.

Asynchronous Data that is sent on the bus network intermittently (as needed) rather than continuously.

Asíncronos Datos que se envían en el bus múltiple de modo intermitente (como sea necesario) en lugar de modo continuo.

Atom The smallest part of a chemical element that still has all the characteristics of that element.

Átomo Partícula más pequeña de un elemento químico que conserva las cualidades íntegras del mismo.

Audio system The sound system for a vehicle; can include radio, cassette player, CD player, amplifier, and speakers.

Sistema de audio El sistema de sonido de un vehículo; puede incluir el radio, el tocacaset, el toca discos compactos, el amplificador, y las bocinas.

Automatic door locks A system that automatically locks all doors through the activation of one switch.

Cerrojos de compuertas automatizados Cerrojos de compuertas automatizados eléctricamente que utilizan o un solenoide o un motor reversible de imán permanente para cerrar y abrir las puertas.

Automatic headlight dimming An electronic feature that automatically switches the headlights from high beam to low beam under two different conditions: light from oncoming vehicles strikes the photocell-amplifier; or light from the taillights of a vehicle that is being passed strikes the photocell-amplifier.

Reducción automática de intensidad luminosa de los faros delanteros Característica electrónica que conmuta los faros delanteros automáticamente de luz larga a luz corta dadas las siguientes circunstancias: la luz de los vehículos que se aproximan alcanza el amplificador de fotocélula, o la luz de los faros traseros de un vehículo que se ha rebasado alcanza el amplificador de fotocélula.

Automatic on/off with time delay Turns on the headlights automatically when ambient light decreases to a predetermined level. Also allows the headlights to remain on for a certain amount of time after the vehicle has been turned off. This system can be used in combination with automatic dimming systems.

Prendido/apagado automático con temporización Prende los faros automáticamente cuando la luz ambiental se oscurece a un nivel predeterminado. Tambien permite que los faros queden prendidos por un tiempo determinado después de que se ha apagado el vehículo. Este sistema se puede utilizar en combinación con los sistemas de regulación de intensidad luminosa automáticos.

Automatic temperature control (ATC) A passenger comfort system that is capable of maintaining a preset temperature level as selected by the operator. Sensors are used to determine the present temperatures, and the system can adjust the level of heating or cooling as required by using actuators to open and close air-blend doors to achieve the desired in-vehicle temperature.

Control automático de la temperatura (CAT) Un sistema de comodidad para el pasajero es capaz de mantener un nivel de temperatura previamente fijo tal como lo selecciona el operador. Los sensores se utilizan para determinar las temperaturas actuales, y el sistema puede ajustar el nivel de calentamiento o enfriamiento como se requiera al usar actuadores para abrir y cerrar las compuertas de recirculación para lograr la temperatura deseada dentro del vehículo.

Automatic traction control A system that prevents slippage of one of the drive wheels. This is done by applying the brake at that wheel and/or decreasing the engine's power output.

Control Automático de Tracción Un sistema que previene el patinaje de una de las ruedas de mando. Esto se efectúa aplicando el freno en esa rueda y/o disminuyendo la salida de potencia del motor.

Avalanche diodes Diodes that conduct in the reverse direction when a reverse bias voltage of about 6.2 volts or higher is applied. Causes the avalanche effect to occur when the reverse electric field moves across the PN junction, causing a wave of ionization that leads to a large current.

Diodo Zener Diodos que hacen conducción en dirección contraria cuando se aplica velocidad invertida de transmisión de baudios de alrededor de 6.2 voltios o más. Produce el efecto Zener o de avalancha cuando el campo eléctrico invertido se mueve a cruzar la unión PN, y causa así una onda de ionización que provoca una gran corriente.

AXODE The model designation given to Ford's electronically shifted automatic transaxle with overdrive.

AXODE La designación del modelo dado al transeje automático de embrague electrónico con sobremarcha.

B circuit A generator regulator circuit that is internally grounded. In the B circuit, the voltage regulator controls the power side of the field circuit.

Circuito B Circuito regulador del generador puesto internamente a tierra. En el circuito B, el regulador de tensión controla el lado de potencia del circuito inductor.

Balanced atom An atom that has an equal number of protons and electrons.

Átomo equilibrado Átomo que tiene el mismo número de protones y de electrones.

Ballast resistor A resistance put in series with a power lead to a component. Its purpose is to reduce the voltage applied to the component and to control the amount of current in the circuit.

Resistencia autorreguladora Una regulación de serie con un conectador de alimentación a un componente. Su propósito es de reducir la tensión que se aplica al componente y controlar la cantidad del corriente en el circuito.

Base The center layer of a bipolar transistor.

Base Capa central de un transistor bipolar.

Base ignition timing The specification for the degree before top dead center at a defined engine speed without any advance that a spark must occur.

Tiempo de encendido fundamento La specificación del grado en el cual tiene que llegar una chispa al punto muerto superior en una velocidad definida sin ser avanzado.

BAT The terminal identifier for the conductor from the generator to the battery positive terminal.

BAT El terminal que identifica el conductor del generador al terminal positivo de la batería.

Battery cables High-current conductors that connect the battery to the vehicle's electrical system.

Cables de batería Conductores de alta corriente que conectan la batería al sistema eléctrico del vehículo.

Battery cell The active unit of a battery.

Acumulador de batería Componente activo de una batería.

Battery holddowns Brackets that secure the battery to the chassis of the vehicle.

Portabatería Los sostenes que fijan la batería al chasis del vehículo.

Battery terminals Terminals at the battery to which the positive and the negative battery cables are connected. The terminals may be posts or threaded inserts.

Bornes de la batería Los bornes en la batería a los cuales se conectan los cables positivos y negativos. Los terminales pueden ser postes o piezas roscadas.

Baud rate The measure of computer data transmission speed in bits per second.

Razón de baúd Medida de la velocidad de la transmisión de datos de una computadora en bits por segundo.

Belt alternator starter (BAS) A high-voltage starter/alternator combination that uses current flow through the stator windings to generate magnetic fields in the rotor, causing the rotor to turn, thus turning the crankshaft and starting the engine magnetic fields in the rotor, the and starting the engine.

Arrancador del alternador por faja (AAF) Una combinación de arrancador / alternador de alto voltaje que utiliza el flujo de corriente a través del devanado estatórico para generar campos magnéticos en el rotor, lo que provoca que el rotor dé vueltas, haciendo girar así el cigüeñal y arrancando el motor.

Belt tension sensor (BTS) A strain gauge-type sensor located on the seat belt anchor of the passenger-side front seat, used to determine if an infant seat is being used.

Sensor de la tensión de la faja (STF) Sensor de tipo medidor de tensiones localizado en el ancla del cinturón del asiento del lado del asiento frontal del pasajero y que se usa para determinar si se usa un asiento para bebés.

Bendix drive A type of starter drive that uses the inertia of the spinning starter motor armature to engage the drive gear to the gears of the flywheel. This type starter drive was used on early models of vehicles and is rarely seen today.

Acoplamiento Bendix Un tipo del acoplamiento del motor de arranque que usa la inercia de la armadura del motor de arranque giratorio para endentar el engranaje de mando con los engranajes del volante. Este tipo de acoplamiento del motor de arranque se usaba en los modelos vehículos antiguos y se ven raramente.

Bias voltage Voltage applied across a diode.

Tensión polarizadora Tensión aplicada a través de un diodo.

Bimetallic strip A metal contact wiper consisting of two different types of metals. One strip will react quicker to heat than the other, causing the strip to flex in proportion to the amount of current flow.

Banda bimetálica Contacto deslizante de metal compuesto de dos tipos de metales distintos. Una banda reaccionará más rápido al calor que la otra, haciendo que la banda se doble en proporción con la cantidad de flujo de corriente.

Binary code A series of numbers represented by 1's and 0's. Any number and word can be translated into a combination of binary 1's and 0's.

Código binario Serie de números representados por unos y ceros. Cualquier número y palabra puede traducirse en una combinación de unos y ceros binarios.

Bipolar The name used for transistors because current flows through the materials of both polarities.

Bipolar Nombre aplicado a los transistores porque la corriente fluye por conducto de materiales de ambas polaridades.

Bit A binary digit.

Bit Dígito binario.

Blend-air door actuator An electric motor that controls the position of the blend air door, in order to supply the in-vehicle temperature the driver selected.

Actuador de puertas por aire mezclado Un motor eléctrico que controla la posición de las puertas por aire mezclado para proporcionar la temperatura dentro del vehículo que seleccione el conductor.

Bluetooth Technology that allows several modules from different manufacturers to be connected using a standardized radio transmission.

Bluetooth Tecnología que permite que se conecten varios módulos de diferente manufactura por medio del uso de transmisión de radio estándar.

Brake assist system (BAS) An additional function of ABS that assists in the application of the brakes during a panic stop situation.

Sistema asistido por freno (SAF) Una función adicional del SFA (sistema de freno asistido) que ayuda a frenar durante una situación de paro forzado.

Brushes Electrically conductive sliding contacts, usually made of copper and carbon.

Escobillas Contactos deslizantes de conduccion eléctrica, por lo general hechos de cobre y de carbono.

Bucking coil One of the coils in a three-coil gauge. It produces a magnetic field that bucks or opposes the low-reading coil.

Bobina compensadora Una de las bobinas de un calibre de tres bobinas. Produce un campo magnético que es contrario o en oposición a la bobina de baja lectura.

Buffer A buffer cleans up a voltage signal. These are used with PM generator sensors to change the AC voltage to a digitalized signal.

Separador Un separador aguza una señal del tensión. Estos se usan con los sensores generadores PM para cambiar la tensión de corriente alterna a una señal digitalizado.

Buffer circuit Changes the AC voltage from the PM generator into a digitalized signal.

Circuito separador Cambia el voltaje de corriente alterna del generador PM a una señal digital.

Bulkhead connector A large connector that is used when many wires pass through the bulkhead or firewall.

Conectador del tabique Un conectador que se usa al pasar muchos alambres por el tabique o mamparo de encendidos.

Bus Used in reference to data transmission since data is being transported from one place to another. The multiplex circuit is often referred to as the bus circuit.

Colectiva Se usa en referir a la transmisión de datos que se estan transportando de un lugar a otro. El circuito multiplex suele referirse como el circuito colectivo.

Bus (+) The bus cirucuit that is most positive when the dominant bit is being transmitted.

Bus positivo (+) Circuito del bus que es más positivo cuando se transmite el bit predominante.

Bus (–) The bus circuit that is most negative when the dominant bit is being transmitted.

Bus negativo (–) Circuito del bus que es más negativo cuando se transmite el bit predominante.

Bus Bar A common electrical connection to which all of the fuses in the fuse box are attached. The bus bar is connected to battery voltage.

Barra colectora Conexión eléctrica común a la que se conectan todos los fusibles de la caja de fusibles. La barra colectora se conecta a la tensión de la batería.

Buzzer An audible warning device that is used to warn the driver of possible safety hazards.

Zumbador Dispositivo audible de advertencia utilizado para prevenir al conductor de posibles riesgos a la seguridad.

Capacitance The ability of two conducting surfaces to store voltage.

Capacitancia Propiedad que permite el almacenamiento de electricidad entre dos conductores aislados entre sí.

Carbon monoxide An odorless, colorless, and toxic gas that is produced as a result of combustion.

Monóxido de carbono Gas inodoro, incoloro y tóxico producido como resultado de la combustión.

Carriers Attached to the shoulder anchor to move or carry the anchor from one end of the track to the other.

Portadoras Conectados al reborde de anclaje para mudar o transportar el anclaje de una extremidad del carril a la otra.

Cartridge fuses *See* maxi-fuse.

Fusibles cartucho *Véase* maxifusible.

Cathode Negatively charged electrode of a voltage cell.

Cátodo Electrodo de carga negativa de un generador de electricidad.

Cathode ray tube Similar to a television picture tube. It contains a cathode that emits electrons and an anode that attracts them. The screen of the tube will glow at the points that are hit by the electrons.

Tubo de rayos catódicos Parecidos a un tubo de pantalla de televisor. Contiene un cátodo que emite los electrones y un ánodo que los atrae. La pantalla del tubo iluminará en los puntos en donde pegan los electrones.

Caustic A material that has the ability to destroy or eat through something. Caustic materials are considered extremely corrosive.

Caustico Una materia que tiene la habilidad de destruir o carcomer algo. Las materias causticas se consideran extremamente corrosivas.

Cell element The assembly of a positive and negative plate in a battery.

Elemento de pila La asamblea de una placa positiva y negativa en una bateria.

Central gateway (CGW) A module on the CAN bus network that is the hub between the different networks.

Puerta central Un módulo de la red de bus CAN que está en el núcleo entre las diferentes redes.

Central processing unit (CPU) The brains of the computer where most calculations take place.

Unidad central de procesamiento (UCP) El cerebro de la computadora en donde se realizan la mayoría de los cálculos.

Charging system Converts the mechanical energy of the engine into electrical energy to recharge the battery and run the electrical accessories.

Sistema de carga Convierte la energía mecánica del motor en energía eléctrica para recargar la batería y hacer trabajar los accesorios eléctricos.

CHMSL The abbreviation for center high-mounted stop light, often referred to as the third brake light.

CHMSL La abreviación para el faro de parada montada alto en el centro que suele referirse como el faro de freno tercero.

Choke An inductor in series with a circuit.

Reactancia Un inductor en serie con un circuit.

Choke coil Fine wire wound into a coil used to absorb oscillations in a switched circuit.

Bobina de inducción Alambre fino devanado en una bobina, utilizado para absorber oscilaciones en un circuito conmutado.

Chrysler Collision Detection (CCD) Chrysler's data bus network first used in 1988. Uses a twisted pair of wires to transmit data.

Detección de colisión de Chrysler Red del bus de datos de Chrysler que se usó primero en 1988. Utilizó un par de hilos torcidos para transmitir datos.

Circuit The path of electron flow consisting of the voltage source, conductors, load component, and return path to the voltage source.

Circuito Trayectoria del flujo de electrones, compuesto de la fuente de tensión, los conductores, el componente de carga y la trayectoria de regreso a la fuente de tensión.

Circuit breaker A mechanical fuse that opens the circuit when amperage is excessive. In most cases, the circuit breaker will reset when the overload is removed.

Interruptor Un fusible mecánico que abre el circuito cuando la intensidad de amperaje es excesiva. En la mayoría de los casos, el interruptor se reengancha al eliminarse la sobrecarga.

Clamping diode A diode that is connected in parallel with a coil to prevent voltage spikes from the coil from reaching other components in the circuit.

Diodo de bloqueo Un diodo que se conecta en paralelo con una bobina para prevenir que los impulsos de tensión lleguen a otros componentes en el circuito.

Climate-controlled seats Vehicle seats that are heated and/or cooled to increase occupant comfort.

Asientos de clima controlado Asientos del vehículo que pueden calentarse y/o enfriarse para aumentar la comodidad del ocupante.

Clock circuit A crystal that electrically vibrates when subjected to current at certain voltage levels. As a result, the chip produces very regular series of voltage pulses.

Circuito de reloj Cristal que vibra electrónicamente cuando está sujeto a una corriente a ciertos niveles de tensión. Como resultado, el fragmento produce una serie sumamente regular de impulsos de tensión.

Clockspring A winding of electrical conducting tape enclosed within a plastic housing. The clockspring maintains continuity between the steering wheel, switches, the air bag, and the wiring harness in all steering wheel positions.

Muelle espiral Una bobina de cinta conductiva eléctrica encerrada en una caja de plástico. El muelle espiral mantiene la corriente continua entre el volante de dirección, los interruptores, la bolsa de aire, y el mazo de alambres en cualquier posición del volante de dirección.

Closed circuit A circuit that has no breaks in the path and allows current to flow.

Circuito cerrado Circuito de trayectoria ininterrumpida que permite un flujo continuo de corriente.

Coil pack A coil assembly that contains two or more coils.

Asamblea de bobina Una asamblea de bobinas que contiene dos bobinas o más.

Cold cranking rating (CCA) Rating indicates the battery's ability to deliver a specified amount of current to start an engine at low ambient temperatures.

Amperios de arranque en frío Tasa indicativa de la capacidad de la batería para producir una cantidad específica de corriente para arrancar un motor a bajas temperaturas ambiente.

Cold engine lock-out switch Prevents blower motor operation until the air entering the passenger compartment reaches a specified temperature.

Interruptor de cierre en motor inactivo Previene la operación del motor del ventilador hasta que el aire que entra en el compartimiento del pasajero alcanza su temperatura específica.

Collector The portion of a bipolar transistor that receives the majority of current carriers.

Dispositivo de toma de corriente Parte del transistor bipolar que recibe la mayoría de los portadores de corriente.

Common connector A connector that is shared by more than one circuit and/or component.

Conector común Un conector que se comparte entre más de un circuito y/o componente.

Commutator A series of conducting segments located around one end of the armature.

Conmutador Serie de segmentos conductores ubicados alrededor de un extremo de la armadura.

Component locator A service manual that lists and describes the exact location of components on a vehicle.

Localizador de componentes Un manual de servicio que cataloga y describe la posición exacta de los componentes en un vehículo.

Composite bulb A headlight assembly that has a replaceable bulb in its housing.

Bombilla compuesta Una asamblea de faros cuyo cárter tiene una bombilla reemplazable.

Composite headlights A halogen headlight system that uses a replacement bulb.

Faros compuestos Un sistema de faros halógenos que usa un foco de recambio.

Compound motor A motor that has the characteristics of a series-wound and a shunt-wound motor.

Motor compuesta Un motor que tiene las características de un motor exitado en serie y uno en derivación.

Computer An electronic device that stores and processes data and is capable of operating other devices.

Computadora Dispositivo electrónico que almacena y procesa datos y que es capaz de ordenar a otros dispositivos.

Concealed headlight System used to help improve fuel economy and styling of the vehicle.

Faros ocultos Un sistema que se usa para mejorar el rendimiento del combustible y el estilo del vehículo.

Condenser A capacitor made from two sheets of metal foil separated by an insulator.

Condensador Capacitador hecho de dos láminas de metal separadas por un medio aislante.

Conduction Bias voltage difference between the base and the emitter has increased to the point that the transistor is switched on. In this condition, the transistor is conducting. Output current is proportional to that of the current through the base.

Conducción La diferencia de la tensión polarizadora entre la base y el emisor ha aumentado hasta el punto que el transistor es conectado. En estas circunstancias, el transistor está conduciendo. La corriente de salida está en proporción con la de la corriente conducida en la base.

Conductor A material in which electrons flow or move easily.

Conductor Una material en la cual los electrones circulen o se mueven fácilmente.

Continuity Refers to the circuit being continuous with no opens.

Continuidad Se refiere al circuito ininterrumpido, sin aberturas.

Control panel assembly Provides for driver input into the automatic temperature control microprocessor.

Asamblea de controles Permite la entrada del conductor al microprocesador del control de temperatura automático. La asamblea de control tambien se refiere como el tablero de instrumentos.

Controller area network (CAN) A two-wire bus network that allows the transfer of data between control modules.

CAN (Red del área del controlador) Red de bus de dos hilos que permite la transferencia de datos entre los módulos de control.

Conventional theory Electrical theory that states current flows from a positive point to a more negative point.

Teoría convencional Teoría de electricidad la cual enuncia que el corriente fluye desde un punto positivo a un punto más negativo.

Cornering lights Lamps that illuminate when the turn signals are activated. They burn steady when the turn signal switch is in a turn position, to provide additional illumination of the road in the direction of the turn.

Faros de viraje Los faros que iluminen cuando se prenden los indicadores de virajes. Quedan prendidos mientras que el indicador de viraje esta en una posición de viraje para proveer mayor iluminción del camino en la dirección del viraje.

Counterelectromotive force (CEMF) An induced voltage that opposes the source voltage.

Fuerza cóntraelectromotriz Tensión inducida en oposición a la tensión fuente.

Courtesy lights Lamps that illuminate the vehicle's interior when the doors are open.

Luces interiores Lámparas que iluminan el interior del vehículo cuando las puertas están abiertas.

Covalent bonding When atoms share valence electrons with other atoms.

Enlace covalente Cuando los átomos comparten electrones de valencia con otros átomos.

Crash sensor Normally open electrical switch designed to close when subjected to a predetermined amount of jolting or impact.

Sensor de impacto Un conmutador normalmente abierto diseñado a cerrarse al someterse a un sacudo de una fuerza predeterminada o un impacto.

Cross-fire The undesired firing of a spark plug that results from the firing of another spark plug. This is caused by electromagnetic induction.

Encendido transversal El encendido no deseable de una bujía que resulta del encendido de otra bujía. Esto se causa por la inducción electromagnética.

Cruise control A system that allows the vehicle to maintain a preset speed with the driver's foot off the accelerator.

Control crucero Un sistema que permite que el vehículo mantenga una velocidad predeterminada sin que el pie del conductor dceprime al accelerador.

Crystal A term used to describe a material that has a definite atom structure.

Cristal Término utilizado para describir un material que tiene una estructura atómica definida.

Current The aggregate flow of electrons through a wire. One ampere represents the movement of 6.25 billion billion electrons (or one coulomb) past one point in a conductor in one second.

Corriente Flujo combinado de electrones a través de un alambre. Un amperio representa el movimiento de 6,25 mil millones de mil millones de electrones (o un colombio) que sobrepasa un punto en un conductor en un segundo.

Cutoff When reverse-bias voltage is applied to the base leg of the transistor. In this condition, the transistor is not conducting and no current will flow.

Corte Cuando se aplica tensión polarizadora inversa a la base del transistor. En estas circunstancias, el transistor no está conduciendo y no fluirá ninguna corriente.

Cycle Completed when the voltage has gone positive, returned to zero, gone negative, and returned to zero.

Ciclo Completado cuando el voltaje ha sido positivo, regresado al cero, ha sido negativo y regresado al cero.

Darlington pair An arrangement of transistors that amplifies current by one transistor acting as a preamplifier that creates a larger base current to the second transistor.

Par Darlington Conjunto de transistores que amplifica la corriente. Un transistor actúa como preamplificador y produce una corriente base más ámplia para el segundo transistor.

D'Arsonval gauge A gauge design that uses the interaction of a permanent magnet and an electromagnet, and the total field effect to cause needle movement.

Calibrador d'Arsonval Calibrador diseñado para utilizar la interacción de un imán permanente y de un electroimán, y el efecto inductor total para generar el movimiento de la aguja.

Daytime running lamps Generally use the high-beam or low-beam headlight system at a reduced intensity to provide additional visibility of the vehicle for other drivers and pedestrians.

Faros diurnos Generalmente usan el sistema de faros de alta y baja intensidad en una intensidad disminuida para proporcionar el vehículo más visibilidad para los otros conductores y los peatones.

DC/DC converter The DC/DC converter is configured to provide a 14-V output from the high voltage input. The 14-V output can be used to supply electrical energy to those components that do not require the high voltage.

Convertidor continua-continua El convertidor continua-continua está configurado para proporcionar una salida de 14 V de una entrada de alto voltaje. La salida de 14 V puede usarse para suplir energía eléctrica a aquellos componentes que no requieran de alto voltaje.

Deep cycling Discharging the battery completely before recharging it.

Operación cíclica completa La descarga completa de la batería previo al recargo.

Delta connection A connection that receives its name from its resemblance to the Greek letter delta (Δ).

Conexión delta Una conexión que recibe su nombre a causa de su aparencia parecida a la letra delta griega.

Delta stator A three-winding AC generator stator with the ends of each winding connected to each other.

Estátor delta Estátor generador de corriente alterna de devanado triple, con los extremos de cada devanado conectados entre sí.

Depletion-type FET Cuts off current flow.

FET tipo agotamiento Corta el flujo del corriente.

Depressed park A system in which the blades drop down below the lower windshield molding to hide them.

Limpiaparabrisas guardadas Un sistema en el cual los brazos se guardan abajo del borde inferior de la parabrisa para asi esconderlos.

Diagnostic module Part of an electronic control system that provides self-diagnostics and/or a testing interface.

Módulo de diagnóstico Parte de un sistema controlado electronicamente que provee autodiagnóstico y/o una interfase de pruebas.

Diaphragm A thin, flexible, circular plate that is held around its outer edge by the horn housing, allowing the middle to flex.

Diagragma Una placa redonda flexible y delgada sostenido en el cárter del claxon por medio de su borde exterior, así permitiendo flexionar la parte central.

Dielectric An insulator material.

Dieléctrico Material aislante.

Digital A voltage signal is either on-off, yes-no, or high-low.

Digital Una señal de tensión está Encendida-Apagada, es Sí-No o Alta-Baja.

Digital instrument clusters Use digital and linear displays to notify the driver of monitored system conditions.

Grupo de instrumentos digitales Usan los indicadores digitales y lineares para notificar el conductor de las condiciones de los sistemas regulados.

Dimmer switch A switch in the headlight circuit that provides the means for the driver to select either high-beam or low-beam operation, and to switch between the two. The dimmer switch is connected in series within the headlight circuit and controls the current path for high and low beam.

Conmutador reductor Conmutador en el circuito para faros delanteros que le permite al conductor que elegir la luz larga o la luz corta, y conmutar entre las dos. El conmutador reductor se conecta en serie dentro del circuito para faros delanteros y controla la trayectoria de la corriente para la luz larga y la luz corta.

DIN The abbreviation for Deutsche Institut füer Normung (German Institute for Standardization) and the recommended standard for European manufacturers to follow.

DIN La abreviatura de Deutsche Institut füer Normung (Normas del Instituto Aleman) y se recomienda que los fabricantes europeos siguen estas normas.

Diode An electrical one-way check valve that will allow current to flow in one direction only.

Diodo Válvula eléctrica de retención, de una vía, que permite que la corriente fluya en una sola dirección.

Diode rectifier bridge A series of diodes that are used to provide a reasonably constant DC voltage to the vehicle's electrical system and battery.

Puente rectificador de diodo Serie de diodos utilizados para proveerles una tensión de corriente continua bastante constante al sistema eléctrico y a la batería del vehículo.

Diode trio Used by some manufacturers to rectify the stator of an AC generator current so that it can be used to create the magnetic field in the field coil of the rotor.

Trío de diodos Utilizado por algunos fabricantes para rectificar el estátor de la corriente de un generador de corriente alterna y poder así utilizarlo para crear el campo magnético en la bobina inductora del rotor.

Direct current (DC) Electric current that flows in one direction.

Corriente continua Corriente eléctrica que fluye en una dirección.

Direct drive A situation where the drive power is the same as the power exerted by the device that is driven.

Transmisión directa Una situación en la cual el poder de mando es lo mismo que la potencia empleada por el dispositivo arrastrado.

Discrete devices Electrical components that are made separately and have wire leads for connections to an integrated circuit.

Dispositivos discretos Componentes eléctricos hechos uno a uno; tienen conductores de alambre para hacer conexiones a un circuito integrado.

Discriminating sensors Part of the air bag circuitry; these sensors are calibrated to close with speed changes that are great enough to warrant air bag deployment. These sensors are also referred to as crash sensors.

Sensores discriminadores Una parte del conjunto de circuitos de Airbag; estos sensores se calibran para cerrar con los cambios de la velocidad que son bastante severas para justificar el despliegue del Airbag. Estos sensores también se llaman los sensores de impacto.

Distributor A mechanism within the ignition system that controls the primary circuit and directs the secondary voltage to the correct spark plug.

Distribuidor Un mecanismo dentro del sistema de encendido que controla el circuito primario y dirije el voltaje secundarios hacia la bujía.

Distributor ignition (DI) Replaces all previous terms for electronically controlled distributor-type ignition systems.

Encendido del distribuidor (DI) Reemplaza todos los téerminos usado previamente para indicar los sistemas de encendido controlados electronicamente por distribuidor.

Doping The addition of another element with three or five valence electrons to a pure semiconductor.

Impurificación La adición de otro elemento con tres o cinco electrones de valencia a un semiconductor puro.

Double-filament lamp A lamp designed to execute more than one function. It can be used in the stoplight circuit, taillight circuit, and the turn signal circuit combined.

Lámpara con filamento doble Lámpara diseñada para llevar a cabo más de una función. Puede utilizarse en una combinación de los circuitos de faros de freno, de faros traseros y de luces indicadoras para virajes.

Double-start override Prevents the starter motor from being energized if the engine is already running.

Sobremarcha de doble marcha Previene que se excita el motor del encendido si ya esta en marcha el motor.

Drain The portion of a field-effect transistor that receives the holes or electrons.

Drenador Parte de un transistor de efecto de campo que recibe los agujeros o electrones.

Drive coil A hollowed field coil used in a positive-engagement starter to attract the movable pole shoe of the starter.

Bobina de excitación Una bobina inductora hueca empleada en un encendedor de acoplamiento directo para atraer la pieza polar móvil del encendedor.

Drive spring Absorbs the initial shock of engagement of the starter.

Resorte de enganche Absorba el choque inicial del enganche.

Dual climate control Provides separate temperature settings for the driver and the front-seat passenger. This system is similar to previous systems except two blend doors are used to control separate temperature settings.

Control de clima doble Provea la regulación de temperatura individual para el conductor ye el pasajero del asiento delantero. Este sistema es parecido a los sistemas anteriores menos que los dos compuertas de mezcla se usan para controlar la regulación individual de la temperatura.

Dual-generator system Uses one generator that operates at 42 volts while another operates at 14 volts.

Sistema de doble voltaje Arquitectura que utiliza dos voltajes separados. Uno es para el sistema de 42 voltios que enciende los accesorios eléctricos que podrían requerir o beneficiarse de mayor voltaje. El resto de las cargas continúan en 14 voltios.

Dual plug Has two spark plugs per cylinder.

Bujía doble Tiene dos bujías por cilindro.

Dual-stator, dual-voltage system Dual voltage (42 volts and 14 volts) is produced from a single alternator that has two output voltages.

Sistema de doble voltaje y doble estator Un alternador sencillo que tiene voltajes de dos salidas y produce un doble voltaje (42 voltios y 14 voltios.)

Dual-voltage system Architecture that uses two separate voltages. One is for the 42-volt system that powers those electrical accessories that would require or benefit from the higher voltage. The remainder of the loads remain on 14 V.

Sistema de doble voltaje Arquitectura que utiliza dos voltajes separados. Uno es para el sistema de 42 voltios que enciende los accesorios eléctricos que podrían requerir o beneficiarse de mayor voltaje. El resto de las cargas continúan en 14 voltios.

Duty cycle The percentage of on time to total cycle time.

Ciclo de trabajo Porcentaje del trabajo efectivo a tiempo total del ciclo.

Dwell The length of time, in degree of distributor shaft rotation, that there is a current flow in the primary circuit prior to firing the spark plug.

Reposo La cantidad del tiempo, medida por grados de rotación del eje del distribuidor, en que hay un flujo de corriente en el circuito primario antes de encender la bujía.

Easy exit An additional function of the memory seat that provides for easier entrance and exit of the vehicle by moving the seat all the way back and down. Some systems also move the steering wheel up and to full retract.

Salida fácil Una función adicional de la memoria del asiento que provee una entrada y salida más fácil del vehículo al mover el asiento hasta su posición más extrema hacia atrás y abajo. Algunos sistemas tambien muevan el volante de dirección hacia arriba y a su posición más alejada.

Eddy currents Small induced currents.

Corriente de Foucault Pequeñas corrientes inducidas.

Electric defoggers Heat the rear window to remove ice and/or condensation. Some vehicles use the same circuit to heat the outside driver's-side mirror.

Desneblador eléctrica Calientan la ventanilla trasxera para remover el hielo y/o la condensación. Algunos vehículos usan el mismo circuito para calentar el espejo lateral del conductor.

Electric vehicle (EV) A vehicle that powers its motor off of a battery pack.

Vehículo eléctrico (VE) Vehículo que apaga su motor por medio de un paquete de baterías.

Electrical accessories Electrical systems or components that provide for additional safety and comfort, including safety accessories such as the horn, windshield wipers, and windshield washers. Comfort accessories include the blower motor, electric defoggers, power mirrors, power windows, power seats, and power door locks.

Accesorios eléctricos Sistemas o componentes eléctricos que proporcionan seguridad y comodidad adicionales, y que incluyen accesorios de seguridad tales como el claxon, limpia brisas y parabrisas. Los accesorios de comodidad incluyen un motor de aire, desnubilizador eléctrico, espejos mecánicos, asientos mecánicos y cierre mecánico de puertas.

Electrical load The working device of the circuit.

Carga eléctrica Dispositivo de trabajo del circuito.

Electrical symbols Used to represent components in the wiring diagram.

Símbolos electrónicos Se usan para representar los componentes en uyn esquema de conexiones.

Electrically Erasable PROM (EEPROM) Memory chip that allows for electrically changing the information one bit at a time.

Capacidad de borrado electrónico PROM Fragmento de memoria que permite el cambio eléctrico de la información un bit a la vez.

Electrochemical The chemical action of two dissimilar materials in a chemical solution.

Electroquímico Acción química de dos materiales distintos en una solución química.

Electrochromic mirror Automatically adjusts to light by using forward and rearward facing photo sensors and a solid-state chip. Based on light intensity differences, the chip applies a small voltage to the silicon layer. As voltage is applied, the molecules of the layer rotate and redirect the light beams. Thus the mirror reflection appears dimmer.

Espejo electrocrómico Se ajusta automáticamente a la luz usando los sensores orientados hacia afrente y atrás juntos con un chip de estado sólido.

Electrolysis The producing of chemical changes by passing electrical current through an electrolyte. The splitting of water into hydrogen and oxygen.

Electrólisis La producción de los cambios químicos al pasar un corriente eléctrico por un electrolito.

Electrolyte A solution of 64% water and 36% sulfuric acid.

Electrolito Solución de un 64% de agua y un 36% de ácido sulfúrico.

Electromagnetic gauge Gauge that produces needle movement by magnetic forces.

Calibrador electromagnético Calibrador que genera el movimiento de la aguja mediante fuerzas magnéticas.

Electromagnetic induction The production of voltage and current within a conductor as a result of relative motion within a magnetic field.

Inducción electromagnética Producción de tensión y de corriente dentro de un conductor como resultado del movimiento relativo dentro de un campo magnético.

Electromagnetic interference (EMI) An undesirable creation of electromagnetism whenever current is switched on and off.

Interferencia electromagnética Fenómeno de electromagnetismo no deseable que resulta cuando se conecta y se desconecta la corriente.

Electromagnetism A form of magnetism that occurs when current flows through a conductor.

Electromagnetismo Forma de magnetismo que ocurre cuando la corriente fluye a través de un conductor.

Electromechanical A device that uses electricity and magnetism to cause a mechanical action.

Electromecánico Un dispositivo que causa una acción mecánica por medio de la electricidad y el magnetismo.

Electromotive force (EMF) *See* voltage.

Fuerza electromotriz *Véase* tensión.

Electron Negative-charged particles of an atom.

Electrón Partículas de carga negativa de un átomo.

Electron theory Defines electrical movement as from negative to positive.

Teoría del electrón Define el movimiento eléctrico como el movimiento de lo negativo a lo positivo.

Electronic ignition (EI) Replaces all previous terms for distributorless ignition systems.

Encendido electrónico (EI) Reemplaza todos los términos previos del sistema de encendido sin distribuidor.

Electronic regulator Uses solid-state circuitry to perform regulatory functions.

Regulador electrónico Usa los circuitos de estado sólido para llevar a cabo los funciones de regulación.

Electronic shift automatic transmissions Computer-controlled automatic transmissions that use several inputs to determine torque converter clutch operation, hydraulic pressure levels, and shift points.

Transmisión de cambio electrónico automático Transmisiones automáticas controladas por computadora que utilizan varias entradas para determinar la operación del embrague del convertidor de par, niveles de la presión hidráulica y los puntos de cambio.

Electronic stability control An additional function of the antilock brake and traction control system that uses additional sensors and inputs to determine if the vehicle is actually moving in the direction intended by the driver, as indicated by steering wheel position sensors, yaw sensors, and lateral sensors. If the actual path is not the intended path, the module will apply the appropriate brake to bring the vehicle back onto the correct path.

Mando electrónico de estabilidad Función adicional de un sistema de antibloqueo de frenos y de control de la tracción que utiliza sensores adicionales y entradas para determinar si efectivamente se mueve el vehículo en la dirección que desea el conductor, tales como indican los sensores de posición del volante, los sensores de guiñada y los sensores laterales. Si el recorrido real no es el corrido deseado, el módulo aplicará el freno apropiado para regresar el vehículo al recorrido correcto.

Electronic suspension systems Electronic suspension systems use a computer to change ride height, shock damping, and spring rates in response to changing road and driving conditions.

Sistemas de suspensión electrónica Los sistemas de suspensión electrónica utilizan una computadora para cambiar la altura de la monta, el amortiguamiento y la relación elástica del muelle como respuesta a los cambios en la carretera y las condiciones de manejo.

Electrostatic field The field that is between the two oppositely charged plates.

Campo electrostático Campo que se encuentra entre las placas de carga opuesta.

Emitter The outer layer of the transistor, which supplies the majority of current carriers.

Emisor Capa exterior del transistor que suministra la mayor parte de los portadores de corriente.

Engine vacuum Formed during the intake stroke of the cylinder. Engine vacuum is any pressure lower than atmospheric pressure.

Vacío del motor Formado durante la carrera de entrada de un cilíndro. El vacío de motor es cualquier presión más baja de la presión atmosférica.

Enhancement-type FET Improves current flow.

FET tipo de acrecentamiento Mejor ael flujo del corriente.

Equivalent series load (equivalent resistance) The total resistance of a parallel circuit, which is equivalent to the resistance of a single load in series with the voltage source.

Carga en serie equivalente (resistencia equivalente) Resistencia total de un circuito en paralelo, equivalente a la resistencia de una sola carga en serie con la fuente de tensión.

Erasable PROM (EPROM) Similar to PROM except that its contents can be erased to allow for new data to be installed. A piece of Mylar tape covers a window. If the tape is removed, the microcircuit is exposed to ultraviolet light and erases its memory.

Capacidad de borrado PROM Parecido al PROM, pero su contenido puede borrarse para permitir la instalación de nuevos datos. Un trozo de cinta Mylar cubre una ventana; si se remueve la cinta, el microcircuito queda expuesto a la luz ultravioleta y borra la memoria.

EVR Stands for external voltage regulator.

EVR Representa el Regulador de Voltaje Externo.

Excitation current Current that magnetically excites the field circuit of the ac generator.

Corriente de excitación Corriente que excita magnéticamente al circuito inductor del generador de corriente alterna.

Face shield A clear plastic shield that protects the entire face.

Máscara protectora Una máscara de plástico transparente que proteje la cara entera.

Feedback 1. Data concerning the effects of the computer's commands are fed back to the computer as an input signal. Used to determine if the desired result has been achieved. 2. A condition that can occur when electricity seeks a path of lower resistance, but the alternate path operates a component other than that intended. Feedback can be classified as a short.

Realimentación 1. Datos referentes a los efectos de las órdenes de la computadora se suministran a la misma como señal de entrada. La realimentación se utiliza para determinar si se ha logrado el resultado deseado. 2. Condición que puede ocurrir cuando la electricidad busca una trayectoria de menos resistencia, pero la trayectoria alterna opera otro componente que aquel deseado. La realimentación puede clasificarse como un cortocircuito.

Fiber optics A medium of transmitting for the transmission of light through polymethyl methacrylate plastic that keeps the light rays parallel even if there are extreme bends in the plastic.

Transmisión por fibra óptica Técnica de transmisión de luz por medio de un plástico de polimetacrilato de metilo que mantiene los rayos de luz paralelos aunque el plástico esté sumamente torcido.

Field coils Heavy copper wire wrapped around an iron core to form an electromagnet.

Bobina del campo El alambre grueso de cobre envuelta alrededor de un núcleo de hierro para formar un electroimán.

Field current The current going to the field windings of a motor or generator.

Corriente inductora El corriente que va a los devanados inductores de un motor o generador.

Field-effect transistor (FET) A unipolar transistor in which current flow is controlled by voltage in a capacitance field.

Transistor de efecto de campo Transistor unipolar en el cual la tensión en un campo de capacitancia controla el flujo de corriente.

Field relay The relay that controls the amount of current going to the field windings of a generator. This is the main output control unit for a charging system.

Relé inductor El relé que controla la cantidad del corriente a los devanados inductores de un generador. Es la unedad principal de potencia de salida de un sistema de carga.

Fifth percentile female The fifth percentile female is determined to be those who weigh less than 100 pounds (45 kg). The fifth percentile female is determined by averaging all potential occupants by size and then plotting the results on a graph.

Quinto percentil femenino La determinan aquellos que pesan menos de 45 kg (100 lbs. Se determina al hacer el promedio de todos los posibles ocupantes por su tamaño, y luego al trazar los resultados en una gráfica.

Firing order Order in which the cylinders of an engine move through the power stroke.

Orden del encendido El orden en el cual los cilindros de un motor cumplen la carrera de potencia.

Fixed resistors Have a set resistance value and are used to limit the amount of current flow in a circuit.

Resistores fijos Tienen un valor de resistencia fijo y se usan para limitar la cantidad de flujo del corriente en un circuito.

Flammable A substance that will support combustion.

Inflamable Una substancia que ampara la combustión.

Flasher Used to open and close the turn signal circuit at a set rate.

Pulsador Se usa para abrir y cerrar el circuito del indicador de vueltas en una velocidad predeterminada.

Floor jack A portable hydraulic tool used to raise and lower a vehicle.

Gato de pie Herramienta hidráulica portátil utilizada para levantar y bajar un vehículo.

Flux density The number of flux lines per square centimeter.

Densidad de flujo Número de líneas de flujo por centímetro cuadrado.

Flux lines Magnetic lines of force.

Líneas de flujo Líneas de fuerza magnética.

Forward bias A positive voltage that is applied to the P-type material and negative voltage to the N-type material of a semiconductor.

Polarización directa Tensión positiva aplicada al material P y tensión negativa aplicada al material N de un semiconductor.

Fuel cell A battery-like component that produces current from hydrogen and aerial oxygen.

Celda de combustible Componente tipo batería que produce corriente del hidrógeno y del oxígeno en el aire.

Fuel pump inertia switch An NC switch that will open if the vehicle is involved in an impact at speeds over 5 mph or if it rolls over. When the switch opens, it turns off power to the fuel pump. This is a safety feature to prevent fuel from being pumped onto the ground or hot engine compartments if the engine dies. The switch has to be manually reset if it is triggered.

Interruptor inercia de la bomba de combustible Un interruptor NC que se abre si el vehículo se involucra en un choque en una velocidad que exceda 5 millas por hora o si se invierte de arriba abajo. Cuando el interruptor se abre, corta la corriente a la bomba del combustible. Este es una precaución de seguridad para prevenir que la bomba vierte el combustibel en el suelo o sobre un compartimento caliente del motor si se muere el motor. El interruptor se tiene que reenganchar a mano si se acciona.

Full field Maximum AC generator output.

Campo completo Salida máxima de un generador de corriente alterna.

Full parallel hybrid Uses an electric motor that is powerful enough to propel the vehicle on its own.

Híbrido en paralelo completo Utiliza un motor eléctrico que es lo suficientemente potente para que el vehículo se impulse por sí mismo.

Full-wave rectification The conversion of a complete AC voltage signal to a DC voltage signal.

Rectificación de onda plena La conversión de una señal completa de tensión de corriente alterna a una señal de tensión de corriente continua.

Fuse A replaceable circuit protection device that will melt should the current passing through it exceed its rating.

Fusible Dispositivo reemplazable de protección del circuito que se fundirá si la corriente que fluye por el mismo excede su valor determinado.

Fuse block The term used to indicate the central location of the fuses contained in a single holding fixture.

Bloque de fusibles El término que su usa para indicar la ubicación central de los fusibles contenidos en una fijación central.

Fuse box A term used to indicate the central location of the fuses contained in a single holding fixture.

Caja de fusibles Término utilizado para indicar la ubicación central de los fusibles contenidos en un solo elemento permanente.

Fusible link A wire made of meltable material with a special heat-resistant insulation. When there is an overload in the circuit, the link melts and opens the circuit.

Cartucho de fusible Alambre hecho de material fusible con aislamiento especial resistente al calor. Cuando ocurre una sobrecarga en el circuito, el cartucho se funde y abre el circuito.

Gain The ratio of amplification in an electronic device.

Ganancia Razón de amplificación en un dispositivo electrónico.

Ganged Refers to a type of switch in which all wipers of the switch move together.

Acoplado en tándem Se refiere a un tipo de conmutador en el cual todos los contactos deslizantes del mismo se mueven juntos.

Gassing The conversion of a battery's electrolyte into hydrogen and oxygen gas.

Burbujeo La conversión del electrolito de una batería al gas de hidrógeno y oxígeno.

Gate The portion of a field-effect transistor that controls the capacitive field and current flow.

Compuerta Parte de un transistor de efecto de campo que controla el campo capacitivo y el flujo de corriente.

Gauge 1. A device that displays the measurement of a monitored system by the use of a needle or pointer that moves along a calibrated scale. 2. The number that is assigned to a wire to indicate its size. The larger the number, the smaller the diameter of the conductor.

Calibrador 1. Dispositivo que muestra la medida de un sistema regulado por medio de una aguja o indicador que se mueve a través de una escala calibrada. 2. El número asignado a un alambre indica su tamaño. Mientras mayor sea el número, más pequeño será el diámetro del conductor.

Gear reduction Occurs when two different sized gears are in mesh and the driven gear rotates at a lower speed than the drive gear but with greater torque.

Desmultiplicación Ocurre cuando dos engranajes de distinctos tamaños se endentan y el engranaje arrastrado gira con una velocidad más baja que el engranaje de mando pero con más par.

Grid growth A condition where the grid grows little metallic fingers that extend through the separators and short out the plates.

Expansión de la rejilla Una condición en la cual la rejilla produce protrusiones metálicas que se extienden por los separadores y causan cortocircuitos en las placas.

Grids The frame structure of a battery that normally has connector tabs at the top. It is generally made of lead alloys.

Rejillas La estructura encuadrador de una batería que normalmente tiene orejas de conexión en la parte superior. Generalmente se fabrica de aleaciones de plomo.

Ground The common negative connection of the electrical system. It is the point of lowest voltage.

Tierra Conexión negativa común del sistema eléctrico. Es el punto de tensión más baja.

Ground side The portion of the circuit that is from the load component to the negative side of the source.

Lado a tierra Parte del circuito que va del componente de carga al lado negativo de la fuente.

Grounded circuit An electrical defect that allows current to return to ground before it has reached the intended load component.

Circuito puesto a tierra Falla eléctrica que permite el regreso de corriente a tierra antes de alcanzar el componente de carga deseado.

Half-field current The current going to the field windings of a motor or generator after it has passed through a resistor in series with the circuit.

Corriente de medio campo El corriente que va a los devanados inductores de un motor o a un generador después de que haya pasado por un resistor conectado en serie con el circuito.

Half-wave rectification Rectification of one-half of an AC voltage.

Rectificación de media onda Rectificación en la que la corriente fluye únicamente durante semiciclos alternados.

Hall-effect switch A sensor that operates on the principle that if a current is allowed to flow through thin conducting material being exposed to a magnetic field, another voltage is produced.

Conmutador de efecto Hall Sensor que funciona basado en el principio de que si se permite el flujo de corriente a través de un material conductor delgado que ha sido expuesto a un campo magnético, se produce otra tensión.

Halogen The term used to identify a group of chemically related nonmetallic elements. These elements include chlorine, fluorine, and iodine.

Halógeno Término utilizado para identificar un grupo de elementos no metálicos relacionados químicamente. Dichos elementos incluyen el cloro, el flúor y el yodo.

Hand tools Tools that use only the force generated from the body to operate. They multiply the force received through leverage to accomplish the work.

Herramientas manuales Herramientas que para funcionar sólo necesitan la fuerza generada por el cuerpo. Para llevar a cabo el trabajo, las herramientas multiplican la fuerza que reciben por medio de la palancada.

Hazardous material Materials that can cause illness, injury, or death or pollute water, air, or land.

Material peligroso Las materias que puedan causar la enfermedad, los daños, la muerte o que puedan contaminar el agua, el aire o la tierra.

Headlight leveling system (HLS) Uses front lighting assemblies with a leveling actuator motor to allow the headlights to be adjusted into different vertical positions to compensate for headlight position that can occur when the vehicle is loaded.

Sistema de nivelación de los faros (SNF) Utiliza el ensamblaje de los faros frontales con un motor accionador de nivelación para permitir el ajuste de los faros en diferentes posiciones verticales para compensar la posición que el faro pudiera tomar si se carga el vehículo.

Head-up display (HUD) Displays images onto the inside of the windshield so the driver can see them without having to take his eyes off the road.

Presentación en pantalla (HUD) Proyecta las imagenes en la parte interior de la parabrisas para que el conductor las pueda ver sin tener que tomar su atención de la pista.

Heat range A statement of how well a spark plug can conduct heat away from its tip.

Factor términa Una indicación de la habilidad de una bujía en alejar el calor de su punta.

Heated windshield system A specialy designed windshield that allows current flow through the glass without interfering with the driver's vision; it is capable of melting ice and frost from the windshield three to five times faster than conventional defroster systems.

Sistema de parabrisas térmico Parabrisas especialmente diseñado para permitir el flujo de la corriente a través del vidrio sin interferir con la visión del conductor; está capacitado para derretir el hielo y la escarcha que haya en el parabrisas de 3 a 5 veces más rápido que los sistemas convencionales anticongelantes.

Heater core flow valve Shuts off the coolant flow through the heater core when the A/C system is in the max air mode.

Válvula del flujo térmico del núcleo Cierra el flujo del enfriador a través del núcleo del calentador cuando el sistema AC está en el mando de aire máximo.

Heat sink An object that absorbs and dissipates heat from another object.

Dispersador térmico Objeto que absorbe y disipa el calor de otro objeto.

H-gate A set of four transistors that can reverse current.

Compuerta H Juego de cuatro transistores que pueden invertir la corriente.

HID High intensity discharge; a lighting system that uses an arc across electrodes instead of a filament.

HID Descarga de Alta Intensidad; un sistema de iluminación que utiliza un arco por dos electrodos en vez de un filamento.

High intensity discharge (HID) Uses an inert gas to amplify the light produced by arcing across two electrodes.

Descarga de alta intensidad (HID) Usa un gas inerte para amplificar la luz producida al conectar dos electrodos con una arca.

High-reading coil Position at a 90-degree angle to the low-reading and bucking coils.

Bobina dfe lectura de alta tensión Posiciona en un ángulo a las bobinas de lectura de tensión baja y las bobinas compensadoras.

High side drivers Control the output device by varying the positive (12-volt) side.

Impulsores del lado de alto potencial Controlan el dispositivo de salida en variar el lado positivo (12 voltíos).

High-voltage ECU (HV ECU) See starter generator control module (SGCM).

UCE de alto voltaje (UCE AV) Vea el módulo de control del generador de arranque (MCGA)

Hoist A lift that is used to raise the entire vehicle.

Elevador Montacargas utilizado para elevar el vehículo en su totalidad.

Holddowns Secure the battery to reduce vibration and to prevent tipping.

Portador Aseguran la batería para disminuir la vibración y prevenir que se vierte.

Hold-in winding A winding that holds the plunger of a solenoid in place after it moves to engage the starter drive.

Devanado de retención Un devanado que posiciona el núcleo móvil de un solenoide después de que mueva para accionar el acoplamiento del motor de arranque.

Hole The absence of an electron in an element's atom. These holes are said to be positively charged since they have a tendency to attract free electrons into the hole.

Agujero Ausencia de un electrón en el átomo de un elemento. Se dice que dichos agujeros tienen una carga positiva puesto que tienden a atraer electrones libres hacia el agujero.

Horn A device that produces an audible warning signal.

Claxon Un dispositivo que produce una señal de advertencia audible.

Hybrid air bag Modules use compressed gas to fill the air bag instead of burning a chemical to produce gas.

Bolsa de aire híbrido Los modulos usan el gas comprimido para llenar la bolsa de aire en vez de quemar una química para producir un gas.

Hybrid battery A battery that combines the advantages of low maintenance and maintenance-free batteries.

Batería híbrida Una batería que combina las ventajas de las baterías de bajo mantenimiento y de no mantenimiento.

Hybrid electric vehicle (HEV) System that has two different power sources. In most hybrid vehicles (HEV), the power sources are a small displacement gasoline or diesel engine and an electric motor.

Vehículo eléctrico híbrido (VEH) Sistema con dos diferentes fuentes de potencia. En la mayoría de los vehículos híbridos (VEH), las fuentes de potencia son un motor de diesel o de gasolina de desplazamiento menor y un motor eléctrico.

Hydrometer A test instrument used to check the specific gravity of the electrolyte to determine the battery's state of charge.

Hidrómetro Instrumento de prueba utilizado para verificar la gravedad específica del electrolito y así determinar el estado de la carga de la batería.

Igniter A combustible device that converts electric energy into thermal energy to ignite the inflator propellant in an air bag system.

Ignitor Un dispositivo combustible que convierte la energía eléctrica a la energía termal para encender el propelente inflador en un sistema Airbag.

Ignition coil A step-up transformer that builds up the low battery voltage of approximately 12.6 volts to a voltage that is high enough to jump across the spark plug gap and ignite the air-fuel mixture.

Bobina de encendido Transformador multiplicador que sube el bajo voltaje de la batería de aproximadamente 12.6 voltios a uno lo suficientemente alto para brincar sobre el huelgo de la bujía y encender la mezcla de aire y combustible.

Ignition switch The power distribution point for most of the vehicle's primary electrical systems.

Selector de encendido Punto de distribución de potencia para la mayoría de los sistemas eléctricos principales del vehículo.

Ignition system Responsible for delivering properly timed high-voltage surges to the spark plugs.

Sistema de encendido Es responsable de llevar subidas de alto voltaje reguladas apropiadamente a las bujías.

Ignition timing Refers to the precise time a spark is sent to the cylinder relative to the piston position.

Tiempo del encendido Refiera al tiempo preciso en el que una chispa es mandada al cilíndro en relación a la posición del piston.

Illuminated entry systems Turn on the courtesy lights before the doors are opened.

Sistemas de entrada iluminada Enciendan las luces de cortesía antes de que se abren las puertas.

Immobilizer system Designed to provide protection against unauthorized vehicle use by disabling the engine if an invalid key is used to start the vehicle or if an attempt to hot-wire the ignition system is made.

Sistema inmovilizante Diseñado para proporcionar protección contra el uso no autorizado del vehículo al desactivar el motor si una llave que no es válida se utiliza para encender el vehículo o para intentar "hacerle el puente" al vehículo.

Incandescence The process of changing energy forms to produce light.

Incandescencia Proceso a través del cual se cambian las formas de energía para producir luz.

Induced voltage Voltage that is produced in a conductor as a result of relative motion within magnetic flux lines.

Tensión inducida Tensión producida en un conductor como resultado del movimiento relativo dentro de líneas de flujo magnético.

Induction The magnetic process of producing a current flow in a wire without any actual contact to the wire. To induce 1 volt, 100 million magnetic lines of force must be cut per second.

Inducción Proceso magnético a través del cual se produce un flujo de corriente en un alambre sin contacto real alguno con el alambre. Para inducir 1 voltio, deben producirse 100 millones de líneas de fuerza magnética por segundo.

Inductive reactance The result of current flowing through a conductor and the resultant magnetic field around the conductor that opposes the normal flow of current.

Reactancia inductiva El resultado de un corriente que circule por un conductor y que resulta en un campo magnético alrededor del conductor que opone el flujo normal del corriente.

Inductive reluctance A statement of a material's ability to strengthen the magnetic field around it.

Reluctancia a la inducción Una indicación de la habilidad de una materia en reenforzar el campo que la rodea.

Inertia The tendency of an object that is at rest and an object that is in motion to stay in motion.

Inercia La tendencia de un objeto que esta en descanso quedarse en descanso y un objeto en movimiento de quedarse en movimiento.

Inertia engagement A type of starter motor that uses rotating inertia to engage the drive pinion with the engine flywheel.

Conexión por inercia Tipo de motor de arranque que utiliza inercia giratoria para engranar el piñon de mando con el volante de la máquina.

Inertia lock retractors Use a pendulum mechanism to lock the belt tightly during sudden movement.

Retractores de cierre tipo inercia Usan un mecanismo de péndulo para enclavar fuertemente la cinta durante un movimiento repentino.

Inflatable knee blocker (IKB) A small air bag that deploys simultaneously with the driver's side airbag to provide upper-leg protection and positioning of the driver.

Bloqueante inflable de la rodilla (BIR) Pequeña bolsa de aire que se despliega simultáneamente con la bolsa de aire lateral del conductor para proporcionar protección a la parte superior de las piernas y el posicionamiento del conductor.

Infrared temperature sensor A senor that measures the surface temperature of an object or person by measuring the intensity of the energy given off by an object.

Sensor infrarrojo para la temperatura Sensor que mide la temperatura de la superficie de un objeto o persona al medir la intensidad de la energía que desprende un objeto.

Installation diagrams Provide a more accurate duplication of where the wire harness, connectors, and components are found on the vehicle.

Esquemas de instalación Proveen una duplicación más precisa de donde se encuentran el cableado preformado, los conectores, y los componentes en el vehículo.

Instrument panel dimming System in which the headlight switch dimming control is used as an input to the computer instead of having direct control of the illumination lights.

Reducción luminosa del tablero de instrumentos Un sistema en el cual el control del interruptor de luminosidad se usa como una entrada a la computadora en vez de tener control directo al luminosidad.

Instrument voltage regulator (IVR) Provides a constant voltage to the gauge, regardless of the voltage output of the charging system.

Instrumento regulador de tensión Le provee tensión constante al calibrador, sin importar cual sea la salida de tensión del sistema de carga.

Insulated side The portion of the circuit from the positive side of the source to the load component.

Lado aislado Parte del circuito que va del lado positivo de la fuente al componente de carga.

Insulator A material that does not allow electrons to flow easily through it.

Aislador Una material que no permite circular fácilmente los electrones.

Integrated circuit (IC chip) A complex circuit of thousands of transistors, diodes, resistors, capacitors, and other electronic devices that are formed onto a small silicon chip. As many as 30,000 transistors can be placed on a chip that is 1/4 inch (6.35 mm) square.

Circuito integrado (Fragmento CI) Circuito complejo de miles de transistores, diodos, resistores, condensadores, y otros dispositivos electrónicos formados en un fragmento pequeño de silicio. En un fragmento de 1/4 de pulgada (6,35 mm) cuadrada, pueden colocarse hasta 30.000 transistores.

Integrated starter generator (ISG) A combination starter generator in one unit that attaches directly to the crankshaft to allow for the automatic stop/start function of an HEV. It can also convert kinetic energy to DC voltage when the vehicle is traveling downhill and there is zero load.

Generador de arranque integrado Combinación de generador de arranque en una unidad que se adhiere directamente al cigüeñal para permitir la función automática de encendido y apagado de un VEH. También puede convertir la energía cinemática a voltaje de corriente continua cuando el vehículo va de bajada y no lleva carga.

Intelligent windshield wipers A wiper system that uses a monitoring system to detect if water is present on the windshield and that automatically turns on the wiper system.

Limpiaparabrisas inteligente Sistema de limpiadores que utiliza un sistema de monitoreo para detectar si hay agua en el parabrisas, y esto hace que automáticamente se encienda el sistema de los limpiadores.

Interface Used to protect the computer from excessive voltage levels and to translate input and output signals.

Interfase Utilizada para proteger la computadora de niveles excesivos de tensión y traducir señales de entrada y salida.

International Standards Organization (ISO) Symbols used to represent the gauge function.

Organización Internacional de Normas (ISO) Los símbolos que se usan para representar la función del indicador.

In-vehicle sensor The in-vehicle sensor contains a temperature-sensing NTC thermistor to measure the average temperature inside the vehicle.

Sensor en el vehículo El sensor dentro del vehículo contiene un termostato de coeficiente de temperatura negativo (CTN) para percibir la temperatura que mide la temperatura promedio dentro del vehículo.

Ion An atom or group of atoms that has an electrical charge.

Ion Átomo o grupo de átomos que poseen una carga eléctrica.

Ionize To electrically charge.

Ionizar Cargar eléctricamente.

ISO An abbreviation for International Standards Organizations.

ISO Una abreviación de las Organizaciones de Normas Internacionales.

ISO 14230-4 A bus data protocol that uses a single-wire bidirectional data line to communicate between the scan tool and the nodes. This data bus is only used for diagnostics and maintains the ISO 9141 protocol with a bud rate of 10.4 Kb/s.

ISO 14230-4 Protocolo de un bus de datos que utiliza una línea de datos en dos direcciones en un hilo sencillo para comunicarse entre el instrumento de exploración y los nodos. Este bus de datos se utiliza solamente para diagnósticos y mantiene el protocolo de ISO 9141 con una velocidad de transmisión de baudios de 10.4 Kb/s.

ISO 9141-2 A class B system with a baud rate of 10.4 Kb/s used only for diagnostic purposes between the nodes on the data bus and an OBD II standardized scan tool.

ISO 9141-2 Sistema B de clase A con una velocidad de transmisión de baudios de 10.4 Kb/s que se utiliza sólo con un propósito de diagnóstico entre los nodos en el bus de datos y un instrumento de exploración estandarizado del sistema de diagnóstico a bordo II o DAB II.

ISO K An adoption of the ISO 9141-2 protocol that allows for bidirectional communication on a single wire. Vehicles that use the ISO-K bus require that the scan tool provide the bias voltage to power up the system.

ISO K La adopción del protocolo del ISO 9141-2 es que permite la comunicación en dos direcciones en un hilo sencillo. Los vehículos que utilizan el bus de ISO-K requieren que el instrumento de exploración proporcione la tensión de polarización para hacer funcionar el sistema.

ISO relays Conform to the specifications of the International Standards Organization (ISO) for common size and terminal patterns.

Relés ISO Conforman a las especificaciones de la Organización Internacional de Normas (ISO) en tamaño normal y conformidades de terminales.

J1850 The bus system that is the class B standard for OBD II. The J1850 standard allows for two different versions based on baud rate. The first supports a baud rate of 41.6 Kb/s that is transmitted by a pulse width modulated (PWM) signal over a twisted pair of wires. The second protocol supports a baud rate of 10.4 Kb/s average that is transmitted by a variable pulse width (VPW) data bus over a single wire.

J1850 El sistema de bus que es el estándar de clase B para el sistema de diagnóstico a bordo II o DAB II. El estándar J1850 permite dos diferentes versiones que se basan en la velocidad de transmisión de baudios. El primero respalda una velocidad de transmisión de baudios de 41.6 Kb/s que transmite una señal de modulación de duración de impulsos (MDI o PWM) mediante un par torcido de hilos. El segundo protocolo respalda una velocidad de transmisión de baudios de 10.4 Kb/s promedio que transmite un bus de datos de anchura variada entre impulsos (AVI o VPW) mediante un hilo sencillo.

Jack stands Support devices used to hold the vehicle off the floor after it has been raised by the floor jack.

Soportes de gato Dispositivos de soporte utilizados para sostener el vehículo sobre el suelo después de haber sido levantado con el gato de pie.

Keyless entry A lock system that allows for locking and unlocking of a vehicle with a touch keypad instead of a key.

Entrada sin llave Un sistema de cerradura que permite cerrar y abrir un vehículo por medio de un teclado en vez de utilizar una llave.

Keyless start system System that allows the vehicle to be started without the use of an ignition key.

Sistema de encendido sin llave Sistema que permite que el vehículo se arranque sin usar una llave de encendido.

K-line One circuit of the ISO 9141-2 data bus that is used for transmitting data from the module to the scan tool. The scan tool provides the bias voltage onto this circuit and the module pulls the voltage low to transmit its data.

Línea K Un circuito del bus de datos ISO 9141-2 que se utiliza para transmitir datos de un módulo a un instrumento de exploración. El instrumento de exploración proporciona la tensión de polarización sobre este circuito, y el módulo baja el voltaje para transmitir sus datos.

Laminated construction Construction of the armature from individual stampings.

Construcción laminada La armadura esta construida de un matrizado individual.

Lamination The process of constructing something with layers of materials that are firmly connected.

Laminación El proceso de construir algo de capas de materiales unidas con mucha fuerza.

Lamp A device that produces light as a result of current flow through a filament. The filament is enclosed within a glass envelope and is a type of resistance wire that is generally made from tungsten.

Lámpara Dispositivo que produce luz como resultado del flujo de corriente a través de un filamento. El filamento es un tipo de alambre de resistencia hecho por lo general de tungsteno, que es encerrado dentro de una bombilla.

Lamp outage module A current-measuring sensor that contains a set of resistors, wired in series with the power supply to the headlights, taillights, and stop lights. If the sensor indicates that a lamp is burned out, the module will alert the driver.

Unidad de avería de la lámpara Sensor para medir corriente que incluye un juego de resistores, alambrado en serie con la fuente de alimentación a los faros delanteros, traseros y a las luces de freno. Si el sensor indica que se ha apagado una lámpara, la unidad le avisará al conductor.

Laser radar sensor Determines the vehicle-to-vehicle distances and relational speeds.

Sensor de radar láser Determina las distancias y las velocidades relacionales entre un vehículo y otro.

Leading edge The edges of the rotating blade that enter the switch in a Hall-effect switch.

Borde anterior Los bordes de la ala giratorio que entran al interruptor en un terruptor efecto Hall.

Lean burn technology Uses lean air-fuel ratios to increase fuel efficiency.

Tecnología de quema limpia Determina las relaciones aire-combustible limpios para aumentar la eficacia del combustible.

Light-emitting diode (LED) A gallium-arsenide diode that converts the energy developed when holes and electrons collide during normal diode operation into light.

Diodo emisor de luz Diodo semiconductor de galio y arseniuro que convierte en luz la energía producida por la colisión de agujeros y electrones durante el funcionamiento normal del diodo.

Lighting system Electrical system that consists of all of the lights used on the vehicle, including headlights, front and rear park lights, front and rear turn signals, side marker lights, daytime running lights, cornering lights, brake lights, back-up lights, instrument cluster backlighting, and interior lighting.

Sistema de iluminación Sistema eléctrico que consta de todas las luces que usa el vehículo, incluyendo los faros, las luces frontales y traseras de estacionamiento, las luces intermitentes frontales y traseras, luces de posición, luces de marcha diurna, luces de esquina?, luces de freno, luces de marcha atrás, iluminación trasera de tablero de controles e iluminación interior.

Limit switch A switch used to open a circuit when a predetermined value is reached. Limit switches are normally responsive to a mechanical movement or temperature changes.

Disyuntor de seguridad Un conmutador que se emplea para abrir un circuito al alcanzar un valor predeterminado. Los disyuntores de seguridad suelen ser responsivos a un movimiento mecánico o a los cambios de temperatura.

Linearity Refers to the sensor signal being as constantly proportional to the measured value as possible. It is an expression of the sensor's accuracy.

Linealidad Significa que la variación del valor de una magnitud es lo más proporcional posible a la variación del valor de otra magnitud. Expresa la precisión del sensor.

Liquid crystal display (LCD) A display that sandwiches electrodes and polarized fluid between layers of glass. When voltage is applied to the electrodes, the light slots of the fluid are rearranged to allow light to pass through.

Visualizador de cristal líquido Visualizador digital que consta de dos láminas de vidrio selladas, entre las cuales se encuentran los electrodos y el fluido polarizado. Cuando se aplica tensión a los electrodos, se rompe la disposición de las moléculas para permitir la formación de caracteres visibles.

L-line One circuit of the ISO 9141-2 data bus that is used by the module to receive data from the scan tool. The module provides the bias onto this circuit and the scan tool pulls the voltage low to communicate.

Línea L Un circuito del bus de datos del ISO 9141-2 que utiliza un módulo que recibe datos de un instrumento de exploración. El módulo proporciona la tensión sobre este circuito y el instrumento de exploración baja el voltaje para comunicarse.

Load device The component that performs some form of work.

Dispositivo de carga El componente que lleva a cabo algun forma de trabajo.

Local interconnect network (LIN) A bus network that was developed to supplement the CAN bus system. The term *local interconnect* refers to all of the modules in the LIN network being located within a limited area.

LIN (Red local de interconexiones) Una red de bus que se desarrolló para complementar el sistema de bus CAN. El término "interconexión local" se refiere a todos los módulos en la red de LIN que se encuentran dentro de un área.

Logic gates Electronic circuits that act as gates to output voltage signals depending on different combinations of input signals.

Compuertas lógicas Circuitos electrónicos que gobiernan señales de tensión de salida, dependiendo de las diferentes combinaciones de señales de entrada.

Look-up tables The part of a microprocessor's memory that indicates how a system should perform in the form of calibrations and specifications.

Tablas de referencia La parte de la memoria de una microprocesora que indica como debe ejecutar las calibraciones y las especificaciones la sistema.

Low-reading coil Wound together with the bucking coil but in the opposite direction.

Bobina de lectura de baja tensión Envueltas juntas con la bobina compensadora pero en una dirección opuesta.

Low side drivers Used to complete the path to ground to turn on an actuator.

Impulsorer del lado a tierra Usados para completar el circuito a tierra para activar un actuador.

LUX The International System unit of measurement of the intensity of light. It is equal to the illumination of a surface one meter away from a single candle (one lumen per square meter).

LUX (lumen por metro cuadrado) Unidad del sistema internacional de medida de la intensidad de la luz. Es igual a la iluminación de una superficie a un metro de distancia de una vela sencilla (un lumen por metro cuadrado.)

Magnetic field The area surrounding a magnet where energy is exerted due to the atoms aligning in the material.

Campo magnético Espacio que rodea un imán donde se emplea la energía debido a la alineación de los átomos en el material.

Magnetic flux density The concentration of the magnetic lines of force.

Densidad de flujo magnético Número de líneas de fuerza magnética.

Magnetic pulse generator Sensor that uses the principle of magnetic induction to produce a voltage signal. Magnetic pulse generators are commonly used to send data to the computer concerning the speed of the monitored component.

Generador de impulsos magnéticos Sensor que funciona según el principio de inducción magnética para producir una señal de tensión. Los generadores de impulsos magnéticos se utilizan comúnmente para transmitir datos a la computadora relacionados a la velocidad del componente regulado.

Magnetism An energy form resulting from atoms aligning within certain materials, giving the materials the ability to attract other metals.

Magnetismo Forma de energía que resulta de la alineación de átomos dentro de ciertos materiales y que le da a éstos la capacidad de atraer otros metales.

Maintenance-free battery A battery that has no provision for the addition of water to the cells. The battery is sealed.

Sin mantención Que no tiene provisión para añadir el agua a la célualas. Es una batería sellada.

Master module Controller on the network that translates messages between different network systems.

Instancia maestra Controlador o entidad única, dentro de una red distribuida, que traduce los mensajes entre los diferentes sistemas de la red.

Material expanders Fillers that can be used in place of the active materials in a battery. They are used to keep the cost of manufacturing low.

Expansores de materias Los rellenos que se pueden usar en vez de las materiales activas de una batería. Se emplean para mantener bajos los costos de la fabricación.

Matrix A rectangular array of grids.

Matriz Red lógica en una rejilla de forma rectangular.

Maxi-fuse A circuit protection device that looks similar to a blade-type fuse except that it is larger and has a higher amperage capacity. Maxi-fuses are used because they are less likely to cause an underhood fire when there is an overload in the circuit. If the fusible link burned in two, it is possible that the "hot" side of the fuse could come into contact with the vehicle frame and the wire could catch on fire.

Maxifusible Dispositivo de protección del circuito parecido a un fusible de tipo de cuchilla, pero más grande y con mayor capacidad de amperaje. Se utilizan maxifusibles porque existen menos probabilidades de que ocasionen un incendio debajo de la capota cuando ocurra una sobrecarga en el circuito. Si el cartucho de fusible se quemase en dos partes, es posible que el lado "cargado" del fusible entre en contacto con el armazón del vehículo y que el alambre se encienda.

Media Oriented System Transport (MOST) A data bus system based on standards established by a cooperative effort between automobile manufacturers, suppliers, and software programmers that resulted in a data system specifically designed for the data transmission of media-oriented data. MOST uses fiber optics to transmit data at a rate up to 25 megabits per second.

MOST Sistema de bus de datos basado en estándares establecidos por un esfuerzo cooperativo entre los fabricantes de vehículos, los proveedores y los programadores de software que resultó en un sistema de datos específicamente diseñado para la transmisión de datos informativos. MOST utiliza fibras ópticas para transmitir datos a una velocidad de 25 megabitios por segundo (25Mb/s.)

Memory seats Power seats that can be programmed to return or adjust to a point designated by the driver.

Asientos con memoria Los asientos automáticos que se pueden programar a regresar o ajustarse a un punto indicado por el conductor.

Metri-pack connector Special wire connectors used in some computer circuits. They seal the wire terminals from the atmosphere, thereby preventing corrosion and other damage.

Conector metri-pack Los conectores de alambres especiales que se emplean en algunos circuitos de computadoras. Impermealizan los bornes de los alambres, así previniendo la corrosión y otros daños.

Mild parallel hybrid Uses an electric motor that is large enough to provide regenerative braking, instant engine startup, and a boost to the combustion engine.

Híbrido de medio paralelo Utiliza un motor eléctrico que es lo suficientemente grande para proveer freno regenerativo, encendido instantáneo del motor y un aumento a la combustión del motor.

Mode door actuator An electric motor that is linked to the mode door to supply airflow to the floor ducts, A/C panel ducts, or defrost ducts.

Actuador de mando puerta Motor eléctrico que está unido al mando puerta para suministrar flujo de aire a los conductos del piso, los conductos del panel de corriente alterna o a los conductos de descongelamiento.

Momentary contact A switch type that operates only when held in position.

Contacto momentáneo Tipo de conmutador que funciona solamente cuando se mantiene en su posición.

MSDS (Material Data Safety Sheet) A fact sheet of hazardous material.

MSDS Una joja de información de materials tóxicos.

Multistage air bags Hybrid air bags that use two squibs to control the rate of inflation.

Bolsas de aire de etapas múltiples Las bolsas de aire híbridas que usan dos petardos para controlar la velocidad de la inflación.

Multiplexer An electronic switch that switches between the different audio sources.

Multiplexor Interruptor electrónico que se usa para hacer cambios entre las diferentes fuentes de audio.

Multiplexing A means of transmitting information between computers. It is a system in which electrical signals are transmitted by a peripheral serial bus instead of conventional wires, allowing several devices to share signals on a common conductor.

Multiplexaje Medio de transmitir información entre computadoras. Es un sistema en el cual las señales eléctricas son transmitidas por una colectora periférica en serie en vez de por líneas convencionales. Esto permite que varios dispositivos compartan señales en un conductor común.

Mutual induction An induction of voltage in an adjacent coil by changing current in a primary coil.

Inducción mutua Una inducción de la tensión en una bobina adyacente que se efectúa al cambiar la tensión en una bobina primaria.

MUX Common acronym for multiplexing.

MUX Una sigla común del proceso de multiplex.

Navigational systems Use satellites to direct the drivers to a desired destinations.

Sistema de navegación Usa los satélites para dirigir el conductor a la destinación deseada.

Negative logic Defines the most negative voltage as a logical 1 in the binary code.

Lógica negativa Define la tensión más negativa como un 1 lógico en el código binario.

Negative temperature coefficient (NTC) thermistors Thermistors that reduce their resistance as the temperature increases.

Termistores con coeficiente negativo de temperatura Termistores que disminuyen su resistencia según aumenta la temperatura.

Nematic Describes a fluid that is a liquid crystal with a threadlike form. It has light slots that can be rearranged by applying small amounts of voltage.

Nemático Describe un flúido que es un cristal líquido con una forma de filamento. Tiene aberturas de luz que se pueden reorganizar por medio de la aplicación de pequeñas cantidades de voltaje.

Neon lights A light that contains a colorless, odorless inert gas called neon. These lamps are discharge lamps.

Luces de neón Una luz que contiene un gas inerto sin color, inodoro llamado neón. Estas lámparas son lámparas de descarga.

Network Incorporating the vehicle's electrical systems together through computers so information gathered by one system can be used by another.

En red Incorporar los sistemas eléctricos del vehículo mediante el uso de computadoras para que la información que obtenga un sistema pueda usarla otro sistema.

Neutral atom *See* balanced atom.

Átomo neutro *Véase* átomo equilibrado.

Neutral junction The center connection to which the common ends of a Y-type stator winding are connected.

Empalme neutro Conexión central a la cual se conectan los extremos comunes de un devanado del estátor de tipo Y.

Neutral safety switch A switch used to prevent the starting of an engine unless the transmission is in PARK or NEUTRAL.

Disyuntor de seguridad en neutral Un conmutador que se emplea para prevenir que arranque un motor al menos de que la transmisión esté en posición PARK o Neutral.

Neutrons Particles of an atom that have no charge.

Neutrones Partículas de un átomo desprovistas de carga.

Node A computer that is connected to a data bus network and capable of sending or receiving messages.

Nodo Computadora conectada a una red de bus de datos y con capacidad de mandar o recibir mensajes.

Nonvolatile RAM memory that will retain its memory if battery voltage is disconnected. NVRAM is a combination of RAM and EEPROM into the same chip. During normal operation, data is written to and read from the RAM portion of the chip. If the power is removed from the chip, or at programmed timed intervals, the data is transferred from RAM to the EEPROM portion of the chip. When the power is restored to the chip, the EEPROM will write the data back to the RAM.

Memoria de acceso aleatorio no volátil [NV RAM] Memoria de acceso aleatorio (RAM) que retiene su memoria si se desconecta la carga de la batería. La NV RAM es una combinación de RAM y EEPROM en el mismo fragmento. Durante el funcionamiento normal, los datos se escriben en y se leen de la parte RAM del fragmento. Si se remueve la alimentación del fragmento, o si se remueve ésta a intervalos programados, se transfieren los datos de la RAM a la parte del EEPROM del fragmento. Cuando se restaura la alimentación en el fragmento, el EEPROM volverá a escribir los datos en la RAM.

Normally closed (NC) switch A switch designation denoting that the contacts are closed until acted upon by an outside force.

Conmutador normalmente cerrado Nombre aplicado a un conmutador cuyos contactos permanecerán cerrados hasta que sean accionados por una fuerza exterior.

Normally open (NO) switch A switch designation denoting that the contacts are open until acted upon by an outside force.

Conmutador normalmente abierto Nombre aplicado a un conmutador cuyos contactos permanecerán abiertos hasta que sean accionados por una fuerza exterior.

N-type material When there are free electrons, the material is called an N-type material. The N means negative and indicates that it is the negative side of the circuit that pushes electrons through the semiconductor and the positive side that attracts the free electrons.

Material tipo N Al material se le llama material tipo N cuando hay electrones libres. La N significa negativo e indica que el lado negativo del circuito empuja los electrones a través del semiconductor y el lado positivo atrae los electrones libres.

Nucleus The core of an atom that contains the protons and neutrons.

Núcleo Parte central de un átomo que contiene los protones y los neutrones.

Occupant classification systems A mandated requirement to reduce the risk of injuries resulting from air bag deployment by determining the weight classification of the front-seat passenger.

Sistemas de clasificación de los ocupantes Mandato para reducir el riesgo de daños que resulten del desarrollo de la bolsa de aire al determinar la clasificación del peso del pasajero del asiento de enfrente.

Occupational safety glasses Eye protection that is designed with special high-impact lens and frames, and provides for side protection.

Gafas de protección para el trabajo Gafas diseñadas con cristales y monturas especiales resistentes y provistas de protección lateral.

OCS service kit Special kit that consists of the seat foam, the bladder, the pressure sensor, the occupant classification module (OCM), and the wiring.

Kit de servicio SCO Kit especial que consiste en hule-espuma del asiento, el depósito, el sensor de presión, el módulo de clasificación del ocupante (MCO) y el alambrado.

OCS validation test A test that confirms that the system can properly classify the occupant.

Prueba de revalidación del SCO Prueba que confirma que el sistema puede clasificar apropiadamente al ocupante.

Odometer A mechanical counter in the speedometer unit indicating total miles accumulated on the vehicle.

Odómetro Aparato mecánico en la unidad del velocímetro con el que se cuentan las millas totales recorridas por el vehículo.

Ohm Unit of measure for resistance. One ohm is the resistance of a conductor such that a constant current of 1 ampere in it produces a voltage of 1 volt between its ends.

Ohmio Unidad de resistencia eléctrica. Un ohmio es la resistencia de un conductor si una corriente constante de 1 amperio en el conductor produce una tensión de 1 voltio entre los dos extremos.

Ohmmeter A test meter used to measure resistance and continuity in a circuit.

Ohmiómetro Instrumento de prueba utilizado para medir la resistencia y la continuidad en un circuito.

Ohm's law Defines the relationship between current, voltage, and resistance.

Ley de Ohm Define la relación entre la corriente, la tensión y la resistencia.

Open circuit A term used to indicate that current flow is stopped. By opening the circuit, the path for electron flow is broken.

Circuito abierto Interrupción en el circuito eléctrico que causa que pare el flujo de corriente.

Optical horn A name Chrysler uses to describe their "flash-to-pass" headlamp system.

Claxón óptico Un nombre que usa Chrysler para describir su sistema de faros "relampaguea para rebasar."

Oscillator Creates a rapid back-and-forth movement of voltage.

Oscilador Crea un movimiento de oscilación rápido de voltaje.

Overload Excess current flow in a circuit.

Sobrecarga Flujo de corriente superior a la que tiene asignada un circuito.

Overrunning clutch A starter drive that uses a roller clutch to transmit torque in one direction only and freewheels in the other direction.

Embrague de sobremarcha Una asamblea de embrague en un acoplamiento del motor de arranque que se emplea para prevenir que el volante del motor dé vueltas al armazón del motor de arranque.

Oversteer The tendency of the back of the vehicle to turn on the vehicle's center of gravity and come around the front of the vehicle.

Tener la dirección muy sensible Tendencia de la parte trasera de un vehículo de dar vuelta en el centro de gravedad del vehículo y de doblar al frente del vehículo.

Oxygen sensor A voltage generating sensor that measures the amount of oxygen present in an engine's exhaust.

Sensor de oxígeno Un sensor generador de tensión que mide la cantidad del oxígeno presente en el gas de escape de un motor.

Parallel circuit A circuit that provides two or more paths for electricity to flow.

Circuito en paralelo Circuito que provee dos o más trayectorias para que circule la electricidad.

Parallel hybrid A hybrid vehicle configuration that has a direct mechanical connection between the engine and the wheels. Both the engine and the electric motor can turn the transmission at the same time.

Híbrido en paralelo completo Utiliza un motor eléctrico que es lo suficientemente potente para que el vehículo se impulse por sí mismo.

Parasitic loads Electrical loads that are still present when the ignition switch is in the OFF position.

Cargas parásitas Cargas eléctricas que todavía se encuentran presente cuando el botón conmutador de encendido está en la posición OFF.

Park contacts Located inside the motor assembly and supply current to the motor after the wiper control switch has been turned to the park position. This allows the motor to continue operating until the wipers have reached their PARK position.

Contactos de Park Ubicado dentro de la asamblea del motor y proveen el corriente al motor después de que el interruptor de control de la limpiaparabrisa se ha puesto en la posición de estacionamiento. Esto permite que el motor continua operando hasta que los brazos de la limpiaparabrisas hayan llegado a su posición de estacionamiento.

Park switch Contact points located inside the wiper motor assembly that supply current to the motor after the wiper control switch has been turned to the PARK position. This allows the motor to continue operating until the wipers have reached their PARK position.

Conmutador PARK Puntos de contacto ubicados dentro del conjunto del motor del frotador que le suministran corriente al motor después de que el conmutador para el control de los frotadores haya sido colocado en la posición PARK. Esto permite que el motor continue su funcionamiento hasta que los frotadores hayan alcanzado la posición original.

Pass key A specially designed vehicle key with a coded resistance value. The term pass is derived from Personal Automotive Security System.

Llave maestra Una llave vehícular de diseño especial que tiene un valor de resistencia codificado. El termino pass se derive de las palabras Personal Automotive Security System (sistema personal de seguridad automotriz).

Passive restraints A passenger restraint system that automatically operates to confine the movement of a vehicle's passengers.

Correas passivas Un sistema de resguardo del pasajero que opera automaticamente para limitar el movimiento de los pasajeros en el vehículo.

Passive suspension systems Use fixed spring rates and shock valving.

Sistemas pasivos de suspensión Utilizan elasticidad de muelle constante y dotación con válvulas amortigadoras.

Peltier element Similar to a bimetal switch, the element consists of two different types of metals that are joined together. This joint area will generate or absorb heat when an electric current is applied to the element at a specified temperature.

Elemento Peltier Similar al interruptor bimetal; consiste de dos diferentes tipos de metal que están unidos. El área de unión generará o absorberá calor cuando se aplique corriente eléctrica al elemento a una temperatura específica.

Permanent magnet gear reduction (PMGR) A starter that uses four or six permanent magnet field assemblies in place of field coils.

Reducción de engranaje de imán permanente (PMGR) Un arrancador que usa cuatro o seis asambleas permanentes de campo magnético en vez de las bobinas de campo.

Permeability Term used to indicate the magnetic conductivity of a substance compared with the conductivity of air. The greater the permeability, the greater the magnetic conductivity and the easier a substance can be magnetized.

Permeabilidad Término utilizado para indicar la aptitud de una sustancia en relación con la del aire, de dar paso a las líneas de fuerza magnética. Mientras mayor sea la permeabilidad, mayor será la conductividad magnética y más fácilmente se comunicará a un cuerpo propiedades magnéticas.

Photo diode Allows current to flow in the opposite direction of a standard diode when it receives a specific amount of light.

Fotodiodo Permite que fluye el corriente en la dirección opuesta de él de un diodo normal al recibir una cantidad específica de luz.

Photocell A variable resistor that uses light to change resistance.

Fotocélula Resistor variable que utiliza luz para cambiar la resistencia.

Phototransistor A transistor that is sensitive to light.

Fototransistor Transistor sensible a la luz.

Photovoltaic diodes Diodes capable of producing a voltage when exposed to radiant energy.

Diodos fotovoltaicos Diodos capaces de generar una tensión cuando se encuentran expuestos a la energía de radiación.

Pickup coil The stationary component of the magnetic pulse generator consisting of a weak permanent magnet that has fine wire wound around it. As the timing disc rotates in front of it, the changes of magnetic lines of force generate a small voltage signal in the coil.

Bobina captadora Componente fijo del generador de impulsos magnéticos compuesta de un imán permanente débil devanado con alambre fino. Mientras gira el disco sincronizador enfrente de él, los cambios de las líneas de fuerza magnética generan una pequeña señal de tensión en la bobina.

Piconets Small transmission cells that assist in the organization of data.

Picoredes Pequeñas células de transmisión que ayudan a organizar los datos.

Piezoelectricity Voltage produced by the application of pressure to certain crystals.

Piezoelectricidad Generación de polarización eléctrica en ciertos cristales a consecuencia de la aplicación de tensiones mecánicas.

Piezoresistive sensor A sensor that is sensitive to pressure changes.

Sensor piezoresistivo Sensor susceptible a los cambios de presión.

Ping (or denotation) A knocking sound that occurs as two flame fronts collide.

Golpeteo (detonación) Un ruido de impacto que ocurre cuando hay colisión entre dos bordes térmicos.

Pinion factor A calculation using the final drive ratio and the tire circumference to obtain accurate vehicle speed signals.

Factor de piñón Una calculación que usa la relación de impulso final y la circunferencia de la llanta para obtener unas señales precisad de la velocidad del vehículo.

Pinion gear A small gear; typically refers to the drive gear of a starter drive assembly or the small drive gear in a differential assembly.

Engranaje de piñón Un engranaje pequeño; tipicamente se refiere al engranaje de arranque de una asamblea de motor de arranque o al engranaje de mando pequeño de la asamblea del diferencial.

Plate straps Metal connectors used to connect the positive or negative plates in a battery.

Abrazaderas de la placa Los conectores metálicos que sirven para conectar las placas positivas o negativas de una batería.

Plates The basic structure of a battery cell; each cell has at least one positive plate and one negative plate.

Placas La estructura básica de una celula de batería; cada celula tiene al menos una placa positiva y una placa negativa.

P-material Silicon or germanium that is doped with boron or gallium to create a shortage of electrons.

Material-P Boro o galio añadidos al silicio o al germanio para crear una insuficiencia de electrones.

PMGR An abbreviation for permanent magnet gear reduction.

PMGR Una abreviación de desmultiplicación del engranaje del imán permanente.

Pneumatic tools Power tools that are powered by compressed air.

Herrimientas neumáticas Herramientas mecánicas accionadas por aire comprimido.

PN junction The point at which two opposite kinds of semiconductor materials are joined together.

Unión pn Zona de unión en la que se conectan dos tipos opuestos de materiales semiconductores.

Polarizers Glass sheets that make light waves vibrate in only one direction. This converts light into polarized light.

Polarizadores Las láminas de vidrio que hacen vibrar las ondas de luz en un sólo sentido. Esto convierte la luz en luz polarizada.

Polarizing The process of light polarization or of setting one end of a field as a positive or negative point.

Polarizadora El proceso de polarización de la luz o de establecer un lado de un campo como un punto positivo o negativo.

Pole The number of input circuits.

Poste El número de los circuitos de entrada.

Pole shoes The components of an electric motor that are made of high-magnetic permeability material to help concentrate and direct the lines of force in the field assembly.

Expansión polar Componentes de un motor eléctrico hechos de material magnético de gran permeabilidad para ayudar a concentrar y dirigir las líneas de fuerza en el conjunto inductor.

Positive engagement starter A type of starter that uses the magnetic field strength of a field winding to engage the starter drive into the flywheel.

Acoplamiento de arranque positivo Un tipo de arrancador que utiliza la fuerza del campo magnético del devanado inductor para accionar el acoplamiento del arrancador en el volante.

Positive plate The plate connected to the positive battery terminal.

Placa positiva La placa conectada al terminal positivo de la batería.

Positive temperature coefficient (PTC) thermistors Thermistors that increase their resistance as the temperature increases.

Termistores con coeficiente positivo de temperatura Termistores que aumentan su resistencia según aumenta la temperatura.

Potential The ability to do something; typically voltage is referred to as the potential. If you have voltage, you have the potential for electricity.

Potencial La capacidad de efectuar el trabajo; típicamente se refiere a la tensión como el potencial. Si tiene tensión, tiene la potencial para la electricidad.

Potentiometer A variable resistor that acts as a circuit divider to provide accurate voltage drop readings proportional to movement.

Potenciómetro Resistor variable que actúa como un divisor de circuito para obtener lecturas de perdidas de tensión precisas en proporcion con el movimiento.

Power The rate of doing electrical work.

Potencia La tasa de habilidad de hacer el trabajo eléctrico.

Power door locks Electric power door locks use either a solenoid or a permanent magnet reversible motor to lock and unlock the door.

Cerradura mecánica de puertas Las cerraduras electro-mecánicos de puertas utilizan o un solenoide o un motor reversible de imanes permanentes para abrir y cerrar la puerta.

Power formula A formula used to calculate the amount of electrical power a component uses. The formula is P = I x E, where P stands for power (measured in watts), I stands for current, and E stands for voltage.

Formula de potencia Una formula que se emplea para calcular la cantidad de potencia eléctrica utilizada por un componente. La formula es P = I x E, en el que el P quiere decir potencia (medida en wats), I representa el corriente y el E representa la tensión.

Power mirrors Outside mirrors that are electrically positioned from the inside of the driver's compartment.

Espejos eléctricos Los espejos exteriores que se ajusten eléctricamente desde el interior del compartimiento del conductor.

Power tools Tools that use forces other than those generated from the body. They can use compressed air, electricity, or hydraulic pressure to generate and multiply force.

Herramientas mecánicas Herramientas que utilizan fuerzas distintas a las generadas por el cuerpo. Dichas fuerzas pueden ser el aire comprimido, la electricidad, o la presión hidráulica para generar y multiplicar la fuerza.

Power windows Windows that are raised and lowered by use of electrical motors.

Ventanillas eléctricas Las ventanillas que se suban y se bajan por medio de los motores eléctricos.

Pressure control solenoid A solenoid used to control the pressure of a fluid, commonly found in electronically controlled transmissions.

Solenoide de control de la presión Un solenoide que controla la presión de un fluido, suele encontrarse en las transmisiones controladas electronicamente.

Pretensioners Used to tighten the seat belt and shoulder harness around the occupant during an accident severe enough to deploy the air bag. Pretensioners can be used on all seat belt assemblies in the vehicle.

Pretensadores Se usan para apretar la cinta de seguridad y el arnés del cuerpo alrededor del ocupante durante un accidente bastante severo como para activar la bolsa de aire. Los pretensadores se pueden usarse en cualquier asamblea de cinta de seguridad del vehículo.

Primary circuit All the components that carry low voltage through the system.

Circuito pirmario Todos los componentes que llevan un voltaje bajo dentro del sistema.

Primary coil winding The second set of winding in the ignition coil. The primary winds will have about 200 turns to create a magnetic field to induce voltage into the secondary winding.

Devanado primario El grupo segundo de devanados en la bobina del encendido. Los devandos primarios tendrán unos 200 vueltas para crear un camp magnético para inducir el voltaje al devanado secundario.

Primary wiring Conductors that carry low voltage and current. The insulation of primary wires is usually thin.

Hilos primarios Hilos conductores de tensión y corriente bajas. El aislamiento de hilos primarios es normalmente delgado.

Printed circuit Made of thin phenolic or fiberglass board with copper deposited on it to create current paths. These are used to simplify the wiring of circuits.

Circuito impreso Un circuito hecho de un tablero de fenólico delgado o de fibra de vidrio el cual tiene depósitos del cobre para crear los trayectorios para el corriente. Estos se emplean para simplificar el cableado de los circuitos.

Prism lens A light lens designed with crystal-like patterns, which distort, slant, direct, or color the light that passes through it.

Lente prismático Un lente de luz con diseños cristalinos que distorcionan, inclinan, dirigen o coloran la luz que lo atraviesa.

Prisms Redirect the light beam and create a broad, flat beam.

Prismas Dirigen un rayo de luz y crean un rayo ancho y plano.

Program A set of instructions that the computer must follow to achieve desired results.

Programa Conjunto de instrucciones que la computadora debe seguir para lograr los resultados deseados.

Program number Represents the amount of heating or cooling required to obtain the temperature set by the driver.

Número de programa Representa la cantidad de calefacción o enfriamiento requerido para obtener la temperatura indicado por el conductor.

Programmable Communication Interface (PCI) A single wire, bidirectional communication bus where each module supplies its own bias voltage and has its own termination resistors. As a message is sent, a variable pulse width modulation (VPWM) voltage between 0 and 7.75 volts is used to represent the 1 and 0 bits.

Interfaz de comunicación programable (PCI) Bus de comunicación en dos direcciones en un hilo sencillo en donde cada módulo proporciona su propia velocidad de transmisión de baudios y tiene sus propias resistencias de unión. Mientras se envía un mensaje, un voltaje de anchura variada entre impulsos entre 0 y 7.75 voltios se utiliza para representar los bits 1 y 0.

Programmer Controls the blower speed, air mix doors, and vacuum motors of the SATC system. Depending on manufacturer, they are also called servo assemblies.

Programador Controla la velocidad del ventilador, las puertas de mezcla de aire y los motores de vacío de un sistema SATC. Según el fabricante, tambien se llaman asambleas servo.

PROM (programmable read only memory) Memory chip that contains specific data that pertains to the exact vehicle in which the computer is installed. This information may be used to inform the CPU of the accessories that are equipped on the vehicle.

PROM (memoria de sólo lectura programable) Fragmento de memoria que contiene datos específicos referentes al vehículo particular en el que se instala la computadora. Esta información puede utilizarse para informar a la UCP sobre los accesorios de los cuales el vehículo está dotado.

Protection device Circuit protector that is designed to "turn off" the system that it protects. This is done by creating an open to prevent a complete circuit.

Dispositivo de protección Protector de circuito diseñado para "desconectar" el sistema al que provee protección. Esto se hace abriendo el circuito para impedir un circuito completo.

Protocol A language used by computers to communicate with each other over a data bus.

Protocolo Lenguaje que se utiliza en computadoras para comunicarse entre sí sobre un mando de bus.

Proton Positively charged particles contained in the nucleus of an atom.

Protón Partículas con carga positiva que se encuentran en el núcleo de todo átomo.

Proton exchange membrane (PEM) Impedes the oxyhydrogen gas reaction in a fuel cell by ensuring that only protons (H+), and not elemental hydrogen molecules (H2), react with the oxygen.

Membrana de intercambio de protones (MIP) Impide la reacción de gas de oxigeno-hidrógeno en una célula de combustible al asegurar que sólo los protones reaccionen con el oxígeno y no las moléculas de hidrógeno elemental.

Prove-out circuit A function of the ignition switch that completes the warning light circuit to ground through the ignition switch when it is in the START position. The warning light will be on during engine cranking to indicate to the driver that the bulb is working properly.

Circuito de prueba Función del botón conmutador de encendido que completa el circuito de la luz de aviso para que se ponga a tierra a través del botón conmutador de encendido cuando éste se encuentra en la posición START. La luz de aviso se encenderá durante el arranque del motor para avisarle al conductor que la bombilla funciona correctamente.

Pull-in windings An electrical coil internal to a solenoid that is energized to create a magnetic field used to move the solenoid plunger to the engaged position.

Devanados de puesta en trabajo Una bobina eléctrica que es íntegra a un solenoide que se excita para crear un campo magnético que sirve para mover el relé de solenoide a la posición de engranaje.

Pulse width The length of time in milliseconds that an actuator is energized.

Duración de impulsos Espacio de tiempo en milisegundos en el que se excita un accionador.

Pulse-width modulation On/off cycling of a component. The period of time for each cycle does not change; only the amount of on time in each cycle changes.

Modulación de duración de impulsos Modulación de impulsos de un componente. El espacio de tiempo de cada ciclo no varía; lo que varía es la cantidad de trabajo efectivo de cada ciclo.

Radial grid A type of battery grid that has its patterns branching out from a common center.

Rejilla radial Un tipo de rejilla de bateria cuyos diseños extienden de un centro común.

Radio choke Absorbs voltage spikes and prevents static in the vehicle's radio.

Impedancia del radio Absorba los impulsos de la tensión y previene la presencia del estático en el radio del vehículo.

Radio frequency interference (RFI) Radio and television interference caused by electromagnetic energy.

Interferencia de frecuencia radioeléctrica Interferencia en la radio y en la televisión producida por energía electromagnética.

RAM (random access memory) Stores temporary information that can be read from or written to by the CPU. RAM can be designed as volatile or nonvolatile.

RAM (memoria de acceso aleatorio) Almacena datos temporales que la UCP puede leer o escribir. La RAM puede ser volátil o no volátil.

Ratio A mathematical relationship between two or more things.

Razón Una relación matemática entre dos cosas o más.

Reactivity A statement of how easily a substance can cause or be a part of a chemical.

Reactividad Una indicación de cuan fácil una sustancia puede causar o ser parte de una química.

Recirc/air inlet door actuator An electric motor that is linked to the recirculation door to provide either outside air or in-vehicle air into the A/C heater case.

Actuador de la compuerta de entrada de recirculación del aire Motor eléctrico que está unido a una compuerta de recirculación para proporcionar ya sea aire de fuera o que provenga del vehículo, dentro de la caja del calentón de corriente alterna.

Recombination battery A type of battery that is sometimes called a dry-cell battery because it does not use a liquid electrolyte solution.

Batería de recombinación Un tipo de batería que a veces se llama una pila seca porque no requiere una solución líquida de electrolita.

Rectification The converting of AC current to DC current.

Rectificación Proceso a través del cual la corriente alterna es transformada en una corriente continua.

Reflectors A device whose surface reflects or radiates light.

Reflectores Un dispositivo cuyo superficie refleja o irradia la luz.

Reformer A high-temperature device that converts hydrocarbon fuels to CO and H2.

Reformador Dispositivo de alta temperatura que convierte los combustibles de hidrocarburo a monóxido de carbono CO y a H2.

Regenerative braking Braking energy is turned back into electricity instead of heat.

Frenado regenerativo La energía de frenado se convierte nuevamente en electricidad en lugar de calor.

Relative compression testing A test that compares the compression between engine cylinders.

Prueba de compresión relativa Prueba que compara la compresión entre los cilindros del motor.

Relay A device that uses low current to control a high-current circuit. Low current is used to energize the electromagnetic coil, while high current is able to pass over the relay contacts.

Relé Dispositivo que utiliza corriente baja para controlar un circuito de corriente alta. La corriente baja se utiliza para excitar la bobina electromagnética, mientras que la corriente alta puede transmitirse a través de los contactos del relé.

Reluctance A term used to indicate a material's resistance to the passage of flux lines.

Reluctancia Término utilizado para señalar la resistencia ofrecida por un circuito al paso del flujo magnético.

Reserve-capacity rating An indicator, in minutes, of how long the vehicle can be driven, with the headlights on, if the charging system should fail. The reserve-capacity rating is determined by the length of time, in minutes, that a fully charged battery can be discharged at 25 amperes before battery cell voltage drops below 1.75 volts per cell.

Clasificación de capacidad en reserva Indicación, en minutos, de cuánto tiempo un vehículo puede continuar siendo conducido, con los faros delanteros encendidos, en caso de que ocurriese una falla en el sistema de carga. La clasificación de capacidad en reserva se determina por el espacio de tiempo, en minutos, en el que una batería completamente cargada puede descargarse a 25 amperios antes de que la tensión del acumulador de la batería disminuya a un nivel inferior de 1,75 amperios por acumulador.

Resistance Opposition to current flow.

Resistencia Oposición que presenta un conductor al paso de la corriente eléctrica.

Resistance wire A special type of wire that has some resistance built into it. These typically are rated by ohms per foot.

Alambre de resistencia Un tipo de alambre especial que por diseño tiene algo de resistencia. Estos tipicamente tienen un valor nominal de ohm por pie.

Resistor block Consists of two or three helically wound wire resistors wired in series.

Bloque de resistencia Consiste de dos o tres cables helicoilades rostáticos intercalados en serie.

Resistive multiplex switch Provides multiple inputs over a single circuit. Since each switch position has a different resistance value, the voltage drop is different. This means a switch can have one power supply wire and one ground wire instead of a separate wire for each switch position.

Interruptor multiplex resistente Provee múltiples entradas de energía sobre un circuito simple. Debido a que la posición de cada interruptor tiene un valor diferente de resistencia, el voltaje de la caída de tensión es diferente. Esto significa que un interruptor puede tener un cable de abastecimiento de energía y un cable conductor a tierra, sin que sea necesario un cable individual para cada posición del interruptor.

Retarded timing Means the spark is late arriving to the combustion chamber.

Sincronización retardada Significa que la chispa se demora en llegar a la cámara de combustión.

Reverse-bias A positive voltage is applied to the N-type material and negative voltage is applied to the P-type material of a semiconductor.

Polarización inversa Tensión positiva aplicada al material N y tensión negativa aplicada al material P de un semiconductor.

Rheostat A two-terminal variable resistor used to regulate the strength of an electrical current.

Reóstato Resistor variable de dos bornes utilizado para regular la resistencia de una corriente eléctrica.

Right-hand rule Identifies the direction of the lines of force of an electromagnet.

Regla de la mano derecha Identifica la dirección de las líneas de fuerza de un electroimán.

ROM (read only memory) Memory chip that stores permanent information. This information is used to instruct the computer on what to do in response to input data. The CPU reads the information contained in ROM, but it cannot write to it or change it.

ROM (memoria de sólo lectura) Fragmento de memoria que almacena datos en forma permanente. Dichos datos se utilizan para darle instrucciones a la computadora sobre cómo dirigir la ejecución de una operación de entrada. La UCP lee los datos que contiene la ROM, pero no puede escribir en ella o puede cambiarla.

Rotor The component of the AC generator that is rotated by the drive belt and creates the rotating magnetic field of the AC generator.

Rotor Parte rotativa del generador de corriente alterna accionada por la correa de transmisión y que produce el campo magnético rotativo del generador de corriente alterna.

Safety goggles Eye protection device that fits against the face and forehead to seal off the eyes from outside elements.

Gafas de seguridad Dispositivo protector que se coloca delante de los ojos para preservarlos de elementos extraños.

Safety stands *See* Jack stands.

Soportes de seguridad *Véase* soportes de gato.

Safing sensor Determines if the collision is severe enough to inflate the air bag.

Monitor de seguridad (safing sensor) Determina si el impacto del choque es lo suficientemente grave para inflar las bolsas de aire.

Satellite radios Provide several commercial-free music and talk show channels using orbiting satellites to provide a digital signal.

Radios por satélite Proporcionan música variada sin comerciales y canales de programas de entrevistas al usar satélites en órbita que dan una señal digital.

Saturation 1. The point at which the magnetic strength eventually levels off, and where an additional increase of the magnetizing force current no longer increases the magnetic field strength. 2. The point where forward-bias voltage to the base leg is at a maximum. With bias voltage at the high limits, output current is also at its maximum.

Saturación 1. Máxima potencia posible de un campo magnético, donde un aumento adicional de la corriente de fuerza magnética no logra aumentar la potencia del campo magnético. 2. La tensión de polarización directa a la base está en su máximo. Ya que polarización directa ha alcanzado su límite máximo, la corriente de salida también alcanza éste.

Schematic An electrical diagram that shows how circuits are connected, but not details such as color codes.

Esquemático Diagrama eléctrico que muestra cómo se conectan los circuitos, pero no los detalles tales como las claves por colores.

Schmitt trigger An electronic circuit used to convert analog signals to digital signals or vice versa.

Disparador de Schmitt Un circuito electrónico que se emplea para convertir las señales análogas en señales digitales o vice versa.

Sealed-beam headlight A self-contained glass unit that consists of a filament, an inner reflector, and an outer glass lens.

Faro delantero sellado Unidad de vidrio que contiene un filamento, un reflector interior y una lente exterior de vidrio.

Seat track position sensor (STPS) A Hall-effect sensor that provides information to the OCM concerning the position of the seat in relation to the air bag.

Sensor de la posición del carril de asiento(SPCA) Sensor de efecto de Hall que proporciona información al MCS (Módulo de control de salida) que concierne a la posición del asiento en relación con la bolsa de aire.

Secondary circuit All the components that carry voltage to the combustion chamber.

Circuito secundario Todos los componentes que conducen el voltaje a la cámara de combustión.

Secondary coil windings One of the two coils in the ignition coil. This winding has several thousand turns and is where low voltage will be transformed to high voltage.

Bobinas secundarias Una o dos bobinas de tensión en la bobina de encendido. Este bobinaje tiene la capacidad de millares de vueltas y de esta forma el voltaje bajo será transformado en voltaje de alta tensión.

Secondary reserve voltage The difference between the required voltage and the maximum available voltage.

Voltaje secundario de la reserva Es la diferencia entre el voltaje requerido y el voltaje máximo disponible.

Secondary wiring Conductors, such as battery cables and ignition spark plug wires, that are used to carry high voltage or high current. Secondary wires have extra thick insulation.

Hilos secundarios Conductores, tales como cables de batería e hilos de bujías del encendido, utilizados para transmitir tensión o corriente alta. Los hilos secundarios poseen un aislamiento sumamente grueso.

Sector gear The section of gear teeth on the regulator.

Engranaje de cables La sección de los dientes de engranaje en el regulador.

Self-induction The generation of an electromotive force by a changing current in the same circuit.

Autoinducción Este es el generamiento de la fuerza electromotriz cuando la corriente cambia en el mismo circuito.

Semiconductor An element that is neither a conductor nor an insulator. Semiconductors are materials that conduct electric current under certain conditions, yet will not conduct under other conditions.

Semiconductor Elemento que no es ni conductor ni aislante. Los semiconductores son materiales que transmiten corriente eléctrica bajo ciertas circunstancias, pero no la transmiten bajo otras.

Sending unit The sensor for the gauge. It is a variable resistor that changes resistance values with changing monitored conditions.

Unidad emisora Sensor para el calibrador. Es un resistor variable que cambia los valores de resistencia según cambian las condiciones reguladas.

Sensing voltage Input voltage to the AC generator.

Detección del voltaje Determina la tensión de entrada de energía al generador de Corriente Alterna AC.

Sensitivity controls A potentiometer that allows the driver to adjust the sensitivity of the automatic dimmer system to surrounding ambient light conditions.

Controles de sensibilidad Un potenciómetro que permite que el conductor ajusta la sensibilidad del sistema de intensidad de iluminación automático a las condiciones de luz ambientales.

Sensor Any device that provides an input to the computer.

Sensor Cualquier dispositivo que le transmite información a la computadora.

Sentry key Describes a sophisticated antitheft system that prevents the engine from starting unless a special key is used.

Llave guardiante centinela Describe un sofisticado sistema de guardia anti-robo el que evita prender el motor, a menos que se haga con una llave específicamente creada.

Separators Normally constructed of glass with a resin coating. These battery plates offer low resistance to electrical flow but high resistance to chemical contamination.

Separadores Normalmente se construyen del vidrio con una capa de resina. Estas placas de la batería ofrecen baja resistencia al flujo de la electricidad pero alta resistencia a la contaminación química.

Sequential logic circuits Flip-flop circuits in which the output is determined by the sequence of inputs. A given input affects the output produced by the next input.

Circuitos de lógica secuenciales Cambia los circuitos en los cuales la salida de energía es determinada según la secuencia de las entradas de corriente. Una entrada de energía afecta la salida de corriente que va ser producida en una próxima entrada de energía.

Sequential sampling The process that the MUX and DEMUX operate on. This means the computer will deal with all of the sensors and actuators one at a time.

Muestreo secuencial La forma en que funcionan los sistemas de las abreviaturas (MUX y DEMUX). Esto significa que la computadora se encargará de que todos los monitores y actuadores funcionen uno por uno.

Series circuit A circuit that provides a single path for current flow from the electrical source through all the circuit's components and back to the source.

Circuito en serie Circuito que provee una trayectoria única para el flujo de corriente de la fuente eléctrica a través de todos los componentes del circuito, y de nuevo hacia la fuente.

Series hybrid Hybrid configuration where propulsion comes directly from the electric motor.

Híbrido en serie Configuración del híbrido en donde la propulsión llega directamente del motor eléctrico.

Series-parallel circuit A circuit that has some loads in series and some in parallel.

Circuito en series paralelas Circuito que tiene unas cargas en serie y otras en paralelo.

Series-wound motor A type of motor that has its field windings connected in series with the armature. This type of motor develops its maximum torque output at the time of initial start. Torque decreases as motor speed increases.

Motor con devanados en serie Un tipo de motor cuyos devanados inductores se conectan en serie con la armadura. Este tipo de motor desarrolla la salida máxima de par de torsión en el momento inicial de ponerse en marcha. El par de torsión disminuye al aumentar la velocidad del motor.

Servomotor An electrical motor that produces rotation of less than a full turn. A feedback mechanism is used to position itself to the exact degree of rotation required.

Servomotor Motor eléctrico que genera rotación de menos de una revolución completa. Utiliza un mecanismo de realimentación para ubicarse al grado exacto de la rotación requerida.

Shell The electron orbit around the nucleus of an atom.

Corteza Órbita de electrones alrededor del núcleo del átomo.

Short An unwanted electrical path; sometimes this path goes directly to ground.

Corto Una trayectoria eléctrica no deseable; a veces este trayectoria viaja directamente a tierra.

Shorted circuit A circuit that allows current to bypass part of the normal path.

Circuito corto Este circuito permite que la corriente pase por una parte del recorrido normal.

Shunt More than one path for current to flow.

Desviación Más de una derivación para que la corriente pueda fluir.

Shunt circuits The branches of the parallel circuit.

Circuitos en derivación Las ramas del circuito en paralelo.

Shunt-wound motor A type of motor whose field windings are wired in parallel to the armature. This type of motor does not decrease its torque as speed increases.

Motor con devanados en derivación Un tipo de motor cuyos devanados inductores se cablean paralelos a la armadura. Este tipo de motor no disminuya su par de torsión al aumentar la velocidad.

Shutter wheel A metal wheel consisting of a series of alternating windows and vanes. It creates a magnetic shunt that changes the strength of the magnetic field from the permanent magnet of the Hall-effect switch or magnetic pulse generator.

Rueda obturadora Rueda metálica compuesta de una serie de ventanas y aspas alternas. Genera una derivación magnética que cambia la potencia del campo magnético, del imán permanente del conmutador de efecto Hall o del generador de impulsos magnéticos.

Sine wave A waveform that shows voltage changing polarity.

Onda senoidal Una forma de onda que muestra un cambio de polaridad en la tensión.

Single phase voltage The sine wave voltage induced in one conductor of the stator during one revolution of the rotor.

Tensión monofásica La tensión en forma de onda senoidal inducida en un conductor del estator durante una revolución del rotor.

Slave module Controller on the network that must communicate through a master controller.

Módulo esclavo Controlador en la red que debe comunicarse por medio del control maestro.

Slip rings Rings that function much like the armature commutator in the starter motor; however, they are smooth.

Anillos colectores Estos funcionan como casi el conmutador inducido en el arranque del motor, excepto que estos anillos son lisos.

Smart sensors Sensors that are capable of sending digital messages on the data bus.

Sensor inteligente Sensores capaces de enviar mensajes digitales en el bus de datos.

Solenoid An electromagnetic device that uses movement of a plunger to exert a pulling or holding force.

Solenoide Dispositivo electromagnético que utiliza el movimiento de un pulsador para ejercer una fuerza de arrastre o de retención.

Sound generator See *buzzer.*

Generador de sonido Vea "timbre."

Source The portion of a field-effect transistor that supplies the current-carrying holes or electrons.

Fuente Terminal de un transistor de efecto de campo que provee los agujeros o electrones portadores de corriente.

Spark plug Electrodes provide gaps inside each combustion chamber across which the secondary current flows to ignite the air-fuel mixture in the combustion chambers.

Enchufe de chispa Electrodos para proveer intervalos dentro de cada cámara de combustión a través de la corriente secundaria que fluye para encender la mezcla de aire/combustible en las cámaras de combustión.

Specific gravity The weight of a given volume of a liquid divided by the weight of an equal volume of water.

Gravedad específica El peso de un volumen dado de líquido dividido por el peso de un volumen igual de agua.

Speedometer An instrument panel gauge that indicates the speed of the vehicle.

Velocímetro Calibrador en el panel de instrumentos que marca la velocidad del vehículo.

Speed-sensitive steering A steering system that allows for high power assist during low-speed maneuvers and increased road feel at high road speeds.

Dirección tacométrica Sistema de dirección que permite la asistencia de alta potencia durante las maniobras de baja velocidad y mayor sensación del camino a altas velocidades del camino.

Squib A pyrotechnic term used for a fire cracker that burns but does not explode. The squib starts the process of air bag deployment.

Mecha Un término pirotécnico usado para prender una pólvora que se quema pero no explota. La mecha inicia el proceso de la salida de la bolsa de aire.

Start/clutch interlock switch Used on vehicles equipped with manual transmissions.

Interruptor de seguridad de embrague Usado en los vehículos equipados con transmisiones manuales.

Starter drive The part of the starter motor that engages the armature to the engine flywheel ring gear.

Transmisión de arranque Parte del motor de arranque que engrana la armadura a la corona del volante de la máquina.

Starter generator control module (SGCM) Also called a high-voltage ECU (HV ECU). Used to control the flow of torque and electrical energy into and from the motor generator of the HEV.

Módulo de control del generador de encendido (MCGE) También se le llama UCE de alto voltaje (UCE AV). Se utiliza para controlar el flujo del par motor y la energía eléctrica dentro y fuera del generador del motor del VHE.

Starting system A combination of mechanical and electrical parts that work together to start the engine by changing the electrical energy that is being supplied by the battery into mechanical energy by use of a starter or cranking motor.

Sistema de encendido Combinación de partes mecánicas y eléctricas que trabajan unidas para encender el motor al cargar la energía eléctrica que proporciona la batería, en energía mecánica mediante el uso de un encendedor o motor de arranque.

State of charge The condition of a battery's electrolyte and plate materials at any given time.

Estado de carga Condición del electrolito y de los materiales de la placa de una batería en cualquier momento dado.

Static electricity Electricity that is not in motion.

Electricidad estática Electricidad que no está en movimiento.

Static neutral point The point at which the fields of a motor are in balance.

Punto neutral estático El punto en que los campos de un motor estan equilibrados.

Stator The stationary coil of the AC generator where current is produced.

Estátor Bobina fija del generador de corriente alterna donde se genera corriente.

Stator neutral junction The common junction of Wye stator windings.

Unión de estátor neutral La unión común de los devanados de un estátor Y.

Stepped resistor A resistor that has two or more fixed resistor values.

Resistor de secciones escalonadas Resistor que tiene dos o más valores de resistencia fija.

Stepper motor An electrical motor that contains a permanent magnet armature with two or four field coils. Can be used to move the controlled device to whatever location is desired. By applying voltage pulses to selected coils of the motor, the armature will turn a specific number of degrees. When the same voltage pulses are applied to the opposite coils, the armature will rotate the same number of degrees in the opposite direction.

Motor de pasos Contiene una armadura magnética permanente con dos, cuatro o más bobinas del campo.

Stranded wire A conductor comprised of many small solid wires twisted together. This type conductor is used to allow the wire to flex without breaking.

Cable trenzado Un conductor que comprende muchos cables sólidos pequeños trenzados. Este tipo de conductor se emplea para permitir que el cable se tuerza sin quebrar.

Sulfation A condition in a battery that reduces its output. The sulfate in the battery that is not converted tends to harden on the plates, and permanent damage to the battery results.

Sulfatación Una condición en una batería que disminuya su potencia de salida. El sulfato en la batería que no se convierte suele endurecerse en las placas y resulta en daños permanentes en la batería.

Supplemental bus networks Bus networks that are on the vehicle in addition to the main bus network.

Redes de bus complementarios Redes de bus que hay en el vehículo aparte de la red del bus principal.

Tachometer An instrument that measures the speed of the engine in revolutions per minute (rpm).

Tacómetro Instrumento que mide la velocidad del motor en revoluciones por minuto (rpm).

Terminals Provide a means of connecting the battery plates to the vehicle's electrical system.

Terminales Provee los medios de conección de las placas de batería al sistema eléctrico del vehículo.

Termination resistors Used to control induced voltages. Since voltage is dropped over resistors, the induced voltage is terminated.

Resistores de terminación Usados para controlar la conducción de los voltajes. Como el voltaje cae sobre los resistores, es así como se determina el voltaje.

Thermistor A solid-state variable resistor made from a semiconductor material that changes resistance in relation to temperature changes.

Termistor Resistor variable de estado sólido hecho de un material semiconductor que cambia su resistencia en relación con los cambios de temperatura.

Three-coil gauge A gauge design that uses the interaction of three electromagnets and the total field effect upon a permanent magnet to cause needle movement.

Calibrador de tres bobinas Calibrador diseñado para utilizar la interacción de tres electroimanes y el efecto inductor total sobre un imán permanente para producir el movimiento de la aguja.

Throw Term used in reference to electrical switches or relays referring to the number of output circuits from the switch.

Posición activa Término utilizado para conmutadores o relés eléctricos en relación con el número de circuitos de salida del conmutador.

Thyristor A semiconductor switching device composed of alternating N and P layers. It can also be used to rectify current from AC to DC.

Tiristor Dispositivo de conmutación del semiconductor compuesto de capas alternas de N y P. Puede utilizarse también para rectificar la corriente de corriente alterna a corriente continua.

Timer circuit Uses a bimetallic strip that opens as a result of the heat being generated by the current flow.

Circuito Sincronizador Consisite de una cinta bimetálica la que se abre, debido al calor generado por el flujo de corriente.

Timer control A potentiometer that is part of the headlight switch in some systems. It controls the amount of time the headlights stay on after the ignition switch is turned off.

Control temporizador Un potenciómetro que es parte del conmutador de los faros en algunos sistemas. Controla la cantidad del tiempo que quedan prendidos los faros después de apagarse la llave del encendido.

Timing disc Known as an armature, reluctor, trigger wheel, pulse wheel, or timing core. It is used to conduct lines of magnetic force.

Disco medidor de tiempo Se conoce como una armadura de inducción, rueda disparadora, rueda de pulsación o un núcleo de tiempo. Este disco es usado para conducir líneas de fuerza magnética.

Tire pressure monitoring systems A safety system that notifies the driver if one or more tires are underinflated or overinflated.

Sistema de monitoreo de la presión de las llantas Sistema de seguridad que le hace saber al conductor si una llanta o más llantas están desinfladas o sobre infladas.

Torque converter A hydraulic device found on automatic transmissions. It is responsible for controlling the power flow from the engine to the transmission; works like a clutch to engage and disengage the engine's power to the drive line.

Convertidor de par Un dispositivo hidráulico en las transmisiones automáticas. Se encarga de controlar el flujo de la potencia del motor a la transmisión; funciona como un embrague para embragar y desembragar la potencia del motor con la flecha motríz.

Total reflection A phenomenon where in a light wave reflects off of the surface 100% when the light wave advances from a medium of high index of refraction to a medium of low index of refraction.

Reflexión total Fenómeno en el que una onda de luz se refleja 100 % de una superficie cuando la onda de luz avanza de un índice medio de refracción de uno alto a uno índice medio de refracción de uno bajo.

Tracer A thin or dashed line of a different color than the base color of the insulation.

Traza líneas Una línea delgada o instrumento de color diferente al color básico de la insulación.

Trailing edge In a Hall-effect switch, the edges of the rotating blade that exit the switch.

Borde de salida Un interruptor de efecto Hall indica los bordes de la paleta giratoria que sale del interruptor.

Transducer A device that changes energy from one form into another.

Transductor Dispositivo que cambia la energía de una forma a otra.

Transistor A three-layer semiconductor used as a very fast switching device.

Transistor Semiconductor de tres capas utilizado como dispositivo de conmutación sumamente rápido.

Trimotor A three-armature motor.

Trimotor Motor de tres armaduras.

Turn-on voltage The voltage required to jump the PN junction and allow current to flow.

Voltaje de conección El voltaje requerido para hacer funcionar el cable de empalme PN y permitir que la corriente fluya.

TVRS An abbreviation for television-radio-suppression cable.

TVRS Una abreviación del cable de supresión del televisión y radio.

Tweeters Smaller speakers that produce the high frequencies of treble.

Tweeters Pequeñas bocinas que producen altas frecuencias agudas.

Two-coil gauge A gauge design that uses the interaction of two electromagnets and the total field effect upon an armature to cause needle movement.

Calibrador de dos bobinas Calibrador diseñado para utilizar la interacción de dos electroimanes y el efecto inductor total sobre una armadura para generar el movimiento de la aguja.

Ultra capacitor A device that stores energy as electrostatic charge. It is the primary device in the power supply during hill climbing, acceleration, and the recovery of braking energy.

Ultra capacitor Dispositivo que guarda energía como carga electrostática. Es el dispositivo primario en la fuente de energía durante una subida, una aceleración y el recobro de la energía de frenado.

Vacuum distribution valve A valve used in vacuum-controlled concealed headlight systems. It controls the direction of vacuum to various vacuum motors or to vent.

Válvula de distribución al vacío Válvula utilizada en el sistema de faros delanteros ocultos controlado al vacío. Regula la dirección del vacío a varios motores al vacío o sirve para dar salida del sistema.

Vacuum fluorescent display (VFD) A display type that uses anode segments coated with phosphor and bombarded with tungsten electrons to cause the segments to glow.

Visualización de fluorescencia al vacío Tipo de visualización que utiliza segmentos ánodos cubiertos de fósforo y bombardeados de electrones de tungsteno para producir la luminiscencia de los segmentos.

Valence ring The outermost orbit of the atom.

Anillo de valencia Órbita más exterior del átomo.

Valve body A unit that consists of many valves and hydraulic circuits. This unit is the central control point for gear shifting in an automatic transmission.

Cuerpo de la válvula Una unedad que consiste de muchas válvulas y circuitos hidráulicos. Esta unedad es el punto central de mando para los cambios de velocidad en una transmisión automática.

Vaporized aluminum Gives a reflecting surface that is comparable to silver.

Aluminio vaporizado Da a la superficie un reflejo brillante como si fuera plata.

Variable resistor A resistor that provides for an infinite number of resistance values within a range.

Resistor variable Resistor que provee un número infinito de valores de resistencia dentro de un margen.

Vehicle instrumentation systems A system that monitors the various vehicle operating systems and provides information to the driver of their correct operation.

Sistemas de instrumentación del vehículo Sistema que monitorea varios sistemas de operación del vehículo y proporciona información de su operación correcta al conductor.

Volatile Easily vaporizes or explodes.

Volátil Vaporiza o explota fácilmente.

Volatile RAM memory that is erased when it is disconnected from its power source. Also known as Keep Alive Memory.

Volátil Memoria RAM cuyos datos se perderán cuando se la desconecta de la fuente de alimentación. Conocida también como memoria de entretenimiento.

Volt The unit used to measure the amount of electrical force.

Voltio Unidad práctica de tensión para medir la cantidad de fuerza eléctrica.

Voltage The difference or potential that indicates an excess of electrons at the end of the circuit the farthest from the electromotive force. It is the electrical pressure that causes electrons to move through a circuit. One volt is the amount of pressure required to move one amp of current through one ohm of resistance.

Tensión Diferencia o potencial que indica un exceso de electrones al punto del circuito que se encuentra más alejado de la fuerza electromotriz. La presión eléctrica genera el movimiento de electrones a través de un circuito. Un voltio equivale a la cantidad de presión requerida para mover un amperio de corriente a través de un ohmio de resistencia.

Voltage drop A resistance in the circuit that reduces the electrical pressure available after the resistance. The resistance can be either the load component, the conductors, any connections, or unwanted resistance.

Caída de tensión Resistencia en el circuito que disminuye la presión eléctrica disponible después de la resistencia. La resistencia puede ser el componente de carga, los conductores, cualquier conexión o resistencia no deseada.

Voltage limiter Connected through the resistor network of a voltage regulator. It determines whether the field will receive high, low, or no voltage. It controls the field voltage for the required amount of charging.

Limitador de tensión Conectado por el red de resistores de un regulador de tensión. Determina si el campo recibirá alta, baja o ninguna tensión. Controla la tensión de campo durante el tiempo indicado de carga.

Voltage regulator Used to control the output voltage of the AC generator, based on charging system demands, by controlling field current.

Regulador de tensión Dispositivo cuya función es mantener la tensión de salida del generador de corriente alterna, de acuerdo a las variaciones en la corriente de carga, controlando la corriente inductora.

Voltmeter A test meter used to read the pressure behind the flow of electrons.

Voltímetro Instrumento de prueba utilizado para medir la presión del flujo de electrones.

Wake-up signal An input signal used to notify the body computer that an engine start and operation of accessories is going to be initiated soon. This signal is used to warm up the circuits that will be processing information.

Señal despertadora Señal de entrada para avisarle a la computadora del vehículo que el arranque del motor y el funcionamiento de los accesorios se iniciarán dentro de poco. Dicha señal se utiliza para calentar los circuitos que procesarán los datos.

Warning lamp A lamp that is illuminated to warn the driver of a possible problem or hazardous condition.

Luz de aviso Lámpara que se enciende para avisarle al conductor sobre posibles problemas o condiciones peligrosas.

Waste spark A spark that occurs during the exhaust stroke of a piston.

Chispa desgastada La chispa que se produce durante el tiempo de escape de un pistón.

Watchdog circuit Supplies a reset voltage to the microprocessor in the event that pulsating output voltages from the microprocessor are interrupted.

Circuito de vigilancia Proporciona un voltaje de reposición para el microprocesador en caso de que se interrumpan los voltajes de potencia útil de pulsaciones del microprocesador.

Watt The unit of measure of electrical power, which is the equivalent of horsepower. One horsepower is equal to 746 watts.

Watio Unidad de potencia eléctrica, equivalente a un caballo de vapor. 746 watios equivalen a un caballo de vapor (CV).

Wattage A measure of the total electrical work being performed per unit of time.

Vataje Medida del trabajo eléctrico total realizado por unidad de tiempo.

Watt-hour rating Equals the battery voltage times ampere-hour rating.

Contador de voltaje-hora nominal Su función es igualar el tiempo del voltaje de la batería con el voltaje-hora nominal.

Weather-pack connector A type of connector that seals the terminal's ends. This type connector is used in electronic circuits.

Conectador impermeable Un tipo de conectador que sella las extremidades de los terminales. Este tipo de conectador se emplea en los circuitos electrónicos.

Wheatstone bridge A series-parallel arrangement of resistors between an input terminal and ground.

Puente de Wheatstone Conjunto de resistores en series paralelas entre un borne de entrada y tierra.

Window regulator Converts the rotary motion of the motor into the vertical movement of the window.

Regulador del vidrio parabrisas Convierte la acción rotatoria del motor en un movimiento vertical sobre el vidrio parabrisas.

Windshield wipers Mechanical arms that sweep back and forth across the windshield to remove water, snow, or dirt.

Limpiador del parabrisas Funciona con brazos mecánicos, de un extremo al otro del vidrio, para remover agua, nieve y suciedad.

Wiper Moveable contact of a variable resistor.

Limpiador Consiste de un movimiento deslizable de resistores variables.

Wireless networks Connection of modules together to transmit information without the use of physical connection by wires.

Redes inalámbricas Conexión entre los módulos para transmitir información sin el uso de conexiones físicas por medio de alambres.

Wiring diagram An electrical schematic that shows a representation of actual electrical or electronic components and the wiring of the vehicle's electrical systems.

Esquema de conexiones Esquema en el que se muestran las conexiones internas de los componentes eléctricos o electrónicos reales y las de los sistemas eléctricos del vehículo.

Wiring harness A group of wires enclosed in a conduit and routed to specific areas of the vehicle.

Cableado preformado Conjunto de alambres envueltos en un conducto y dirigidos hacia áreas específicas del vehículo.

Woofers Large speakers that produce the low frequencies of midrange and bass.

Woofers Bocinas grandes que producen bajas frecuencias de media distancia y bajo.

Worm gear A type of gear whose teeth wrap around the shaft. The action of the gear is much like that of a threaded bolt or screw.

Engranaje de tornillo sin fin Un tipo de engranaje cuyos dientes se envuelven alrededor del vástago. El movimiento del engranaje es muy parecido a un perno enroscado o una tuerca.

Wye wound connection A type of stator winding in which one end of the individual windings are connected at a common point. The structure resembles the letter Y.

Conexión Y Un tipo de devanado estátor en el cual una extremidad de los devanados individuales se conectan en un punto común. La estructura parece la letra "Y."

Yaw The tendency for the vehicle to rotate around its center of gravity.

Giro longitudinal Tendencia de un vehículo a girar sobre su propio eje de gravedad.

Y-type stator A three-winding AC generator that has one end of each winding connected at the neutral junction.

Estátor de tipo Y Generador de corriente alterna de devanado triple; un extremo de cada devanado se conecta al empalme neutro.

Zener diode A diode that allows reverse current to flow above a set voltage limit.

Diodo Zener Diodo que permite que el flujo de corriente en dirección inversa sobrepase el límite de tensión determinado.

Zener voltage The voltage that is reached when a diode conducts in reverse direction.

Tensión de Zener Tensión alcanzada cuando un diodo conduce en una dirección inversa.

INDEX